Graduate Texts in Mathematics

68

Joachim Weidmann

Linear Operators in Hilbert Spaces

Translated by
Joseph Szücs

Springer-Verlag
New York Heidelberg Berlin

Joachim Weidmann

Mathematisches Seminar der
Johann-Wolfgang-Goethe-Universität
Institut für Angewandte Mathematik
Robert-Mayer-Strasse 10
6 Frankfurt a. M.
Federal Republic of Germany

Joseph Szücs

American Mathematical Society
P.O. Box 6248
Providence, RI 02940
USA

AMS Subject Classifications: 47Axx', 47B05, 47B10, 47B15, 47B20, 47B25, 47B30, 47E05, 81A09, 81A10, 81A45

With 1 Figure.

Library of Congress Cataloging in Publication Data

Weidmann, Joachim.
Linear operators in Hilbert spaces.

(Graduate texts in mathematics; 68)
Translation of Lineare Operatoren in Hilberträumen.
Bibliography: p.
Includes index.
1. Linear operators. 2. Hilbert space. I. Title. II. Series.
QA329.2.W4413 515'.72 79-12649

Exclusively authorized English translation of the original German edition *Lineare Operatoren in Hilberträumen* published in the series Mathematische Leitfäden edited by G. Köthe © by B. G. Teubner, Stuttgart, 1976.

9 8 7 6 5 4 3 2

ISBN-13: 978-1-4612-6029-5 e-ISBN-13: 978-1-4612-6027-1
DOI: 10.107/978-1-4612-6027-1

To the memory of
Konrad Jörgens

Preface to the English edition

This English edition is almost identical to the German original *Lineare Operatoren in Hilberträumen*, published by B. G. Teubner, Stuttgart in 1976. A few proofs have been simplified, some additional exercises have been included, and a small number of new results has been added (e.g., Theorem 11.11 and Theorem 11.23). In addition a great number of minor errors has been corrected.

Frankfurt, January 1980 J. Weidmann

Preface to the German edition

The purpose of this book is to give an introduction to the theory of linear operators on Hilbert spaces and then to proceed to the interesting applications of differential operators to mathematical physics. Besides the usual introductory courses common to both mathematicians and physicists, only a fundamental knowledge of complex analysis and of ordinary differential equations is assumed. The most important results of Lebesgue integration theory, to the extent that they are used in this book, are compiled with complete proofs in Appendix A. I hope therefore that students from the fourth semester on will be able to read this book without major difficulty. However, it might also be of some interest and use to the teaching and research mathematician or physicist, since among other things it makes easily accessible several new results of the spectral theory of differential operators.

In order to limit the length of the text, I present the results of abstract functional analysis only insofar as they are significant for this book. I prove those theorems (for example, the closed graph theorem) that also hold in more general Banach spaces by Hilbert space methods whenever this leads to simplification. The typical concepts of Hilbert space theory, "orthogonal" and "self-adjoint," stand clearly at the center. The spectral theorem for self-adjoint operators and its applications are the central topics of this book. A detailed exposition of the theory of expansions in terms of generalized eigenfunctions and of the spectral theory of ordinary differential operators (Weyl–Titchmarsh–Kodaira) was not possible within the framework of this book.

In the first three chapters pre-Hilbert spaces and Hilbert spaces are introduced, and their basic geometric and topologic properties are proved. Chapters 4 and 5 contain the fundamentals of the theory of (not necessarily bounded) linear operators on Hilbert spaces, including general spectral theory. Besides the numerous examples scattered throughout the text, in Chapter 6 certain important classes of linear operators are studied in detail. Chapter 7 contains the spectral theory of self-adjoint operators (first for compact operators, and then for the general case), as well as some important consequences and a detailed characterization of the spectral points. In Chapter 8 von Neumann's extension theory for symmetric operators is developed and is applied to, among other things, the Sturm-Liouville operators. Chapter 9 provides some important results of perturbation theory for self-adjoint operators. Chapter 10 begins with proofs of the most significant facts about Fourier transforms in $L_2(\mathbb{R}^m)$, applications to partial differential operators, in particular to Schrödinger and Dirac operators, follow. Finally, Chapter 11 gives a short introduction to (time dependent) scattering theory with some typical results; to my regret, I could only touch upon the far reaching results of recent years.

Exercises are not used later in the text, with a few exceptions. They mainly serve to deepen understanding of the material and give opportunity for practice; however, I often use them to formulate further results which I cannot treat in the text. The level of difficulty of the exercises varies widely. Because I give many exercises with detailed hints, they can be solved in general without much difficulty.

Now I want to very heartily thank all those who helped me with the production of this book. Mrs. Hose turned my notes into an excellent typed manuscript with infinite diligence. Messrs. R. Hollstein, D. Keim and H. Küch spent much time reading the whole manuscript and discussing with me their suggestions for improvement. Messrs. R. Colgen and W. Stork helped me with the proofreading. I thank the publisher and the editors for their pleasant cooperation.

My teacher Konrad Jörgens inspired me to study this subject; he influenced the present exposition in several ways. I dedicate this volume to his memory.

Hattersheim am Main, the summer of 1976 Joachim Weidmann

Contents

Vector spaces with a scalar product, pre-Hilbert spaces 1

In what follows we consider vector spaces over a field $\mathbb{K}$, where $\mathbb{K}$ is either the field $\mathbb{C}$ of complex numbers or the field $\mathbb{R}$ of real numbers; accordingly, we speak of a complex or a real vector space. For every $c \in \mathbb{K}$ let c^* be the complex conjugate of c; so for $c \in \mathbb{R}$ the star has no significance.

As a rule, we assume the most important notions and results of linear algebra to be known.

1.1 Sesquilinear forms

Let H be a vector space over $\mathbb{K}$. A mapping $s: H \times H \to \mathbb{K}$ is called a *sesquilinear form* on H if for all $f, g, h \in H$ and $a, b \in \mathbb{K}$ we have

$$s(f, ag+bh) = as(f,g) + bs(f,h), \tag{1.1}$$

$$s(af+bg, h) = a^*s(f,h) + b^*s(g,h). \tag{1.2}$$

If (1.2) holds without stars, then s is called a bilinear form on H; in particular every sesquilinear form on a real vector space is a bilinear form.

Property (1.1) is obviously equivalent to the two properties

$$s(f, g+h) = s(f,g) + s(f,h), \tag{1.1'}$$

$$s(f, ag) = as(f,g). \tag{1.1''}$$

Similarly, (1.2) is equivalent to

$$s(f+g, h) = s(f,h) + s(g,h), \tag{1.2'}$$

$$s(af, g) = a^*s(f,g). \tag{1.2''}$$

If s is a sesquilinear form on H, then the mapping $q : H \to \mathbb{K}$ that is defined by $q(f) = s(f, f)$ for each $f \in H$ is called the *quadratic form* on H generated or induced by s. For each quadratic form q we obviously have

$$q(af) = |a|^2 q(f) \quad \text{for all} \quad f \in H, a \in \mathbb{K}; \tag{1.3}$$

so we have, in particular, $q(af) = q(f)$ for every $a \in \mathbb{K}$ with $|a| = 1$.

The following theorem shows that in a complex vector space the generating sesquilinear form is uniquely determined by the quadratic form; for real vector spaces this is not necessarily true in general; see Exercise 1.2.

Theorem 1.1 (Polarization identity). *Let H be a complex vector space, s a sesquilinear form on H, and q the quadratic form generated by s. Then for all $f, g \in H$ we have*

$$s(f, g) = \tfrac{1}{4}\{q(f+g) - q(f-g) + iq(f-ig) - iq(f+ig)\}. \tag{1.4}$$

The proof of this identity may be given by calculating the right side of (1.4) according to the rules (1.1) and (1.2).

Theorem 1.2 (Parallelogram law). *Let s be a sesquilinear form on a vector space H, and let q be the corresponding quadratic form on H. Then for all $f, g \in H$ we have*

$$q(f+g) + q(f-g) = 2[q(f) + q(g)]. \tag{1.5}$$

PROOF. For every $f, g \in H$ we have

$$\begin{aligned} q(f+g) + q(f-g) &= s(f, f) + s(f, g) + s(g, f) + s(g, g) \\ &\quad + s(f, f) - s(f, g) - s(g, f) + s(g, g) \\ &= 2q(f) + 2q(g). \end{aligned}$$

□

A sesquilinear form s on H is said to be *Hermitian* provided that for every $f, g \in H$ we have

$$s(f, g) = s(g, f)^*. \tag{1.6}$$

A Hermitian bilinear form on a real vector space is said to be *symmetric*.

If s is a Hermitian sesquilinear form, and q the quadratic form generated by s, then we obviously have $q(f) \in \mathbb{R}$ for all $f \in H$; we say briefly that q is *real*. The following theorem shows, among other things, that Hermitian sesquilinear forms can be characterized by this property of their associated quadratic forms. We also obtain that symmetric bilinear forms are uniquely determined by the corresponding quadratic forms.

Theorem 1.3. *Let H be a vector space over $\mathbb{K}$, s a sesquilinear form on H, and q the quadratic form generated by s.*

(a) *If* $\mathbb{K}=\mathbb{C}$, *then the following statements are equivalent*:

(i) *s is symmetric*,

(ii) *q is real*,

(iii) *for all* $f, g \in H$ *we have*

$$\operatorname{Re} s(f, g)=\tfrac{1}{4}\{q(f+g)-q(f-g)\}, \tag{1.7}$$

(iv) *for all* $f, g \in H$ *we have*

$$\operatorname{Im} s(f, g)=\tfrac{1}{4}\{q(f-ig)-q(f+ig)\}. \tag{1.7'}$$

(b) *If* $\mathbb{K}=\mathbb{R}$, *then the following statements are equivalent*:

(i) *s is symmetric*,

(ii) *for all* $f, g \in H$ *we have*

$$s(f, g)=\tfrac{1}{4}\{q(f+g)-q(f-g)\}. \tag{1.8}$$

PROOF.

(a) (ii) *follows from* (i): $q(f)^*=s(f,f)^*=s(f,f)=q(f)$, i.e., $q(f)$ is real.

(iii) *follows from* (ii): Because $q(h)\in\mathbb{R}$ for all $h \in H$, it follows from (1.4) that

$$\begin{aligned}\operatorname{Re} s(f, g)&=\tfrac{1}{4}\operatorname{Re}\{q(f+g)-q(f-g)+iq(f-ig)-iq(f+ig)\}\\&=\tfrac{1}{4}\{q(f+g)-q(f-g)\}.\end{aligned}$$

(iv) *follows from* (iii): Because of (iii) we have

$$\begin{aligned}\operatorname{Im} s(f, g)&=-\operatorname{Re}\{is(f, g)\}\\&=\operatorname{Re} s(f,-ig)=\tfrac{1}{4}\{q(f-ig)-q(f+ig)\}.\end{aligned}$$

(i) *follows from* (iv):

$$\begin{aligned}s(g, f)^*=\operatorname{Re} s(g, f)-i\operatorname{Im} s(g, f)&=\operatorname{Im} s(g, if)-i\operatorname{Im} s(g, f)\\&=\tfrac{1}{4}\{q(g+f)-q(g-f)-iq(g-if)+iq(g+if)\}\\&=\tfrac{1}{4}\{q(f+g)-q(f-g)+iq(f-ig)-iq(f+ig)\}=s(f, g);\end{aligned}$$

here we have used (1.3) with $a=-1$, $a=i$, and $a=-i$.

(b) (ii) *follows from* (i) by calculating the right side of (ii) while using the symmetry of s.

(i) *follows from* (ii):

$$\begin{aligned}s(g, f)&=\tfrac{1}{4}\{q(g+f)-q(g-f)\}\\&=\tfrac{1}{4}\{q(f+g)-q(f-g)\}=s(f, g).\end{aligned}$$ □

A Hermitian sesquilinear form is said to be *non-negative* when

$$s(f, f) \geq 0 \quad \text{for all} \quad f \in H; \tag{1.9}$$

it is said to be *positive* when

$$s(f,f) > 0 \quad \text{for all} \quad f \in H \quad \text{with} \quad f \neq 0. \tag{1.10}$$

Since we have $s(0, 0) = 0$, every positive sesquilinear form is non-negative. We also say that the corresponding quadratic forms are *non-negative*, respectively *positive*. (Because of Theorem 1.3, the word "Hermitian" may be omitted from this definition in the complex case; this does not hold in the real case, cf. Exercise 1.3.)

Theorem 1.4. *If s is a non-negative sesquilinear form on H, and q denotes the quadratic form generated by s, then for every $f, g \in H$ we have the Schwarz inequality*

$$|s(f, g)| \leqslant [q(f)q(g)]^{1/2}. \tag{1.11}$$

If s is positive, then the equality sign in (1.11) *holds if and only if f and g are linearly dependent; the equality $s(f, g) = [q(f)q(g)]^{1/2}$ holds if and only if there exists a $c \geqslant 0$ such that $f = cg$ or $g = cf$.*

Proof. Let $f, g \in H$. For all $t \in \mathbb{R}$ we have

$$0 \leqslant q(f+tg) = q(f) + 2t \operatorname{Re}(f, g) + t^2 q(g).$$

This second degree polynomial in t has either no root or a double root. Since this holds for a polynomial $at^2 + 2bt + c$ if and only if $b^2 - ac \leqslant 0$, it follows that

$$[\operatorname{Re} s(f, g)]^2 \leqslant q(f)q(g). \tag{1.12}$$

If one chooses $a \in \mathbb{K}$ such that $|a| = 1$ and $as(f, g) = |s(f, g)|$ holds, then it follows from (1.12) with $h = ag$ that

$$\begin{aligned} |s(f, g)|^2 &= [\operatorname{Re} as(f, g)]^2 = [\operatorname{Re} s(f, h)]^2 \\ &\leqslant q(f)q(h) = q(f)q(ag) = q(f)q(g); \end{aligned}$$

this is the *Schwarz inequality*.

Let s now be positive and let $s(f, g) = [q(f)q(g)]^{1/2}$ be true. If $g = 0$, then the equality $g = 0f$ proves the assertion. Consequently, let $g \neq 0$. Because of the equality $[\operatorname{Re} s(f, g)]^2 - q(f)q(g) = 0$, the polynomial considered above has a double root t_0; hence we have $q(f + t_0 g) = 0$ i.e., $f = -t_0 g$. From $-t_0 s(g, g) = s(f, g) \geqslant 0$ it follows that $-t_0 \geqslant 0$. If we have $|s(f, g)| = [q(f)q(g)]^{1/2}$ and choose a and h as above, then $s(f, h) = [q(f)q(h)]^{1/2}$ follows. According to the part just proved we then have either $g = 0 = 0f$, or there exists a $c > 0$ such that $f = ch = acg$. In both cases f and g are linearly dependent. One can verify the converses of the last two assertions by simple calculation. □

Example 1. For each $m \in \mathbb{N}$ ($\mathbb{N}$ denotes the set $\{1, 2, 3, \ldots\}$ of positive integers) let $\mathbb{C}^m$ be the complex vector space of the m-tuples $f =$

$(f_1, f_2, \ldots, f_m)$, $g = (g_1, g_2, \ldots, g_m)$, ... of complex numbers with the addition

$$f + g = (f_1 + g_1, f_2 + g_2, \ldots, f_m + g_m)$$

and multiplication by $a \in \mathbf{C}$

$$af = (af_1, af_2, \ldots, af_m).$$

If $(a_{jk})_{j,k=1,\ldots,m}$ is a complex $m \times m$ matrix, then

$$s(f,g) = \sum_{j,k=1}^{m} a_{jk} f_j^* g_k \quad \text{for} \quad f, g \in \mathbf{C}^m$$

defines a sesquilinear form on $\mathbf{C}^m$. s is Hermitian if and only if the matrix (a_{jk}) is Hermitian, i.e., if for every $j, k = 1, 2, \ldots, m$ we have $a_{jk} = a_{kj}^*$. s is non-negative (positive) if, for example, (a_{jk}) is a diagonal matrix with non-negative (positive) entries in the diagonal. An important special case of a positive sesquilinear form on $\mathbf{C}^m$ occurs when (a_{jk}) is the unit matrix. Then

$$s(f,g) = \sum_{j=1}^{m} f_j^* g_j.$$

EXAMPLE 2. On the real vector space $\mathbf{R}^m$ (symmetric, non-negative, positive) bilinear forms can be given accordingly.

EXAMPLE 3. Let $C[0, 1]$ be the complex vector space of complex-valued continuous functions defined on $[0, 1]$ with the addition

$$(f + g)(x) = f(x) + g(x)$$

and multiplication by $a \in \mathbf{C}$

$$(af)(x) = af(x).$$

If $r : [0, 1] \to \mathbf{C}$ is continuous, then by

$$s(f,g) = \int_0^1 f(x)^* g(x) r(x)\, dx \qquad f, g \in C[0,1]$$

a sesquilinear form is defined on $C[0, 1]$. It is Hermitian if and only if r is real-valued; it is non-negative if and only if $r(x) \geq 0$ for all $x \in [0, 1]$; it is positive if and only if $r(x) \geq 0$ for all $x \in [0, 1]$ and r does not vanish identically on any non-trivial interval.

EXAMPLE 4. Let $C_{\mathbf{R}}[0, 1]$ be the real vector space of real-valued continuous functions defined on $[0, 1]$. For each continuous function $r : [0, 1] \to \mathbf{R}$ the bilinear form

$$s(f,g) = \int_0^1 f(x) g(x) r(x)\, dx \qquad f, g \in C[0,1]$$

is symmetric. Concerning non-negativity and positivity the same assertions hold as in Example 3.

EXAMPLE 5. If $k:[0, 1]\times[0, 1]\to\mathbb{C}$ is continuous, then by

$$s(f, g) = \int_0^1 \int_0^1 k(x, y) f(x)^* g(y)\, dy\, dx$$

a sesquilinear form is defined on $C[0, 1]$. This is Hermitian if and only if the kernel k is Hermitian, i.e., if for every $x, y \in [0, 1]$ we have $k(x, y) = k(y, x)^*$.

EXERCISES

1.1. Prove the assertions given in Examples 1–5.

1.2. The matrix

$$\begin{pmatrix} 0 & 1 \\ -1 & 0 \end{pmatrix}$$

generates a non-zero sesquilinear form on $\mathbb{R}^2$ (cf. Example 2), the quadratic form of which vanishes. Consequently, in a real vector space sesquilinear forms are not determined uniquely by the corresponding quadratic forms.

1.3. Let s be the sesquilinear form on $\mathbb{R}^2$ generated by the matrix

$$\begin{pmatrix} 1 & a \\ 0 & 1 \end{pmatrix}.$$

If $|a| < 2$ ($|a| \leq 2$), then we have $s(f, f) > 0$ for all $f \in \mathbb{R}^2$ such that $f \neq 0$ ($s(f, f) \geq 0$ for all $f \in \mathbb{R}^2$). If $a \neq 0$, then s is not symmetric.

1.4. Let s be a non-negative sesquilinear form on H, q the quadratic form generated by s, and $N = \{f \in H : q(f) = 0\}$. Show that
(a) N is a subspace (sub-vectorspace) of H.
(b) If $f \in N$ and $g \in H$, then we have $s(f, g) = 0$ and $q(f+g) = q(g)$.
(c) In the Schwarz inequality the equality sign holds if and only if f and g are linearly dependent modulo N, i.e., if there are numbers $a, b \in \mathbb{K}$ not vanishing simultaneously and such that $af + bg \in N$.
(d) We have $s(f, g) = [q(f)q(g)]^{1/2}$ if and only if there is a $c \geq 0$ such that $f - cg \in N$ or $g - cf \in N$.

1.5. Prove the *Cauchy inequality*

$$\left|\sum_{j=1}^m f_j^* g_j\right|^2 \leq \sum_{j=1}^m |f_j|^2 \sum_{j=1}^m |g_j|^2$$

with the aid of Example 1 and the Schwarz inequality.

1.2 Scalar products and norms

A positive sesquilinear form on H is called a *scalar product* (or *inner product*) on H. In what follows scalar products will be denoted mostly by $\langle .,. \rangle$ and occasionally they will be given an index in order to distinguish between them. A non-negative sesquilinear form is called a *semi-scalar product*. Examples for (semi-) scalar products may be obtained from the exercises in Section 1.1.

The mapping $s : H \times H \to \mathbb{K}$ *is a scalar product if and only if for all* $f, g, h \in H$ *and* $a \in \mathbb{K}$ *we have*

(i) $s(f, g+h) = s(f, g) + s(f, h)$,
(ii) $s(f, ag) = as(f, g)$,
(iii) $s(f, g) = s(g, f)^*$, (1.13)
(iv) $s(f, f) \geq 0$,
(v) $s(f, f) > 0$ if $f \neq 0$.

For the *proof* we only have to observe that the properties (1.1) and (1.2) follow from (i), (ii) and (iii). Similarly, *a mapping* $s : H \times H \to \mathbb{K}$ *is a semi-scalar product if and only if s satisfies properties* [(1.13) (i-iv)].

A mapping $p : H \to \mathbb{R}$ is called a *norm* on H if for all $f, g \in H$ and $a \in \mathbb{K}$ we have

(i) $p(f) \geq 0$,
(ii) $p(af) = |a| p(f)$, (1.14)
(iii) $p(f+g) \leq p(f) + p(g)$ (*triangle inequality*),
(iv) $p(f) > 0$ provided $f \neq 0$.

A mapping $p : H \to \mathbb{R}$ is called a *seminorm* on H if it satisfies the properties [(1.14) (i–iii)]. In what follows norms will mostly be denoted by $\| \, . \, \|$ and for more precise distinctions they will occasionally be given different indices.

Remark. If p is a seminorm on H, then for all $f, g \in H$ we have

$$p(f \pm g) \geq |p(f) - p(g)|.$$

Proof. The triangle inequality implies

$$p(f) = p(f - g + g) \leq p(f-g) + p(g),$$

thus

$$p(f) - p(g) \leq p(f-g).$$

Similarly, $p(g) = p(g - f + f) \leq p(g-f) + p(f)$; thus

$$-(p(f) - p(g)) \leq p(f-g).$$

From these two inequalities $p(f-g) \geq |p(f) - p(g)|$ follows. One can show the inequality $p(f+g) \geq |p(f) - p(g)|$ in a similar way. □

Example 1. In $\mathbb{C}^m$ (or $\mathbb{R}^m$) let us define two norms by

$$\|f\|_1 = \sum_{j=1}^{m} |f_j| \quad \text{and} \quad \|f\|_\infty = \max\{|f_j| : j = 1, \ldots, m\}.$$

If $c_j > 0$ for $j = 1, 2, \ldots, m$, then by

$$p_1(f) = \sum_{j=1}^{m} c_j |f_j| \quad \text{and} \quad p_\infty(f) = \max\{c_j |f_j| : j = 1, \ldots, m\}$$

two seminorms are defined. These seminorms are norms if all the c_j are positive.

EXAMPLE 2. If r is a non-negative continuous function on [0, 1], then by

$$p_1(f) = \int_0^1 r(x)|f(x)|\,dx$$

and

$$p_\infty(f) = \max\{r(x)|f(x)| : 0 \leq x \leq 1\}$$

two seminorms are defined on $C[0, 1]$. These are norms if r does not vanish identically on any non-trivial interval. For $r(x) \equiv 1$ these norms will be denoted by $\|\,.\,\|_1$ and $\|\,.\,\|_\infty$, respectively:

$$\|f\|_1 = \int_0^1 |f(x)|\,dx,$$

$$\|f\|_\infty = \max\{|f(x)| : 0 \leq x \leq 1\}.$$

A large number of norms can be generated with the aid of scalar products because of the following theorem.

Theorem 1.5. *If s is a semi-scalar product on H, then $p(f) = [s(f, f)]^{1/2}$ defines a seminorm on H.*

If $\langle .\,,.\rangle$ is a scalar product on H, then $\|f\| = \langle f, f\rangle^{1/2}$ defines a norm on H.

PROOF. Property [(1.14) (i)] follows immediately from [(1.13) (iv)]; [(1.14) (iv)] follows from [(1.13) (v)]. It is sufficient to prove the remaining properties for the first case. Because of [(1.13) (ii)] and [(1.13) (iii)] we have

$$p(af) = [s(af, af)]^{1/2} = [|a|^2 s(f, f)]^{1/2} = |a|[s(f, f)]^{1/2} = |a|p(f),$$

which is [(1.14) (ii)]. With the aid of the Schwarz inequality it follows that

$$\begin{aligned} p(f+g)^2 &= p(f)^2 + 2\,\mathrm{Re}\, s(f, g) + p(g)^2 \\ &\leq p(f)^2 + 2|s(f, g)| + p(g)^2 \leq p(f)^2 + 2p(f)p(g) + p(g)^2 \\ &= (p(f) + p(g))^2, \end{aligned}$$

which is the triangle inequality [(1.14) (iii)]. □

From the Schwarz inequality for non-negative sesquilinear forms we obtain for the norm $\|\,.\,\|$ (seminorm p) induced by a scalar product $\langle .\,,.\rangle$ (semi-scalar product s) that

$$|\langle f, g\rangle| \leq \|f\|\|g\|, \tag{1.15}$$

$$|s(f, g)| \leq p(f)p(g). \tag{1.15'}$$

Proposition. *If* $\langle.,.\rangle$ *is a scalar product on* H *and* $\|.\|$ *denotes the norm generated by it (cf. Theorem 1.5), then* $\|f+g\|=\|f\|+\|g\|$ *if and only if there exists an* $a\geqslant 0$ *such that* $f=ag$ *or* $g=af$.

Proof. If $f=ag$ with $a\geqslant 0$, then we have

$$\|f+g\|=\|(1+a)g\|=(1+a)\|g\|=\|ag\|+\|g\|=\|f\|+\|g\|$$

(this part of the assertion holds for any norm). Conversely, if $\|f+g\|=\|f\|+\|g\|$ then

$$\|f\|^2+2\|f\|\,\|g\|+\|g\|^2=\|f+g\|^2=\|f\|^2+2\,\mathrm{Re}\,\langle f,g\rangle+\|g\|^2,$$

thus $\mathrm{Re}\,\langle f,g\rangle=\|f\|\,\|g\|$. Using (1.15) this implies $\langle f,g\rangle=\|f\|\,\|g\|$. Now Theorem 1.4 gives the assertion. □

For a norm $\|.\|$ (seminorm p) induced by a scalar product (semi-scalar product) the *parallelogram identity*

$$\|f+g\|^2+\|f-g\|^2=2(\|f\|^2+\|g\|^2), \tag{1.16}$$

respectively

$$p(f+g)^2+p(f-g)^2=2(p(f)^2+p(g)^2). \tag{1.16'}$$

follows from Theorem 1.2.

If one considers a (semi-)norm as the length of a vector, then these equalities have the following geometric meaning: In a parallelogram the sum of the squares of the diagonals equals the sum of the squares of the sides. According to (1.4) [respectively (1.8)] the scalar product $\langle.,.\rangle$ (respectively semi-scalar product s) which we started with is given by the *polarization identity*

$$\langle f,g\rangle=\begin{cases}\frac{1}{4}\{\|f+q\|^2-\|f-g\|^2+i\|f-ig\|^2-i\|f+ig\|^2\}, & \mathbb{K}=\mathbb{C},\\ \frac{1}{4}\{\|f+g\|^2-\|f-g\|^2\}, & \mathbb{K}=\mathbb{R},\end{cases} \tag{1.17}$$

respectively

$$s(f,g)=\begin{cases}\frac{1}{4}\{p(f+g)^2-p(f-g)^2+ip(f-ig)^2-ip(f+ig)^2\}, & \mathbb{K}=\mathbb{C},\\ \frac{1}{4}\{p(f+g)^2-p(f-g)^2\}, & \mathbb{K}=\mathbb{R}.\end{cases} \tag{1.17'}$$

The following theorem enables us to decide if a given (semi-)norm is generated by a (semi-)scalar product.

Theorem 1.6 (Jordan and von Neumann). *A norm* $\|.\|$ *on a vector space* H *is generated by a scalar product* $\langle.,.\rangle$ *in the sense of Theorem 1.5 if and only if the parallelogram identity (1.16) is satisfied. If this is so then the scalar product* $\langle.,.\rangle$ *is given by (1.17). A corresponding statement holds true for seminorms and semi-scalar products.*

PROOF. If the norm $\|\,.\,\|$ is induced by the scalar product $\langle .\,,.\rangle$, then (1.16) holds true and the scalar product can be recaptured from the norm by means of (1.17). It remains to be shown that if $\|\,.\,\|$ satisfies the parallelogram identity and $\langle .\,,.\rangle$ is defined by (1.17), then $\langle .\,,.\rangle$ is a scalar product and generates the norm $\|\,.\,\|$. We restrict ourselves to the proof in the complex case; the real case goes analogously and is even a little simpler.

Let $\langle .\,,.\rangle$ be defined by (1.17). We show that $\langle .\,,.\rangle$ is a scalar product.

[(1.13) (iv–v)]: For all $f\in H$ by virtue of the definition of $\langle .\,,.\rangle$ we have

$$\begin{aligned}\langle f,f\rangle &= \tfrac{1}{4}\{\|f+f\|^2-\|0\|^2+i\|f-if\|^2-i\|f+if\|^2\}\\ &= \tfrac{1}{4}\{4\|f\|^2-0+2i\|f\|^2-2i\|f\|^2\} = \|f\|^2.\end{aligned}$$

The properties [(1.13) (iv–v)] of $\langle .\,,.\rangle$ now follow from the corresponding properties of the norm $\|\,.\,\|$. At the same time we obtain that $\|\,.\,\|$ is generated by $\langle .\,,.\rangle$.

[(1.13) (iii)]: For all $f, g\in H$ we have

$$\begin{aligned}\langle g,f\rangle^* &= \tfrac{1}{4}\{\|g+f\|^2-\|g-f\|^2+i\|g-if\|^2-i\|g+if\|^2\}^*\\ &= \tfrac{1}{4}\{\|f+g\|^2-\|f-g\|^2-i\|f+ig\|^2+i\|f-ig\|^2\}\\ &= \langle f,g\rangle.\end{aligned}$$

[(1.13) (i)]: For all $f, g, h\in H$ because of (1.16) we have

$$\begin{aligned}&\langle f,g\rangle+\langle f,h\rangle\\ &= \tfrac{1}{4}\{\|f+g\|^2-\|f-g\|^2+i\|f-ig\|^2-i\|f+ig\|^2\\ &\qquad+\|f+h\|^2-\|f-h\|^2+i\|f-ih\|^2-i\|f+ih\|^2\}\\ &= \tfrac{1}{4}\Big\{\Big\|\Big(f+\frac{g+h}{2}\Big)+\frac{g-h}{2}\Big\|^2+\Big\|\Big(f+\frac{g+h}{2}\Big)-\frac{g-h}{2}\Big\|^2\\ &\qquad-\Big\|\Big(f-\frac{g+h}{2}\Big)+\frac{g-h}{2}\Big\|^2-\Big\|\Big(f-\frac{g+h}{2}\Big)-\frac{g-h}{2}\Big\|^2\\ &\qquad+i\Big\|\Big(f-i\frac{g+h}{2}\Big)+i\frac{g-h}{2}\Big\|^2+i\Big\|\Big(f-i\frac{g+h}{2}\Big)-i\frac{g-h}{2}\Big\|^2\\ &\qquad-i\Big\|\Big(f+i\frac{g+h}{2}\Big)+i\frac{g-h}{2}\Big\|^2-i\Big\|\Big(f+i\frac{g+h}{2}\Big)-i\frac{g-h}{2}\Big\|^2\Big\}\\ &= \frac{1}{2}\Big\{\Big\|f+\frac{g+h}{2}\Big\|^2+\Big\|\frac{g-h}{2}\Big\|^2-\Big\|f-\frac{g+h}{2}\Big\|^2-\Big\|\frac{g-h}{2}\Big\|^2\\ &\qquad+i\Big\|f-i\frac{g+h}{2}\Big\|^2+i\Big\|\frac{g-h}{2}\Big\|^2-i\Big\|f+i\frac{g+h}{2}\Big\|^2-i\Big\|\frac{g-h}{2}\Big\|^2\Big\}\\ &= 2\Big\langle f,\frac{g+h}{2}\Big\rangle. \end{aligned}\tag{1.18}$$

Since by (1.17) we obviously have $\langle f,0\rangle=0$, from (1.18) it follows by substituting $h=0$ that

$$2\langle f,\tfrac{1}{2}g\rangle=\langle f,g\rangle.\tag{1.19}$$

From (1.18) and (1.19) it follows that

$$\langle f, g\rangle + \langle f, h\rangle = 2\left\langle f, \frac{g+h}{2}\right\rangle = \langle f, g+h\rangle,$$

which is the required property.

[(1.13) (ii)]: We already know that $\langle f, g\rangle = 2\langle f, g/2\rangle$. From this and from property [(1.13) (i)] we obtain by induction that

$$2^{-n}m\langle f, g\rangle = \langle f, 2^{-n}mg\rangle \quad \text{for all} \quad n, m \in \mathbb{N}_0$$

($\mathbb{N}_0$ is the set of non-negative integers $\{0, 1, 2, \ldots\}$). If $a \geq 0$, then there exist numbers $a_k = 2^{-n(k)}m(k)$ such that $a_k \to a$ as $k \to \infty$. By the proposition preceding Example 1 we have

$$|\|f \pm a_k g\| - \|f \pm ag\|| \leq |a_k - a| \, \|g\|,$$
$$\|f \pm ia_k g\| - \|f \pm iag\|| \leq |a_k - a| \, \|g\|,$$

therefore because of (1.17)

$$\langle f, a_k g\rangle \to \langle f, ag\rangle \quad \text{as} \quad k \to \infty.$$

From this it follows that

$$a\langle f, g\rangle = \lim_{k\to\infty} a_k\langle f, g\rangle = \lim_{k\to\infty} \langle f, a_k g\rangle = \langle f, ag\rangle.$$

Furthermore, we have

$$\begin{aligned}\langle f, -g\rangle &= \tfrac{1}{4}\{\|f-g\|^2 - \|f+g\|^2 + i\|f+ig\|^2 - \|f-ig\|^2\} \\ &= -\langle f, g\rangle;\end{aligned}$$

consequently $\langle f, ag\rangle = a\langle f, g\rangle$ for all $a \in \mathbb{R}$. As we also have

$$\begin{aligned}\langle f, ig\rangle &= \tfrac{1}{4}\{\|f+ig\|^2 - \|f-ig\|^2 + i\|f+g\|^2 - i\|f-g\|^2\} \\ &= i\langle f, g\rangle.\end{aligned}$$

The equality $\langle f, ag\rangle = a\langle f, g\rangle$ follows for all $a \in \mathbb{C}$. The proof for semi-norms is completely analogous. □

If H is a (complex or real) vector space and $\langle . , . \rangle$ is a scalar product on H, then we call the pair $(H, \langle . , . \rangle)$ a *vector space with scalar product* or a *pre-Hilbert space*. If it is clear which scalar product is meant on H, then we shall briefly write H for the pair mentioned. If $\| . \|$ is a norm on H, then we call the pair $(H, \| . \|)$ a normed space. Here we shall also only write H in most cases. By Theorem 1.5 the norm $\|f\| = \langle f, f\rangle^{1/2}$ is defined in a natural way on every pre-Hilbert space. Therefore in what follows we shall consider every pre-Hilbert space as a normed space.

EXAMPLE 3. On $\mathbb{C}^m$ respectively $\mathbb{R}^m$ by

$$\langle f, g\rangle = \sum_{j=1}^{m} f_j^* g_j$$

a scalar product is defined. The corresponding norm

$$\|f\| = \left\{ \sum_{j=1}^{m} |f_j|^2 \right\}^{1/2}$$

is the *Euclidean length* of the vector f, thus $\|f-g\|$ is the *Euclidean distance* of the points f and g.

EXAMPLE 4. On $C[0, 1]$ by

$$\langle f, g\rangle = \int_0^1 f(x)^* g(x)\,\mathrm{d}x, \quad \|f\| = \left\{ \int_0^1 |f(x)|^2\,\mathrm{d}x \right\}^{1/2}$$

a scalar product and the corresponding norm are defined.

EXAMPLE 5. Let l_2 be the *Hilbert sequence space*, i.e., the set of (real or complex) sequences $f = (f_n) = (f_1, f_2, \dots)$ for which $\Sigma_{n=1}^{\infty}|f_n|^2 < \infty$. Then l_2 will be a (real or complex) vector space if one defines addition and multiplication as follows:

$$f + g = (f_n + g_n),\ af = (af_n) \quad \text{for} \quad f, g \in l_2 \quad \text{and} \quad a \in \mathbb{K}.$$

It is clear that this definition of multiplication is meaningful since along with $\Sigma_{n=1}^{\infty}|f_n|^2 < \infty$ we also have $\Sigma_{n=1}^{\infty}|af_n|^2 < \infty$. If f and g are in l_2, then for every $N \in \mathbb{N}$ we have

$$\sum_{n=1}^{N} |f_n + g_n|^2 \leqslant 2\left\{ \sum_{n=1}^{N} |f_n|^2 + \sum_{n=1}^{N} |g_n|^2 \right\} \leqslant 2\left\{ \sum_{n=1}^{\infty} |f_n|^2 + \sum_{n=1}^{\infty} |g_n|^2 \right\},$$

consequently we also have

$$\sum_{n=1}^{\infty} |f_n + g_n|^2 \leqslant 2\left\{ \sum_{n=1}^{\infty} |f_n|^2 + \sum_{n=1}^{\infty} |g_n|^2 \right\} < \infty,$$

i.e., $f + g \in l_2$. It is easy to see that by

$$\langle f, g\rangle = \sum_{n=1}^{\infty} f_n^* g_n, \quad f, g \in l_2$$

a scalar product is defined on l_2; the series converges, because $|f_n^* g_n| \leqslant (|f_n|^2 + |g_n|^2)/2$. The induced norm is

$$\|f\| = \left\{ \sum_{n=1}^{\infty} |f_n|^2 \right\}^{1/2}$$

Unless otherwise stated, in what follows l_2 will always denote the complex sequence space.

Exercises

1.6. The norms in Examples 1 and 2 are not generated by scalar products.

1.7. The proposition after Theorem 1.5 does not hold true in general for norms that are not generated by scalar products.

1.8. Let p be a seminorm on H generated by a semi-scalar product and let $N=\{f\in H: p(f)=0\}$. We have $p(f+g)=p(f)+p(g)$ if and only if there exists an $a>0$ such that $f-ag\in N$ or $g-af\in N$.

1.9. (a) Let A^2 be the set of functions f holomorphic on $\mathbb{C}_1=\{z\in\mathbb{C}: |z|<1\}$ for which

$$\int_{\mathbb{C}_1}|f(x+iy)|^2\,dx\,dy<\infty$$

(the integral can be understood as an improper Riemann integral or as a Lebesgue integral). A^2 is a vector space. By

$$\langle f,g\rangle_1=\int_{\mathbb{C}_1}f(x+iy)^*g(x+iy)\,dx\,dy,\ \|f\|_1^2=\int_{\mathbb{C}_1}|f(x+iy)|^2\,dx\,dy$$

a scalar product and the corresponding norm are defined on A^2.

(b) Let H^2 be the set of functions f holomorphic on $\mathbb{C}_1$ for which the limit

$$\lim_{r\to 1}\int_0^{2\pi}|f(re^{it})|^2\,dt$$

is finite. H^2 is a vector space (*Hardy-class*). By

$$\|f\|_2=\left\{\lim_{r\to 1}\int_0^{2\pi}|f(re^{it})|^2\,dt\right\}^{1/2}$$

a norm is defined on H^2. This norm is generated by the scalar product

$$\langle f,g\rangle_2=\lim_{r\to 1}\int_0^{2\pi}f(re^{it})^*g(re^{it})\,dt.$$

(c) If $f(z)=\sum_{n=0}^{\infty}f_nz^n$, $g(z)=\sum_{n=0}^{\infty}g_nz^n$ are the Taylor series of f and g, then we have

$$\langle f,g\rangle_1=\pi\sum_{n=0}^{\infty}\frac{1}{n+1}f_n^*g_n,\quad \langle f,g\rangle_2=2\pi\sum_{n=0}^{\infty}f_n^*g_n.$$

(d) H^2 is a subspace of A^2 and we have $\|f\|_1^2\leq\frac{1}{2}\|f\|_2^2$ for $f\in H^2$.

(e) For all $f\in H^2$ we have

$$\|f\|_2^2=\sup\left\{\int_0^{2\pi}|f(re^{it})|^2\,dt: 0<r<1\right\}.$$

1.10. Let A be an arbitrary set, let $\mu: A\to(0,\infty)$, and let $l_2(A;\mu)$ be the set of functions $f: A\to\mathbb{C}$ that vanish outside a countable set (that may vary with f) and for which $\sum_{\alpha\in A}\mu(\alpha)|f(\alpha)|^2<\infty$.

(a) $(l_2A;\mu)$ is a subspace of the space of all complex valued functions on A.

(b) By

$$\langle f,g\rangle=\sum_{\alpha\in A}\mu(\alpha)f(\alpha)^*g(\alpha),\quad f,g\in l_2(A;\mu),$$

a scalar product is defined on $l_2(A;\mu)$.

1.11. Let A be an arbitrary set; for each $\alpha \in A$ let $(H_\alpha, \langle .\,,.\rangle_\alpha)$ be a pre-Hilbert space. Then

$$H = \Big\{ f = (f_\alpha)_{\alpha \in A} \in \prod_{\alpha \in A} H_\alpha : \\ f_\alpha \neq 0 \text{ for at most countably many } \alpha \in A, \ \sum_{\alpha \in A} \|f_\alpha\|_\alpha^2 < \infty \Big\}$$

is a vector space (with componentwise addition and multiplication). By

$$\langle (f_\alpha), (g_\alpha) \rangle = \sum_{\alpha \in A} \langle f_\alpha, g_\alpha \rangle_\alpha, \quad (f_\alpha), (g_\alpha) \in H,$$

a scalar product is defined on H, i.e., $(H, \langle .\,,.\rangle)$ is a pre-Hilbert space.

Hilbert spaces 2

2.1 Convergence and completeness

Let $(H, \|\,.\,\|)$ be a normed space. A sequence (f_n) in H is said to be *convergent* if there exists an $f \in H$ such that $\|f_n - f\| \to 0$ as $n \to \infty$. There exists at most one $f \in H$ with $\|f_n - f\| \to 0$; since from $\|f_n - f\| \to 0$ and $\|f_n - g\| \to 0$ it follows that $\|f - g\| \leqslant \|f - f_n\| + \|f_n - g\| \to 0$, thus $f = g$. We say that the sequence (f_n) *tends* to f and call f the *limit* of the sequence (f_n). In symbols we write $f = \lim_{n\to\infty} f_n$ or $f_n \to f$ as $n \to \infty$. If no confusion is possible, we shall occasionally abbreviate these by writing $f = \lim f_n$, or $f_n \to f$.

Proposition.
(a) *From $f_n \to f$ it follows that $\|f_n\| \to \|f\|$; the sequence $(\|f_n\|)$ is bounded.*
(b) *If $(H, \langle . , . \rangle)$ is a pre-Hilbert space, then we also have that $f_n \to f$ and $g_n \to g$ imply $\langle f_n, g_n \rangle \to \langle f, g \rangle$.*

PROOF.
(a) By the proposition preceding Example 1 of Section 1.2, we have $|\|f_n\| - \|f\|| \leqslant \|f_n - f\|$; from this the assertion follows.
(b) We have $|\langle f_n, g_n \rangle - \langle f, g \rangle| \leqslant |\langle f_n, g_n \rangle - \langle f_n, g \rangle| + |\langle f_n, g \rangle - \langle f, g \rangle| \leqslant \|f_n\|\,\|g_n - g\| + \|f_n - f\|\,\|g\| \to 0$, since the sequence $(\|f_n\|)$ is bounded on account of (a). □

A sequence (f_n) in H is called a *Cauchy sequence* if for each $\epsilon > 0$ there exists an $n_0 \in \mathbb{N}$ such that for $n, m \geqslant n_0$ we have $\|f_n - f_m\| \leqslant \epsilon$. In what follows, we shall briefly write for this $\|f_n - f_m\| \to 0$ as $n, m \to \infty$. *Every convergent sequence is a Cauchy sequence*: if f is the limit of the sequence (f_n), then $\|f_n - f_m\| \leqslant \|f_n - f\| + \|f - f_m\| \to 0$ as $n, m \to \infty$. Conversely, in an arbitrary normed space (or pre-Hilbert space) not every Cauchy sequence is convergent, as Example 1 below shows.

Proposition.
(a) *If* (f_n) *is a Cauchy sequence, then the sequnce* $(\|f_n\|)$ *is convergent (thus it is bounded).*
(b) *If* $(H, \langle .,. \rangle)$ *is a pre-Hilbert space and* (f_n), (g_n) *are Cauchy sequences, then the sequence* $(\langle f_n, g_n \rangle)$ *is convergent.*

PROOF.
(a) As $|\|f_n\| - \|f_m\|| \leq \|f_n - f_m\|$, the sequence $(\|f_n\|)$ is a Cauchy sequence in $\mathbb{R}$, thus it is convergent and bounded.
(b) By (a) there exists a $c > 0$ such that $\|f_n\| \leq c$ and $\|g_m\| \leq c$ for all $n, m \in \mathbb{N}$. Since

$$|\langle f_n, g_n \rangle - \langle f_m, g_m \rangle| \leq |\langle f_n, g_n - g_m \rangle| + |\langle f_n - f_m, g_m \rangle| \leq c(\|g_n - g_m\| + \|f_n - f_m\|),$$

the sequence $(\langle f_n, g_n \rangle)$ is also a Cauchy sequence. □

EXAMPLE 1. Let $(C[0, 1], \langle .,. \rangle)$ be the pre-Hilbert space introduced in Section 1.2, Example 4. We show that not every Cauchy sequence is convergent. For this let the sequence (f_n) in $C[0, 1]$ be defined in the following way: $f_1(x) = 1$ for all $x \in [0, 1]$, and

$$f_n(x) = \begin{cases} 1 & \text{for } 0 \leq x \leq \frac{1}{2} \\ 1 - \left(x - \frac{1}{2}\right)n & \text{for } \frac{1}{2} < x < \frac{1}{2} + \frac{1}{n}, \\ 0 & \text{for } \frac{1}{2} + \frac{1}{n} \leq x \leq 1, \end{cases}$$

for $n = 2, 3, \ldots$. This sequence is a Cauchy sequence, since for $2 \leq n \leq m$ we have

$$\|f_n - f_m\|^2 \leq \int_{1/2}^{1/2+1/n} |f_n(x) - f_m(x)|^2 \, dx \leq \frac{1}{n}.$$

To prove that the sequence (f_n) is not convergent let us assume that there exists an $f \in C[0, 1]$ such that $f_n \to f$, i.e., $\|f_n - f\| \to 0$. Then for $2 \leq n \leq m$ we have

$$\int_0^{1/2} |f(x) - 1|^2 \, dx + \int_{(1/2)+(1/n)}^1 |f(x)|^2 \, dx$$
$$= \int_0^{1/2} |f(x) - f_m(x)|^2 \, dx + \int_{(1/2)+(1/n)}^1 |f(x) - f_m(x)|^2 \, dx$$
$$\leq \int_0^1 |f(x) - f_m(x)|^2 \, dx.$$

Since the right-hand side tends to 0 as $m \to \infty$, we have

$$\int_0^{1/2} |f(x) - 1|^2 \, dx + \int_{(1/2)+(1/n)}^1 |f(x)|^2 \, dx = 0 \quad \text{for} \quad n \geqslant 2.$$

Since f is continuous, it follows from this that

$$f(x) = 1 \quad \text{for} \quad x \in \left[0, \tfrac{1}{2}\right],$$
$$f(x) = 0 \quad \text{for} \quad x \in \left(\tfrac{1}{2}, 1\right].$$

However, this contradicts the continuity of f. Therefore the sequence (f_n) cannot be convergent in $C[0, 1]$.

A normed space $(H, \| \cdot \|)$ is said to be *complete* if every Cauchy sequence is convergent. A complete normed space is called a *Banach space*; a complete pre-Hilbert space is called a *Hilbert space*.

EXAMPLE 2. The space $C[0, 1]$ becomes a Banach space with the norm (cf. Section 1.2, Example 2)

$$\|f\|_\infty = \max\{|f(x)| : 0 \leqslant x \leqslant 1\}.$$

(By Exercise 1.6 it is not a Hilbert space.) For suppose (f_n) is a Cauchy sequence, i.e., assume that for every $\epsilon > 0$ there exists an $n_0 \in \mathbb{N}$ such that for all $n, m \geqslant n_0$ and for all $x \in [0, 1]$ we have $|f_n(x) - f_m(x)| \leqslant \epsilon$. Then $(f_n(x))$ is convergent for every $x \in [0, 1]$; let $f(x) = \lim f_n(x)$. First we show that this f is continuous. For $\epsilon > 0$ let n_0 be chosen as above and for this n_0 let $\delta > 0$ be chosen so that for $|x_1 - x_2| \leqslant \delta$ we have $|f_{n_0}(x_1) - f_{n_0}(x_2)| \leqslant \epsilon$. From this it follows for $|x_1 - x_2| \leqslant \delta$ that

$$\begin{aligned} |f(x_1) - f(x_2)| &\leqslant |f(x_1) - f_{n_0}(x_1)| + |f_{n_0}(x_1) - f_{n_0}(x_2)| + |f_{n_0}(x_2) - f(x_2)| \\ &= \lim_{m\to\infty} |f_m(x_1) - f_{n_0}(x_1)| + |f_{n_0}(x_1) - f_{n_0}(x_2)| \\ &\quad + \lim_{m\to\infty} |f_{n_0}(x_2) - f_m(x_2)| \leqslant 3\epsilon. \end{aligned}$$

This proves the continuity of f. Now we show that $f_n \to f$. For $\epsilon > 0$ let n_0 be chosen again as above. Then for $n \geqslant n_0$ we have

$$\begin{aligned} \|f_n - f\|_\infty &= \max\{|f_n(x) - f(x)| : 0 \leqslant x \leqslant 1\} \\ &= \max\left\{ \lim_{m\to\infty} |f_n(x) - f_m(x)| : 0 \leqslant x \leqslant 1 \right\} \leqslant \epsilon, \end{aligned}$$

consequently $f_n \to f$.

EXAMPLE 3. $\mathbb{C}^m$ and $\mathbb{R}^m$ are Banach spaces with the norms $\| \cdot \|_1$, $\| \cdot \|_\infty$ and $\| \cdot \|$ from Section 1.2, Examples 1 and 3. This follows easily from the fact that a sequence (f_n) is a Cauchy sequence (convergent sequence) in $\mathbb{C}^m$ or $\mathbb{R}^m$ if and only if it converges componentwise. (The proof can also be obtained as a special case of Example 4.) $\mathbb{C}^m$ (respectively $\mathbb{R}^m$) is therefore a Hilbert space with the scalar product

$$\langle f, g\rangle = \sum_{j=1}^{m} f_j^* g_j.$$

EXAMPLE 4. Let the scalar product and the norm in l_2 be defined by

$$\langle f, g\rangle = \sum_{n=1}^{\infty} f_n^* g_n, \quad \|f\| = \left\{ \sum_{n=1}^{\infty} f_n^2 \right\}^{1/2},$$

as in Section 1.2, Example 5. We show that l_2 is complete, therefore it is a Hilbert space. Let $(f^{(n)})$ be a Cauchy sequence, $f^{(n)} = (f_{1,n}, f_{2,n}, f_{3,n}, \ldots)$. As

$$|f_{j,n} - f_{j,m}| \leqslant \|f^{(n)} - f^{(m)}\|,$$

the sequence $(f_{j,n})_{n\in\mathbb{N}}$ is a Cauchy sequence for each $j \in \mathbb{N}$, i.e., there are numbers $f_j \in \mathbb{C}$ such that $f_{j,n} \to f_j$ as $n \to \infty$. It remains to prove that $f = (f_j) \in l_2$ and $f^{(n)} \to f$ as $n \to \infty$. For $\epsilon > 0$ let $n_0 \in \mathbb{N}$ be chosen so that for $n, m \geqslant n_0$ we have $\|f^{(n)} - f^{(m)}\| \leqslant \epsilon$. Then for all $k \in \mathbb{N}$ we have

$$\sum_{j=1}^{k} |f_{j,n} - f_j|^2 = \lim_{m\to\infty} \sum_{j=1}^{k} |f_{j,n} - f_{j,m}|^2 \leqslant \limsup_{m\to\infty} \|f^{(n)} - f^{(m)}\|^2 \leqslant \epsilon^2,$$

therefore also

$$\sum_{j=1}^{\infty} |f_{j,n} - f_j|^2 \leqslant \epsilon^2 \quad \text{for} \quad n \geqslant n_0.$$

It follows from this that $f^{(n)} - f \in l_2$, thus $f \in l_2$, also, and $\|f^{(n)} - f\| \leqslant \epsilon$ for $n \geqslant n_0$, i.e., $f^{(n)} \to f$.

EXAMPLE 5. The *Lebesgue space* $L_2(M)$ for a Lebesgue measurable subset M of $\mathbb{R}^m$: For the concepts and results of this example a knowledge of Lebesgue's integration theory is needed (cf. Appendix A). This will be assumed in what follows. The notions of "measurable," "almost everywhere," and "integrable" refer to Lebesgue measure in $\mathbb{R}^m$.

Let M be a measurable subset of $\mathbb{R}^m$. First we treat the function space

$$\mathfrak{L}_2(M) = \left\{ f : f \text{ measurable complex-valued on } M, \int_M |f(x)|^2 \, dx < \infty \right\}.$$

$\mathfrak{L}_2(M)$ is a vector space, since with $f, g \in \mathfrak{L}_2(M)$, $a \in \mathbb{C}$ the functions af and $f + g$ are also measurable and because of

$$|af(x)| = |a|\,|f(x)| \quad \text{and} \quad |f(x) + g(x)|^2 \leqslant 2|f(x)|^2 + 2|g(x)|^2$$

we have

$$\int_M |af(x)|^2 \, dx < \infty \quad \text{and} \quad \int_M |f(x) + g(x)|^2 \, dx < \infty.$$

It is obvious that by $s(f, g) = \int_M f^*(x)g(x)\,dx$, $f, g \in \mathfrak{L}_2(M)$ a semi-scalar product is defined on $\mathfrak{L}_2(M)$. Let

$$\mathfrak{N}(M) = \{f : f \text{ measurable complex-valued function on } M,\ f(x) = 0 \text{ almost everywhere on } M\}.$$

Then $\mathfrak{N}(M)$ is a subspace of $\mathfrak{L}_2(M)$ and we have

$$\mathfrak{N}(M) = \{f \in \mathfrak{L}_2(M) : s(f, f) = 0\}.$$

Now we define

$$L_2(M) = \mathfrak{L}_2(M)/\mathfrak{N}(M);$$

Thus we build equivalence classes in $\mathfrak{L}_2(M)$ by placing two functions in the same class if they coincide almost everywhere. Addition of these classes and multiplication by a complex number are defined via representatives: If $\tilde{f}$ and $\tilde{g}$ are the equivalence classes of f and g and $a \in \mathbb{C}$, then

$$a\tilde{f} = (af)^{\sim}, \quad a\tilde{f} + b\tilde{g} = (af + bg)^{\sim}.$$

The scalar product of two equivalence classes $\tilde{f}$ and $\tilde{g}$ is defined by

$$\langle \tilde{f}, \tilde{g} \rangle = s(f, g) = \int_M f^*(x)g(x)\,dx,$$

where f and g are representatives of $\tilde{f}$ and $\tilde{g}$. It is evident that this definition does not depend on the choice of f and g. From $\langle \tilde{f}, \tilde{f} \rangle = 0$ it follows that the representatives of $\tilde{f}$ vanish almost everywhere, i.e., $\tilde{f}$ is the zero element of $\mathfrak{L}_2(M)/\mathfrak{N}(M)$. Therefore $\langle . , . \rangle$ is actually a scalar product, and $L_2(M)$ is thus a pre-Hilbert space. In what follows we shall denote the functions $f \in \mathfrak{L}_2(M)$ and the corresponding equivalence classes $\tilde{f} \in L_2(M)$ by the same symbol f. The function f is then always an arbitrary representative of the corresponding equivalence class.

Theorem 2.1. ($L_2(M)$, $\langle . , . \rangle$) *is complete, thus it is a Hilbert space. If $f_n \to f$, then there is a subsequence (f_{n_j}) of (f_n) such that*

$$f_{n_j}(x) \to f(x) \quad \textit{as} \quad j \to \infty, \quad \textit{almost everywhere in} \quad M$$

(here $f_{n_j}(\,.\,)$ and $f(\,.\,)$ are arbitrary representatives of f_{n_j}, respectively f).

Proof. Let (f_n) be a Cauchy sequence in $L_2(M)$. For each $j \in \mathbb{N}$ there exists an n_j such that

$$\|f_m - f_n\| < 2^{-j} \quad \text{for} \quad n, m > n_j.$$

Without loss of generality we may assume that $n_{j+1} > n_j$ for all $j \in \mathbb{N}$. Then we have in particular $\|f_{n_{j+1}} - f_{n_j}\| < 2^{-j}$. In what follows let $f_n(\,.\,)$ be an arbitrary (however, in the rest of the proof fixed) representative of f_n.

For all $k \in \mathbb{N}$ let $g_k : M \to \mathbb{R}$ be defined by the equality

$$g_k(x) = \sum_{j=1}^{k} |f_{n_{j+1}}(x) - f_{n_j}(x)|.$$

The sequence $(g_k^2(\,.\,))$ is non-decreasing, and

$$\int g_k(x)^2\, dx = \|g_k\|^2 \leqslant \left(\sum_{j=1}^{k} 2^{-j}\right)^2 < 1$$

for all $k \in \mathbb{N}$. By B. Levi's theorem (Theorem A 7) the sequence g_k^2, and thus also the sequence (g_k), is convergent almost everywhere. Then the sequence of the functions

$$f_{n_k} = f_{n_1} + \sum_{j=1}^{k-1} (f_{n_{j+1}} - f_{n_j})$$

also converges almost everywhere to a measurable function $f(\,.\,)$. We show that $f(\,.\,) \in \mathfrak{L}_2(M)$ and that in the sense of $L_2(M)$ we have $f_n \to f$ as $n \to \infty$. For each $\epsilon > 0$ let $n(\epsilon)$ and $j(\epsilon)$ be chosen so that for $n \geqslant n(\epsilon)$ and $j \geqslant j(\epsilon)$ we have

$$\int_M |f_{n_j}(x) - f_n(x)|^2\, dx = \|f_{n_j} - f_n\|^2 \leqslant \epsilon.$$

The functions $|f_{n_j}(\,.\,) - f_n(\,.\,)|^2$ are non-negative, their integrals are bounded by ϵ and for $j \to \infty$ we have

$$|f_{n_j}(x) - f_n(x)|^2 \to |f(x) - f_n(x)|^2 \quad \text{almost everywhere in } M.$$

By Fatou's lemma it follows from this that $|f(\,.\,) - f_n(\,.\,)|^2$ is integrable and that we have

$$\int_M |f(x) - f_n(x)|^2\, dx \leqslant \epsilon \quad \text{for} \quad n \geqslant n(\epsilon).$$

Therefore $f(\,.\,) - f_n(\,.\,) \in \mathfrak{L}_2(M)$ and, consequently, $f(\,.\,) \in \mathfrak{L}_2(M)$. Besides, we have $\|f - f_n\|^2 \leqslant \epsilon$ for $n \geqslant n(\epsilon)$, i.e., $f_n \to f$ in the sense of $L_2(M)$. The second part of the assertion is proved by the fact that $f_{n_j}(x) \to f(x)$ almost everywhere. □

If we look only at real valued functions in this example, then we obtain the real Hilbert space $L_{2,\mathbb{R}}(M)$.

EXAMPLE 6. All the reasoning of Example 5 can be carried out analogously if ρ is a measure generated by a regular interval function on $\mathbb{R}^m$ (cf. Appendix A), M is a ρ-measurable subset of $\mathbb{R}^m$ and $L_2(M\,;\rho)$ is the corresponding space of square integrable functions with respect to ρ. Theorem 2.1 holds true for $L_2(M\,;\rho)$ also. We omit the details here.

EXERCISES

2.1. Let (f_n) be a sequence in the normed space $(H, \|\cdot\|)$ with $\Sigma_{n=1}^{\infty}\|f_n\| < \infty$.
(a) $f_n \to 0$ and the sequence $(\Sigma_{j=1}^{n} f_j)$ is a Cauchy sequence.
(b) If H is a Banach space, then the sequence $(\Sigma_{j=1}^{n} f_j)$ is convergent; we write $\Sigma_{j=1}^{\infty} f_j$ for the limit of this sequence.

2.2. (a) In Exercise 1.11 H is a Hilbert space if and only if all H_α are Hilbert spaces.
(b) The space $l_2(A;\mu)$ of Exercise 1.10 is a Hilbert space.
(c) The spaces A^2 and H^2 of Exercise 1.9 are Hilbert spaces.
Hint: This can be proved with the aid of Exercise 1.9(c) or the mean value property of holomorphic functions.

2.3. (a) Let $C^k[0, 1]$ be the vector space of k times continuously differentiable complex (or real) valued functions defined on $[0, 1]$. By

$$\langle f, g\rangle_k = \sum_{j=0}^{k} \int_0^1 f^{(j)}(x)^* g^{(j)}(x)\,dx$$

a scalar product is defined on $C^k[0, 1]$. The space $(C^k[0, 1], \langle.,.\rangle_k)$ is not complete.
(b) Let $W_{2,k}(0, 1)$ be the space of those complex-valued functions on $[0, 1]$ that are $k-1$ times continuously differentiable, whose $(k-1)$th derivative is absolutely continuous (cf. Appendix A 5) and whose kth derivate is in $L_2(0, 1)$. By

$$\langle f, g\rangle_k = \sum_{j=0}^{k} \int_0^1 f^{(j)}(x)^* g^{(j)}(x)\,dx$$

a scalar product is defined on $W_{2,k}(0, 1)$. The pair $(W_{2,k}(0, 1), \langle.,.\rangle_k)$ is a Hilbert space.
(c) $C^k[0, 1]$ is a subspace of $W_{2,k}(0, 1)$. For each $f \in W_{2,k}(0, 1)$ there exists a sequence (f_n) from $C^k[0, 1]$ such that $f_n \to f$ in the sense of $W_{2,k}(0, 1)$.

2.2 Topological notions

Let $(H, \|\cdot\|)$ be a normed space. A subset A of H is said to be *open* if for each $f \in A$ there exists an $\epsilon > 0$ such that the *ball*

$$K(f, \epsilon) = \{g \in H : \|g - f\| < \epsilon\}$$

lies in A.

EXAMPLE 1. For each $r > 0$ and each $h \in H$ the ball $K(h, r) = \{g \in H : \|g - h\| < r\}$ is open. It will be called the *open ball around h with radius r*. The assertion is obvious for $r = 0$, as $K(h, r)$ is then empty (the empty set is open). Now let $r > 0$, $g \in K(h, r)$, then we have

$$\epsilon = r - \|g - h\| > 0$$

and for each $f \in K(g, \epsilon)$

$$\|h-f\| \leqslant \|h-g\|+\|g-f\| < \|h-g\|+\epsilon = r,$$

i.e., $K(g, \epsilon) \subset K(h, r)$.

A subset A of H is said to be *closed* if $\complement A = H \setminus A$, the *complement* of A, is open.

EXAMPLE 2. For each $f \in H$ and each $r > 0$ the ball $\overline{K}(f, r) = \{g \in H: \|g-f\| \leqslant r\}$ is closed, because for $g \in \complement\overline{K}(f, r)$ we have $\|g-f\|-r>0$ and $K(g, \|g-f\|-r) \subset \complement\overline{K}(f, r)$. The set $\overline{K}(f, r)$ is called the *closed ball around f with radius r*.

Closed sets can be characterized in another way. For this we mention another definition. An element $f \in H$ is called a *contact point* of the subset A of H if for each $\epsilon > 0$ there exists a $g \in A$ such that $\|g-f\| < \epsilon$. The set of all contact points of A will be denoted by $\overline{A}$. We obviously have $A \subset \overline{A}$.

Proposition.
(1) *$A \subset B$ implies $\overline{A} \subset \overline{B}$.*
(2) *We have $f \in \overline{A}$ if and only if there exists a sequence (f_n) in A such that $f_n \to f$.*
(3) *We have $\overline{\overline{A}} = A$.*

PROOF. (1) and (2) are clear.
(3) Let $f \in \overline{\overline{A}}$, $\epsilon > 0$. Then there exists a $g \in \overline{A}$ such that $\|g-f\| < \epsilon/2$ and for this g there exists an $h \in A$ such that $\|h-g\| < \epsilon/2$; consequently $\|h-f\| < \epsilon$. Therefore $f \in \overline{A}$ holds, i.e., $\overline{\overline{A}} \subset \overline{A}$. Since $\overline{A} \subset \overline{\overline{A}}$, it follows that $\overline{A} = \overline{\overline{A}}$. □

Theorem 2.2. *$\overline{A}$ is closed. A is closed if and only if $A = \overline{A}$. The set $\overline{A}$ is the smallest closed subset of H that contains A.*

PROOF. First we show that $\overline{A}$ is closed, i.e., $\complement\overline{A}$ is open. Let $f \in \complement\overline{A}$. Since $\overline{A} = \overline{\overline{A}}$, then we have $f \in \complement\overline{\overline{A}}$, i.e., f is not a contact point of $\overline{A}$. Therefore there is an $\epsilon > 0$ such that $K(f, \epsilon) \cap \overline{A} = \varnothing$ and consequently $K(f, \epsilon) \subset \complement\overline{A}$. If $A = \overline{A}$, then A is closed by the first part of our theorem. If A is closed, then $\complement A$ is open, i.e., for each $f \in \complement A$ there exists an $\epsilon > 0$ such that $K(f, \epsilon) \cap A = \varnothing$. However, this means that no element f of $\complement A$ is a contact point of A, therefore $\overline{A} \subset A$ and thus $A = \overline{A}$.

If $B \subset H$ is closed and $A \subset B$, then it follows that $\overline{A} \subset \overline{B} = B$, therefore $\overline{A} \subset B$. □

On the basis of Theorem 2.2 it is justified to call $\overline{A}$ the *closure* (*closed hull*) of A.

EXAMPLE 3. For $r>0$ the closed ball $\overline{K}(f, r)$ is the closure of $K(f, r)$. For if $g \in \overline{K}(f, r)$, then for all $n \in \mathbb{N}$ the element $g_n = f + \left(1 - \frac{1}{n}\right)(g-f)$ belongs to $K(f, r)$ and we have $g_n \to g$. Hence $\overline{K}(f, r) \subset \overline{K(f, r)}$. As $\overline{K}(f, r)$ is closed, we also have $\overline{K(f, r)} \subset \overline{K}(f, r)$.

Theorem 2.3. *The closure of a subspace of H is a subspace.*

PROOF. Let T be a subspace of H, let $f, g \in \overline{T}$ and let $a, b \in \mathbb{K}$. Then there are sequences (f_n) and (g_n) in T such that $f_n \to f$, $g_n \to g$. It follows that

$$af + bg = a \lim f_n + b \lim g_n = \lim (af_n + bg_n).$$

As $af_n + bg_n \in T$, it follows that $af + bg \in \overline{T}$. □

If $(H, \|\,.\,\|)$ is a normed space and T is a subspace of H, then the restriction of $\|\,.\,\|$ to T defines a norm on T. Thus T becomes a normed space $(T, \|\,.\,\|)$ in a natural way. Analogously, if $(H, \langle .\,, .\rangle)$ is a pre-Hilbert space, then we can consider T as a pre-Hilbert space $(T, \langle .\,, .\rangle)$.

Theorem 2.4. *A subspace T of a Banach space $(H, \langle\,.\,\rangle)$ (respectively a Hilbert space $(H, \langle .\,, .\rangle)$) is closed if and only if $(T, \|\,.\,\|)$ is a Banach space (respectively $(T, \langle .\,, .\rangle)$ is a Hilbert space).*

PROOF. If T is closed, and (f_n) is a Cauchy sequence in T, then there exists an $f \in H$ such that $f_n \to f$; therefore $f \in \overline{T} = T$, i.e., T is complete. If T is complete and $f \in \overline{T}$, then there exists a sequence (f_n) from T such that $f_n \to f$; as (f_n) is a Cauchy sequence, (f_n) is convergent in T, i.e., $f \in T$. □

Let A and B now be subsets of a normed space H. The set A is said to be *dense relative to* B if $B \subset \overline{A}$ holds. If, in addition, $A \subset B$, then we say that A is a *dense subset* of B (or briefly A is *dense in* B). If A is dense relative to H, then we say briefly that A is *dense*.

Proposition. *If A_1 is dense relative to A_2 and A_2 is dense relative to A_3, then A_1 is dense relative to A_3.*

PROOF. From $A_3 \subset \overline{A}_2$ and $A_2 \subset \overline{A}_1$ it follows that $A_3 \subset \overline{\overline{A}}_1 = \overline{A}_1$. □

EXAMPLE 4. A sequence of complex numbers $f = (f_n)$ is said to be *finitary* if only finitely many members f_n are different from zero, i.e., $f = (f_1, f_2, \ldots, f_{n_0}, 0, 0, \ldots)$. The set of finitary sequences is a subspace $l_{2,0}$ of l_2. We show that $l_{2,0}$ is dense (in l_2). Let $f = (f_n)$ be an arbitrary element of l_2. Then for each $j \in \mathbb{N}$ we have $f^{(j)} = (f_1, \ldots, f_j, 0, 0, \ldots) \in l_{2,0}$ and

$$\|f - f^{(j)}\|^2 = \sum_{n=j+1}^{\infty} |f_n|^2 \to 0 \quad \text{as} \quad j \to \infty.$$

Consequently $f^{(j)} \to f$, i.e., $f \in \overline{l_{2,0}}$.

EXAMPLE 5. Let P be the vector space of polynomials in one variable. P can be considered as a subspace of the pre-Hilbert space $(C[0, 1], \langle .\,,.\rangle)$ from Section 1.2, Example 4. P is dense in $C[0, 1]$: By Weierstrass' approximation theorem (cf. *Hewitt-Stromberg* [18], (7.31)) for each continuous f and for every $\epsilon > 0$ there exists a polynomial p such that $\max\{|f(x) - p(x)| : x \in [0, 1]\} \leq \epsilon$. We also have then that $\|f - p\| \leq \epsilon$, i.e., f is a contact point of P.

EXAMPLE 6. Let M be a measurable subset of $\mathbb{R}^m$. Let

$$\begin{aligned} L_{2,0}(M) = \{f \in L_2(M):\ & \text{there exists a } K > 0 \text{ such that} \\ & |f(x)| \leq K \text{ almost everywhere in } M, \\ & \text{and } f(x) = 0 \text{ almost everywhere in } \{x \in M : |x| > K\}\}. \end{aligned}$$

$L_{2,0}(M)$ is dense in $L_2(M)$. For let f be an element of $L_2(M)$ and for each $n \in \mathbb{N}$ let

$$f_n(x) = \begin{cases} f(x) & \text{if } \quad |x| \leq n \quad \text{and} \quad f(x) \leq n \\ 0 & \text{otherwise.} \end{cases}$$

Then we have $|f_n(x)| \leq |f(x)|$ for all $n \in \mathbb{N}$ and all $x \in M$, and $f_n(x) \to f(x)$ as $n \to \infty$. By Lebesgue's dominated convergence theorem it now follows that

$$\|f_n - f\|^2 = \int_M |f_n(x) - f(x)|^2 \, dx \to 0 \quad \text{as} \quad n \to \infty.$$

Therefore $f_n \to f$. Since $f_n \in L_{2,0}(M)$, the assertion follows.

EXAMPLE 7. A subset J of $\mathbb{R}^m$ of the form

$$J = \{x = (x_1, \ldots, x_m) \in \mathbb{R}^m : a_j \lesseqgtr x_j \lesseqgtr b_j, j = 1, 2, \ldots, m\}$$

with $a_j, b_j \in \mathbb{R}$, is called an *interval* in $\mathbb{R}^m$; here any combination of the signs $<$ and $\leq$ is permitted. A function $f : \mathbb{R}^m \to \mathbb{C}$ is called a step function if there are finitely many intervals $J_1, \ldots, J_n$ and complex numbers $c_1, \ldots, c_n$ such that

$$f(x) = \sum_{j=1}^{n} c_j \chi_{J_j}(x),$$

where χ_A denotes the *characteristic function* of A, i.e.,

$$\chi_A(x) = \begin{cases} 1 & \text{for} \quad x \in A \\ 0 & \text{otherwise.} \end{cases}$$

The set $T(\mathbb{R}^m)$ of *step functions* on $\mathbb{R}^m$ is obviously a vector space (the linear operations are defined as usual). We show that $T(\mathbb{R}^m)$ *is a dense subspace of* $L^2(\mathbb{R}^m)$. To *prove* this it is enough to show that $T(\mathbb{R}^m)$ is dense in $L_{2,0}(\mathbb{R}^m)$. It is obvious that $T(\mathbb{R}^m) \subset L_{2,0}(\mathbb{R}^m)$. Let $f \in L_{2,0}(\mathbb{R}^m)$. Then f is integrable and there exists (cf. Theorem A6) a sequence (f_n) from $T(\mathbb{R}^m)$

such that $f_n(x)\to f(x)$ almost everywhere in $\mathbb{R}^m$ and

$$\int |f_n(x)-f(x)|\,dx \to 0 \quad \text{as } n\to\infty.$$

(Integrals for which no domain of integration is given are always taken over the whole space $\mathbb{R}^m$.) If for $K>0$ we have $|f(x)|\leqslant K$ almost everywhere, then we may assume that $|f_n(x)|\leqslant K$ for all $x\in\mathbb{R}^m$, and for all $n\in\mathbb{N}$. Consequently, we have

$$\|f_n-f\|^2\leqslant 2K\int |f_n(x)-f(x)|\,dx\to 0 \quad \text{as } n\to\infty,$$

i.e., $f_n\to f$.

EXAMPLE 8. Let $C_0^\infty(\mathbb{R}^m)$ be the space of infinitely many times differentiable complex-valued functions with compact support (i.e., for every $f\in C_0^\infty(\mathbb{R}^m)$ there exists a compact subset K in $\mathbb{R}^m$ such that f vanishes outside K; the smallest set K of this kind is called the *support* of f, in symbols $\operatorname{supp} f$). We show: *$C_0^\infty(\mathbb{R}^m)$ is a dense subspace of $L_2(\mathbb{R}^m)$*. For the *proof* it is enough to show that $C_0^\infty(\mathbb{R}^m)$ is dense relative to $T(\mathbb{R}^m)$. To prove this it is enough to show that for every interval J the characteristic function χ_J is a contact point of $C_0^\infty(\mathbb{R}^m)$. For this, let us define $\delta_\epsilon\in C_0^\infty(\mathbb{R}^m)$ by

$$\tilde{\delta}_\epsilon(x)=\begin{cases}\exp\left\{(|x|^2-\epsilon^2)^{-1}\right\} & \text{for } |x|<\epsilon,\\ 0 & \text{for } |x|\geqslant\epsilon,\end{cases}$$

and

$$\delta_\epsilon=\left\{\int \tilde{\delta}_\epsilon(x)\,dx\right\}^{-1}\tilde{\delta}_\epsilon.$$

The reader can verify himself that $\delta_\epsilon\in C_0^\infty(\mathbb{R}^m)$ and $\operatorname{supp}\delta_\epsilon=\{x\in\mathbb{R}^m : |x|\leqslant\epsilon\}$ hold. If J is now an interval in $\mathbb{R}^m$ and for $n\in\mathbb{N}$ we define

$$f_n(x)=\int \delta_{1/n}(x-y)\chi_J(y)\,dy,\quad x\in\mathbb{R}^m,$$

then we have $f_n\in C_0^\infty(\mathbb{R}^m)$,

$$f_n(x)=\begin{cases}1 \ \text{ for } \ x\in\mathbb{R}^m \ \text{ with } \ d(x,\complement J)\geqslant\dfrac{1}{n},\\ 0 \ \text{ for } \ x\in\mathbb{R}^m \ \text{ with } \ d(x,J)\geqslant\dfrac{1}{n},\end{cases}$$

and $0\leqslant f_n(x)\leqslant 1$ for all $x\in\mathbb{R}^m$ (here $d(x,A)$ stands for the Euclidean distance of the point x from the set A). We have $f_n(x)\to\chi_J(x)$ for all x that do not lie on the boundary of J. Therefore $f_n(x)\to\chi_J(x)$ almost everywhere. Thus by the Lebesgue dominated convergence theorem it follows that

$$\|f_n - \chi_J\|^2 = \int |f_n(x) - \chi_J(x)|^2 \, dx \to 0,$$

i.e., $f_n \to \chi_J$ in the sense of $L_2(\mathbb{R}^m)$.

A subset A of a normed space is said to be *separable* if there exists an at most countable subset B of A which is dense in A. If T is a subspace of H and B is a subset of H, then B is said to be *total with respect to* T if the linear hull $L(B)$ (the set of finite linear combinations of elements of B, or, in other words, the smallest subspace of H that contains B) is dense relative to T. We use the concept *total in* if $B \subset T$, and *total* if $T = H$.

EXAMPLE 9. The spaces $\mathbb{C}^m$ and $\mathbb{R}^m$ are separable, as the set of elements with rational components (in $\mathbb{C}^m$ this means that the real and imaginary parts of the components are rational) is enumerable and dense.

Theorem 2.5. *Let* $(H, \| \, . \, \|)$ *be a normed space.*
(a) *If* A *is a separable subset of* H, *then* $\bar{A}$ *is separable, also.*
(b) *If* A *is separable and* $A_1 \subset A$, *then* A_1 *is separable, too.*
(c) *A subspace* T *of* H *is separable if and only if there exists an at most countable subset* A *of* H *that is total with respect to* T.

PROOF.
(a) Let B be at most countable and dense in A. Since A is dense in $\bar{A}$, the set B is also dense in $\bar{A}$.
(b) Let $B = \{f_n : n \in \mathbb{N}\}$ be an at most countable set that is dense relative to A. Let J be the set of those pairs $(n, m) \in \mathbb{N} \times \mathbb{N}$ for which there exists an $f \in A_1$ such that $\|f - f_n\| \leqslant 1/m$. For every $n, m \in J$ let us choose a $g_{nm} \in A_1$ such that $\|g_{nm} - f_n\| \leqslant 1/m$. The set $B_1 = \{g_{nm} : (n, m) \in J\}$ is then at most countable. We show that it is dense with respect to A_1, i.e., A_1 is separable. Let $f \in A_1$. As B is dense in A, B is also dense relative to A_1. Therefore for every $k \in \mathbb{N}$ there exists an $n(k)$ such that $\|f_{n(k)} - f\| \leqslant 1/k$. Hence $(n(k), k) \in J$ and we have

$$\|g_{n(k),k} - f\| \leqslant \|g_{n(k),k} - f_{n(k)}\| + \|f_{n(k)} - f\| \leqslant \frac{2}{k},$$

i.e., $g_{n(k),k} \to f$ as $k \to \infty$. Hence B_1 is a dense subset of A_1.
(c) If T is separable, then there exists an at most countable subset B which is dense in T. Since $L(B) \supset B$, the set $L(B)$ is dense in T, too. Let B now be at most countable and let $T \subset \overline{L(B)}$. Then $\overline{L(B)}$ is separable, for the set $L_r(B)$ of finite linear combinations of elements of B with rational coefficients is dense in $\overline{L(B)}$ and $L_r(B)$ is countable. Since $T \subset \overline{L(B)}$, the subspace T is separable, also.

EXAMPLE 10. l_2 is separable, as the set of unit vectors $\{e_n = (\delta_{nj})_{j \in \mathbb{N}} : n \in \mathbb{N}\}$ is total in l_2: the linear hull of the unit vectors is $l_{2,0}$.

EXAMPLE 11. $L_2(\mathbb{R}^m)$ is separable. By Example 7 it is enough to show that $T(\mathbb{R}^m)$ is separable. Let S_0 be the set of characteristic functions of intervals with rational end points. The set S_0 is countable and obviously dense in the set S of characteristic functions of all intervals, therefore $\overline{L(S_0)} \supset L(S)$. Because $L(S) = T(\mathbb{R}^m)$, it follows from this that $\overline{L(S_0)} \supset T(\mathbb{R}^m)$, i.e., $T(\mathbb{R}^m)$ is separable.

EXAMPLE 12. For every measurable subset M of $\mathbb{R}^m$ the space $L_2(M)$ is separable. The space $L_2(M)$ may be considered as a subspace of $L_2(\mathbb{R}^m)$ provided we identify each $f \in L_2(M)$ with the element $\tilde{f} \in L_2(\mathbb{R}^m)$ defined by

$$\tilde{f}(x) = \begin{cases} f(x) & \text{for } x \in M, \\ 0 & \text{for } x \notin M. \end{cases}$$

EXAMPLE 13. Let ρ be a measure on $\mathbb{R}^m$ (cf. Appendix A) and let M be a ρ-measurable subset of $\mathbb{R}^m$. *The Hilbert space* $L_2(M, \rho)$ (cf. Section 2.1, Example 6) *is separable*. This can be proved for $M = \mathbb{R}^m$ as in Example 11 (in the course of the proof of $S \subset \bar{S}_0$ one has to notice that the boundaries of intervals in general have measures different from zero). For a general M we obtain the assertion by considering $L_2(M, \rho)$ as a subspace of $L_2(\mathbb{R}^m, \rho)$. *The space* $T(\mathbb{R}^m)$ *of step functions is dense in* $L_2(\mathbb{R}^m, \rho)$.

EXERCISES

2.4. A subset A of a normed space H is separable if and only if its closed linear hull $\overline{L(A)}$ is separable.

2.5. Prove that the function δ_ε from Example 8 is infinitely many times differentiable.

2.6. Let G be an open subset of $\mathbb{R}^m$, let ρ be a measure on $\mathbb{R}^m$, and let $L_2(G, \rho)$ be defined as in Section 2.1, Example 6.
(a) If $L_{2,0}(G, \rho)$ is the subspace of $L_2(G, \rho)$ consisting of all bounded functions with compact support in G, then $L_{2,0}(G, \rho)$ is dense in $L_2(G, \rho)$.
(b) If $T(G)$ is the space of step functions whose supports are contained in G (these are then compact subsets of G!), then $T(G)$ is dense in $L_2(G, \rho)$.
(c) $C_0^\infty(G)$, the space of infinitely many times continuously differentiable functions with compact support in G, is dense in $L_2(G, \rho)$.

2.7. The spaces A^2 and H^2 of Exercise 1.9 are separable (cf. also Exercise 2.2c).

2.8. (a) Prove the separability of $L_2(a, b)$ for $-\infty < a < b < \infty$ with the aid of the Weierstrass approximation theorem (cf. Example 5).
(b) With the aid of (a), prove the separability of $L_2(\mathbb{R})$.
(c) Prove the separability of $L_2(\mathbb{R}^m)$ analogously.

2.9. (a) A subset A of a normed space is *not* separable if and only if there exists an $a > 0$ and an uncountable subset B of A such that for all $f, g \in B, f \neq g$ we have $\|f - g\| > a$.

(b) If a subspace T is not separable, then for each $a > 0$ there exists such a set B.

(c) $l_2(A\,;\,\mu)$ is separable if and only if A is at most countable.

2.10. Let H be a pre-Hilbert space and let T be a dense subspace of H.

(a) The closures of $\{f \in T : \|f\| \leqslant 1\}$ and of $\{f \in T : \|f\| < 1\}$ are equal to $\bar{K}(0, 1)$.

(b) For every $f \in H$ we have $\|f\| = \sup\{|\langle f, g\rangle| : g \in T, \|g\| \leqslant 1\} = \sup\{|\langle f, g\rangle| : g \in T, \|g\| < 1\}$.

Orthogonality 3

3.1 The projection theorem

Let $(H, \langle .\,,.\rangle)$ be a pre-Hilbert space. Two elements $f, g \in H$ are said to be *orthogonal* (in symbols $f \perp g$) if $\langle f, g\rangle = 0$. If $f \perp g$, then we obviously have $\|f+g\|^2 = \|f\|^2 + \|g\|^2$; this formula often is referred to as the *Pythagorean theorem*. An element $f \in H$ is said to be *orthogonal* to the subset A of H (in symbols $f \perp A$), if $f \perp g$ for all $g \in A$. Two subsets A and B of H are said to be orthogonal (in symbols $A \perp B$) if $\langle f, g\rangle = 0$ for all $f \in A$, $g \in B$. If A is a subset of H, then the set $A^{\perp} = \{f \in H : f \perp A\}$ is called the *orthogonal complement* of A.

Proposition.
(a) *We have* $\{0\}^{\perp} = H$, $H^{\perp} = \{0\}$, i.e., 0 *is the only element orthogonal to every element.*
(b) *For every subset* A *of* H *the set* $A^{\perp}$ *is a closed subspace of* H.
(c) $A \subset B$ *implies* $B^{\perp} \subset A^{\perp}$.
(d) *We have* $A^{\perp} = L(A)^{\perp} = \overline{L(A)}^{\perp}$.

PROOF.
(a) For every $f \in H$ we have $\langle 0, f\rangle = 0$. If $f \neq 0$, then $\langle f, f\rangle \neq 0$, i.e., f is not orthogonal to H.
(b) If $f, g \in A^{\perp}$ and $a, b \in \mathbb{K}$, then for all $h \in A$ it follows that

$$\langle af + bg, h\rangle = a^*\langle f, h\rangle + b^*\langle g, h\rangle = 0,$$

i.e., $af + bg \in A^{\perp}$. Therefore $A^{\perp}$ is a subspace. It remains to prove that $\overline{A^{\perp}} \subset A^{\perp}$. Let $f \in \overline{A^{\perp}}$, and let (f_n) be a sequence from $A^{\perp}$ such that

$f_n \to f$. Then we have for all $h \in A$ that

$$\langle f, h\rangle = \lim\langle f_n, h\rangle = 0,$$

consequently $f \in A^\perp$.

(c) If $f \in B^\perp$, then we have $\langle f, h\rangle = 0$ for all $h \in B$, therefore also for all $h \in A$, and thus $f \in A^\perp$.

(d) Since $A \subset L(A) \subset \overline{L(A)}$, from (c) it follows that

$$\overline{L(A)}^\perp \subset L(A)^\perp \subset A^\perp.$$

It remains to prove that $A^\perp \subset \overline{L(A)}^\perp$. If $f \in A^\perp$, then we have $\langle f, h\rangle = 0$ for all $h \in A$, and therefore for all $h \in L(A)$, as well. If $h \in \overline{L(A)}$, then there exists a sequence (h_n) from $L(A)$ such that $h_n \to h$. Consequently, we have

$$\langle f, h\rangle = \lim\langle f, h_n\rangle = 0.$$

i.e., $f \in \overline{L(A)}^\perp$. □

In order to prove the projection theorem we need an approximation theorem, which we prove with somewhat more generality than we actually need. A subset A of a vector space is said to be *convex* if from $x, y \in A$ and $0 \leq a \leq 1$ it follows that $ax + (1-a)y \in A$. Any subspace is obviously convex.

Theorem 3.1. *Let H be a Hilbert space and let A be a non-empty closed convex subset of H. Then for each $f \in H$ there exists a unique $g \in A$ such that*

$$\|f - g\| = d(f, A) = \inf\{\|f - h\| : h \in A\}.$$

PROOF. There always exists a sequence (g_n) of elements of A such that $\|g_n - f\| \to d = d(f, A)$. If we replace f by $g_n - f$ and g by $g_m - f$ in the parallelogram identity (1.16), then on account of the inequality $\|f - h\| \geq d$ for all $h \in A$, we have

$$\begin{aligned}\|g_n - g_m\|^2 &= 2\|g_n - f\|^2 + 2\|g_m - f\|^2 - 4\|f - \tfrac{1}{2}(g_n + g_m)\|^2\\ &\leq 2\|g_n - f\|^2 + 2\|g_m - f\|^2 - 4d^2 \to 0\end{aligned}$$

as $n, m \to \infty$ (here we have used the fact that $(g_n + g_m)/2$ lies in A, since A is convex). Hence (g_n) is a Cauchy sequence. As H is a Hilbert space, there exists a $g \in H$ such that $g_n \to g$. We have $g \in A$, since A is closed. Moreover, we have

$$\|g - f\| = \lim\|g_n - f\| = d.$$

It remains to prove that g is uniquely defined. If $g, h \in A$ are such that $\|f - g\| = \|f - h\| = d$, then for the sequence $(g_n) = (g, h, g, h, g, \ldots)$ we obviously have $\|g_n - f\| = d$. By the above reasoning (g_n) is a Cauchy sequence, i.e. we have $g = h$. □

Theorem 3.2 (*Projection theorem*). *Let H be a Hilbert space, and let T be a closed subspace of H. Then we have $T^{\perp\perp}=T$. Each $f\in H$ can be uniquely decomposed in the form $f=g+h$ with $g\in T$ and $h\in T^{\perp}$. This g is called the (orthogonal) projection of f onto T.*

PROOF. As T is convex and closed, by Theorem 3.1 there exists a $g\in T$ such that $\|f-g\|=d(f, T)$. Let us set $h=f-g$.

$h\in T^{\perp}$: We have to prove that for all $w\in T$ we have $\langle w, h\rangle=0$. For $w=0$ this is clear, so let $w\in T$, $w\neq 0$. Then for all $a\in\mathbb{K}$ the element $g+aw$ also belongs to T. Therefore

$$\begin{aligned} d^2 = d(f, T)^2 &\leqslant \|f-(g+aw)\|^2 = \|h-aw\|^2 \\ &= \|h\|^2 - 2\operatorname{Re}(a\langle h, w\rangle) + |a|^2\|w\|^2 \\ &= d^2 - 2\operatorname{Re}(a\langle h, w\rangle) + |a|^2\|w\|^2. \end{aligned}$$

With $a=\|w\|^{-2}\langle w, h\rangle$ it follows from this that

$$\|w\|^{-2}|\langle w, h\rangle|^2 \leqslant 0,$$

so $\langle w, h\rangle=0$.

In order to prove the uniqueness of the representation $f=g+h$ let us assume that $f=g+h=g'+h'$ with $g, g'\in T$, and $h, h'\in T^{\perp}$. Then we have $g-g'\in T$ and $h'-h\in T^{\perp}$, therefore

$$g-g' = h'-h \in T\cap T^{\perp} = \{0\}.$$

It follows from this that $g=g'$ and $h=h'$.

It remains to prove that $T=T^{\perp\perp}$.

$T\subset T^{\perp\perp}$: If $f\in T$, then by the definition of $T^{\perp}$ we have $\langle f, g\rangle=0$ for all $g\in T^{\perp}$, i.e., f is orthogonal to $T^{\perp}$, $f\in T^{\perp\perp}$.

$T^{\perp\perp}\subset T$: Let $f\in T^{\perp\perp}$. On the basis of what we have already proved the element f may be represented in the form $f=g+h$ with $g\in T\subset T^{\perp\perp}$, $h\in T^{\perp}$. From this it follows that $h=f-g\in T^{\perp}\cap T^{\perp\perp}$, hence $h=0$, i.e., $f=g\in T$. □

Proposition.

(a) *Let H be a Hilbert space. For every subset A of H we have $A^{\perp\perp}=\overline{L(A)}$, i.e., $A^{\perp\perp}$ is the smallest closed subspace containing A.*

(b) *In a Hilbert space H we have $A^{\perp}=\{0\}$ if and only if $\overline{L(A)}=H$ holds, i.e., if A is total.*

PROOF.

(a) Since $A^{\perp}=\overline{L(A)}^{\perp}$, the projection theorem shows that $\overline{L(A)}=\overline{L(A)}^{\perp\perp}=A^{\perp\perp}$.

(b) If $A^{\perp}=\{0\}$, then we have $\overline{L(A)}=A^{\perp\perp}=\{0\}^{\perp}=H$. If $A^{\perp\perp}=H$, then we have $A^{\perp}=A^{\perp\perp\perp}=H^{\perp}=\{0\}$, as $A^{\perp}$ is a closed subspace. □

If T_1 and T_2 are subspaces of a vector space such that $T_1\cap T_2=\{0\}$, then

$$T_1+T_2 = \{f+g : f\in T_1, g\in T_2\}$$

is a *direct sum* (consequently we write $T_1 \dot{+} T_2$), i.e., each element from $T_1 + T_2$ has exactly one representation of the form $f+g$ with $f \in T_1$ and $g \in T_2$. If T_1 and T_2 are subspaces of a pre-Hilbert space with $T_1 \perp T_2$, then we have $T_1 \cap T_2 = \{0\}$. In this case we call the direct sum $T_1 \dot{+} T_2$ an *orthogonal sum* and we denote it by $T_1 \oplus T_2$.

Theorem 3.3.

(a) *Let H be a pre-Hilbert space, and let T_1 and T_2 be orthogonal subspaces. If $T_1 \oplus T_2$ is closed, then T_1 and T_2 are closed.*

(b) *If H is a Hilbert space and T_1, T_2 are closed orthogonal subspaces, then $T_1 \oplus T_2$ is closed.*

(c) *If H is a Hilbert space and T and T_1 are closed subspaces such that $T_1 \subset T$, then there exists exactly one closed subspace T_2 such that $T_2 \subset T$, $T_2 \perp T_1$ and $T = T_1 \oplus T_2$.*

For the subspace T_2, defined uniquely by part (c) of this theorem, we write briefly $T_2 = T \ominus T_1$. The subspace T_2 is called the *orthogonal complement* of T_1 *with respect to* T. For $T = H$ we obtain that $H \ominus T_1 = T_1^{\perp}$.

PROOF.

(a) We show that T_1 is closed (the proof for T_2 goes the same way). Let $f \in \overline{T}_1$ and let (f_n) be a sequence from T_1 such that $f_n \to f$. Since $T_1 \subset T_1 \oplus T_2$, we have $\overline{T}_1 \subset \overline{T_1 \oplus T_2} = T_1 \oplus T_2$. Hence $f \in T_1 \oplus T_2$ and thus we have $f = g_1 + g_2$ with $g_1 \in T_1$, $g_2 \in T_2$. On the other hand, it follows from $f_n \in T_1$ that $f_n \perp T_2$ and so $f \perp T_2$, and, consequently, $g_2 = f - g_1 \in T_2 \cap T_2^{\perp}$. Therefore $g_2 = 0$. From this it follows that $f = g \in T_1$.

(b) We have to prove that $\overline{T_1 \oplus T_2} \subset T_1 \oplus T_2$. Let $f \in \overline{T_1 \oplus T_2}$; then there exists a sequence $(f_{1,n} + f_{2,n}) \in T_1 \oplus T_2$ with $f_{1,n} \in T_1$, $f_{2,n} \in T_2$ and $f_{1,n} + f_{2,n} \to f$. Since

$$\|f_{1,n} + f_{2,n} - f_{1,m} - f_{2,m}\|^2 = \|f_{1,n} - f_{1,m}\|^2 + \|f_{2,n} - f_{2,m}\|^2,$$

the sequences $(f_{1,n})$ and $(f_{2,n})$ are Cauchy sequences. Consequently $f_{1,n} \to f_1 \in T_1$, $f_{2,n} \to f_2 \in T_2$. From this it follows that

$$f = \lim (f_{1,n} + f_{2,n}) = f_1 + f_2 \in T_1 \oplus T_2.$$

(c) By Theorem 2.4 T is a Hilbert space. Without loss of generality we may assume that $T = H$. In this case let us set $T_2 = T_1^{\perp}$. Then by the projection theorem (Theorem 3.2) we have $H = T_1 \oplus T_2$. In order to prove uniqueness, let us choose an arbitrary subspace T_2' such that $H = T_1 \oplus T_2'$. Then we surely have $T_2' \subset T_1^{\perp}$. If $f \in T_1^{\perp}$, then $f = f_1 + f_2$ with $f_1 \in T_1$, $f_2 \in T_2'$. Here we must have $f_1 = 0$, since $0 = \langle f_1, f \rangle = \langle f_1, f_1 \rangle = \|f_1\|^2$. Therefore we have $f = f_2 \in T_2'$, i.e., $T_1^{\perp} \subset T_2'$ and thus $T_1^{\perp} = T_2'$. □

EXAMPLE 1. Let $-\infty \leq a \leq c \leq b \leq \infty$. In $L_2(a, b)$ by

$$T_1 = \{f \in L_2(a, b) : f(x) = 0 \quad \text{almost everywhere in} \quad (a, c)\}$$
$$T_2 = \{f \in L_2(a, b) : f(x) = 0 \quad \text{almost everywhere in} \quad (c, b)\}$$

two subspaces are defined, and we have $L_2(a, b) = T_1 \oplus T_2$: For $g \in T_1$ and $h \in T_2$ we obviously have $\langle g, h \rangle = 0$. Moreover, for each $f \in L_2(a, b)$ we have $g = \chi_{(a, c)} f \in T_1$, $h = \chi_{c, b} f \in T_2$ and $f = g + h$.

EXAMPLE 2. In $L_2(-a, a)$ by

$$T_\pm = \{f \in L_2(-a, a) : f(x) = \pm f(-x) \quad \text{almost everywhere in} \quad (-a, a)\}$$

two subspaces are defined and $L_2(-a, a) = T_+ \oplus T_-$: For $g \in T_+$ and $h \in T_-$ we have

$$\langle g, h \rangle = \int_{-a}^{a} g(x)^* h(x)\, dx = \int_{-a}^{0} g(x)^* h(x)\, dx - \int_{-a}^{0} g(x)^* h(x)\, dx = 0.$$

Let us set $f_\pm(x) = \frac{1}{2}(f(x) \pm f(-x))$. Then for each $f \in L_2(-a, a)$ we have $f_\pm \in T_\pm$ and $f = f_+ + f_-$. The subspace T_+ is the space of *even functions*. T_- is the space of *odd functions*.

If $T_1, \ldots, T_n$ are mutually orthogonal subspaces of H, then we call the (direct) sum of these spaces an *orthogonal sum* and we write

$$\bigoplus_{j=1}^{n} T_j = T_1 \oplus T_2 \oplus \cdots \oplus T_n.$$

Parts (a) and (b) of Theorem 3.3 can be extended to this case. For infinitely many subspaces see Exercise 3.3.

If A is an arbitrary set and $(H_\alpha, \langle \cdot\,, \cdot \rangle_\alpha)$ is a pre-Hilbert space for each $\alpha \in A$, then by Exercise 1.11 the space

$$H = \Big\{ f = (f_\alpha)_{\alpha \in A} \in \prod_{\alpha \in A} H_\alpha :$$
$$f_\alpha \neq 0 \quad \text{for at most countably many} \quad \alpha \in A, \quad \text{and} \quad \sum_{\alpha \in A} \|f_\alpha\|^2 < \infty \Big\}$$

is a pre-Hilbert space with the scalar product

$$\langle (f_\alpha), (g_\alpha) \rangle = \sum_{\alpha \in A} \langle f_\alpha, g_\alpha \rangle_\alpha \quad \text{for} \quad (f_\alpha), (g_\alpha) \in H.$$

By Exercise 2.2a the space H is a Hilbert space if and only if all H_α are Hilbert spaces. If we identify H_α with the subspace of elements $(f_\beta)_{\beta \in A}$ such that $f_\beta = 0$ for $\beta \neq \alpha$, then the spaces H_α become pairwise orthogonal subspaces of H. Therefore $(H, \langle \cdot\,, \cdot \rangle)$ is called the *orthogonal sum* of the spaces $(H_\alpha, \langle \cdot\,, \cdot \rangle_\alpha)$ in symbols $H = \oplus_{\alpha \in A} H_\alpha$ (if A is finite, then $H = \prod_{\alpha \in} H_\alpha$).

Theorem 3.4. *Let H be a Hilbert space. If T is a closed subspace and S is a finite dimensional subspace, then $T+S$ is closed.*

PROOF. The problem can be reduced by induction, to the case where S is one dimensional; $S=L(f)$. If we write $f=f_1+f_2$ with $f_1 \in T$ and $f_2 \in T^{\perp}$, then we have $T+S=T \oplus L(f_2)$. Therefore $T+S$ is closed by Theorem 3.3(b). □

EXERCISES

3.1. Let H be the pre-Hilbert space $\{f \in C[0,1] : f(1)=0\}$ with the scalar product $\langle f, g\rangle = \int_0^1 f(t)^* g(t)\,dt$. The subspace $T=\{f \in H : \int_0^1 f(t)\,dt=0\}$ is closed, $T \neq H$, and $T^{\perp}=\{0\}$.

3.2. Let H be the pre-Hilbert space $C[-1,1]$ with the scalar product $\langle f, g\rangle = \int_{-1}^{1} f(t)^* g(t)\,dt$. The subspaces $T_1=\{f \in H : f(t)=0$ for $t \leqslant 0\}$ and $T_2=\{f \in H : f(t)=0$ for $t \geqslant 0\}$ are closed and such that $T_1 \perp T_2$. The orthogonal sum $T_1 \oplus T_2$ is not closed (cf. Theorem 3.3(b)).

3.3. Let H be a Hilbert space, and let $\{T_\alpha : \alpha \in A\}$ be a family of pairwise orthogonal subspaces of H.
(a) If $(f_\alpha) \in \prod_{\alpha \in A} T_\alpha$, and $f_\alpha \neq 0$ for at most countably many α, and $\sum_{\alpha \in A} \|f_\alpha\|^2 < \infty$, then $f=\sum_{\alpha \in A} f_\alpha$ can be defined. The subspace T of all f of this form is called the *orthogonal sum* of T_α, in symbols $T=\bigoplus_{\alpha \in A} T_\alpha$.
Hint: Build a sequence (α_n) from those α for which $f_\alpha \neq 0$ and define $\sum_{\alpha \in A} f_\alpha$ as $\sum_{n=1}^{\infty} f_{\alpha_n}$. This definition is independent of the choice of the sequence (α_n).
(b) T is closed if and only if all T_α are closed.
(c) If all T_α are different from $\{0\}$, then T is separable if and only if A is countable and all T_α are separable.

3.4. Let H be a pre-Hilbert space, let D be a dense subspace of H, and let N be a finite dimensional subspace of H. Then $D \cap N^{\perp}$ is dense in $N^{\perp}$.
Hint: By induction on $n=\dim N$ we can reduce the problem to the case $n=1$, i.e., $N=L(g)$, $g \neq 0$. Then there exists an $h \in D$ such that $\langle g, h\rangle = 1$. If $f \in N^{\perp}$, then there exists a sequence (f_n) from D such that $f_n \to f$. For the sequence $f_n' = f_n - \langle g, f_n\rangle h$ we then have $f_n' \in D \cap N^{\perp}$ and $f_n' \to f$.

3.5. Let H be a pre-Hilbert space and let T_1 and T_2 be subspaces of H such that $T_1 \perp T_2$. Then we have $\overline{T_1 \oplus T_2} \supset \overline{T}_1 \oplus \overline{T}_2$. If H is a Hilbert space, then we have $\overline{T_1 \oplus T_2} = \overline{T}_1 \oplus \overline{T}_2$.

3.2 Orthonormal systems and orthonormal bases

Let $(H, \langle .\,,.\rangle)$ be a pre-Hilbert space. A family $M=\{e_\alpha : \alpha \in A\}$ of elements from H is called an *orthonormal system* (ONS) if we have

$$\langle e_\alpha, e_\beta\rangle = \delta_{\alpha\beta} \quad \text{for} \quad \alpha, \beta \in A$$

($\delta_{\alpha\beta}$ denotes the Kronecker delta, i.e., $\delta_{\alpha\alpha}=1$ for all $\alpha\in A$ and $\delta_{\alpha\beta}=0$ for $\alpha\neq\beta$). An orthonormal system M is called an *orthonormal basis* (ONB) *of the subspace* T if M is total in T (i.e., $M\subset T$ and $\overline{L(M)}\supset T$). If M is an ONB of H, then M is called an *orthonormal basis*.

Proposition.

(a) *Each ONS is linearly independent (i.e., every finite subsystem is linearly independent).*

(b) *Each ONB M is a maximal ONS (i.e., if M' is an ONS such that $M\subset M'$, then we have $M'=M$).*

(c) *If H is a Hilbert space, then each maximal ONS is an ONB.*

PROOF.

(a) If $\{e_1,\ldots,e_n\}$ is a finite subsystem of an ONS, then $\{e_1,\ldots,e_n\}$ is also an ONS. If $\sum_{j=1}^n a_je_j=0$, then it follows that

$$0=\left\|\sum_{j=1}^n a_je_j\right\|^2=\sum_{j=1}^n |a_j|^2.$$

Therefore $a_j=0$ for all j.

(b) Let M be an ONB. If M were not maximal, then there would be an $e\in H$ such that $e\perp M$; consequently $e\perp\overline{L(M)}$, this contradicts $\overline{L(M)}=H$.

(c) Let M be a maximal ONS in the Hilbert space H. If M were not total, i.e., if we had $L(M)^\perp\neq\{0\}$, then there would be an $e\in L(M)^\perp$ such that $\|e\|=1$. Hence $M'=M\cup\{e\}$ would be a larger ONS which contradicts the maximality of M. □

EXAMPLE 1. The set of unit vectors $\{e_1,\ldots,e_m\}$ is an ONB in $\mathbb{K}^m$ (e_j is the vector with 1 at the jth place and zero otherwise).

EXAMPLE 2. The set of unit vectors $\{e_k=(\delta_{kn})_{n\in\mathbb{N}}: k\in\mathbb{N}\}$ is an ONB in l_2.

EXAMPLE 3. An ONB in A^2 is $\{e_n: n\in\mathbb{N}_0\}$ with $e_n(z)=[(n+1)/\pi]^{1/2}z^n$. An ONB of H^2 is $\{f_n: n\in\mathbb{N}_0\}$ with $f_n(z)=[2\pi]^{-1/2}z^n$. This follows immediately from Exercise 1.9, in particular part (c).

EXAMPLE 4. In $L_2(0,1)$ the set $M=\{e_n: n\in\mathbb{Z}\}$ with $e_n(x)=\exp(2i\pi nx)$ is an ONS, as one can verify by a simple calculation. We show that M is an ONB, i.e., that M is total. For this let $\hat{C}[0,1]=\{f\in C[0,1]: f(0)=f(1)\}$. For each $f\in\hat{C}[0,1]$ by *Fejér*'s theorem there exists a sequence (f_n) of trigonometric polynomials ($f_n\in L(M)$) such that f_n uniformly tends to f. We also have then that $f_n\to f$ in the sense of $L_2(0,1)$, i.e., M is total in $\hat{C}[0,1]$. If we also prove that $\hat{C}[0,1]$ is dense in $C[0,1]$, then everything will be proved. For $f\in C[0,1]$ and $n\in\mathbb{N}$ let us define

$$f_n(x)=\begin{cases} f(x) & \text{for}\quad 0\leqslant x\leqslant 1-\dfrac{1}{n}\\[2mm] f(0)+(1-x)n\left(f\left(1-\dfrac{1}{n}\right)-f(0)\right) & \text{for}\quad 1-\dfrac{1}{n}<x\leqslant 1.\end{cases}$$

Then we obviously have $f_n\in \hat{C}[0,1]$ and $f_n\to f$.

EXAMPLE 5. Let $F_0=L\{e_\lambda:\lambda\in\mathbb{R}\}$ with $e_\lambda:\mathbb{R}\to\mathbb{C}$, $e_\lambda(x)=\exp(i\lambda x)$. On F_0 by

$$\langle f,g\rangle=\lim_{T\to\infty}\frac{1}{2T}\int_{-T}^{T}f(x)^*g(x)\,dx$$

a scalar product is defined. For the proof of the existence of this limit it is sufficient to treat $f=e_\lambda$ and $g=e_\mu$. For these we have

$$\begin{aligned}\langle f,g\rangle &= \lim_{T\to\infty}\frac{1}{2T}\int_{-T}^{T}e^{i(\mu-\lambda)x}\,dx=\lim_{T\to\infty}\frac{1}{2T}\frac{-i}{\mu-\lambda}\left[e^{i(\mu-\lambda)T}-e^{-i(\mu-\lambda)T}\right]\\ &=0\quad\text{for}\quad \mu\neq\lambda.\end{aligned}$$

$$\langle f,g\rangle=\lim_{T\to\infty}\frac{1}{2T}\int_{-T}^{T}dx=1\quad\text{for}\quad\mu=\lambda.$$

The properties [(1.13)(i-iv)] of scalar products are obviously satisfied. If

$$f=\sum_{k=1}^{n}a_ke_{\lambda_k}\quad\text{then}\quad\langle f,f\rangle=\sum_{k=1}^{n}|a_k|^2.$$

Therefore we have $\langle f,f\rangle=0$ if and only if $f=0$, which is property [(1.13)(v)]. By construction, $M=\{e_\lambda:\lambda\in\mathbb{R}\}$ is total in F_0, i.e., M is an ONB in F_0. The space F_0 is not a Hilbert space. For if (λ_k) is a sequence of mutually distinct real numbers and (a_k) is a sequence of complex numbers such that $a_k\neq 0$ for all k and $\Sigma|a_k|^2<\infty$, then the sequence (f_n) with $f_n=\Sigma_{k=1}^{n}a_ke_{\lambda_k}$ is a Cauchy sequence that is not convergent.

The following theorem, known as the Gram-Schmidt *orthogonalization process* enables us to generate orthonormal systems and (in separable spaces) orthonormal bases.

Theorem 3.5. *Let H be a pre-Hilbert space. For each finite or countably infinite set $F=\{f_n\}$ from H there exists a finite or countably infinite orthonormal system $M=\{e_n\}$ such that $L(F)=L(M)$. If F is linearly independent, then we can also insure $L(f_1,\ldots,f_n)=L(e_1,\ldots,e_n)$ for all n.*[1] *If we require that in the representation $e_n=\Sigma_{j=1}^{n}a_jf_j$ the coefficient a_n is positive, then M is uniquely determined.*

[1]In the sequel we write $L(e_1,\ldots,e_n)$ in place of $L(\{e_1,\ldots,e_n\})$.

(In what follows we shall always use the Gram-Schmidt orthogonalization process with this additional requirement.)

Proof. It is obviously enough to prove only the last part of the assertion. Every normed element from $L(f_1)$ has the form $b_1\|f_1\|^{-1}f_1$ with $|b_1|=1$. The additional condition $a_1=b_1\|f_1\|^{-1}>0$ gives $b_1=1$. That is, $e_1=\|f_1\|^{-1}f_1$. So we obviously have $L(e_1)=L(f_1)$, as well. Let us now suppose that $e_1,\ldots,e_n$ are determined in such a way that $L(e_1,\ldots,e_n)=L(f_1,\ldots,f_n)$. For every

$$g=\sum_{j=1}^{n} b_je_j+b_{n+1}f_{n+1}\in L(e_1,\ldots,e_n,f_{n+1})=L(f_1,\ldots,f_{n+1})$$

such that $g\perp L(e_1,\ldots,e_n)$ we then have

$$0=\langle e_i,g\rangle=b_i+b_{n+1}\langle e_i,f_{n+1}\rangle,\quad i=1,\ldots,n,$$

thus $b_i=-b_{n+1}\langle e_i,f_{n+1}\rangle$. Consequently, e_{n+1} necessarily has the form

$$e_{n+1}=\frac{b_{n+1}}{|b_{n+1}|}\left\|f_{n+1}-\sum_{i=1}^{n}\langle e_i,f_{n+1}\rangle e_i\right\|^{-1}\left(f_{n+1}-\sum_{i=1}^{n}\langle e_i,f_{n+1}\rangle e_i\right).$$

From the requirement,

$$a_{n+1}=\frac{b_{n+1}}{|b_{n+1}|}\left\|f_{n+1}-\sum_{i=1}^{n}\langle e_i,f_{n+1}\rangle e_i\right\|^{-1}>0$$

it follows that $b_{n+1}>0$, therefore $b_{n+1}|b_{n+1}|^{-1}=1$. Consequently,

$$e_{n+1}=\left\|f_{n+1}-\sum_{i=1}^{n}\langle e_i,f_{n+1}\rangle e_i\right\|^{-1}\left(f_{n+1}-\sum_{i=1}^{n}\langle e_i,f_{n+1}\rangle e_i\right).$$

By construction, we have $e_{n+1}\in L(f_1,\ldots,f_{n+1})$, hence $L(e_1,\ldots,e_{n+1})\subset L(f_1,\ldots,f_{n+1})$. From the formula for e_{n+1} it follows that $f_{n+1}\in L(e_1,\ldots,e_{n+1})$, therefore we also have $L(f_1,\ldots,f_{n+1})\subset L(e_1,\ldots,e_{n+1})$. □

Example 6. In $L_2(-1,1)$ the set $F=\{f_n: n\in\mathbb{N}_0\}$ with $f_n(x)=x^n$ is a linearly independent system. The application of Schmidt's process provides an ONS $M=\{p_n: n\in\mathbb{N}_0\}$, where $p_n(x)=\sum_{j=0}^{n}a_{nj}x^j$ holds with $a_{nn}>0$; i.e., p_n is a polynomial of degree n with a positive leading coefficient. These polynomials are called the *Legendre polynomials*. As F is total, the Legendre polynomials constitute an ONB in $L_2(-1,1)$. The polynomial p_n can be given explicitly:

$$p_n(x)=(2^nn!)^{-1}\left(\frac{2n+1}{2}\right)^{1/2}\frac{d^n}{dx^n}(x^2-1)^n,\quad n\in\mathbb{N}_0.$$

In order to prove this formula it is sufficient to show that the expression given for p_n is a polynomial of degree n whose leading coefficient is positive and that $\langle p_n, p_m\rangle = \delta_{nm}$. The first assertion is obvious. For $j<m$ we obtain by a $(j+1)$-fold integration by parts (the integrated terms vanish) that

$$\int_{-1}^{1} x^j p_m(x)\,dx = C_m \int_{-1}^{1} x^j \frac{d^m}{dx^m}(x^2-1)^m\,dx$$
$$= C_m(-1)^{j+1}\int_{-1}^{1} \frac{d^{j+1}}{dx^{j+1}}(x^j)\frac{d^{m-j-1}}{dx^{m-j-1}}(x^2-1)^m\,dx = 0.$$

This implies that $\langle p_n, p_m\rangle = 0$ for $n\neq m$. It remains to prove that $\|p_n\| = 1$. By integration by parts we obtain that

$$\int_{-1}^{1}\left[\frac{d^n}{dx^n}(x^2-1)^n\right]^2 dx = (-1)^n\int_{-1}^{1}(x^2-1)^n\frac{d^{2n}}{dx^{2n}}(x^2-1)^n\,dx$$
$$= (2n)!\int_{-1}^{1}(1-x)^n(1+x)^n\,dx$$
$$= (2n)!\frac{n}{n+1}\int_{-1}^{1}(1-x)^{n-1}(1+x)^{n+1}\,dx$$
$$= (2n)!\frac{n(n-1)}{(n+1)(n+2)}\int_{-1}^{1}(1-x)^{n-2}(1+x)^{n+2}\,dx = \dots$$
$$= (2n)!\frac{n(n-1)\dots 1}{(n+1)(n+2)\dots 2n}\int_{-1}^{1}(1+x)^{2n}\,dx$$
$$= (n!)^2\frac{1}{2n+1}\left[(1+x)^{2n+1}\right]_{-1}^{1} = (n!)^2(2n+1)^{-1}2^{2n+1}.$$

From this it follows that $\|p_n\| = 1$.

We can see in an entirely analogous way that the generalized Legendre polynomials

$$p_{a,b,n}(x) = \left[(b-a)^n n!\right]^{-1}\left[\frac{2n+1}{b-a}\right]^{1/2}\frac{d^n}{dx^n}\left[(x-a)(x-b)\right]^n, \quad n\in\mathbb{N}$$

constitute an ONB in $L_2(a, b)$.

Theorem 3.6. *Let H be a pre-Hilbert space.*

(a) *If $\{e_1, \dots, e_n\}$ is a (finite) ONS in H, then for each $f\in H$ there exists a $g\in L(e_1, \dots, e_n)$ such that $\|g-f\| = d(f, L(e_1, \dots, e_n))$; we have*

$$g = \sum_{j=1}^{n}\langle e_j, f\rangle e_j.$$

(b) *Let $\{e_\alpha : \alpha\in A\}$ be an ONS in H and let $f\in H$. Then at most countably many of the numbers $\langle e_\alpha, f\rangle$ are different from zero, and we have the Bessel inequality*

$$\|f\|^2 \geq \sum_{\alpha\in A}|\langle e_\alpha, f\rangle|^2.$$

(c) *An ONS $\{e_\alpha : \alpha \in A\}$ is an ONB if and only if for all $f \in H$ the Parseval equality*

$$\|f\|^2 = \sum_{\alpha \in A} |\langle e_\alpha, f\rangle|^2$$

holds. Then we also have

$$f = \sum_{\alpha \in A} \langle e_\alpha, f\rangle e_\alpha \quad \textit{for all} \quad f \in H.$$

PROOF.

(a) For all $c_1, \ldots, c_n \in \mathbb{K}$ we have

$$\left\|f - \sum_{j=1}^{n} c_j e_j\right\|^2 = \|f\|^2 - \sum_{j=1}^{n} c_j\langle f, e_j\rangle - \sum_{j=1}^{n} c_j^*\langle e_j, f\rangle + \sum_{j=1}^{n} |c_j|^2$$

$$= \|f\|^2 - \sum_{j=1}^{n} |\langle e_j, f\rangle|^2 + \sum_{j=1}^{n} |c_j - \langle e_j, f\rangle|^2.$$

Therefore $\|f - \sum_{j=1}^{n} c_j e_j\|$ is minimal if and only if $c_j = \langle e_j, f\rangle$.

(b) For every finite set $\{\alpha_1, \ldots, \alpha_n\} \subset A$ we have

$$\|f\|^2 = \left\|f - \sum_{j=1}^{n} \langle e_{\alpha_j}, f\rangle e_{\alpha_j}\right\|^2 + \sum_{j=1}^{n} |\langle e_{\alpha_j}, f\rangle|^2,$$

by part (a). Hence

$$\sum_{j=1}^{n} |\langle e_{\alpha_j}, f\rangle|^2 \leqslant \|f\|^2.$$

From this the assertion follows because for every $\epsilon > 0$ only finitely many $j \in \mathbb{N}$ exist with the property $|\langle e_{\alpha_j}, f\rangle|^2 \geqslant \epsilon$.

(c) Let us assume the Parseval equality for all $f \in H$. Let $f \in H$, and let (α_j) be the sequence of those α for which $\langle e_\alpha, f\rangle \neq 0$. Then we have

$$\left\|f - \sum_{j=1}^{n} \langle e_{\alpha_j}, f\rangle e_{\alpha_j}\right\|^2 = \|f\|^2 - \sum_{j=1}^{n} |\langle e_{\alpha_j}, f\rangle|^2 \to 0$$

as $n \to \infty$. Consequently, $f \in \overline{L(M)}$, i.e., $\{e_\alpha : \alpha \in A\}$ is an ONB. Moreover, it follows that

$$f = \sum_{j=1}^{\infty} \langle e_{\alpha_j}, f\rangle e_{\alpha_j} = \sum_{\alpha \in A} \langle e_\alpha, f\rangle e_\alpha.$$

Let $\{e_\alpha : \alpha \in A\}$ now be an ONB, i.e., let $H = \overline{L(M)}$. For every $f \in H$ and for every $\epsilon > 0$ there exist $n \in \mathbb{N}$, $\alpha_1, \ldots, \alpha_n \in A$ and $c_1, \ldots, c_n \in \mathbb{K}$ such that

$$\left\|f - \sum_{j=1}^{n} c_j e_{\alpha_j}\right\|^2 < \epsilon.$$

By parts (a) and (b) it follows from this that

$$0 \leq \|f\|^2 - \sum_{\alpha\in A} |\langle e_\alpha, f\rangle|^2 \leq \|f\|^2 - \sum_{j=1}^{n} |\langle e_{\alpha_j}, f\rangle|^2$$

$$= \left\|f - \sum_{j=1}^{n} \langle e_{\alpha_j}, f\rangle e_{\alpha_j}\right\|^2 \leq \left\|f - \sum_{j=1}^{n} c_j e_{\alpha_j}\right\|^2 < \epsilon.$$

and consequently that

$$\|f\|^2 = \sum_{\alpha\in A} |\langle e_\alpha, f\rangle|^2.$$

Theorem 3.7 (Expansion theorem). *Let H be a pre-Hilbert space and let $M = \{e_\alpha : \alpha \in A\}$ be an ONS in H.*

(a) *If (α_n) is a sequence of pairwise different elements from A, (c_n) a sequence from $\mathbb{K}$, and the series $\sum c_n e_{\alpha_n}$ is convergent (i.e., $\lim_{m\to\infty} \sum_{n=1}^{m} c_n e_{\alpha_n}$ exists), then we have $(c_n) \in l_2$. If H is a Hilbert space, then this series is convergent if and only if $(c_n) \in l_2$.*

(b) *If $g = \sum c_n e_{\alpha_n}$, then we have*

$$c_n = \langle e_{\alpha_n}, g\rangle \quad \text{for all} \quad n \in \mathbb{N},$$

$$\|g\|^2 = \sum |c_n|^2$$

and

$$\langle g, f\rangle = \sum \langle g, e_{\alpha_n}\rangle\langle e_{\alpha_n}, f\rangle \quad \text{for all} \quad f \in H.$$

(c) *The set of all elements from H which can be represented by a convergent sum $\sum c_n e_{\alpha_n}$ equals $\overline{L(M)}$.*

PROOF.

(a) The sequence $(\sum_{n=1}^{m} c_n e_{\alpha_n})_{m\in\mathbb{N}}$ is a Cauchy sequence if and only if we have

$$\sum_{n=m+1}^{k} |c_n|^2 = \left\| \sum_{n=m+1}^{k} c_n e_{\alpha_n}\right\|^2 =$$

$$\left\| \sum_{n=1}^{m} c_n e_{\alpha_n} - \sum_{n=1}^{k} c_n e_{\alpha_n}\right\|^2 \to 0 \quad \text{for} \quad m < k \quad \text{and} \quad m, k \to \infty.$$

This holds true if and only if $\sum |c_n|^2 < \infty$.

(b) We have

$$\|g\|^2 = \lim_{m\to\infty} \left\| \sum_{n=1}^{m} c_n e_{\alpha_n}\right\|^2 = \lim_{m\to\infty} \sum_{n=1}^{m} |c_n|^2 = \sum |c_n|^2$$

and

$$\langle e_{\alpha_n}, g\rangle = \lim_{m\to\infty} \left\langle e_{\alpha_n}, \sum_{j=1}^{m} c_j e_{\alpha_j}\right\rangle = \lim_{m\to\infty} \langle e_{\alpha_n}, c_n e_{\alpha_n}\rangle = c_n,$$

In the same manner, it follows for all $f \in H$ that

$$\langle g, f\rangle = \lim_{m\to\infty}\left\langle \sum_{n=1}^{m} c_n e_{\alpha_n}, f\right\rangle = \lim_{m\to\infty}\sum_{n=1}^{m}\langle g, e_{\alpha_n}\rangle\langle e_{\alpha_n}, f\rangle$$
$$= \sum \langle g, e_{\alpha_n}\rangle\langle e_{\alpha_n}, f\rangle.$$

(c) Each element $\sum c_n e_{\alpha_n}$ is in $\overline{L(M)}$, by construction. The converse follows from Theorem 3.6(c), as M is an ONB in $\overline{L(M)}$. □

Proposition. *Let H be a Hilbert space, and let $M = \{e_\alpha : \alpha \in A\}$ be an ONS. For each $f \in H$ the vector $\sum_{\alpha\in A}\langle e_\alpha, f\rangle e_\alpha$ is the orthogonal projection of f onto $\overline{L(M)}$.*

PROOF. If $\sum_{j=1}^{n} c_j e_{\alpha_j}$ is an arbitrary element from $L(M)$, then by Theorem 3.6(a) we have

$$\left\| f - \sum_{\alpha\in A}\langle e_\alpha, f\rangle e_\alpha\right\|^2 = \|f\|^2 - \sum_{\alpha\in A}|\langle e_\alpha, f\rangle|^2$$
$$\leq \|f\|^2 - \sum_{j=1}^{n}|\langle e_{\alpha_j}, f\rangle|^2$$
$$= \left\| f - \sum_{j=1}^{n}\langle e_{\alpha_j}, f\rangle e_{\alpha_j}\right\|^2 \leq \left\| f - \sum_{j=1}^{n} c_j e_{\alpha_j}\right\|^2,$$

consequently

$$\left\| f - \sum_{\alpha\in A}\langle e_\alpha, f\rangle e_\alpha\right\|^2 = \inf\{\|f-g\|^2 : g \in L(M)\} = \inf\{\|f-g\|^2 : g \in \overline{L(M)}\}.$$

Theorem 3.8. *Let M_1 and M_2 be measurable subsets of $\mathbb{R}^p$ and $\mathbb{R}^q$, respectively and let $\{e_n : n \in \mathbb{N}\}$ and $\{f_m : m \in \mathbb{N}\}$ be orthonormal bases of $L_2(M_1)$ and $L_2(M_2)$, respectively. If we define $g_{nm} \in L_2(M_1 \times M_2)$ by $g_{nm}(x, y) = e_n(x)f_m(y)$ for $x \in M_1, y \in M_2$, then $\{g_{nm} : (n, m) \in \mathbb{N}\times\mathbb{N}\}$ is an orthonormal basis of $L_2(M_1 \times M_2)$.*

PROOF. It is obvious that the functions g_{nm} are in $L_2(M_1 \times M_2)$ and they form an orthonormal system. It remains to prove that $\{g_{nm} : (n, m) \in \mathbb{N}\times\mathbb{N}\}$ is total. Let h be an element of $L_2(M_1 \times M_2)$ such that $h \perp g_{nm}$ for all n, m. For all $x \in M_1$ let $h_x(y) = h(x, y)$. By Fubini's theorem we have $h_x \in L_2(M_2)$ for all $x \in M_1 \setminus N$ with some set N of measure zero, and we have (Parseval's equality)

$$\|h\|^2 = \int_{M_1}\|h_x\|^2\,dx = \int_{M_1}\sum_{m=1}^{\infty}|\langle f_m, h_x\rangle|^2\,dx$$
$$= \int_{M_1}\sum_{m=1}^{\infty}\left|\int_{M_2} h(x, y)f_m(y)^*\,dy\right|^2 dx$$
$$= \sum_{m=1}^{\infty}\int_{M_1}\left|\int_{M_2} h(x, y)f_m(y)^*\,dy\right|^2 dx.$$

Let us define k_m by

$$k_m(x) = \int_{M_2} h(x, y) f_m(y)^* \, dy = \langle f_m, h_x \rangle, \; x \in M_1 \setminus N.$$

Then k_m also belongs to $L_2(M_1)$. Another application of Parseval's equality shows that

$$\int_{M_1} \left| \int_{M_2} h(x, y) f_m(y)^* \, dy \right|^2 dx = \|k_m\|^2 = \sum_{n=1}^{\infty} |\langle e_n, k_m \rangle|^2$$

$$= \sum_{n=1}^{\infty} \left| \int_{M_1} k_m(x) e_n(x)^* \, dx \right|^2 = \sum_{n=1}^{\infty} \left| \int_{M_1} \int_{M_2} h(x, y) f_m(y)^* e_n(x)^* \, dy \, dx \right|^2$$

$$= \sum_{n=1}^{\infty} \left| \int_{M_1 \times M_2} h(x, y) g_{nm}(x, y)^* \, dx \, dy \right|^2 = \sum_{n=1}^{\infty} |\langle g_{nm}, h \rangle|^2 = 0.$$

By summing up, it follows that $\|h\| = 0$. □

EXERCISES

3.6. If $T = \bigoplus_{\alpha \in A} T_\alpha$ and M_α are orthonormal bases in T_α, then $M = \bigcup_{\alpha \in A} M_\alpha$ is an ONB in T.

3.7. Let H be an infinite dimensional Hilbert space, and let M be an ONB of H. The cardinality of every dense subset of H is at least that of M. There exists a dense subset of H with cardinality equal to that of M.

3.3 Existence of orthonormal bases, dimension of a Hilbert space

Up to now we have always assumed the existence of orthonormal bases; only in examples did we see that in certain spaces orthonormal bases exist. The question is then whether all Hilbert spaces or pre-Hilbert spaces have orthonormal bases. It is relatively easy to show that each separable pre-Hilbert space has an ONB. For non-separable spaces it is a little harder to answer this question.

Theorem 3.9. *Let H be a separable pre-Hilbert space.*

(a) *H possesses an ONB.*

(b) *If M_1 is a finite ONS in H, then there exists an ONB in H such that $M \supset M_1$.*

(c) *If H is a Hilbert space and M_1 is an ONS in H, then there exists an ONB M in H such that $M \supset M_1$.*

(d) *H is m-dimensional ($m < \infty$) if and only if there exists an ONB containing m elements. Then each ONB in H has exactly m elements.*

(e) *H is infinite dimensional (i.e., not finite-dimensional) if and only if there exists an ONB containing a countable infinity of elements. Then each ONB in H is enumerably infinite.*

PROOF.
(a) follows from (b) if we choose $M_1 = \varnothing$.
(b) Let $H \neq \{0\}$ and let $M_1 = \{e_1, \ldots, e_n\}$ be an ONS. As H is separable, there exists a countable dense subset $A = \{f_n : n \in \mathbb{N}\}$ in H. We define the elements $g_1, g_2, \ldots,$ from A recursively in the following way: let $g_1 = f_{j_1}$, where j_1 is the smallest index for which $\{e_1, \ldots, e_n, f_{j_1}\}$ is linearly independent. If $g_1, \ldots, g_k$ are defined, then let $g_{k+1} = f_{j_{k+1}}$, where j_{k+1} is the smallest index for which $\{e_1, \ldots, e_m, g_1, \ldots, g_k, f_{j_{k+1}}\}$ is linearly independent. We obviously have $j_{k+1} > j_k$. With $B = \{e_1, \ldots, e_n, g_1, g_2, \ldots\}$ we then have $L(A) \subset L(B)$, i.e., B is total. If we apply Gram-Schmidt's orthogonalization method to B, we obtain an ONS M, the first n elements of which coincide with $e_1, \ldots, e_n$ (as these are already orthonormal). We have $\overline{L(M)} = \overline{L(B)} = H$, i.e., M is an ONB in H with $M_1 \subset M$.
(c) Let H be a separable Hilbert space and let M_1 be an ONS. Then $L(M_1)^{\perp}$ is also separable. Therefore, by part (a), there exists an ONB M_2 of $L(M_1)^{\perp}$. The set $M = M_1 \cup M_2$ is then an ONS, and $L(M_1 \cup M_2) = L(M_1) \oplus L(M_2)$. Consequently, by Exercise 3.5

$$\overline{L(M_1 \cup M_2)} = \overline{L(M_1) \oplus L(M_2)} = \overline{L(M_1)} \oplus \overline{L(M_2)} = H$$

hence M is an ONB of H such that $M_1 \subset M$.
(d) Let H be m-dimensional, i.e., assume that the maximal number of linearly independent elements equals m. As every ONS is linearly independent, it consists of at most m elements. If $M = \{e_1, \ldots, e_k\}$ is an ONS with less than m elements, then we have $\dim L(M) < \dim H$, therefore $L(M) \neq H$. Thus there exists an $f \in H$ such that $\{e_1, \ldots, e_k, f\}$ is linearly independent. The Schmidt orthogonalization process provides an ONS $M' = \{e_1, \ldots, e_k, e_{k+1}\}$ such that $M \subset M'$, i.e., M is no ONB. Hence every orthonormal basis has exactly m elements.
(e) If H is infinite dimensional, then every ONB has at least a countable infinity of elements, for otherwise H would be finite dimensional by part (d). It remains to prove that each ONB $M = \{e_\alpha : \alpha \in A\}$ is countable. Let $N = \{f_n : n \in \mathbb{N}\}$ be a countable dense subset. For each $\alpha \in A$ there exists an $n(\alpha) \in \mathbb{N}$ such that $\|f_{n(\alpha)} - e_\alpha\| < \frac{1}{2}$. Because $\|e_\alpha - e_\beta\| = \sqrt{2}$ for $\alpha \neq \beta$, we have

$$\begin{aligned} \|f_{n(\alpha)} - f_{n(\beta)}\| &> \|e_\alpha - e_\beta\| - \|f_{n(\alpha)} - e_\alpha\| - \|e_\beta - f_{n(\beta)}\| \\ &> \sqrt{2} - 1 > 0 \quad \text{for} \quad \alpha \neq \beta. \end{aligned}$$

This means that the mapping $\alpha \mapsto n(\alpha) : A \to \mathbb{N}$ is injective. Consequently A is countable. Conversely, if an orthonormal basis in H is infinite, then H is not finite dimensional. □

Proposition. *Every finite dimensional pre-Hilbert space is complete; in particular, every finite dimensional subspace of a pre-Hilbert space is a closed subspace.*

PROOF. Let H be an m-dimensional pre-Hilbert space. Then there exists an orthonormal basis $\{e_1, \ldots, e_m\}$. Let (f_n) be a Cauchy sequence in H with $f_n = \sum_{j=1}^m a_{jn} e_j$. Then we have

$$\|f_n - f_p\|^2 = \sum_{j=1}^m |a_{jn} - a_{jp}|^2,$$

thus for each j the sequence $(a_{jn})_{n \in N}$ is a Cauchy sequence as $n \to \infty$. Therefore $a_{jn} \to a_j$ as $n \to \infty$ with some a_j. Putting

$$f = \sum_{j=1}^m a_j e_j,$$

we have

$$\|f - f_n\|^2 = \left\| \sum_{j=1}^m (a_j - a_{jn}) e_j \right\|^2 = \sum_{j=1}^m |a_j - a_{jn}|^2 \to 0$$

as $n \to \infty$. Consequently, (f_n) is convergent in H. □

Proposition. *A pre-Hilbert space is separable if and only if it possesses an at most countable ONB.*

PROOF. By Theorem 3.9 each separable pre-Hilbert space possesses an at most countable ONB. If M is an at most countable ONB in H, then the set $L_r(M)$ of linear combinations of elements of M with rational coefficients is dense in $L(M)$, and thus it is also dense in H. As $L_r(M)$ is at most countable, H is separable. □

Theorem 3.10. *Let H be a Hilbert space.*

(a) *H possesses an ONB.*

(b) *If M_0 is an ONS, then there exists an ONB M in H such that $M \supset M_0$.*

(c) *All ONB of H have the same cardinality.*

REMARK. Theorem 3.10(a) and (b) do not hold for (non-separable) pre-Hilbert spaces; cf., for example, N. Bourbaki [2], Chapter 5, §2, Exercise 2.

PROOF. Part (a) follows from part (b) by choosing $M_0 = \emptyset$.

(b) Let $\mathfrak{M}$ be the set of all those ONS which contain M_0. $\mathfrak{M}$ is *partially ordered* by the inclusion "$\subset$" (i.e., we have $M \subset M$ for all $M \in \mathfrak{M}$; from $M_1 \subset M_2$, $M_2 \subset M_3$, it follows that $M_1 \subset M_3$; from $M_1 \subset M_2$, $M_2 \subset M_1$ it follows that $M_1 = M_2$). If $\mathfrak{N}$ is a *linearly ordered* subset of $\mathfrak{M}$ (i.e., for $M_1, M_2 \in \mathfrak{N}$ we have $M_1 \subset M_2$ or $M_2 \subset M_1$), then $\mathfrak{N}$ has an *upper bound* $M \in \mathfrak{M}$ (i.e., for every $M' \in \mathfrak{N}$ we have $M' \subset M$); for the upper bound M we may take the union of all $N \in \mathfrak{N}$.

This M *is an* ONS: If $f_1, f_2 \in M$, then there exist $M_1, M_2 \in \mathfrak{N}$ such that $f_1 \in M_1, f_2 \in M_2$. Since $M_1 \subset M_2$ or $M_2 \subset M_1$ holds, we have $f_1, f_2 \in M_2$ or $f_1, f_2 \in M_1$. Therefore $f_1 \perp f_2$.

As M contains all $N \in \mathfrak{N}$, M is an upper bound of $\mathfrak{N}$. By Zorn's lemma this implies the existence of at least one *maximal element* $M_{\max} \in \mathfrak{M}$ (i.e., for each $M \in \mathfrak{M}$ such that $M_{\max} \subset M$ we have $M_{\max} = M$).

This $M_{\max}$ *is an* ONB: If we had $\overline{L(M_{\max})} \neq H$, then, as H is a Hilbert space, there would be (cf. part (b) of the proposition preceding Theorem 3.3) an $f \in L(M_{\max})^{\perp}$ such that $\|f\| = 1$, i.e., $M_{\max} \cup \{f\}$ would be an ONS such that $M_{\max} \subset M_{\max} \cup \{f\}$ and $M_{\max} \neq M_{\max} \cup \{f\}$; this would contradict the maximality of $M_{\max}$. The requirement $M_0 \subset M_{\max}$ is obviously satisfied.

(c) Let M_1, M_2 be ONB of H. If $|M_1| = m < \infty$ (we write $|M|$ for the cardinality of M), then by the proposition preceding Theorem 3.10, the space H is separable and by Theorem 3.9(d) we have $\dim H = m = |M_2|$.

Now let $|M_1| \geq |\mathbb{N}|$. For each $f \in M_1$ let $K(f) = \{g \in M_2 : \langle g, f \rangle \neq 0\}$. By Theorem 3.6(b) we have $|K(f)| \leq |\mathbb{N}|$ for all $f \in M_1$. We have $\cup \{K(f) : f \in M_1\} = M_2$, since if $g \in M_2 \setminus \cup \{K(f) : f \in M_1\}$ we would have $g \perp M_1$, therefore $g = 0$ (as M_1 is total); however, this is impossible because all elements of M_2 have norm 1. Consequently, it follows that

$$|M_2| \leq \sum_{f \in M_1} |K(f)| \leq |M_1|\,|\mathbb{N}| \leq |M_1|.$$

On the other hand, $|M_2| \geq |\mathbb{N}|$ is true, for otherwise M_1 would be finite, also. We can therefore prove that $|M_1| \leq |M_2|$ in the same way. □

The *algebraic dimension* of a vector space is the cardinality of a maximal set of linearly independent elements (*algebraic basis*). In Hilbert spaces it is useful to introduce another notion of dimension. The *dimension* (more precisely, the *Hilbert space dimension*) of a Hilbert space H is the cardinality of an ONB of H. By Theorem 3.10(c) this dimension does not depend on the choice of the ONB. By Theorem 3.9(d) for finite dimensional Hilbert spaces the two definitions of dimension coincide; for infinite dimensional spaces this is not the case, cf. Exercise 3.8.

Proposition. *There exist Hilbert spaces of arbitrary (Hilbert space) dimension.*

PROOF. Let κ be an arbitrary cardinal number, and let A be a set of cardinality κ. Let $l_2(A)$ be the Hilbert space $l_2(A; \mu)$ with $\mu(\alpha) = 1$ for all $\alpha \in A$ (cf. Exercise 1.10 and 2.2(b)). The dimension of $l_2(A)$ equals $\kappa = |A|$, as $M = \{f_\alpha : \alpha \in A\}$, where $f_\alpha(\beta) = \delta_{\alpha\beta}$, is an ONB. □

Theorem 3.11. *If H is a Hilbert space and S and T are closed subspaces of H such that $S \cap T^{\perp} = \{0\}$, then we have* $\dim S \leqslant \dim T$ (dim = *Hilbert space dimension*).

PROOF. Let us distinguish between two different cases.

(a) $\dim T = k < \infty$: Assume that $\dim S > \dim T$ holds. If $\{e_1, \ldots, e_k\}$ is an ONB of T and $\{f_1, \ldots, f_{k+1}\}$ is an ONS in S, then the system of homogeneous equations (k equations, $k+1$ unknowns)

$$\sum_{j=1}^{k+1} c_j \langle e_m, f_j \rangle = 0, \quad m = 1, \ldots, k$$

has a non-trivial solution. Therefore there exists a non-zero element $f = \sum_{j=1}^{k+1} c_j f_j$ of $S \cap T^{\perp}$, which contradicts the assumption.

(b) $\dim T \geqslant |\mathbb{N}|$. Let M_1 and M_2 be orthonormal bases of T and S, respectively. For each $e \in M_1$ let $K(e) = \{f \in M_2 : \langle e, f \rangle \neq 0\}$. We have $\cup_{e \in M_1} K(e) = M_2$, because for $f \in M_2 \setminus \cup_{e \in M_1} K(e)$ we would have $f \perp M_1$, thus $f \perp T$; which would contradict the assumption. Since for each $e \in M_1$ the set $K(e)$ is at most countable, it follows that $|M_2| \leqslant |\mathbb{N}|\,|M_1| = |M_1|$. □

EXERCISES

3.8. Let H be a Hilbert space and let A be a countable subset of H such that $L(A) = H$. Then H is finite dimensional, i.e., no Hilbert space of algebraic dimension $|\mathbb{N}|$ exists.
Hint: Apply the Schmidt orthogonalization process to A; for the resulting ONB M we have $L(M) = H$. (It can actually be proved that no infinite dimensional Hilbert space can have an algebraic dimension smaller than the cardinality of the continuum; cf. N. Bourbaki [2], Chap. 5, §2, Exercise 1.)

3.9. (a) Let $(H, \langle.\,,.\rangle)$ be a pre-Hilbert space. For any n elements $f_1, \ldots, f_n$ of H the *Gram determinant* is defined by $D(f_1, \ldots, f_n) = \det(\langle f_j, f_k \rangle)$. We have $D(f_1, \ldots, f_n) \geqslant 0$; the equality sign holds if and only if the elements $f_1, \ldots, f_n$ are linearly dependent (in the case $n = 2$ this is Schwarz' inequality.)
Hint: Use induction on n. In going from $n-1$ to n use the fact that the value of the determinant does not change if the first column is replaced by $\langle f_j, f_1 - P_n(f_1) \rangle$, where $P_n(g)$ denotes the orthogonal projection of g onto $L(f_2, \ldots, f_n)$.

(b) Prove the same assertion by using the fact that the matrix $(\langle f_j, f_k \rangle)_{j,k=1,\ldots,n}$ is the product of the matrices

$$(\langle f_j, e_l \rangle)_{\substack{j=1,\ldots,n \\ l=1,\ldots,m}} \quad \text{and} \quad (\langle e_l, f_k \rangle)_{\substack{l=1,\ldots,m, \\ k=1,\ldots,n}}$$

where $\{e_1, \ldots, e_m\}$ is an ONB of $L\{f_1, \ldots, f_n\}$.

(c) Prove an analogous statement for semi-scalar products.

3.10. Part (c) of the proposition preceding Example 1 of Section 3.2 does not hold in pre-Hilbert spaces.
Hint: In l_2 let $f=(1/n)$, $H=l_{2,0}$, $H_1=l_{2,0}\cap\{f\}^{\perp}$. By Exercise 3.4 we have $\bar{H}_1=\{f\}^{\perp}$. If M is an ONB of H, then M is a maximal ONS in $l_{2,0}$, without being an ONB in $l_{2,0}$.

3.4 Tensor products of Hilbert spaces

Let H_1 and H_2 be vector spaces over $\mathbb{K}$. We denote by $F(H_1, H_2)$ the vector space of *formal* linear combinations of the pairs (f, g) with $f\in H_1$, $g\in H_2$, i.e.,

$$F(H_1, H_2)=\left\{\sum_{j=1}^{n} c_j(f_j, g_j): c_j\in\mathbb{K}, f_j\in H_1, g_j\in H_2, j=1, 2, \ldots, n;\ n\in\mathbb{N}\right\}.$$

Let N be the subspace of $F(H_1, H_2)$ spanned by the elements of the form

$$\sum_{j=1}^{n}\sum_{k=1}^{m} a_j b_k(f_j, g_k)-1\times\left(\sum_{j=1}^{n} a_j f_j, \sum_{k=1}^{m} b_k g_k\right). \tag{3.1}$$

The quotient space

$$H_1\otimes H_2=F(H_1, H_2)/N$$

is called the *algebraic tensor product* of H_1 and H_2.

The product $H_1\times H_2$ can be considered as a subset of $F(H_1, H_2)$, if one identifies $(f, g)\in H_1\times H_2$ with $1(f, g)\in F(H_1, H_2)$. The equivalence class from $H_1\otimes H_2$ defined by (f, g) will be denoted by $f\otimes g$; these elements are called *simple tensors*. Each element of $H_1\otimes H_2$ is representable as a finite linear combination of simple tensors. Such a linear combination of simple tensors is equal to zero if and only if it is a finite linear combination of elements of the form

$$\sum_{j=1}^{n}\sum_{k=1}^{m} a_j b_k f_j\otimes g_k-\left(\sum_{j=1}^{n} a_j f_j\right)\otimes\left(\sum_{k=1}^{m} b_k g_k\right). \tag{3.2}$$

In particular, we have

$$\left(\sum_{j=1}^{n} a_j f_j\right)\otimes\left(\sum_{k=1}^{m} b_k g_k\right)=\sum_{j=1}^{n}\sum_{k=1}^{m} a_j b_k f_j\otimes g_k. \tag{3.3}$$

If $(H_1, \langle.\,,.\rangle_1)$ and $(H_2, \langle.\,,.\rangle_2)$ are Hilbert spaces over $\mathbb{K}$, then

$$s\left(\sum_{j=1}^{n} c_j(f_j, g_j), \sum_{k=1}^{m} c_k'(f_k', g_k')\right)=\sum_{j=1}^{n}\sum_{k=1}^{m} c_j^* c_k' \langle f_j, f_k'\rangle_1\langle g_j, g_k'\rangle_2$$

defines a sesquilinear form on $F(H_1, H_2)$. For arbitrary $f\in N$ and $g\in$

$F(H_1, H_2)$ we have $s(f, g) = s(g, f) = 0$, as one can verify by simple calculation. Consequently, by

$$\left\langle \sum_{j=1}^{n} c_j f_j \otimes g_j, \sum_{j=1}^{n} c'_k f'_k \otimes g'_k \right\rangle = s\left(\sum_{j=1}^{n} c_j (f_j, g_j), \sum_{k=1}^{m} c'_k (f'_k, g'_k) \right)$$

a sesquilinear form is defined on $H_1 \otimes H_2$.

We show that $\langle . , . \rangle$ is a scalar product on $H_1 \otimes H_2$. In order to prove this it is enough to show that $\langle f, f \rangle > 0$ holds for all $f \in H_1 \otimes H_2$, $f \neq 0$. Indeed, let $f = \sum_{j=1}^{n} c_j f_j \otimes g_j \neq 0$. If $\{e_k\}$ and $\{e'_k\}$ are orthonormal bases of $L\{f_1, \ldots, f_n\}$ and $L\{g_1, \ldots, g_n\}$, respectively, then

$$f = \sum_{k,l} c_{kl} e_k \otimes e'_l \quad \text{with} \quad c_{kl} = \sum_j c_j \langle e_k, f_j \rangle \langle e'_l, g_j \rangle, \tag{3.4}$$

and thus

$$\langle f, f \rangle = \sum_{k,l} |c_{kl}|^2 > 0.$$

Therefore $(H_1 \otimes H_2, \langle . , . \rangle)$ is a pre-Hilbert space. The completion of this pre-Hilbert space (cf. Section 4.3) will be denoted by $H_1 \hat{\otimes} H_2$ and called the (*complete*) *tensor product* of the Hilbert spaces H_1 and H_2.

From (3.4) for each $f = \sum_{j=1}^{n} c_j f_j \otimes g_j \in H_1 \otimes H_2$ it follows by means of (3.3) that

$$f = \sum_l f'_l \otimes e'_l = \sum_k e_k \otimes g'_k \tag{3.5}$$

where $\{e_k\}$ and $\{e'_l\}$ are orthonormal systems in $L\{f_1, \ldots, f_n\}$ and $L\{g_1, \ldots, g_n\}$, respectively; the elements f'_l and g'_k are contained in $L\{f_1, \ldots, f_n\}$ and $L\{g_1, \ldots, g_n\}$, respectively.

EXAMPLE 1. Let ρ_1 and ρ_2 be measures on $\mathbb{R}$, and let $H_1 = L_2(\mathbb{R}, \rho_1)$, $H_2 = L_2(\mathbb{R}, \rho_2)$. By (3.1) an element $\sum_{j=1}^{n} c_j (f_j, g_j)$ from $F(H_1, H_2)$ is in N if and only if the function

$$(x, y) \mapsto \sum_{j=1}^{n} c_j f_j(x) g_j(y), \quad (x, y) \in \mathbb{R}^2$$

vanishes almost everywhere with respect to the product measure $\rho_1 \times \rho_2$. The algebraic tensor product $H_1 \otimes H_2$ is thus composed of equivalence classes of functions, square integrable on $\mathbb{R}^2$ with respect to $\rho_1 \times \rho_2$. For $f, g \in H_1 \otimes H_2$ we have

$$\langle f, g \rangle = \int f(x, y)^* g(x, y) \, d\rho_1(x) \, d\rho_2(y).$$

As $H_1 \otimes H_2$ obviously contains all step functions on $\mathbb{R}^2$, the space $H_1 \hat{\otimes} H_2$ is isomorphic to $L_2(\mathbb{R}^2, \rho_1 \times \rho_2)$.

Theorem 3.12. *Let H_1 and H_2 be Hilbert spaces.*

(a) *If M_1 and M_2 are total subsets of H_1 and H_2, respectively, then the set $\{f \otimes g : f \in M_1, g \in M_2\}$ is total in $H_1 \hat{\otimes} H_2$.*

(b) *If $\{e_\alpha : \alpha \in A\}$ and $\{f_\beta : \beta \in B\}$ are orthonormal bases of H_1 and H_2, respectively, then $\{e_\alpha \otimes f_\beta : \alpha \in A, \beta \in B\}$ is an orthonormal basis of $H_1 \hat{\otimes} H_2$.*

PROOF.

(a) Let $\sum_{j=1}^n f_j \otimes g_j \in H_1 \otimes H_2$, $\epsilon > 0$. For each $j \in \{1, 2, \ldots, n\}$ there exist elements $f_j' \in L(M_1)$ and $g_j' \in L(M_2)$ such that $\|f_j - f_j'\| \, \|g_j\| < \epsilon/2n$ and $\|g_j - g_j'\| \, \|f_j\| < \epsilon/2n$. Then we have

$$\|f_j \otimes g_j - f_j' \otimes g_j'\| = \|(f_j - f_j') \otimes g_j + f_j' \otimes (g_j - g_j')\| < \epsilon/n$$

and consequently

$$\left\| \sum_{j=1}^n f_j \otimes g_j - \sum_{j=1}^n f_j' \otimes g_j' \right\| < \epsilon.$$

Because

$$\sum_{j=1}^n f_j' \otimes g_j' \in L(M_1) \otimes L(M_2) = L\{f \otimes g : f \in M_1, g \in M_2\},$$

the assertion is proved, since $H_1 \otimes H_2$ is dense in $H_1 \hat{\otimes} H_2$.

(b) By part (a) the set $\{e_\alpha \otimes f_\beta : \alpha \in A, \beta \in B\}$ is total in $H_1 \otimes H_2$. Moreover, we have

$$\langle e_\alpha \otimes f_\beta, e_{\alpha'} \otimes f_{\beta'} \rangle = \delta_{\alpha\alpha'} \delta_{\beta\beta'} \quad \text{for all} \quad \alpha \in A, \beta \in B,$$

i.e., $\{e_\alpha \otimes f_\beta : \alpha \in A, \beta \in B\}$ is an orthonormal basis. □

EXERCISES

3.11. Two non-zero tensors $f_1 \otimes g_1$ and $f_2 \otimes g_2$ are equal to each other if and only if there exists a $c \in \mathbb{K}$, $c \neq 0$ that satisfies $f_2 = cf_1$, $g_2 = c^{-1} g_1$.

3.12. Let H_1 and H_2 be Hilbert spaces.

(a) We have $\dim [H_1 \hat{\otimes} H_2] = (\dim H_1)(\dim H_2)$ (Hilbert space dimensions).

(b) If H_1 and H_2 are different from $\{0\}$, then $H_1 \hat{\otimes} H_2$ is separable if and only if H_1 and H_2 are both separable.

(c) $(H_1 \otimes H_2, \langle . , . \rangle)$ is complete if and only if H_1 or H_2 is finite dimensional.

4 Linear operators and their adjoints

4.1 Basic notions

Let H_1 and H_2 be vector spaces over $\mathbb{K}$. A *linear operator* T from H_1 into H_2 is, by definition, a linear mapping of a subspace $D(T)$ of H_1 into H_2. The subspace $D(T)$ is called the *domain* of T. The image $R(T) = T(D(T)) = \{Tf : f \in D(T)\}$ is called the *range* of T. Since we only treat linear operators here, we shall speak only about *operators from H_1 into H_2*. If $H_1 = H_2 = H$, then T is called an *operator on H*. A linear operator from H into $\mathbb{K}$ is called a *linear functional*. The range of an operator T from H_1 into H_2 is a subspace of H_2. An operator is injective if and only if $Tf = 0$ implies $f = 0$. In this case the *inverse* T^{-1} of T is defined by

$$D(T^{-1}) = R(T),\ T^{-1}g = f \quad \text{for} \quad g = Tf \in R(T).$$

T^{-1} is a (linear) operator from H_2 into H_1. For an operator T from H_1 into H_2 and for $a \in \mathbb{K}$ the operator aT is defined by

$$D(aT) = D(T) \quad \text{and} \quad (aT) = a(Tf) \quad \text{for} \quad f \in D(aT).$$

For two operators S and T from H_1 into H_2 the *sum* $S+T$ is defined by

$$D(S+T) = D(S) \cap D(T),\ (S+T)f = Sf + Tf \quad \text{for} \quad f \in D(S+T).$$

If T is an operator from H_1 into H_2 and S is an operator from H_2 into H_3, then the *product* ST is defined by

$$D(ST) = \{f \in D(T) : Tf \in D(S)\},\ (ST)f = S(Tf) \quad \text{for} \quad f \in D(ST).$$

If D is a subspace of H_1, then the set of those operators from H_1 into H_2 whose domain is D is a vector space over $\mathbb{K}$; the zero element is the operator whose domain is D and which sends all elements of D to 0. Let S

and T be operators from H_1 into H_2. An operator T is called an *extension* of S (or S is a *restriction* of T) if we have

$$D(S) \subset D(T) \quad \text{and} \quad Sf = Tf \quad \text{for} \quad f \in D(S).$$

For this we write $S \subset T$ or $T \supset S$.

EXAMPLE 1. Let M be a measurable subset of $\mathbb{R}^m$ and let $t: M \to \mathbb{C}$ be a measurable function on M. The *maximal operator of multiplication* by t on $L_2(M)$ is defined by

$$D(T) = \{ f \in L_2(M) : tf \in L_2(M) \}, \; Tf = tf \quad \text{for} \quad f \in D(T).$$

The set $D(T)$ is obviously a subspace of $L_2(M)$ and T is an operator on $L_2(M)$.

(4.1) *$D(T)$ is dense.*

PROOF. For each $n \in \mathbb{N}$ let $M_n = \{x \in M : |t(x)| < n\}$. Then $M_n \subset M_{n+1}$ and $\cup_{n=1}^{\infty} M_n = M$. For each $f \in L_2(M)$ the function $f_n = \chi_{M_n} f$ belongs to $D(T)$ and we have $f_n \to f$. □

(4.2) *The following statements are equivalent*:
(a) *$R(T)$ is dense,*
(b) *$t(x) \neq 0$ almost everywhere in M,*
(c) *T is injective.*
If one of these assumptions is satisfied, then T^{-1} is the multiplication operator defined by the function

$$t_1(x) = \begin{cases} t(x)^{-1} & \text{for} \quad x \in M \quad \text{such that} \quad t(x) \neq 0 \\ 0 & \text{for} \quad x \in M \quad \text{such that} \quad t(x) = 0. \end{cases}$$

PROOF. (b) *follows from* (a): Each $f \in L_2(M)$, that vanishes outside the set $M_1 = \{x \in M : t(x) = 0\}$, is orthogonal to $R(T)$. Therefore $L_2(M_1) = \{0\}$, i.e., M_1 has measure 0.

(a) *follows from* (b): Let $M_n = \{x \in M : |t(x)| > 1/n\}$; then $M_n \subset M_{n+1}$ and $M \setminus \cup_{n=1}^{\infty} M_n$ has measure zero. For every $g \in L_2(M)$ the function $g_n = \chi_{M_n} g$ belongs to $R(T)$ and $g_n \to g$ holds.

(c) *follows from* (b): If $Tf = 0$, then $t(x)f(x) = 0$ almost everywhere in M. Therefore $f(x) = 0$ almost everywhere, too, and thus $f = 0$.

(b) *follows from* (c): If $M_1 = \{x \in M : t(x) = 0\}$, then for all $f \in L_2(M)$, vanishing outside M_1, we have $Tf = 0$; therefore $f = 0$. From this it follows that M_1 has measure zero.

If one of the above conditions is satisfied, then T is injective and $t(x) \neq 0$ almost everywhere in M. Hence we have

$$\begin{aligned} D(T^{-1}) = R(T) &= \{ g \in L_2(M) : \text{there exists an } f \in L_2(M) \text{ such that } g = tf \} \\ &= \{ g \in L_2(M) : t_1 g \in L_2(M) \}, \end{aligned}$$

and for $g \in D(T^{-1})$ and $f \in L_2(M)$ such that $g = tf$ we have

$$T^{-1}g = f = t_1 g;$$

consequently T^{-1} is the multiplication operator induced by t_1. □

(4.3) *We have $D(T) = L_2(M)$ if and only if a C exists for which $|t(x)| \leq C$ almost everywhere in M. We have $R(T) = L_2(M)$ if and only if a $c > 0$ exists for which $|t(x)| \geq c$ almost everywhere in M.*

Proof. If $|t(x)| \leq C$ almost everywhere, then $tf \in L_2(M)$ for all $f \in L_2(M)$. Therefore $D(T) = L_2(M)$. Conversely, let $D(T) = L_2(M)$. Let us assume that no C exists for which $|t(x)| \leq C$ almost everywhere. For $n \in \mathbb{N}$ set $M_n = \{x \in M : |t(x)| \geq n\}$, $N_n = M_{n-1} \setminus M_n$ with $M_0 = M$. Then all M_n have positive measures, and the intersection $\cap_{n=1}^{\infty} M_n$ has measure zero. Therefore there exists a subsequence (n_k) of $\mathbb{N}$ such that all N_{n_k} have positive measures. We have $|t(x)| \geq n_k - 1$ for $x \in N_{n_k}$. For all $k \in \mathbb{N}$ let us choose $f_k \in L_2(M)$ in such a way that f_k vanish outside N_{n_k} and $\|f_k\| = 1/k$. Since the functions f_k are mutually orthogonal, we have

$$f = \sum_{k=1}^{\infty} f_k \in L_2(M).$$

However, tf is not in $L_2(M)$, i.e., $f \notin D(T)$, this contradicts the fact that $D(T) = L_2(M)$. (A simpler proof of this can be found in Exercise 5.5.)

If $|t(x)| \geq c > 0$ almost everywhere, then T^{-1} exists, and for the inducing function t_1 we have $|t_1(x)| \leq c^{-1}$ almost everywhere. Therefore $R(T) = D(T^{-1}) = L_2(M)$. If $R(T) = L_2(M)$, then by (4.2) the operator T is injective and $D(T^{-1}) = L_2(M)$. For the inducing function t_1 we have $|t_1(x)| \leq C$ almost everywhere, thus $|t(x)| \geq C^{-1}$ almost everywhere. □

If t, s are measurable functions on M and T, S are the multiplication operators induced by t, s, then $T + S$ is a restriction of the multiplication operator induced by $t + s$, as from $tf \in L_2(M)$ and $sf \in L_2(M)$ it obviously follows that $(t + s)f \in L_2(M)$.

Example 2. If $\psi : \mathbb{R} \to \mathbb{C}$ is continuous, then by

$$Tf = \int \psi(x) f(x)\, dx, f \in L_{2,0}(\mathbb{R})$$

a linear functional is defined with $D(T) = L_{2,0}(\mathbb{R})$. If $\psi \in L_2(\mathbb{R})$, then T can be defined on the whole space $L_2(\mathbb{R})$.

The subset $N(T) = \{f \in D(T) : Tf = 0\}$ is called the *kernel* of T. For every operator T from H_1 into H_2 the set $N(T)$ is a subspace of $D(T)$ (consequently of H_1, as well).

Theorem 4.1. *Let H be a vector space over $\mathbb{K}$, and let $T_1, \ldots, T_n, T$ be linear functionals such that $D(T_1) = \ldots = D(T_n) = D(T) = H$. If we have $N(T) \supset \cap_{i=1}^n N(T_i)$, then there exist $a_1, \ldots, a_n \in \mathbb{K}$ such that*

$$T = \sum_{i=1}^{n} a_i T_i.$$

Proof. We prove this by induction on n. Let n be equal to 1. If $T_1 = 0$, then $N(T) \supset N(T_1) = H$, therefore $T = 0$. If $T_1 \neq 0$, then there exists an $f_0 \in H$ such that $T_1 f_0 = 1$. For every $f \in H$ we then have $f - T_1(f) f_0 \in N(T_1) \subset N(T)$. Consequently, $Tf = T(f_0) T_1 f$, i.e., $T = T(f_0) T_1$.

Let us now assume that the assertion is true for $n-1$ $(n > 1)$. Assume that $N(T) \supset \cap_{i=1}^n N(T_i)$. If $N(T_n) \supset \cap_{i=1}^{n-1} N(T_i)$, then it follows that $N(T) \supset \cap_{i=1}^{n-1} N(T_i)$. Therefore by the induction hypothesis we have $T = \Sigma_{i=1}^{n-1} c_i T_i$. So in this case the assertion holds. If $N(T_n) \not\supset \cap_{i=1}^{n-1} N(T_i)$, then there exists an $f_0 \in \cap_{i=1}^{n-1} N(T_i)$ such that $T_n f_0 = 1$. Let

$$T_0 = T - T(f_0) T_n.$$

For all $f \in \cap_{i=1}^{n-1} N(T_i)$ we have

$$T_i(f - T_n(f) f_0) = T_i(f) - T_n(f) T_i(f_0) = 0 - 0 = 0, \quad i = 1, 2, \ldots, n-1$$

and

$$T_n(f - T_n(f) f_0) = T_n(f) - T_n(f) T_n(f_0) = T_n(f) - T_n(f) = 0,$$

i.e.,

$$f - T_n(f) f_0 \in \bigcap_{i=1}^{n} N(T_i) \subset N(T),$$

and thus

$$T_0(f) = T(f) - T(f_0) T_n(f) = T(f - T_n(f) f_0) = 0.$$

Therefore we have

$$N(T_0) \supset \bigcap_{i=1}^{n-1} N(T_i).$$

By the induction hypothesis it follows from this that

$$T - T(f_0) T_n = T_0 = \sum_{i=1}^{n-1} c_i T_i,$$

i.e., T is a linear combination of $T_1, \ldots, T_n$. □

Let H_1 and H_2 now be normed spaces. An operator T from H_1 into H_2 is said to be *continuous at the point* $f \in D(T)$ if for every sequence (f_n) from $D(T)$ such that $f_n \to f$ we have $Tf_n \to Tf$. The operator T is *continuous* by definition if T is continuous at each point of $D(T)$. The operator T is said to be *bounded* if there exists a $C > 0$ such that $\|Tf\| \leq C\|f\|$ for all $f \in D(T)$. Any such C is called a *bound* of T.

Theorem 4.2. *Let T be an operator from H_1 into H_2. Then the following assertions are equivalent*:
(a) *T is continuous*,
(b) *T is continuous at* 0,
(c) *T is bounded.*

PROOF. (b) obviously follows from (a).

(c) *follows from* (b): Let us assume that T is not bounded. Then for every $n \in \mathbb{N}$ there exists an $f_n \in D(T)$ such that $\|Tf_n\| > n\|f_n\|$. From this it follows that, in particular, $f_n \neq 0$; without loss of generality we may assume that $\|f_n\| = 1/n$. Consequently we have $f_n \to 0$ and $\|Tf_n\| > n(1/n) = 1$, which contradicts the continuity of T at zero.

(a) *follows from* (c): Assume $\|Tf\| \leqslant C\|f\|$ for all $f \in D(T)$. If $f \in D(T)$ and (f_n) is a sequence from $D(T)$ such that $f_n \to f$, then we have

$$\|Tf_n - Tf\| = \|T(f_n - f)\| \leqslant C\|f_n - f\| \to 0.$$

i.e., $Tf_n \to Tf$, which proves the continuity of T. □

For a bounded operator T from H_1 into H_2 the *norm* $\|T\|$ is defined by

$$\|T\| = \inf\{C > 0 : \|Tf\| \leqslant C\|f\| \quad \text{for all} \quad f \in D(T)\} \tag{4.4}$$

(in Section 4.2 we shall justify the word "norm"). Since for every $\epsilon > 0$ we have

$$\|Tf\| \leqslant (\|T\| + \epsilon)\|f\| \quad \text{for all} \quad f \in D(T),$$

the norm $\|T\|$ is a bound for T, thus

$$\|Tf\| \leqslant \|T\|\,\|f\| \quad \text{for all} \quad f \in D(T). \tag{4.5}$$

EXAMPLE 1 (Continued). A function $s : M \to \mathbb{R}$ is said to be *essentially bounded from above* if there exists a $C \in \mathbb{R}$ such that $s(x) \leqslant C$ almost everywhere in M. Each C of this kind is called an *essential upper bound* of s. The greatest lower bound of all essential upper bounds is called the *essential supremum* of s, in symbols ess sup s. It is itself an essential upper bound for s. Indeed, if C_0 denotes this greatest lower bound, then for every $n \in \mathbb{N}$ the number $C_0 + (1/n)$ is an essential upper bound, i.e., $s(x) - C_0 - (1/n) \leqslant 0$ holds almost everywhere. By letting $n \to \infty$ it follows that $s(x) - C_0 \leqslant 0$ almost everywhere. Analogously, we may define the concepts of *essentially bounded from below*, *essential lower bound*, and *essential infimum*. A complex-valued function s is said to be *essentially bounded*, if $|s|$ is essentially bounded from above.

(4.6) *The operator T from Example* 1 *is bounded if and only if t is essentially bounded. We have* $\|T\| = \text{ess sup}\,|t|$.

PROOF. If t is essentially bounded, and $C = \text{ess sup}\,|t|$, then we have $|t(x)| \leqslant C$ almost everywhere. Therefore

$$\|Tf\|^2 = \int_M |t(x)f(x)|^2\,dx \leqslant C^2 \int_M |f(x)|^2\,dx = C^2\|f\|^2,$$

i.e., T is bounded and $\|T\| \leqslant C$. If $C=0$, then we have $\|T\|=0$. If $C>0$, then for every $\epsilon \in (0, C)$ the set $M_\epsilon = \{x \in M : |t(x)| \geqslant C-\epsilon\}$ has a positive measure and for all $f \in L_2(M)$, that vanish outside M_ϵ, we have

$$\|Tf\|^2 = \int_M |t(x)f(x)|^2 \, dx \geqslant (C-\epsilon)^2 \int_{M_\epsilon} |f(x)|^2 \, dx = (C-\epsilon)^2 \|f\|^2.$$

Therefore $\|T\| \geqslant C-\epsilon$, and thus $\|T\| = C$.

If t is not essentially bounded, then for every $n \in \mathbb{N}$ the set $M_n = \{x \in M : t(x) \geqslant n\}$ has a positive measure and for every $f \in D(T)$, that vanishes outside M_n, we have

$$\|Tf\| \geqslant n\|f\|.$$

Therefore T is not bounded. □

EXAMPLE 2 (Continued). The functional T of Example 2 is bounded if $\psi \in L_2(\mathbb{R})$, since then

$$|Tf| \leqslant \|\psi\| \, \|f\|.$$

From Theorem 4.8 (theorem of Riesz) it will follow that T is continuous if and only if $\psi \in L_2(\mathbb{R})$.

EXAMPLE 3. Let M_1 and M_2 be measurable subsets of $\mathbb{R}^p$ and $\mathbb{R}^m$, respectively. Then $M_2 \times M_1$ is a measurable subset of $\mathbb{R}^{m+p}$. The points of $\mathbb{R}^{m+p}$ can be written in the form (x, y) with $x \in \mathbb{R}^m$, $y \in \mathbb{R}^p$. Assume $k \in L_2(M_2 \times M_1)$. By Fubini's theorem

$$\int_{M_1} |k(x,y)|^2 \, dy < \infty \quad \text{almost everywhere in } M_2,$$

i.e., we have $k(x, .) \in L_2(M_1)$ almost everywhere in M_2. Consequently, for all $f \in L_2(M_1)$ we can define

$$(Kf)(x) = \int_{M_1} k(x,y) f(y) \, dy \quad \text{almost everywhere in } M_2.$$

Then we have

$$|(Kf)(x)| \leqslant \|f\| \left\{ \int_{M_1} |k(x,y)|^2 \, dy \right\}^{1/2}. \tag{4.7}$$

For every $g \in L_2(M_2)$ the function h defined by $h(x,y) = k(x,y)f(y)g(x)$ is integrable on $M_2 \times M_1$. Therefore by Fubini's theorem the function

$$g(x)(Kf)(x) = g(x) \int_{M_1} k(x,y) f(y) \, dy = \int_{M_1} h(x,y) \, dy$$

is a measurable function on M_2. If we put $M_{2,n} = \{x \in M_2 : |x| \leqslant n\}$ and $g_n = \chi_{M_{2,n}}$, then we can see that $\chi_{M_{2,n}} Kf$ is measurable for every $n \in \mathbb{N}$. Consequently, Kf is measurable.

Because of (4.7) we have $Kf \in L_2(M_2)$ and

$$\|Kf\| \leq \left\{ \int_{M_2} \int_{M_1} |k(x,y)|^2 \, dy \, dx \right\}^{1/2} \|f\|.$$

Since the mapping $f \mapsto Kf$ is obviously linear, we have defined a continuous operator K from $L_2(M_1)$ into $L_2(M_2)$ such that $D(K) = L_2(M_1)$ and

$$\|K\| \leq \left\{ \int_{M_2} \int_{M_1} |k(x,y)|^2 \, dy \, dx \right\}^{1/2}.$$

Such an operator is called a *Hilbert-Schmidt operator* (cf. also Section 6.2).

Exercises

4.1. (a) The reasoning of Example 1 can be carried out completely analogously if M is replaced by an arbitrary σ-finite measure space $(X, \mathfrak{B}, \mu)$. (A measure space $(X, \mathfrak{B}, \mu)$ is said to be *σ-finite* if X can be written as the union of countably many subsets of finite measure.)

(b) In Example 3 replace M_1 and M_2 by two arbitrary σ-finite measure spaces $(X_1, \mathfrak{B}_1, \mu_1)$ and $(X_2, \mathfrak{B}_2, \mu_2)$.
Hint: Observe that the set $X_{2,0} = \{x \in X_2 : \int |k(x,y)|^2 \, d\mu_1(y) > 0\}$ is the union of the countably many sets $X_{2,n} = \{x \in X_2 : \int |k(x,y)|^2 \, d\mu_1(y) > 1/n\}$ $(n \in \mathbb{N})$ with finite measures, and $\chi_{X_{2,n}} \in L_2(X_2, \mathfrak{B}_2, \mu_2)$.

4.2. Let $(X_1, \mathfrak{B}_1, \mu_1)$ and $(X_2, \mathfrak{B}_2, \mu_2)$ be σ-finite measure spaces. Let $k : X_2 \times X_1 \to \mathbb{C}$ be $\mu_2 \times \mu_1$-measurable and let $k(x,.) \in L_2(X_1, \mathfrak{B}_1, \mu_1)$ almost everywhere in X_2. Then for each $f \in L_2(X_1, \mathfrak{B}_1, \mu_1)$ the function $K_0 f$ defined by

$$K_0 f(x) = \int_{X_1} k(x,y) f(y) \, d\mu_1(y)$$

is μ_2-measurable and we have $|K_0 f(x)| \leq \|k(x,.)\| \, \|f\|$. By

$$D(K) = \{f \in L_2(X_1, \mathfrak{B}_1, \mu_1) : K_0 f \in L_2(X_2, \mathfrak{B}_2, \mu_2)\}$$
$$Kf = K_0 f \quad \text{for} \quad f \in D(K)$$

an operator is defined from $L_2(X_1, \mathfrak{B}_1, \mu_1)$ into $L_2(X_2, \mathfrak{B}_2, \mu_2)$ (such an operator is called a *Carleman operator*; cf. also Section 6.2). There are functions k of this kind for which $D(K) = \{0\}$.

4.2 Bounded linear operators and functionals

Theorem 4.3. *Let H_1 and H_2 be normed spaces. Let T be an operator from H_1 into H_2.*

(a) *We have*

$$\sup\{\|Tf\| : f \in D(T), \|f\| \leq 1\} = \sup\{\|Tf\| : f \in D(T), \|f\| = 1\}$$
$$= \sup\{\|Tf\| : f \in D(T), \|f\| < 1\}$$

(where the value ∞ is allowed). T is bounded if and only if one of these values is finite; if one is finite then the others are finite, also, and they are equal to $\|T\|$.

(b) *If H_2 is a pre-Hilbert space and M is a subspace of H_2 such that $R(T)\subset\overline{M}$, then T is bounded if and only if*

$$\sup\{|\langle Tf, g\rangle| : f\in D(T), g\in M, \|f\|=\|g\|=1\}$$

is finite. This number is then equal to $\|T\|$.

(c) *If T is an operator on a pre-Hilbert space H and D(T) is dense, then T is bounded if and only if*

$$\sup\{|\langle Tf, g\rangle| : f, g\in D(T), \|f\|=\|g\|=1\}$$

is finite. This number is then equal to $\|T\|$.

PROOF.

(a) If we define $\|T\|$ to be equal to ∞ if T is unbounded, then we only have to prove that the three values, which we denote by c_1, c_2, c_3, respectively, are all equal to $\|T\|$. As $\|T\|$ is a bound for T, we surely have $\|T\| \geqslant c_1$. The inequality $c_1 \geqslant c_2$ is obvious. If $f=0$, then $\|Tf\|=0 \leqslant c_2$. If $0<\|f\|<1$, then with $g=\|f\|^{-1}f$ we have

$$\|g\| = 1 \quad\text{and}\quad \|Tg\| = \|f\|^{-1}\|Tf\| \geqslant \|Tf\|.$$

Consequently $c_2 \geqslant c_3$. What remains is to prove that $c_3 \geqslant \|T\|$. This is evident if $\|T\|=0$. Therefore suppose $\|T\|>0$. For every $\epsilon\in(0, 1)$ there exists an $f\in D(T)$ such that $\|Tf\|>(1-\epsilon)\|T\|\ \|f\|$. Hence $f\neq 0$ and for $g=[(1+\epsilon)\|f\|]^{-1}f$ we have

$$\|g\| < 1 \quad\text{and}\quad \|Tg\| \geqslant \frac{1-\epsilon}{1+\epsilon}\|T\|.$$

From this it follows that $c_3 \geqslant (1-\epsilon)/(1+\epsilon)\|T\|$ for all $\epsilon\in(0, 1)$ and thus $c_3 \geqslant \|T\|$.

(b) By (a) we have $\|T\|=\sup\{\|Tf\| : f\in D(T), \|f\|=1\}$. On the other hand,

$$\|Tf\| = \sup\{|\langle Tf, g\rangle| : g\in M, \|g\|=1\}.$$

Indeed, if $Tf=0$, this is obvious. If $Tf\neq 0$, then $\|Tf\| \geqslant |\langle Tf, g\rangle|$ for all $g\in M$ such that $\|g\|=1$, and there exists a sequence (g_n) from M such that $\|g_n\|=1$ and $g_n\to\|Tf\|^{-1}Tf$, therefore $\langle Tf, g_n\rangle\to\|Tf\|$. These arguments together give the assertion.

(c) follows from (b) if we choose $M=D(T)$. □

An operator T is said to be *densely defined* if $D(T)$ is dense. An operator T, which is densely defined on a pre-Hilbert space H, is said to be *symmetric* if for all $f, g\in D(T)$ we have $\langle Tf, g\rangle=\langle f, Tg\rangle$.

Theorem 4.4. *Let T be a densely defined operator on a complex pre-Hilbert space, or a symmetric operator defined on an arbitrary pre-Hilbert space. T is bounded if and only if*

$$C = \sup\{|\langle f, Tf\rangle| : f \in D(T), \|f\| \leq 1\} < \infty.$$

If T is bounded, then we have
(a) $\|T\| \leq 2C$, *if H is a complex Hilbert space,*
(b) $\|T\| = C$, *if T is symmetric.*

PROOF. Let us set $\|T\| = \infty$ for an unbounded T. Then we only have to prove inequality (a) and equality (b).
(a) For all $f, g \in D(T)$ such that $\|f\| \leq 1$ and $\|g\| \leq 1$, from (1.4) with $s(f, g) = \langle f, Tg\rangle$ and from (1.16) it follows that

$$\begin{aligned} 2C &\geq \tfrac{1}{4}C\{4\|f\|^2 + 4\|g\|^2\} \\ &= \tfrac{1}{4}C\{\|f+g\|^2 + \|f-g\|^2 + \|f+ig\|^2 + \|f-ig\|^2\} \\ &\geq \tfrac{1}{4}|\langle f+g, T(f+g)\rangle - \langle f-g, T(f-g)\rangle - i\langle f+ig, T(f+ig)\rangle \\ &\quad + i\langle f-ig, T(f-ig)\rangle| = |\langle f, Tg\rangle|. \end{aligned}$$

The assertion follows from this by Theorem 4.3.
(b) By (1.7) and (1.8) with $s(f, g) = \langle f, Tg\rangle$, it follows for all $f, g \in D(T)$ such that $\|f\| \leq 1$ and $\|g\| \leq 1$ that

$$\begin{aligned} C &\geq \tfrac{1}{2}C\{\|f\|^2 + \|g\|^2\} = \tfrac{1}{4}C\{\|f+g\|^2 + \|f-g\|^2\} \\ &\geq \tfrac{1}{4}|\langle f+g, T(f+g)\rangle - \langle f-g, T(f-g)\rangle| = |\mathrm{Re}\,\langle f, Tg\rangle|. \end{aligned}$$

If we choose $a \in \mathbb{K}$ so that $|a| = 1$ and $a\langle f, Tg\rangle = |\langle f, Tg\rangle|$ hold, then it follows (with $h = a^*f$) that

$$|\langle f, Tg\rangle| = \langle h, Tg\rangle = |\mathrm{Re}\,\langle h, Tg\rangle| \leq C.$$

By Theorem 4.3(c) it follows from this that $\|T\| \leq C$. The inequality $C \leq \|T\|$ is evident by Theorem 4.3(c). □

Theorem 4.5. *Let T be a bounded operator from a normed space H_1 into a Banach space H_2. Then there exists a unique bounded extension S of T such that $D(S) = \overline{D(T)}$. We have $\|S\| = \|T\|$.*

PROOF. *Uniqueness*: Assume S is a continuous extension of T such that $D(S) = \overline{D(T)}$. If $f \in D(S)$, then there exists a sequence (f_n) from $D(T)$ such that $f_n \to f$. As S is continuous, we have $Sf = \lim Sf_n = \lim Tf_n$, i.e., S is (if it exists at all) determined by T uniquely.

Existence: Assume that $f \in \overline{D(T)}$ and (f_n) is a sequence from $D(T)$ such that $f_n \to f$. Then (f_n) is a Cauchy sequence. Since T is bounded, the sequence (Tf_n) is a Cauchy sequence, also, for we have $\|Tf_n - Tf_m\| \leq \|T\|\,\|f_n - f_m\|$. Therefore there exists a $g \in H_2$ such that $Tf_n \to g$. This g is independent of the choice of the sequence (f_n) from $D(T)$ with $f_n \to f$.

Indeed, if (f_n') is another sequence of this kind, then the sequence $(f_1, f_1', f_2, f_2', \ldots)$ converges to f, also. Hence the sequence $(Tf_1, Tf_1', Tf_2, Tf_2' \ldots)$ is also convergent; the limit has to be equal to g. Consequently, (Tf_n') tends to g, as well. Let us define $Sf = g$.

S is linear: If $f_1, f_2 \in \overline{D(T)}$ and $(f_{1,n}), (f_{2,n})$ are sequences in $D(T)$ such that $f_{1,n} \to f_1, f_{2,n} \to f_2$, then for all $a, b \in \mathbb{K}$ it follows that

$$S(af_1 + bf_2) = \lim T(af_{1,n} + bf_{2,n}) = \lim (aTf_{1,n} + bTf_{2,n}) = aSf_1 + bSf_2.$$

S is bounded and $\|S\| = \|T\|$: For $f \in \overline{D(T)}$ and (f_n) from $D(T)$ such that $f_n \to f$ we have $\|f_n\| \to \|f\|$ and

$$\|Sf\| = \lim \|Tf_n\| \leqslant \lim \|T\| \, \|f_n\| = \|T\| \, \|f\|.$$

Therefore $\|S\| \leqslant \|T\|$. As $\|S\| \geqslant \|T\|$ obviously holds, the assertion follows. □

The set of those bounded operators from H_1 into H_2, whose domain is H_1, will be denoted by $B(H_1, H_2)$. By Section 4.1 the set $B(H_1, H_2)$ is a vector space.

Theorem 4.6. *Let* $\| \, . \, \|$ *be defined as in* (4.4). *Then* $(B(H_1, H_2), \| \, . \, \|)$ *is a normed space. If* H_2 *is a Banach space, then* $(B(H_1, H_2), \| \, . \, \|)$ *is a Banach space, too.*

PROOF. It is clear that $\| \, . \, \|$ is a semi-norm. If $\|T\| = 0$, then $\|Tf\| = 0$ for all $f \in H_1$ such that $\|f\| \leqslant 1$; therefore $Tf = 0$ for all $f \in H_1$, and thus $T = 0$, the zero element in $B(H_1, H_2)$. Consequently, $\| \, . \, \|$ is a norm. Assume now that H_2 is a Banach space. If (T_n) is a Cauchy sequence in $B(H_1, H_2)$, then for every $f \in H_1$ the sequence $(T_n f)$ is a Cauchy sequence and, consequently, a convergent sequence. Let us define: $Tf = \lim T_n f$. Then T is linear, because for $f, g \in H_1$ and $a, b \in \mathbb{K}$ we have

$$T(af + bg) = \lim T_n(af + bg) = \lim (aT_n f + bT_n g) = aTf + bTg.$$

As (T_n) is a Cauchy sequence, $(\|T_n\|)$ is convergent, say $\|T_n\| \to C$. For all $f \in H_1$ we have

$$\|Tf\| = \lim \|T_n f\| \leqslant \lim \|T_n\| \, \|f\| = C\|f\|,$$

i.e., $T \in B(H_1, H_2)$. What remains is to prove that $T_n \to T$. For every $\epsilon > 0$ there exists an $n(\epsilon) \in \mathbb{N}$ such that $\|T_n - T_m\| \leqslant \epsilon$ for $n, m \geqslant n(\epsilon)$. Therefore, for $n \geqslant n(\epsilon)$ and for all $f \in H_1$ we have

$$\|(T_n - T)f\| = \lim_{m \to \infty} \|(T_n - T_m)f\| \leqslant \epsilon\|f\|,$$

i.e., $\|T_n - T\| \leqslant \epsilon$ for $n \geqslant n(\epsilon)$. Hence $T_n \to T$. □

Theorem 4.7.

(a) *Let H_1 and H_2 be normed spaces. If $T \in B(H_1, H_2)$, then $N(T)$ is a closed subspace of H_1.*

(b) *Let T be a linear functional on a Hilbert space H such that $D(T) = H$. Then T is continuous if and only if $N(T)$ is closed.*

PROOF.

(a) Let $f \in \overline{N(T)}$. Then there exists a sequence (f_n) from H_1 such that $f_n \to f$ and $Tf_n = 0$. Since T is continuous, it follows that $Tf = \lim Tf_n = 0$, i.e., $f \in N(T)$.

(b) If T is continuous, then $N(T)$ is closed by part (a). Let $N(T)$ now be closed. If $N(T) = H$, then $T = 0$; consequently T is continuous. If $N(T) \neq H$, then $N(T)^\perp \neq \{0\}$. Therefore there exists a $g \in N(T)^\perp$ such that $g \neq 0$. Because $\langle g, f \rangle = 0$ for all $f \in N(T)$, we have $N(T_g) \supset N(T)$ for the functional $T_g f = \langle g, f \rangle$. By Theorem 4.1 this implies that $T = cT_g$ with some $c \in \mathbb{K}$. Consequently T is continuous. (The proof can also be carried out analogously to the second part of the proof of Theorem 4.8, without using Theorem 4.1.) □

REMARK. If H_1 and H_2 are pre-Hilbert spaces, one may expect that $(B(H_1, H_2), \|\,.\,\|)$ is a pre-Hilbert space, also, i.e., the norm is induced by a scalar product. This holds true if $H_1 = \mathbb{K}$ or $H_2 = \mathbb{K}$. However, this is not the case if dim $H_1 \geqslant 2$ and dim $H_2 \geqslant 2$; cf. Exercise 4.3.

For $T \in B(H_1, H_2)$ and $S \in B(H_2, H_3)$ the product ST is in $B(H_1, H_3)$, since we have

$$D(ST) = \{f \in H_1 : Tf \in D(S) = H_2\} = H_1$$

and

$$\|STf\| \leqslant \|S\| \, \|Tf\| \leqslant \|S\| \, \|T\| \, \|f\| \quad \text{for all} \quad f \in H_1,$$

i.e., we have $ST \in B(H_1, H_3)$ and

$$\|ST\| \leqslant \|S\| \, \|T\|. \tag{4.8}$$

We write $B(H)$ for $B(H, H)$. For $S, T_1, T_2 \in B(H)$ we have

$$S(T_1 + T_2) = ST_1 + ST_2, \; (T_1 + T_2)S = T_1 S + T_2 S.$$

The operator I with $D(I) = H$ and $If = f$ for all $f \in H$ obviously belongs to $B(H)$ and we have

$$\|I\| = 1, \tag{4.9}$$

and $IT = TI = T$ for all $T \in B(H)$. The operator I is called the *identity operator* on H. The set $B(H)$ is thus an algebra with an *identity element*. As the norm on $B(H)$ satisfies relations (4.8) and (4.9), the algebra $B(H)$ is a *normed algebra* with an identity element. If H is a Banach space, then $B(H)$ is complete; we call $B(H)$ a *Banach algebra* (with an identity element).

Now we shall study the set of continuous linear functionals defined on a Hilbert space H, i.e., the set $B(H, \mathbb{K})$ for a Hilbert space H over $\mathbb{K}$. Every element $g \in H$ defines a linear functional T_g such that $D(T_g) = H$ by the formula

$$T_g(f) = \langle g, f\rangle.$$

As $|T_g(f)| \leq \|g\|\,\|f\|$, the functional T_g is bounded, and $\|T_g\| \leq \|g\|$. Since for $f = g$ we have $|T_g(g)| = \|g\|^2$, it follows that $\|T_g\| = \|g\|$. Every continuous linear functional defined on H is actually of this form.

Theorem 4.8 (F. Riesz). *Let H be a Hilbert space. Every $g \in H$ induces a continuous linear functional on H by $T_g(f) = \langle g, f\rangle$. We have $\|T_g\| = \|g\|$. This mapping of H onto $B(H, \mathbb{K})$ is bijective and antilinear, i.e., we have $T_{ag+bh} = a^* T_g + b^* T_h$.*

PROOF. The first part has already been shown. The antilinearity follows from

$$T_{ag+bh}(f) = \langle ag + bh, f\rangle = a^*\langle g, f\rangle + b^*\langle h, f\rangle = a^* T_g(f) + b^* T_h(f).$$

As $\|T_g\| = \|g\|$, the mapping $g \mapsto T_g$ is injective. What remains is to prove that it is also surjective. Let $T \in B(H, \mathbb{K})$; we construct a $g \in H$ such that $T = T_g$.

If $T = 0$, then we can choose $g = 0$. If $T \neq 0$, then the kernel $N(T) = \{f \in H : T(f) = 0\}$ is a closed subspace of H, different from H, i.e., $N(T)^{\perp} \neq \{0\}$. Let $g \in N(T)^{\perp}$ such that $\|g\| = 1$. Let $a = T(g)$. For every $f \in H$ we obviously have $T(f)g - T(g)f \in N(T)$. Therefore $T(f)g - T(g)f$ is orthogonal to g, i.e.,

$$0 = \langle g, T(f)g - T(g)f\rangle = T(f) - a\langle g, f\rangle,$$

and thus

$$T(f) = T(g)\langle g, f\rangle = \langle a^* g, f\rangle.$$

Consequently, we have $T = T_{a^*g}$. (We can also prove this last part with the aid of Theorem 4.1, cf. the proof of Theorem 4.7(b).) □

EXAMPLE 1 (Continuation of Example 2 from Section 4.1). We can now show that the continuous function $\psi : \mathbb{R} \to \mathbb{C}$ induces a continuous functional on $L_{2,0}(\mathbb{R})$ by $T(f) = \int \psi(x) f(x)\,dx$ if and only if $\psi \in L_2(\mathbb{R})$. If T is continuous, then it can be extended uniquely to a continuous functional on $L_2(\mathbb{R})$ (which we denote by T, as well). By Theorem 4.8 we have $T = T_g$ with some $g \in L_2(\mathbb{R})$. Therefore, we have for all $f \in L_{2,0}(\mathbb{R})$ that

$$0 = T(f) - T_g(f) = \int (\psi(x) - g^*(x)) f(x)\,dx.$$

For an arbitrary $n \in \mathbb{N}$ let us define

$$f_n(x) = \begin{cases} \psi^*(x) - g(x) & \text{for } |x| \leq n, \\ 0 & \text{for } |x| > n. \end{cases}$$

Then for all n it follows that

$$\int_{-n}^{n} |\psi(x) - g^*(x)|^2 \, dx = \int (\psi(x) - g^*(x)) f_n(x) \, dx = 0,$$

hence

$$\int |\psi(x) - g^*(x)|^2 \, dx = 0.$$

From this we can infer that $\psi = g^* \in L_2(\mathbb{R})$.

Exercises

4.3. (a) If H is a Hilbert space, then the norm in $B(H, \mathbb{K})$ is defined by a scalar product (i.e., $B(H, \mathbb{K})$ is a Hilbert space).
Hint: If T_1, T_2 are the functionals induced by $g_1, g_2 \in H$, then let $\langle T_1, T_2 \rangle = \langle g_2, g_1 \rangle$.
(b) If H_1 and H_2 are (pre-) Hilbert spaces, then the norm in $B(H_1, H_2)$ is induced by a scalar product if and only if $\dim H_1 = 1$ or $\dim H_2 = 1$.
Hint: If $\dim H_1 \geqslant 2$ and $\dim H_2 \geqslant 2$, then let $f_1, f_2 \in H_1$, $g_1, g_2 \in H_2$ be such that $\langle f_i, f_j \rangle = \langle g_i, g_j \rangle = \delta_{ij}$ and $T_j f = \langle f_j, f \rangle g_j$ for $f \in H_1$, $j = 1, 2$. For these two operators the parallelogram identity does not hold.

4.4. For each $x_0 \in [0, 1]$ there exists exactly one $g \in W_{2,1}(0, 1)$ such that for all $f \in W_{2,1}(0, 1)$ we have

$$f(x_0) = \int_0^1 \{ g^*(x) f(x) + g'^*(x) f'(x) \} \, dx$$

(cf. Exercise 2.3(b)).
Hint: The functional $Tf = f(x_0)$ is continuous on $W_{2,1}(0, 1)$.

4.5. The set of Hilbert-Schmidt operators on $L_2(M)$ (cf. Section 4.1, Example 3) is a sub-algebra of $B(L_2(M))$. It is a Banach algebra with the *Hilbert-Schmidt norm*

$$|||K||| = \left\{ \int_M \int_M |k(x, y)|^2 \, dx \, dy \right\}^{1/2}.$$

Hint: If the Hilbert-Schmidt operators K and H are induced by the kernels k and h, then $L = HK$ is induced by the kernel

$$l(x, y) = \int_M h(x, z) k(z, y) \, dz.$$

4.6. (a) Theorem 4.4 does not hold true for non-symmetric operators in real (pre-) Hilbert spaces. There are non-vanishing operators T such that the quadratic form $q_T(f) = \langle f, Tf \rangle$ vanishes on $D(T)$.
(b) Show that the constant $2C$ in Theorem 4.4(a) is optimal.
(c) In a complex (pre-) Hilbert space H the quantity $\|T\|_q = \sup \{ |\langle f, Tf \rangle| : f \in D(T), \|f\| \leqslant 1 \}$ is a norm on $B(H)$. We have $\|T\| \leqslant 2\|T\|_q \leqslant 2\|T\|$.

4.7. Let T be a bounded operator from a Hilbert space H_1 into a Banach space H_2. Then there exists an extension $S \in B(H_1, H_2)$ of T such that $\|S\| = \|T\|$.
Hint: Define $Sf = 0$ for $f \in D(T)^{\perp}$.

4.8. If H is an infinite dimensional Hilbert space, then $B(H)$ is not separable.
Hint: Let $\{e_n : n \in \mathbb{N}\}$ be an ONS in H; for every sequence $\alpha = (\alpha_n)$ from $\{0, 1\}$ let $T_\alpha \in B(H)$ be defined by $T_\alpha f = \Sigma \alpha_n \langle e_n, f \rangle e_n$; this is an uncountable set of operators such that $\|T_\alpha - T_\beta\| = 1$ for $\alpha \neq \beta$.

4.9. Let H be a Banach space. There are no operators $A, B \in B(H)$ such that $AB - BA = I$.
Hint: From $AB - BA = I$ it follows that $(n+1)B^n = AB^{n+1} - B^{n+1}A$ for all $n \in \mathbb{N}$, therefore $\|B^n\| \leq 2/(n+1)\|A\| \, \|B\| \, \|B^n\|$, i.e., $B^n = 0$ for large n; this implies that $0 = B^n = B^{n-1} = \ldots = B^0 = I$.

4.3 Isomorphisms, completion

Let H_1 and H_2 be normed spaces. An operator U from H_1 into H_2 is called an *isometry*, if $D(U) = H_1$ and $\|Uf\|_2 = \|f\|_1$ for all $f \in H_1$. An isometry U from H_1 into H_2 is called an *isomorphism* of H_1 onto H_2 if $R(U) = H_2$. Every isomorphism U of H_1 onto H_2 is injective and U^{-1} is an isomorphism of H_2 onto H_1.

If H_1 and H_2 are pre-Hilbert spaces and U is an isomorphism of H_1 onto H_2, then it follows from the polarization identity that $\langle Uf, Ug \rangle_2 = \langle f, g \rangle_1$ for all $f, g \in H_1$. (The subscripts of the norms and scalar products will be omitted in the sequel, as it will be always clear from the context, to which spaces the elements belong.) Two normed spaces H_1 and H_2 are said to be *isomorphic* (or *equivalent*) if there exists an isomorphism of H_1 onto H_2.

Theorem 4.9. *Let H_1 and H_2 be isomorphic normed spaces. H_1 is a Banach space (Hilbert space) if and only if H_2 is a Banach space (Hilbert space).*

PROOF. Let H_1 be a Banach space and let U be an isomorphism of H_1 onto H_2. If (f_n) is a Cauchy sequence in H_2, then $(U^{-1}f_n)$ is a Cauchy sequence in H_1; hence there exists a $g \in H_1$ such that $U^{-1}f_n \to g$. With $f = Ug \in H_2$ we have $f_n \to f$, i.e., H_2 is complete. If H_1 is a Hilbert space, then H_2 is complete and since $\|U^{-1}f\| = \|f\|$ for all $f \in H_2$, the parallelogram identity holds in H_2, i.e., the norm of H_2 is defined by a scalar product. As U^{-1} is also an isomorphism, we can prove analogously the reverse direction. □

Theorem 4.10. *Let H be a Hilbert space, and let A be a set, the cardinality of which equals the (Hilbert space) dimension of H. Then H is isomorphic to $l_2(A)$. In particular, all infinite dimensional separable Hilbert spaces are isomorphic to l_2. Hilbert spaces having the same dimension are isomorphic to each other.*

PROOF. Let $\{e_\alpha : \alpha \in A\}$ be an ONB of H. For every $f = \Sigma f_\alpha e_\alpha \in H$ let Uf be the function $A \to \mathbb{C}$ with $(Uf)(\alpha) = f_\alpha$. It is easy to see that U is an isomorphism of H onto $l_2(A)$. All the other assertions are obvious consequences of this. □

It is often useful to know that every normed space (pre-Hilbert space) can be considered as a dense subspace of a Banach space (Hilbert space). If H is a normed space (pre-Hilbert space), and $\hat{H}$ is a Banach space (Hilbert space), then $\hat{H}$ is called a *completion* of H provided that H is isomorphic to a dense subspace of $\hat{H}$.

Theorem 4.11. *For each normed space (pre-Hilbert space) H there exists a completion $\hat{H}$. Two arbitrary completions are isomorphic.*

PROOF. We construct a completion $\hat{H}$. For this let $\mathcal{H}$ be the set of all Cauchy sequences in H. Two Cauchy sequences (f_n) and (g_n) from $\mathcal{H}$ are considered *equivalent* (in symbols $(f_n)\sim(g_n)$), if $\|f_n - g_n\|\to 0$. This obviously defines an *equivalence relation* (since we have $(f_n)\sim(f_n)$, from $(f_n)\sim(g_n)$ it follows that $(g_n)\sim(f_n)$, and $(f_n)\sim(g_n)$ and $(g_n)\sim(h_n)$ imply $(f_n)\sim(h_n)$). Let $\hat{H}$ be the set of all equivalence classes. The elements of $\hat{H}$ will be denoted by $\hat{f}, \hat{g}, \ldots$. We shall write in particular $\hat{f}=[(f_n)]$ if (f_n) belongs to the equivalence class $\hat{f}$ and $\hat{f}=[f]$ if $\hat{f}$ is the equivalence class of the sequences that converge to $f\in H$ (notice that $f_n\to f$ and $(f_n)\sim(g_n)$ imply $g_n\to f$).

With $a[(f_n)]+b[(g_n)]=[(af_n+bg_n)]$ the set $\hat{H}$ becomes a vector space; the zero element is $\hat{0}=[0]$. We show that a norm can be introduced on $\hat{H}$ by putting

$$\|[(f_n)]\| = \lim \|f_n\|.$$

For this, we have to notice that the sequence $(\|f_n\|)$ is convergent (cf. the proposition preceding Example 1 of Section 2.1) and the limit does not depend on the choice of representatives, since for $(f_n)\sim(g_n)$ we have $|\|f_n\| - \|g_n\|| \leqslant \|f_n - g_n\|\to 0$. The properties of a semi-norm obviously hold. If $\|[(f_n)]\|=0$, then $f_n\to 0$, i.e., we have $[(f_n)]=\hat{0}$; consequently, we have defined a norm.

If H is a pre-Hilbert space, then we define a scalar product by

$$\langle[(f_n)],[(g_n)]\rangle = \lim \langle f_n, g_n\rangle.$$

This is obviously a semi-scalar product; since it induces the above norm, it is a scalar product. Therefore $\hat{H}$ is a normed space or a pre-Hilbert space, respectively.

Now let $[H]=\{[f]\in\hat{H} : f\in H\}$. The set $[H]$ is obviously a subspace of $\hat{H}$. The space H is isomorphic to $[H]$, since by $Uf=[f]$ an isomorphism of H onto $[H]$ is defined.

$[H]$ is dense in $\hat{H}$: Let $\hat{f}=[(f_n)]\in\hat{H}$. For each $\epsilon>0$ there exists an $n(\epsilon)\in\mathbb{N}$ such that $\|f_n - f_m\|<\epsilon$ for $n, m\geqslant n(\epsilon)$. Therefore for $m\geqslant n(\epsilon)$ we have

$$\|\hat{f}-[f_m]\| = \lim_{n\to\infty}\|f_n - f_m\| \leqslant \epsilon,$$

i.e., $\hat{f}$ is a contact point of $[H]$.

It remains to prove that $\hat{H}$ is complete. Let $(\hat{f}_k)$ be a Cauchy sequence in $\hat{H}$. Since $[H]$ is dense in $\hat{H}$, for every $n \in \mathbb{N}$ there exists a $g_n \in H$ such that $\|\hat{f}_n - [g_n]\| \leq 1/n$. Since we have

$$\begin{aligned} \|g_n - g_m\| &= \|[g_n] - [g_m]\| \\ &\leq \|[g_n] - \hat{f}_n\| + \|\hat{f}_n - \hat{f}_m\| + \|\hat{f}_m - [g_m]\| \to 0 \end{aligned}$$

as $n, m \to \infty$, the sequence (g_n) is a Cauchy sequence. We have

$$\begin{aligned} \|\hat{f}_k - [(g_n)]\| &\leq \|\hat{f}_k - [g_k]\| + \|[g_k] - [(g_n)]\| \\ &\leq \frac{1}{k} + \lim_{n\to\infty} \|g_k - g_n\| \to 0 \end{aligned}$$

as $k \to \infty$; consequently, $\hat{f}_k \to [(g_n)]$.

Now let $\hat{H}$ and $\tilde{H}$ be two completions of H and let $\hat{U}$ and $\tilde{U}$ be the corresponding isomorphisms of H onto the dense subspaces $\hat{U}(H)$ and $\tilde{U}(H)$ of $\hat{H}$ and $\tilde{H}$, respectively. Then $V_0 = \hat{U}\tilde{U}^{-1}$ is an isomorphism of $\tilde{U}(H)$ onto $\hat{U}(H)$. By Theorem 4.5 V_0 can be extended to an element V of $B(\tilde{H}, \hat{H})$. For every $\tilde{f} \in \tilde{H}$ there exists a sequence $(\tilde{f}_n)$ from $\tilde{U}(H)$ such that $\tilde{f}_n \to \tilde{f}$. We have

$$\|V\tilde{f}\| = \lim \|V\tilde{f}_n\| = \lim \|V_0\tilde{f}_n\| = \lim \|\tilde{f}_n\| = \|\tilde{f}\|,$$

i.e., V is an isometry. In order to prove that V is an isomorphism of $\tilde{H}$ onto $\hat{H}$ we have to show that $R(V) = \hat{H}$. Let $\hat{f} \in \hat{H}$. Then there exists a sequence $(\hat{f}_n)$ from $\hat{U}(H)$ such that $\hat{f}_n \to \hat{f}$. If we put $\tilde{f}_n = \tilde{U}\hat{U}^{-1}\hat{f}_n$, then $(\tilde{f}_n)$ is a Cauchy sequence in $\tilde{H}$. Therefore there exists an $\tilde{f} \in \tilde{H}$ such that $\tilde{f}_n \to \tilde{f}$. Then we have

$$\hat{f} = \lim \hat{f}_n = \lim \hat{U}\tilde{U}^{-1}\tilde{f}_n = \lim V\tilde{f}_n = V\tilde{f}.$$

Thus $\hat{f} \in R(V)$, i.e., $\tilde{H}$ and $\hat{H}$ are isomorphic. □

Proposition. *Let H_1 and H_2 be normed spaces (pre-Hilbert spaces) and let H_1 be isomorphic to a dense subspace of H_2. If $\hat{H}_1$ and $\hat{H}_2$ are completions of H_1 and H_2, respectively, then $\hat{H}_1$ and $\hat{H}_2$ are isomorphic.*

PROOF. Let U be an isomorphism of H_1 onto the dense subspace $U(H_1)$ of H_2, and let V be an isomorphism of H_2 onto the dense subspace $V(H_2)$ of $\hat{H}_2$. Then $VU(H_1)$ is a dense subspace of $\hat{H}_2$, hence H_1 is isomorphic to a dense subspace of $\hat{H}_2$, i.e., $\hat{H}_2$ is a completion of H_1 and, consequently, it is isomorphic to $\hat{H}_1$. □

Theorem 4.12. *Let H be a pre-Hilbert space, and let $T_1, \ldots, T_n$ be linear functionals such that $D(T_j) = H$ and $L(T_1, \ldots, T_n) \cap B(H, K) = \{0\}$. Then*

$$M = \bigcap_{j=1}^{n} N(T_j) = \{f \in H : T_j f = 0 \ \text{for}\ j = 1, \ldots, n\}$$

is a dense subspace of H. In particular, the kernel of an unbounded functional is dense.

PROOF. Without loss of generality we may assume that H is a dense subspace of a Hilbert space $(H_0, \langle .\,,.\rangle)$. (If $\hat{H}$ is a completion of H and U is an isomorphism of H onto a dense subspace of $\hat{H}$, then we replace H by $U(H)$ and T_j by T_jU^{-1}; the following proof shows that $U(M)$ is dense in $U(H)$; consequently M is dense in H.)

Since M is surely a subspace, it is enough to show that M is dense in H_0, i.e., $M^{\perp}=\{0\}$. Let $g\in M^{\perp}$. Then for the continuous functional $T_g : H\to \mathbb{K}$, $T_gf=\langle g,f\rangle$ we have $M\subset N(T_g)$. Hence by Theorem 4.1 we have $T_g\in L(T_1,\dots,T_n)$. By assumption $T_g=0$; consequently $g=0$. □

EXERCISES

4.10. Let H be a pre-Hilbert space with an orthonormal basis $\{e_\alpha : \alpha\in A\}$. Then $l_2(A)$ is a completion of H. (In particular, we obtain that l_2 is a completion of any infinite dimensional separable pre-Hilbert space; in proving this we do not need Theorem 4.11).

4.11. Let H be a pre-Hilbert space over $\mathbb{K}$. A mapping $A : H\to\mathbb{K}$ is said to be an *antilinear functional*, if for all $f, g\in H$ and $a, b\in\mathbb{K}$ we have $A(af+bg)=a^*Af+b^*Ag$. The functional A is said to be *bounded*, if there exists a $C\geqslant 0$ such that $\|Af\|\leqslant C\|f\|$ for all $f\in H$. Let H^+ be the set of bounded antilinear functionals on H.
(a) An antilinear functional is continuous if and only if it is bounded.
(b) H^+ becomes a Banach space with the norm $\|A\|=\sup\{|Af| : f\in H, \|f\|\leqslant 1\}$.
(c) To each $g\in H$ there corresponds an $A_g\in H^+$ defined by $A_gf=\langle f,g\rangle$. The mapping $E : H\to H^+$, $g\mapsto A_g$ is isometric.
(d) $\overline{E(H)}$ is a completion of H.
(e) We have $\overline{E(H)}=H^+$.
Hint: Use Theorem 4.5 and 4.8. (This exercise provides a completion for *all* pre-Hilbert spaces without reference to Theorem 4.11.)

4.12. (a) Let H_1 and H_2 be isomorphic normed spaces. Then H_1 is separable if and only if H_2 is separable.
(b) A normed space H is separable if and only if one (and then each) of its completions is separable.
(c) Every infinite dimensional separable pre-Hilbert space is isomorphic to a dense subspace of l_2 (cf. also Exercise 4.10).

4.13. Let H_1 be a normed space, H_2 a Banach space and $\hat{H}_1$ a completion of H_1. Then $B(H_1, H_2)$ and $B(\hat{H}_1, H_2)$ are isomorphic.

4.14. Let H_1 and H_2 be Hilbert spaces. If U is an isomorphism of H_1 onto H_2, then for every subset M of H_1 we have $U(M^{\perp})=(UM)^{\perp}$.

4.15. Let $G\subset\mathbb{R}^m$ be open and let $L_{2,0}(G)$ be as in Exercise 2.6(a). Assume that $\psi_1,\dots,\psi_n : G\to\mathbb{C}$ are locally square integrable (i.e., square integrable on each compact subset of G) and that $L(\psi_1,\dots,\psi_n)\cap L_2(G)=\{0\}$. Then

$M = \{f \in L_{2,0}(G) : \int_G \psi_j(x) f(x)\,dx = 0$ for $j = 1, \ldots, n\}$ is a dense subspace of $L_2(G)$.

4.16. Let T be a bounded operator from a pre-Hilbert space H_1 into a Banach space H_2. Then there exists an extension $S \in B(H_1, H_2)$ of T such that $\|S\| = \|T\|$.
Hint: Use Exercise 4.7.

4.17. Let $\Omega_m = \{x \in \mathbb{R}^m : |x| = 1\}$ be the unit sphere in $\mathbb{R}^m$ and let $C(\Omega_m)$ be the space of continuous functions on Ω_m.

(a) By $\langle f, g \rangle = \int_\Omega f(\omega) g(\omega)\,d\sigma_m(\omega)$ a scalar product is defined on $C(\Omega_m)$ (here $d\sigma_m(\omega)$ denotes the surface element of Ω_m). Let this pre-Hilbert space be denoted by $C_2(\Omega_m)$.

(b) Let the measure ρ_m (cf. Appendix A) be defined for every interval $J \subset \mathbb{R}^m$ by $\rho_m(J)$ = the surface of that part of Ω_m which lies in J. The space $L_2(\mathbb{R}^m, \rho_m)$ is a completion of $C_2(\Omega_m)$; we shall denote it simply by $L_2(\Omega_m)$.

(c) $L_2(\Omega_m)$ is separable.

(d) The space of infinitely many times continuously differentiable functions (i.e., the set of the restrictions of infinitely many times continuously differentiable functions defined on $\mathbb{R}^m$) is dense in $L_2(\Omega_m)$.

4.4 Adjoint operator

Assume that H_1 and H_2 are Hilbert spaces, T is an operator from H_1 into H_2, and S is an operator from H_2 into H_1. The operator S is called a *formal adjoint* of T if we have

$$\langle g, Tf \rangle = \langle Sg, f \rangle \quad \text{for all} \quad f \in D(T), g \in D(S).$$

T is then a formal adjoint of S, also. We say that S and T are formal adjoints of each other. The operator S_0 such that $D(S_0) = \{0\}$ is a formal adjoint of every operator from H_1 into H_2.

If S is a formal adjoint of T, then for every $g \in D(S)$ the linear functional L_g with

$$D(L_g) = D(T),\ L_g f = \langle g, Tf \rangle$$

is continuous, since for all $f \in D(L_g)$ we have

$$L_g f = \langle g, Tf \rangle = \langle Sg, f \rangle,$$

i.e., L_g is the restriction, to $D(T)$, of the continuous functional T_{Sg} induced by Sg.

If $D(T)$ is dense, and the functional L_g is continuous, then by Theorem 4.5 this functional can be extended to $H_1 = \overline{D(T)}$ in a unique way, i.e., there exists an element $h_g \in H_1$, uniquely determined by g and T via

$$\langle g, Tf \rangle = L_g f = \langle h_g, f \rangle \quad \text{for all} \quad f \in D(T).$$

If S is a formal adjoint operator of T, and $g \in D(S)$, then we surely have $Sg = h_g$. Therefore in this case every formally adjoint operator of T is a restriction of the *adjoint operator* T^* to be defined below.

Let T be a densely defined operator from H_1 into H_2, and let

$$\begin{aligned} D^* &= \{ g \in H_2 : \text{the functional } f \mapsto \langle g, Tf \rangle \text{ is continuous on } D(T)\} \\ &= \{ g \in H_2 : \text{there exists an } h_g \in H_1 \text{ such that } \langle h_g, f\rangle = \langle g, Tf\rangle \\ &\qquad \text{for all } f \in D(T)\}. \end{aligned}$$

The element h_g *is uniquely determined*: If $\langle h_1, f\rangle = \langle g, Tf\rangle = \langle h_2, f\rangle$ for all $f \in D(T)$, then $h_1 - h_2 \in D(T)^\perp = \{0\}$, consequently, $h_1 = h_2$.

D^* is a *subspace* of H_1 and the correspondence $D^* \to H_1$, $g \mapsto h_g$ is a *linear transformation*, since for $g_1, g_2 \in D^*$ and $a, b \in \mathbb{K}$ we obviously have

$$h_{ag_1 + bg_2} = ah_{g_1} + bh_{g_2}.$$

Thus by $D(T^*) = D^*$, $T^*g = h_g$ for $g \in D(T^*)$ a linear operator T^* from H_1 into H_2 is defined. The operator T^* is a formal adjoint of T and is an extension of all formal adjoints of T.

Theorem 4.13. *Let T be a densely defined operator from H_1 into H_2.*
(a) *If T^* is also densely defined, then T^{**} is an extension of T.*
(b) *We have $N(T^*) = R(T)^\perp$.*

PROOF.
(a) As T and T^* are formal adjoints of each other, T is a restriction of the adjoint operator T^{**} of T^*.
(b) We have $g \in N(T^*)$ if and only if $g \in D(T^*)$ and $T^*g = 0$ hold. Since $D(T)$ is dense, this is equivalent to the relation

$$\langle Tf, g\rangle = \langle f, T^*g\rangle = 0 \quad \text{for all} \quad f \in D(T).$$

This holds if and only if $g \in R(T)^\perp$. □

Theorem 4.14. *Let T be a densely defined operator from H_1 into H_2.*
(a) *T is bounded if and only if $T^* \in B(H_2, H_1)$.*
(b) *If T is bounded, then $\|T\| = \|T^*\|$.*
(c) *If T is bounded, then T^{**} is the (by Theorem 4.5 uniquely determined) continuous extension of T to the whole space H_1. For $T \in B(H_1, H_2)$ we have $T^{**} = T$.*

PROOF.
(a) *and* (b): Let T be bounded. Then for all $g \in H_2$ and $f \in D(T)$ we have

$$|L_g f| = |\langle g, Tf\rangle| \leqslant \|g\| \, \|T\| \, \|f\|,$$

i.e., L_g is continuous for all $g \in H_2$. Therefore $D(T^*) = H_2$.

By Theorem 4.3(b) we have

$$\|T^*\| = \sup\{|\langle T^*g, f\rangle| : f \in D(T), g \in H_2, \|f\| = 1, \|g\| = 1\}$$
$$= \sup\{|\langle g, Tf\rangle| : f \in D(T), g \in H_2, \|f\| = 1, \|g\| = 1\} = \|T\|.$$

If $T^* \in B(H_2, H_1)$, then T^{**} belongs to $B(H_1, H_2)$. Hence the restriction T of T^{**} is also bounded.

(c) By Theorem 4.13 (a) we have $T \subset T^{**}$. By part (a) we have $T^{**} \in B(H_1, H_2)$. As T is densely defined and continuous, the assertion follows from Theorem 4.5.

EXAMPLE 1. Let M_1 and M_2 be measurable subsets of $\mathbb{R}^m$ and $\mathbb{R}^n$, respectively. Let K denote the Hilbert-Schmidt operator from $L_2(M_1)$ into $L_2(M_2)$, induced by $k \in L_2(M_2 \times M_1)$. (cf. Section 4.1, Example 3):

$$(Kf)(x) = \int_{M_1} k(x,y)f(y)\,\mathrm{d}y \quad \text{for} \quad f \in L_2(M_1).$$

For all $f \in L_2(M_1)$ and $g \in L_2(M_2)$ the function $g(x)k(x,y)f(y)$ is integrable on $M_2 \times M_1$. Therefore by Fubini's theorem we have

$$\langle g, Kf\rangle = \int_{M_2} g(x)^* \left\{ \int_{M_1} k(x,y)f(y)\,\mathrm{d}y \right\} \mathrm{d}x$$

$$= \int_{M_1} f(y) \left\{ \int_{M_2} k(x,y)^* g(x)\,\mathrm{d}x \right\}^* \mathrm{d}y = \langle Hg, f\rangle,$$

where H is the Hilbert-Schmidt operator induced by the kernel $h(y,x) = k(x,y)^*$. If we define the *adjoint kernel* k^* of the kernel k by

$$k^*(y,x) = k(x,y)^*,$$

then K^* is the operator induced by k^*.

EXAMPLE 2. Let T be a continuous linear functional on a Hilbert space H, i.e., a continuous operator from H into $\mathbb{K}$. We want to compute T^*. There exists a uniquely determined $g \in H$ such that

$$Tf = \langle g, f\rangle \quad \text{for all} \quad f \in H.$$

Hence for all $z \in \mathbb{K}$ and all $f \in H$ we have

$$z^* Tf = \langle zg, f\rangle,$$

i.e., $T^*z = zg$ for all $z \in \mathbb{K}$.

EXAMPLE 3. This example shows that $D(T^*) = \{0\}$ may be true. To prove this, for every $k \in \mathbb{N}$ let the sequence $(n_{k,l})_{l \in \mathbb{N}}$ of positive integers be chosen in such a way that

$$\{n_{k,l} : l \in \mathbb{N}\} \cap \{n_{j,l} : l \in \mathbb{N}\} = \varnothing \quad \text{for} \quad j \neq k,$$

$$\bigcup_{k \in \mathbb{N}} \{n_{k,l} : l \in \mathbb{N}\} = \mathbb{N}$$

(we leave the construction of such sequences to the reader). With these sequences let us define the operator T on l_2 by

$$D(T) = l_{2,0},$$

$$Tf = \left(\sum_{l=1}^{\infty} f_{n_{k,l}}\right)_{k \in \mathbb{N}} = \left(\sum_{l=1}^{\infty} f_{n_{1,l}}, \sum_{l=1}^{\infty} f_{n_{2,l}}, \ldots\right).$$

Let us observe that here all the sums occurring are finite. Moreover, we have $Tf \in l_{2,0}$. Therefore the operator T is well-defined.

We show that $g \in D(T^*)$ implies $g = 0$. Let $g = (g_n) \in D(T^*)$ and $h = (h_n) = T^*g$. Then for all $f \in D(T) = l_{2,0}$ we have

$$\sum_k \sum_l g_k^* f_{n_{k,l}} = \langle g, Tf \rangle = \langle T^*g, f \rangle = \sum_n h_n^* f_n = \sum_k \sum_l h_{n_{k,l}}^* f_{n_{k,l}}$$

(here one should notice that all sums are finite). If we choose f equal to the unit vector $e_{n_{k,l}}$ (thus $f_{n_{k,l}} = 1$ and $f_n = 0$ for $n \neq n_{k,l}$), then it follows that

$$h_{n_{k,l}} = g_k \quad \text{for all} \quad l \in \mathbb{N}, k \in \mathbb{N}.$$

As $h \in l_2$, this is only possible if $h = 0$. From this it follows that $g = 0$.

Let T be an operator from H_1 into H_2. The *graph* of T is the subset

$$G(T) = \{(f, Tf) : f \in D(T)\}$$

of $H_1 \times H_2$, where $H_1 \times H_2$ can be considered as a Hilbert space in the sense of Section 3.1: $H_1 \times H_2 = H_1 \oplus H_2$.

Theorem 4.15. *A subset G of $H_1 \times H_2$ is the graph of an operator from H_1 into H_2 if and only if G is a subspace possessing the following property: $(0, g) \in G$ implies $g = 0$. Each subspace of a graph is a graph.*

PROOF. If T is an operator from H_1 into H_2, then $G(T)$ is obviously a subspace, as for $(f_i, g_i) \in G(T)$, $a_i \in \mathbb{K}$ $(i = 1, 2)$ we have

$$\begin{aligned} a_1(f_1, g_1) + a_2(f_2, g_2) &= a_1(f_1, Tf_1) + a_2(f_2, Tf_2) \\ &= (a_1 f_1 + a_2 f_2, T(a_1 f_1 + a_2 f_2)) \in G(T). \end{aligned}$$

If $(0, g) \in G(T)$, then it follows that $g = T0 = 0$.

Let G now be a subspace of $H_1 \times H_2$ having the above mentioned property. We construct an operator T for which $G = G(T)$ holds. For this, let

$$D(T) = \{f \in H_1 : \text{there exists a } g \in H_2 \text{ such that } (f, g) \in G\}.$$

For every $f \in D(T)$ there exists exactly one $g \in H_2$ such that $(f, g) \in G$, as $(f, g_1) \in G$ and $(f, g_2) \in G$ imply $(0, g_1 - g_2) \in G$ (by G being a subspace), consequently, $g_1 - g_2 = 0$. Therefore we can define a mapping T from $D(T)$ into H_2 by

$$Tf = g \quad \text{for} \quad (f, g) \in G.$$

T is linear: If $f_1, f_2 \in D(T)$ and $a_1, a_2 \in \mathbb{K}$, then we have $(f_i, Tf_i) \in G$ for $i = 1, 2$. Hence (as G is a subspace) $(a_1 f_1 + a_2 f_2, a_1 Tf_1 + a_2 Tf_2) \in G$. By the definition of T we have

$$T(a_1 f_1 + a_2 f_2) = a_1 Tf_1 + a_2 Tf_2.$$

By construction, we also have $G = G(T)$. The last assertion can be obtained from this immediately. □

In the sequel we shall use the mappings

$$U : H_1 \times H_2 \to H_2 \times H_1, \; U(f_1, f_2) = (f_2, -f_1)$$
$$V : H_1 \times H_2 \to H_2 \times H_1, \; V(f_1, f_2) = (f_2, f_1).$$

U and V are obviously isomorphisms of $H_1 \oplus H_2$ onto $H_2 \oplus H_1$. The inverse operators U^{-1} and V^{-1} are given by

$$U^{-1} : H_2 \times H_1 \to H_1 \times H_2, \; U^{-1}(f_2, f_1) = (-f_1, f_2),$$
$$V^{-1} : H_2 \times H_1 \to H_1 \times H_2, \; V^{-1}(f_2, f_1) = (f_1, f_2).$$

Theorem 4.16. *Let T be a densely defined operator from H_1 into H_2. Then we have*

$$G(T^*) = U(G(T)^\perp) = (UG(T))^\perp$$

(here the symbol $\perp$ has to be understood in the sense of $H_1 \oplus H_2$, respectively $H_2 \oplus H_1$).

PROOF. By the definition of T^* we have

$$\begin{aligned} G(T^*) &= \{(g, h) \in H_2 \times H_1 : \langle g, Tf \rangle_2 = \langle h, f \rangle_1 \quad \text{for all} \quad f \in D(T)\} \\ &= \{(g, h) \in H_2 \times H_1 : \langle (g, h), (Tf, -f) \rangle = 0 \quad \text{for all} \quad (f, Tf) \in G(T)\} \\ &= (UG(T))^\perp = U(G(T)^\perp). \end{aligned}$$

The last equality follows simply from the definition of U (cf. Exercise 4.14). □

Theorem 4.17. *Let T be a densely defined injective operator from H_1 into H_2.*
(a) *We have $G(T^{-1}) = VG(T)$.*
(b) *If $R(T)$ is dense, then T^* is also injective, and we have $T^{*-1} = T^{-1*}$.*

PROOF. Part (a) is obvious.

(b) By Theorem 4.13(b) we have $N(T^*) = R(T)^\perp = \{0\}$, i.e., T^* is injective. As $G(T^{-1}) = VG(T)$, it follows (cf. Exercise 4.14) that

$$\begin{aligned} G(T^{-1*}) &= U^{-1}(G(T^{-1})^\perp) = U^{-1}V(G(T)^\perp) \\ &= V^{-1}U(G(T)^\perp) = V^{-1}G(T^*) = G(T^{*-1}). \qquad \square \end{aligned}$$

An operator T on a Hilbert space H is said to be *Hermitian*, if it is a formal adjoint of itself, i.e., if we have

$$\langle Tf, g\rangle = \langle f, Tg\rangle \quad \text{for all} \quad f, g \in D(T).$$

An operator T on H is *symmetric* (cf. Section 4.2) if it is Hermitian and densely defined. Since a densely defined operator T is Hermitian if and only if it is a restriction of T^*, we have: *an operator T is symmetric if and only if T is densely defined and $T \subset T^*$.* An operator T on H is said to be *self-adjoint*, if T is densely defined and $T = T^*$.

REMARK. For operators from $B(H)$ the notions of Hermitian, symmetric, and self-adjoint are equivalent.

Theorem 4.18. *An operator T on a complex Hilbert space H is Hermitian if and only if the quadratic form $q(f) = \langle f, Tf\rangle$ defined on $D(T)$ is real.*

PROOF. By definition, T is Hermitian if and only if the sesquilinear form $s(f, g) = \langle f, Tg\rangle$ is Hermitian on $D(T)$. The assertion follows from this by Theorem 1.3(a). $\square$

A characterization of symmetric and self-adjoint operators may be obtained immediately from Theorem 4.16; where U is defined by $U(f, g) = (g, -f)$ on $H \oplus H$.

Proposition. *Let T be a densely defined operator on the Hilbert space H.*
(a) *T is symmetric if and only if*

$$G(T) \subset U(G(T)^\perp) \quad \textit{or} \quad UG(T) \subset G(T)^\perp.$$

(b) *T is self-adjoint if and only if*

$$G(T) = U(G(T)^\perp) \quad \textit{or} \quad UG(T) = G(T)^\perp,$$

i.e.,

$$G(T) \perp UG(T) \quad \textit{and} \quad G(T) \oplus UG(T) = H \oplus H.$$

Proposition. *If T and S are densely defined operators from H_1 into H_2 and $T \subset S$, then we have $S^* \subset T^*$.*

Theorem 4.19. *Let T_1 and T_2 be densely defined operators from H_1 into H_2 and from H_2 into H_3, respectively.*

(a) *If T_2T_1 is densely defined, then we have $T_1^*T_2^* \subset (T_2T_1)^*$.*
(b) *If $T_2 \in B(H_2, H_3)$, then we have $(T_2T_1)^* = T_1^*T_2^*$.*

PROOF.
(a) We have to show that the operators $T_1^*T_2^*$ and T_2T_1 are formal adjoints of each other. Let $f \in D(T_1^*T_2^*)$ and $g \in D(T_2T_1)$. Then $f \in D(T_2^*)$, $T_2^*f \in D(T_1^*)$, $g \in D(T_1)$, and $T_1g \in D(T_2)$. Consequently, from the definition of the adjoint operator it follows that

$$\langle T_1^*T_2^*f, g\rangle = \langle T_2^*f, T_1g\rangle = \langle f, T_2T_1g\rangle.$$

(b) Because of part (a) we only have to prove that $D((T_2T_1)^*) \subset D(T_1^*T_2^*)$. Let $f \in D((T_2T_1)^*)$. As $T_2^* \in B(H_3, H_2)$, for all $g \in D(T_2T_1) = D(T_1)$, we have

$$\langle (T_2T_1)^*f, g\rangle = \langle f, T_2T_1g\rangle = \langle T_2^*f, T_1g\rangle.$$

By the definition of the adjoint operator it follows from this that $T_2^*f \in D(T_1^*)$, i.e., $f \in D(T_1^*T_2^*)$. □

Theorem 4.20. *Let S and T be operators from H_1 into H_2.*
(a) *If T is densely defined, then we have $(aT)^* = a^*T^*$ for all $a \in \mathbb{K}$ such that $a \neq 0$.*
(b) *If $T + S$ is densely defined, then $(T+S)^* \supset T^* + S^*$.*
(c) *If $S \in B(H_1, H_2)$ and T is densely defined, then we have $(T+S)^* = T^* + S^*$.*

PROOF.
(a) is evident (it follows from Theorem 4.19).
(b) Let $f \in D(T^* + S^*) = D(T^*) \cap D(S^*)$. Then for all $g \in D(T+S) = D(T) \cap D(S)$ we have by the definition of the adjoint operator that

$$\begin{aligned}\langle (T^*+S^*)f, g\rangle &= \langle T^*f, g\rangle + \langle S^*f, g\rangle = \langle f, Tg\rangle + \langle f, Sg\rangle \\ &= \langle f, (T+S)g\rangle,\end{aligned}$$

i.e., $f \in D((T+S)^*)$ and $(T+S)^*f = T^*f + S^*f$.
(c) Because of part (b) we only have to prove that $D((T+S)^*) \subset D(T^* + S^*) = D(T^*)$. Let $f \in D((T+S)^*)$. Then for all $g \in D(T+S) = D(T)$ we have

$$\langle [(T+S)^* - S^*]f, g\rangle = \langle f, (T+S)g\rangle - \langle f, Sg\rangle = \langle f, Tg\rangle.$$

From this it follows that $f \in D(T^*)$. □

Theorem 4.21. *Let T be self-adjoint and injective. Then T^{-1} is self-adjoint, too.*

PROOF. $R(T)$ is dense, since we have $\{0\} = N(T) = N(T^*) = R(T)^\perp$. Thus the assertion follows from Theorem 4.17(b). □

EXERCISES

4.18. Let S and T be densely defined operators from H_1 into H_2, and from H_2 into H_3, respectively. Assume that TS is densely defined, S is injective and $S^{-1} \in B(H_2, H_1)$. Then we have $(TS)^* = S^*T^*$.
Hint: By Theorem 4.19(a) we only have to show that $D((TS)^*) \subset D(S^*T^*)$ holds; for all $f \in D((TS)^*)$ and $g \in D(TS) = S^{-1}D(T)$ we have $\langle f, TSg \rangle = \langle (S^{-1})^*(TS)^*f, Sg \rangle$.

4.19. Let $T \in B(H_1, H_2)$.
(a) We have $T^*T \in B(H_1)$, $TT^* \in B(H_2)$ and $\|T^*T\| = \|TT^*\| = \|T\|^2$.
(b) T^*T and TT^* are self-adjoint.

4.5 The theorem of Banach-Steinhaus, strong and weak convergence

We first prove the theorem of *Banach-Steinhaus*, which is also known as the *uniform boundedness principle*.

Theorem 4.22. *Let H_1 and H_2 be Banach spaces, and let M be a subset of $B(H_1, H_2)$. If M is pointwise bounded (i.e., for each $f \in H_1$ there exists a $C(f) \geq 0$ such that $\|Tf\| \leq C(f)$ for all $T \in M$), then M is bounded (i.e., there exists a $C \geq 0$ such that $\|T\| \leq C$ for all $T \in M$).*

PROOF. *1. step.* It is enough to show that there exist an $f_0 \in H_1$, a $\rho > 0$, and a $C' \geq 0$ such that $\|Tf\| \leq C'$ for all $f \in K(f_0, \rho)$ and for all $T \in M$. Indeed, if f_0, ρ, C' have these properties, then for all $g \in K(0, \rho)$ and for all $T \in M$ we have

$$\|Tg\| = \|T(f_0 + g - f_0)\| \leq \|T(f_0 + g)\| + \|Tf_0\| \leq C' + C(f_0) = C'',$$

since $f_0 + g \in K(f_0, \rho)$. Consequently, for all $g \in K(0, 1)$ and $T \in M$ we have

$$\|Tg\| \leq \rho^{-1}C'' = C,$$

i.e., $\|T\| \leq C$ for all $T \in M$.
2. step. What remains is to prove the existence of f_0, ρ, and C' with the above properties. We assume that no such elements exist, i.e., for each $f_0 \in H_1$ and for each $\rho > 0$ the set $\{\|Tf\| : T \in M, f \in K(f_0, \rho)\}$ is unbounded. In particular, the set $\{\|Tf\| : T \in M, f \in K(0, 1)\}$ is unbounded. Therefore there exist an $f_1 \in K(0, 1)$ and a $T_1 \in M$ such that $\|T_1 f_1\| > 1$. Since T_1 is continuous, there exists a ρ_1, $0 < \rho_1 < 2^{-1}$ such that

$$\overline{K}(f_1, \rho_1) \subset K(0, 1) \quad \text{and} \quad \|T_1 f\| > 1 \quad \text{for all} \quad f \in \overline{K}(f_1, \rho_1).$$

Since $\{\|Tf\| : T \in M, f \in K(f_1, \rho_1)\}$ is unbounded, there exist an $f_2 \in K(f_1, \rho_1)$ and a $T_2 \in M$ such that $\|T_2 f_2\| > 2$. As T_2 is continuous, there exists a ρ_2, $0<\rho_2<2^{-2}$ such that

$$\overline{K}(f_2, \rho_2) \subset K(f_1, \rho_1) \quad \text{and} \quad \|T_2 f\| > 2 \quad \text{for all} \quad f \in \overline{K}(f_2, \rho_2).$$

In this way, by induction we obtain a sequence (f_n) from H_1, (T_n) from M, and (ρ_n) from $(0, 1)$ such that $\overline{K}(f_{n+1}, \rho_{n+1}) \subset K(f_n, \rho_n)$, $\rho_n<2^{-n}$, and $\|T_n f\|>n$ for all $f \in \overline{K}(f_n, \rho_n)$. In particular, we have

$$\|f_n - f_m\| < \rho_{n_0} \quad \text{for} \quad n, m > n_0.$$

Since $\rho_n < 2^{-n}$, it follows from this that (f_n) is a Cauchy sequence. Thus there exists an $f \in H_1$ such that $f_n \to f$. Since for $n > m$ we have $f_n \in K(f_m, \rho_m)$, it follows that $f \in \overline{K}(f_m, \rho_m)$ for all $m \in \mathbb{N}$. Consequently $\|T_m f\| > m$ is true for $m \in \mathbb{N}$, which contradicts $\|T_m f\| \leq C(f)$. □

Let H_1 and H_2 be normed spaces. A sequence (T_n) from $B(H_1, H_2)$ is said to be *strongly convergent* to $T \in B(H_1, H_2)$, if for all $f \in H_1$ we have $Tf = \lim T_n f$. For this we shall write $T = s-\lim T_n$ or $T_n \xrightarrow{s} T$. The operator T is called the *strong limit* of the sequence (T_n). It is obvious that every sequence (T_n) in $B(H_1, H_2)$ has at most one strong limit. A sequence (T_n) from $B(H_1, H_2)$ is said to be a *strong Cauchy sequence*, if for every $f \in H_1$ the sequence $(T_n f)$ is a Cauchy sequence in H_2. Every strongly convergent sequence is a strong Cauchy sequence.

Theorem 4.23. *Let H_1 and H_2 be normed spaces.*

(a) *If (T_n) is a strongly convergent sequence in $B(H_1, H_2)$ and $T = s-\lim T_n$, then $\|T\| \leq \liminf \|T_n\|$.*

(b) *If the sequence (T_n) from $B(H_1, H_2)$ is bounded and $(T_n g)$ is a Cauchy sequence for every g in a dense subset M of H_1, then (T_n) is a strong Cauchy sequence.*

(c) *If H_1 is a Banach space, then every strong Cauchy sequence in $B(H_1, H_2)$ is bounded.*

(d) *If H_1 and H_2 are Banach spaces and (T_n) is a strong Cauchy sequence in $B(H_1, H_2)$, then there exists a $T \in B(H_1, H_2)$ such that $T_n \xrightarrow{s} T$.*

PROOF.

(a) Let $C = \liminf \|T_n\|$. Then there exists a subsequence (T_{n_k}) of (T_n) such that $\|T_{n_k}\| \to C$ as $k \to \infty$. Hence for all $f \in H_1$ we have

$$\|Tf\| = \lim \|T_{n_k} f\| \leq \lim \|T_{n_k}\| \, \|f\| = C\|f\|,$$

i.e., $\|T\| \leq C$.

(b) Let $f \in H_1$, $\epsilon > 0$. We have to show that there exists an $n_0 \in \mathbb{N}$ such that $\|T_n f - T_m f\| \leq \epsilon$ for all $m, n \geq n_0$. As M is dense, there exists a $g \in M$ such that $\|f - g\| \leq \epsilon/3C$ (with $C = \sup\{\|T_n\| : n \in \mathbb{N}\}$). If we now choose n_0 in such a way that $\|T_n g - T_m g\| \leq \epsilon/3$ for all $n, m \geq n_0$, then we have

$$\|T_nf - T_mf\| \leq \|T_n(f-g)\| + \|T_ng - T_mg\| + \|T_m(g-f)\|$$
$$\leq \epsilon \quad \text{for} \quad n, m \geq n_0.$$

(c) For every $f \in H_1$ the sequence (T_nf) is a Cauchy sequence and thus it is bounded. Consequently, by Theorem 4.22 there exists a C such that $\|T_n\| \leq C$ for all $n \in \mathbb{N}$.

(d) For every $f \in H_1$ the sequence (T_nf) is a Cauchy sequence, so it is convergent in H_2. Let us define T by $Tf = \lim T_nf$. Then $D(T) = H_1$. The operator T is linear, since for all $f, g \in H_1$ and $a, b \in \mathbb{K}$ we have

$$T(af+bg) = \lim T_n(af+bg) = \lim(aT_nf+bT_ng) = aTf + bTg.$$

By part (c) there exists a $C \geq 0$ such that $\|T_n\| \leq C$ for all $n \in \mathbb{N}$. Consequently, we have

$$\|Tf\| = \lim \|T_nf\| \leq C\|f\|,$$

i.e., $T \in B(H_1, H_2)$. By construction, we obviously have $T_n \overset{s}{\to} T$. □

EXAMPLE 1. Let the operators $T_n \in B(l_2)$ be defined by

$$T_n(f_1, f_2, f_3, \ldots) = (f_{n+1}, f_{n+2}, \ldots).$$

Then for all $f \in l_2$ we obviously have $T_nf \to 0$, i.e., $T_n \overset{s}{\to} 0$. For all $f \in l_2$ we have $\|T_nf\| \leq \|f\|$, consequently $\|T_n\| \leq 1$. Moreover, for $e_j = (\delta_{j,n})$ we have $T_je_{j+1} = e_1$. Therefore $\|T_n\| = 1$. From this it follows that strong convergence does not in general imply convergence in the norm of $B(H_1, H_2)$.

Let H be a pre-Hilbert space. A sequence (f_n) from H is said to *converge weakly* to $f \in H$ if for all $g \in H$ we have $\langle f_n, g\rangle \to \langle f, g\rangle$. For this we write $f = w - \lim f_n$ or $f_n \overset{w}{\to} f$. The element f is called the *weak limit* of the sequence (f_n). Every sequence has at most one weak limit. A sequence (f_n) from H is called a *weak Cauchy sequence*, if for every $g \in H$ the sequence $(\langle f_n, g\rangle)$ is a Cauchy sequence in $\mathbb{K}$. Every weakly convergent sequence is a weak Cauchy sequence.

Theorem 4.24. *Let H be a pre-Hilbert space.*

(a) *If (f_n) is a weakly convergent sequence in H, and $f = w - \lim f_n$, then we have $\|f\| \leq \liminf \|f_n\|$.*

(b) *If the sequence (f_n) is bounded in H and $(\langle f_n, g\rangle)$ is a Cauchy sequence for all g from a dense subset M of H, then (f_n) is a weak Cauchy sequence.*

(c) *If H is a Hilbert space, then every weak Cauchy sequence is bounded in H.*

(d) *If H is a Hilbert space and (f_n) is a weak Cauchy sequence in H, then there exists an $f \in H$ such that $f_n \overset{w}{\to} f$.*

The *proof* immediately follows from Theorem 4.23, if we notice that the weak convergence of (f_n) is equivalent to the strong convergence of the sequence (T_{f_n}) of the linear functionals induced by f_n. In part (d) we have to use Riesz' theorem (Theorem 4.8). The details are left to the reader.

Theorem 4.25. *Let H be a Hilbert space. Every bounded sequence (f_n) in H contains a weakly convergent subsequence (f_{n_k}).*

PROOF. Let $M = L\{f_n : n \in \mathbb{N}\}$. Then $M \oplus M^\perp$ is dense in H. For every $k \in \mathbb{N}$ the sequence $(\langle f_n, f_k \rangle)_{n \in \mathbb{N}}$ is bounded. Consequently, by induction we can find, for all $j \in \mathbb{N}$, a subsequence $(f_{n_{j,l}})_{l \in \mathbb{N}}$ of (f_n) such that $(f_{n_{j+1,l}})_{l \in \mathbb{N}}$ is a subsequence of $(f_{n_{j,l}})_{l \in \mathbb{N}}$ and $(\langle f_{n_{j,l}}, f_j \rangle)_{l \in \mathbb{N}}$ is convergent. With the *diagonal sequence* $(f_{n_l}) = (f_{n_{l,l}})$ the sequence $(\langle f_{n_l}, f_j \rangle)_{l \in \mathbb{N}}$ is then convergent for all $j \in \mathbb{N}$. Since for all $f \in M^\perp$ we have $\langle f_{n_l}, f \rangle = 0$ for all $l \in \mathbb{N}$, the sequence $(\langle f_{n_l}, f \rangle)$ is convergent for all f from the dense subspace $M \oplus M^\perp$. Therefore by Theorem 4.24(b) and (d) (f_{n_l}) is weakly convergent. □

EXAMPLE 2. Every orthonormal sequence (f_n) weakly converges to zero. This follows from the Bessel inequality $\|f\|^2 \geq \sum |\langle f_n, f \rangle|^2$. In particular, the sequence of unit vectors $(e_j = (\delta_{jn}))$ in l_2 tends to zero weakly. This example also shows that weak convergence does not imply strong convergence in general.

EXAMPLE 3. For every $f = \sum_{j=1}^{\infty} f_j e_j \in l_2$ let the sequence $(f^{(n)})$ be defined by $f^{(n)} = \sum_{j=1}^{\infty} f_j e_{j+n}$. Then $(f^{(n)})$ converges weakly to zero, since for all $g = \sum g_j e_j \in l_2$ we have

$$|\langle f^{(n)}, g \rangle|^2 = \left| \sum_{j=1}^{\infty} f_j^* g_{j+n} \right|^2 \leq \|f\|^2 \sum_{j=n+1}^{\infty} |g_j|^2 \to 0$$

as $n \to \infty$.

EXAMPLE 4. In the pre-Hilbert space $l_{2,0}$ the sequence (ke_k) (with $e_k = (\delta_{nk})$) weakly converges to 0. However, it is unbounded.

Let H_1 and H_2 be pre-Hilbert spaces. A sequence (T_n) from $B(H_1, H_2)$ is said to *converge* to $T \in B(H_1, H_2)$ *weakly*, if for all $f \in H_1$ the sequence $(T_n f)$ in H_2 weakly converges to Tf, i.e., $\langle T_n f, g \rangle \to \langle Tf, g \rangle$ for all $f \in H_1$ and $g \in H_2$. In this case we shall write $T = w - \lim T_n$ or $T_n \xrightarrow{w} T$, and call T the *weak limit* of the sequence (T_n). A sequence (T_n) from $B(H_1, H_2)$ is said to be a *weak Cauchy sequence* if $(T_n f)$ is a weak Cauchy sequence in H_2 for each $f \in H_1$.

Theorem 4.26. *Let H_1 and H_2 be pre-Hilbert spaces.*
(a) *If (T_n) is a weakly convergent sequence in $B(H_1, H_2)$ and $T = w - \lim T_n$, then we have $\|T\| \leq \liminf \|T_n\|$.*

(b) *If the sequence* (T_n) *from* $B(H_1, H_2)$ *is bounded and* $(\langle T_n f, g\rangle)$ *is a Cauchy sequence for every* $f \in M_1$ *and every* $g \in M_2$, *where* M_1 *and* M_2 *are dense subsets of* H_1 *and* H_2, *respectively, then* (T_n) *is a weak Cauchy sequence.*

(c) *If* H_1 *and* H_2 *are Hilbert spaces, then every weak Cauchy sequence* (T_n) *from* $B(H_1, H_2)$ *is bounded.*

(d) *If* H_1 *and* H_2 *are Hilbert spaces, and* (T_n) *is a weak Cauchy sequence in* $B(H_1, H_2)$, *then there exists a* $T \in B(H_1, H_2)$ *such that* $T_n \xrightarrow{w} T$.

PROOF.

(a) Let $C = \liminf \|T_n\|$ and let (T_{n_k}) be a subsequence such that $\|T_{n_k}\| \to C$. Then for all $f \in H_1$ and $g \in H_2$ we have

$$|\langle Tf, g\rangle| = \lim |\langle T_{n_k} f, g\rangle| \leqslant \lim \|T_{n_k}\| \, \|f\| \, \|g\| = C\|f\| \, \|g\|.$$

The assertion follows from this by Theorem 4.3.

(b) We have to prove that for arbitrary $f \in H_1$, $g \in H_2$, and $\epsilon > 0$ there exists an n_0 such that

$$|\langle (T_n - T_m) f, g\rangle| \leqslant \epsilon \quad \text{for} \quad n, m \geqslant n_0.$$

Let $C = 1 + (1 + \|f\| + \|g\|) \sup \{\|T_n\| : n \in \mathbb{N}\}$. Since the sets M_j are dense in H_j $(j = 1, 2)$, there exist $f_0 \in M_1$ and $g_0 \in M_2$ such that

$$\|f - f_0\| \leqslant \frac{\epsilon}{5C}, \ \|g - g_0\| \leqslant \frac{\epsilon}{5C}, \ \|f_0\| \leqslant \|f\| + 1.$$

If we choose $n_0 \in \mathbb{N}$ in such a way that for $n, m \geqslant n_0$ we have $|\langle (T_n - T_m) f_0, g_0\rangle| \leqslant \epsilon/5$, then for $n, m \geqslant n_0$ it follows that

$$\begin{aligned} |\langle (T_n - T_m) f, g\rangle| &\leqslant |\langle T_n (f - f_0), g\rangle| \\ &\quad + |\langle T_n f_0, g - g_0\rangle| + |\langle (T_n - T_m) f_0, g_0\rangle| \\ &\quad + |\langle T_m f_0, g_0 - g\rangle| + |\langle T_m (f_0 - f), g\rangle| \\ &\leqslant \epsilon. \end{aligned}$$

(c) For every $f \in H_1$ the sequence $(T_n f)$ is a weak Cauchy sequence in H_2. Since H_2 is a Hilbert space, the sequence $(T_n f)$ is bounded by Theorem 4.24(c). The boundedness of the sequence $(\|T_n\|)$ follows from this via Theorem 4.22, as H_1 is also a Hilbert space.

(d) For every $f \in H_1$ the sequence $(T_n f)$ is a weak Cauchy sequence, therefore by Theorem 4.24(d) it is weakly convergent in H_2. We define T by $Tf = w - \lim T_n f$ for all $f \in H_1$. We can prove the linearity of T, as in the proof of Theorem 4.23(d). By part (c) there exists a $C \geqslant 0$ such that $\|T_n\| \leqslant C$ for all $n \in \mathbb{N}$. It follows that

$$|\langle Tf, g\rangle| = \lim |\langle T_n f, g\rangle| \leqslant C\|f\| \, \|g\| \quad \text{for all} \quad f \in H_1, g \in H_2.$$

Consequently T is bounded. By construction, we obviously have $T_n \xrightarrow{w} T$. □

Proposition.
(a) $T_n \to T$ *implies* $T_n \xrightarrow{s} T$; $T_n \xrightarrow{s} T$ *implies* $T_n \xrightarrow{w} T$.
(b) *If* H_1 *and* H_2 *are Hilbert spaces, then* $T_n \xrightarrow{w} T$ *is equivalent to* $T_n^* \xrightarrow{w} T^*$.

EXAMPLE 5. Let us consider the operators T_n from Example 1. The adjoint operators T_n^* are obviously defined by

$$T_n^*\left(\sum_{j=1}^{\infty} f_j e_j\right) = \sum_{j=1}^{\infty} f_j e_{n+j} \quad \text{for all} \quad \sum_{j=1} f_j e_j \in l_2.$$

By Example 3 we have $T_n^* \xrightarrow{w} 0$. However, we do not have $T_n^* \xrightarrow{s} 0$. Therefore, strong convergence does not imply the strong convergence of the adjoint operators.

Theorem 4.27. *Let* H_1, H_2 *be Hilbert spaces, and let* T *be an operator from* H_1 *into* H_2 *such that* $D(T) = H_1$. *Then the following assertions are equivalent*:
(i) T *is bounded (i.e.,* $f_n \to f$ *implies* $Tf_n \to Tf$),
(ii) $f_n \xrightarrow{w} f$ *implies* $Tf_n \xrightarrow{w} Tf$,
(iii) $f_n \to f$ *implies* $Tf_n \xrightarrow{w} Tf$.

PROOF. (i) *implies* (ii): If $f_n \xrightarrow{w} f$, then for every $g \in H_2$ (notice that T^* exists and $T^* \in B(H_2, H_1)$) we have

$$\langle g, Tf_n\rangle = \langle T^*g, f_n\rangle \to \langle T^*g, f\rangle = \langle g, Tf\rangle,$$

i.e., $Tf_n \xrightarrow{w} Tf$.

(ii) *implies* (iii): This is obvious, as $f_n \to f$ implies $f_n \xrightarrow{w} f$.

(iii) *implies* (i): Let us assume that T is not bounded, i.e. there exists a sequence (f_n) from H_1 such that $\|f_n\| \leq 1$ and $\|Tf_n\| \geq n^2$. Then we have $(1/n)f_n \to 0$. Therefore from (iii) it follows that $(1/n)Tf_n \xrightarrow{w} 0$. By Theorem 4.24(c) the sequence $((1/n)Tf_n)$ is thus bounded. This contradicts the fact that

$$\left\|\frac{1}{n} Tf_n\right\| = \frac{1}{n}\|Tf_n\| \geq n. \qquad \square$$

Theorem 4.28. *Let* H *be a Hilbert space and let* (T_n) *be a bounded sequence of symmetric operators from* $B(H)$.
(a) *If* $T_n \xrightarrow{w} T$ *for some* $T \in B(H)$, *then* T *is also symmetric.*
(b) *If the sequence* $(\langle f, T_n f\rangle)$ *is non-decreasing for every* $f \in H$, *then there exists a* $T \in B(H)$ *such that* $T_n \xrightarrow{s} T$. *The same holds true if the sequence* $(\langle f, T_n f\rangle)$ *is non-increasing for every* $f \in H$.

PROOF.
(a) For all $f, g \in H$ we have

$$\langle f, Tg\rangle = \lim \langle f, T_n g\rangle = \lim \langle T_n f, g\rangle = \langle Tf, g\rangle.$$

Therefore, T is symmetric.

(b) The sequence $(\langle f, T_n f\rangle)$ is non-decreasing and bounded for every $f \in H$, consequently it is convergent. If $C = 2 \sup \{\|T_n\| : n \in \mathbb{N}\}$, then $\|T_n - T_m\| \leq C$ for all $n, m \in \mathbb{N}$. The Schwarz inequality applied to the non-negative sesquilinear form $s(g, f) = \langle g, (T_n - T_m)f\rangle$ shows that for all $f \in H$ we have

$$\begin{aligned}\|(T_n - T_m)f\| &= \langle (T_n - T_m)f, (T_n - T_m)f\rangle^{1/2} = \{s((T_n - T_m)f, f)\}^{1/2}\\ &\leq \{s((T_n - T_m)f, (T_n - T_m)f)s(f, f)\}^{1/4}\\ &= \{\langle (T_n - T_m)f, (T_n - T_m)^2 f\rangle \langle f, (T_n - T_m)f\rangle\}^{1/4}\\ &\leq \|(T_n - T_m)f\|^{1/4}\|(T_n - T_m)^2 f\|^{1/4}\langle f, (T_n - T_m)f\rangle^{1/4}\\ &\leq C^{3/4}\|f\|^{1/2}\langle f, (T_n - T_m)f\rangle^{1/4} \to 0 \quad \text{as} \quad n, m \to \infty.\end{aligned}$$

So $(T_n f)$ is a Cauchy sequence for every $f \in H$, consequently (T_n) is strongly convergent. □

Exercises

4.20. Let H_1, H_2 and H_3 be normed spaces, and let $S_n, S \in B(H_2, H_3)$, $T_n, T \in B(H_1, H_2)$, $S_n \xrightarrow{s} S$, $T_n \xrightarrow{s} T$.
(a) If the sequence (S_n) is bounded, then $S_n T_n \xrightarrow{s} ST$.
(b) If H_2 is a Banach space, then $S_n T_n \xrightarrow{s} ST$.

4.21. (a) Let H be a Hilbert space. If $f_n \xrightarrow{w} f$ in H and $\|f\| \geq \limsup \|f_n\|$, then $f_n \to f$.
Hint: Treat $\|f_n - f\|^2$.
(b) Let H_1 and H_2 be Hilbert spaces, and let T_n and T be isometries from H_1 into H_2 such that $T_n \xrightarrow{w} T$. Then we have $T_n \xrightarrow{s} T$.

4.22. Let H_1, H_2 and H_3 be Hilbert spaces, and assume $S_n, S \in B(H_2, H_3)$, $T_n, T \in B(H_1, H_2)$.
(a) $S_n \xrightarrow{w} S$, $T_n \xrightarrow{s} T$ imply $S_n T_n \xrightarrow{w} ST$.
(b) $S_n \xrightarrow{s} S$, $T_n \xrightarrow{w} T$ do *not* imply $S_n T_n \xrightarrow{w} ST$.
Hint: Let $H_1 = H_2 = H_3 = l_2$, $T_n(\sum f_j e_j) = \sum f_j e_{j+n}$, $S_n = T_n^*$. Then we have $S_n \xrightarrow{s} 0$, $T_n \xrightarrow{w} 0$, $S_n T_n \to I$.
(c) $S_n \to S$, $T_n \xrightarrow{w} T$ imply $S_n T_n \xrightarrow{w} ST$.

4.23. Let H_1 and H_2 be Hilbert spaces, and take A_n, A, from $B(H_1, H_2)$ and B from $B(H_2, H_1)$. Then $A_n \xrightarrow{w} A$, $A_n^* \xrightarrow{s} B$ imply $B = A^*$, therefore that $A_n^* \xrightarrow{s} A^*$.

4.24. (a) If H is a finite dimensional Hilbert space, then $f_n \xrightarrow{w} f$ is equivalent to $f_n \to f$.
(b) If H_1 and H_2 are Hilbert spaces, and H_2 is finite dimensional, then for T_n, T from $B(H_1, H_2)$ the statements $T_n \xrightarrow{w} T$ and $T_n \xrightarrow{s} T$ are equivalent; this holds true in particular for linear functionals on a Hilbert space.
(c) If H_1 and H_2 are finite dimensional, then for T_n, T from $B(H_1, H_2)$ the statements $T_n \xrightarrow{w} T$, $T_n \xrightarrow{s} T$, $T_n \to T$ are equivalent.

4.25. Let H be a Hilbert space.

(a) If M is a closed subspace of H and (f_n) is a sequence in M such that $f_n \xrightarrow{w} f$, then we have $f \in M$ (we say that M is *weakly closed*).

(b) If (f_n) is a sequence in H such that $f_n \xrightarrow{w} f$, then there exists a subsequence (f_{n_k}) of (f_n) such that $(1/m)\Sigma_{k=1}^{m} f_{n_k} \to f$ as $m \to \infty$.
Hint: Treat the case $f_n \xrightarrow{w} 0$ first and choose the subsequence (f_{n_k}) such that for $k<j$ we have $|\langle f_{n_k}, f_{n_j}\rangle| \leq j^{-1}$.

(c) A convex subset M of H is closed if and only if it is weakly closed.

4.26. Let H_1, H_2 be Hilbert spaces and let H_1 be separable. If (T_n) is a bounded sequence from $B(H_1, H_2)$, then there exists a weakly convergent subsequence (T_{n_k}).

4.27. Let H be a complex Hilbert space and assume that (T_n) is a sequence in $B(H)$ such that $(\langle f, T_n f\rangle)$ is bounded for every $f \in H$. Then the sequence (T_n) is bounded in $B(H)$. (For symmetric operators T_n this holds also in real Hilbert spaces.)

4.28. Let H_1 and H_2 be Hilbert spaces. Assume that (T_n) is a bounded sequence from $B(H_1, H_2)$, $T \in B(H_1, H_2)$, and M_1 and M_2 are dense subsets of H_1 and H_2, respectively. If $T_n f \to Tf$ for all $f \in M_1$ (respectively, $\langle g, T_n f\rangle \to \langle g, Tf\rangle$ for all $f \in M_1$ and $g \in M_2$), then it follows that $T_n \xrightarrow{s} T$ (respectively $T_n \xrightarrow{w} T$).

4.6 Orthogonal projections, isometric and unitary operators

Let H be a Hilbert space and let M be a closed subspace of H. By Theorem 3.2 every $f \in H$ can be uniquely represented in the form $f = g + h$ with $g \in M$ and $h \in M^{\perp}$; g is called the orthogonal projection of f onto M. If we define the mapping P_M by $D(P_M) = H$ and $P_M f = g$, then P_M is a linear operator on H such that $D(P_M) = H$, since for $f_1 = g_1 + h_1$ and $f_2 = g_2 + h_2$ with $g_j \in M$ and $h_j \in M^{\perp}$ we have $af_1 + bf_2 = (ag_1 + bg_2) + (ah_1 + bh_2)$ with $ag_1 + bg_2 \in M$, $ah_1 + bh_2 \in M^{\perp}$, therefore $P_M(af_1 + bf_2) = ag_1 + bg_2 = aP_M f_1 + bP_M f_2$. The operator P_M is called the *orthogonal projection onto* M.

Because $\|f\|^2 = \|g\|^2 + \|h\|^2$, we have $\|P_M f\| = \|g\| \leq \|f\|$ for all $f \in H$, i.e., we have $\|P_M\| \leq 1$. If $M = \{0\}$, then it is obvious that $P_M = 0$. If $M \neq \{0\}$, and $f \in M$, $f \neq 0$, then $P_M f = f$, hence $\|P_M\| = 1$. As $P_M f = f$ for all $f \in M$, it follows that $P_M^2 = P_M P_M = P_M$, i.e., P_M is idempotent. We have $P_M f \in M$ for all $f \in H$ and $P_M f = f$ for all $f \in M$, therefore $R(P_M) = M$. As $P_M f = 0$ if and only if $f \in M^{\perp}$, we see that $N(P_M) = M^{\perp}$. An operator P on H is called an *orthogonal projection* if there exists a closed subspace M such that $P = P_M$.

Theorem 4.29. *For an operator $P \in B(H)$ the following statements are equivalent:*

(i) *P is an orthogonal projection,*

(ii) *$I-P$ is an orthogonal projection,*
(iii) *P is idempotent and $R(P)=N(P)^{\perp}$,*
(iv) *P is idempotent and self-adjoint.*
We have $R(P)=N(I-P)$ and $N(P)=R(I-P)$.

PROOF. (i) *and* (ii) *are equivalent*: From the definition of orthogonal projections it follows immediately that P is the orthogonal projection onto M if and only if $I-P$ is the orthogonal projection onto $M^{\perp}$. From this it also follows that $R(P)=M=M^{\perp\perp}=N(I-P)$, $N(P)=M^{\perp}=R(I-P)$.

(i) *and* (iii) *are equivalent*: Using the above reasoning we only have to show that (iii) implies (i). As $R(P)=N(P)^{\perp}$, the range $R(P)$ is a closed subspace. For all $g\in R(P)$ we have $Pg=g$, as P is idempotent. If we write $f\in H$ in the form $f=g+h$, where $g\in R(P)$ and $h\in R(P)^{\perp}=N(P)$, then we have $Pf=Pg+Ph=g$, i.e., P is the orthogonal projection onto $R(P)$.

(iv) *follows from* (i): P is idempotent, as (iii) follows from (i). For all $f_1=g_1+h_1, f_2=g_2+h_2$ with $g_j\in R(P)$, $h_j\in R(P)^{\perp}$ we have

$$\begin{aligned}\langle Pf_1,f_2\rangle &= \langle g_1,g_2+h_2\rangle=\langle g_1,g_2\rangle=\langle g_1+h_1,g_2\rangle\\ &=\langle f_1,Pf_2\rangle,\end{aligned}$$

i.e., P is self-adjoint.

(iii) *follows from* (iv): We only have to prove that $R(P)=N(P)^{\perp}$. If $f\in R(P), f=Pg$, then $(I-P)f=0$, consequently $f\in N(I-P)$. If $f\in N(I-P)$, then $f-Pf=0$, consequently $f\in R(P)$. Therefore we have $R(P)=N(I-P)$ and thus $R(P)$ is closed. From this it follows that $R(P)=R(P)^{\perp\perp}=N(P^*)^{\perp}=N(P)^{\perp}$. □

Theorem 4.30. *Let M and N be closed subspaces of a Hilbert space H, and let P_M and P_N be the orthogonal projections onto M and N, respectively.*

(a) *$P=P_MP_N$ is an orthogonal projection if and only if $P_MP_N=P_NP_M$ holds; then we have $P=P_{M\cap N}$. We have $M\perp N$ if and only if $P_MP_N=0$ (or $P_NP_M=0$).*

(b) *$Q=P_M+P_N$ is an orthogonal projection if and only if $M\perp N$, then we have $Q=P_{M\oplus N}$.*

(c) *$R=P_M-P_N$ is an orthogonal projection if and only if $N\subset M$; we then have $R=P_{M\ominus N}$.*

PROOF.

(a) If $P=P_MP_N$ is an orthogonal projection, then P is self-adjoint. Consequently, $P_MP_N=P=P^*=(P_MP_N)^*=P_N^*P_M^*=P_NP_M$. Let $P_MP_N=P_NP_M$ hold. Then it follows that $P^2=(P_MP_N)^2=P_MP_NP_MP_N=P_M^2P_N^2=P_MP_N=P$ and $P^*=(P_MP_N)^*=(P_NP_M)^*=P_M^*P_N^*=P_MP_N=P$. Therefore P is an orthogonal projection by Theorem 4.29. Since $P=P_MP_N=P_NP_M$, we have $R(P)\subset R(P_M)\cap R(P_N)=M\cap N$. On the other hand, if $f\in M\cap N$, then we have $Pf=P_MP_Nf=P_Mf=f$, therefore $M\cap N\subset R(P)$. Consequently, $M\cap N=R(P)$. It is obvious that $P_MP_N=0$ if and only if $g\in R(P_M)^{\perp}=M^{\perp}$ holds for all $g\in R(P_N)=N$, i.e., if $N\perp M$. The other assertion follows similarly.

(b) If $Q=P_M+P_N$ is an orthogonal projection, then $\|f\|^2 \geq \|Qf\|^2 = \langle Qf, f\rangle = \langle P_Mf, f\rangle + \langle P_Nf, f\rangle = \|P_Mf\|^2 + \|P_Nf\|^2$. For $f=P_Mg$ it follows that $\|P_Mg\|^2 \geq \|P_Mg\|^2 + \|P_NP_Mg\|^2$. Therefore $P_NP_Mg=0$ for all $g \in H$, i.e., $P_NP_M=0$. It follows analogously that $P_MP_N=0$. Consequently, by part (a), $M \perp N$ holds. We obviously have $R(Q) \subset R(P_M) + R(P_N) = M \oplus N$. Conversely, if $f=g+h \in M \oplus N$ with $g \in M$, $h \in N$, then $Qf = Qg + Qh = P_Mg + P_Nh = g+h=f$. Therefore $R(Q) = M \oplus N$. If $M \perp N$, then by part (a), we have $P_MP_N = P_NP_M = 0$, consequently $Q^2 = (P_M+P_N)^2 = P_M^2 + P_N^2 = P_M + P_N = Q$. As the operators P_M and P_N are self-adjoint, Q is also self-adjoint, consequently it is an orthogonal projection.

(c) If $R=P_M-P_N$ is the orthogonal projection onto the subspace L, then, because of the equality $P_M = P_L + P_N$, by part (b) we have $L \perp N$ and $M = L \oplus N \supset N$. Therefore $L = M \ominus N$, i.e., R is the orthogonal projection onto $M \ominus N$. Conversely, if $N \subset M$ and $L = M \ominus N$, then by part (b) we have $P_M = P_L + P_N$, hence $R = P_M - P_N = P_L$ is an orthogonal projection. □

For two symmetric operators $A, B \in B(H)$ we write $A \leq B$ (or $B \geq A$) if for all $f \in H$ we have $\langle Af, f\rangle \leq \langle Bf, f\rangle$ (by Theorem 4.18 $\langle Af, f\rangle$ and $\langle Bf, f\rangle$ are real). A is said to be *non-negative* if $A \geq 0$.

Theorem 4.31. *Let M and N be closed subspaces of the Hilbert space H, and let P_M and P_N be the orthogonal projections onto M and N, respectively.*
(a) *We have $0 \leq P_M \leq I$.*
(b) *The following statements are equivalent*:
(i) $P_M \leq P_N$, (iii) $P_NP_M = P_M$,
(ii) $M \subset N$, (iv) $P_MP_N = P_M$.

PROOF.
(a) For all $f \in H$ we have $\langle Of, f\rangle = 0 \leq \|P_Mf\|^2 = \langle P_Mf, f\rangle \leq \|f\|^2 = \langle If, f\rangle$.
(b) (i) *implies* (ii): If $P_M \leq P_N$, then $\|P_Mf\|^2 = \langle P_Mf, f\rangle \leq \langle P_Nf, f\rangle = \|P_Nf\|^2$ for all $f \in H$. Therefore $N(P_N) \subset N(P_M)$ and thus $M = R(P_M) = N(P_M)^\perp \subset N(P_N)^\perp = R(P_N) = N$.
(ii) *implies* (iii): By Theorem 4.20 (a) and (b), with $L = N \ominus M$ we have $P_NP_M = (P_M + P_L)P_M^2 = P_M = P_M$.
(iii) *implies* (iv): As P_NP_M is an orthogonal projection (namely P_M), by Theorem 4.30 (a) we have $P_MP_N = P_NP_M = P_M$.
(iv) *implies* (i): Because of the equality $P_MP_N = P_M$ we have, for all $f \in H$, that

$$\langle P_Mf, f\rangle = \|P_Mf\|^2 = \|P_MP_Nf\|^2 \leq \|P_Nf\|^2 = \langle P_Nf, f\rangle. \quad \square$$

A sequence (T_n) of symmetric operators $T_n \in B(H)$ is said to be *monotone* (*non-decreasing* or *non-increasing*, respectively) if for every $f \in H$ the sequence $(\langle T_n f, f\rangle)$ is monotone (non-decreasing or non-increasing, respectively). Theorem 4.28(b) says that any bounded non-decreasing sequence of symmetric operators is strongly convergent. For sequences of orthogonal projections we have

Theorem 4.32.

(a) *If (P_n) is a monotone sequence of orthogonal projections acting on the Hilbert space H, then there exists an orthogonal projection P acting on H such that $P_n \xrightarrow{s} P$.*

(b) *If (P_n) is non-decreasing* (i.e., $P_n \leqslant P_{n+1}$), *then P is the orthogonal projection onto $\overline{\bigcup_{n\in\mathbb{N}} R(P_n)}$.*

(c) *If (P_n) is non-increasing* (i.e., $P_n \geqslant P_{n+1}$), *then P is the orthogonal projection onto $\bigcap_{n\in\mathbb{N}} R(P_n)$.*

PROOF.

(a) Because of Theorem 4.28(b) there exists a self-adjoint operator $P \in B(H)$ such that $P_n \xrightarrow{s} P$. As $\langle P^2 f, g\rangle = \langle Pf, Pg\rangle = \lim \langle P_n f, P_n g\rangle = \lim \langle P_n f, g\rangle = \langle Pf, g\rangle$ for all $f, g \in H$, the operator P is idempotent and thus it is an orthogonal projection.

(b) If $f \perp \overline{\bigcup_{n\in\mathbb{N}} R(P_n)}$, then $P_n f = 0$ for all $n \in \mathbb{N}$, consequently $Pf = \lim P_n f = 0$. If $f \in \bigcup_{n\in\mathbb{N}} R(P_n)$, then $f \in R(P_{n_0})$ for some $n_0 \in \mathbb{N}$. Since $R(P_{n_0}) \subset R(P_n)$ for all $n \geqslant n_0$, we have $P_n f = P_{n_0} f$ for $n \geqslant n_0$. Therefore $Pf = f$, i.e., $\bigcup_{n\in\mathbb{N}} R(P_n) \subset R(P)$. As $R(P)$ is closed, it follows that $\overline{\bigcup_{n\in\mathbb{N}} R(P_n)} = R(P)$.

(c) The sequence (Q_n) with $Q_n = I - P_n$ is non-decreasing; $Q = \lim Q_n$ is therefore the orthogonal projection onto $\overline{\bigcup_{n\in\mathbb{N}} R(Q_n)} = \overline{\bigcup_{n\in\mathbb{N}} N(P_n)}$. Then $P = I - Q$ is the orthogonal projection onto $\bigcup_{n\in\mathbb{N}} N(P_n)^{\perp} = \bigcap_{n\in\mathbb{N}} N(P_n)^{\perp} = \bigcap_{n\in\mathbb{N}} R(P_n)$. □

In the calculation of the norm of the difference of two orthogonal projections the following theorem is often useful.

Theorem 4.33. *Let P_1 and P_2 be orthogonal projections acting on the Hilbert space H. Then we have*

$$\|P_1 - P_2\| = \max\{\rho_{12}, \rho_{21}\}.$$

where

$$\rho_{jk} = \sup\left\{\|P_j h\| : h \in R(P_k)^{\perp}, \|h\| \leqslant 1\right\}.$$

PROOF.

(a) By the definition of the norm of an operator we have (notice that $R(P_k)^{\perp} = N(P_k)$)

$$\begin{aligned}\|P_1 - P_2\| &= \sup\{\|(P_1 - P_2)f\| : f \in H, \|f\| \leqslant 1\}\\ &\geqslant \sup\left\{\|(P_1 - P_2)f\| : f \in R(P_1)^{\perp}, \|f\| \leqslant 1\right\}\\ &= \sup\left\{\|P_2 f\| : f \in R(P_1)^{\perp}, \|f\| \leqslant 1\right\} = \rho_{21}.\end{aligned}$$

The inequality $\|P_1 - P_2\| \geq \rho_{12}$ follows in a similar way.

(b) We have $P_1 - P_2 = P_1(I - P_2) - (I - P_1)P_2$. As $(I - P_2)f \in R(P_2)^\perp$, for all $f \in H$ it follows that

$$\|P_1(I - P_2)f\| \leq \rho_{12}\|(I - P_2)f\|.$$

Moreover, since $(I - P_1)P_2 f \in R(P_1)^\perp$, we have

$$\begin{aligned}\|(I - P_1)P_2 f\|^2 &= \langle (I - P_1)P_2 f, (I - P_1)P_2 f\rangle = \langle P_2(I - P_1)P_2 f, P_2 f\rangle \\ &\leq \|P_2(I - P_1)P_2 f\|\,\|P_2 f\| \leq \rho_{21}\|(I - P_1)P_2 f\|\,\|P_2 f\|,\end{aligned}$$

consequently

$$\|(I - P_1)P_2 f\| \leq \rho_{21}\|P_2 f\|.$$

This implies for all $f \in H$

$$\begin{aligned}\|(P_1 - P_2)f\|^2 &= \|P_1(I - P_2)f - (I - P_1)P_2 f\|^2 \\ &= \|P_1(I - P_2)f\|^2 + \|(I - P_1)P_2 f\|^2 \\ &\leq \rho_{12}^2\|(I - P_2)f\|^2 + \rho_{21}^2\|P_2 f\|^2 \\ &\leq \max\{\rho_{12}^2, \rho_{21}^2\}(\|(I - P_2)f\|^2 + \|P_2 f\|^2) \\ &= \max\{\rho_{12}^2, \rho_{21}^2\}\|f\|^2,\end{aligned}$$

therefore

$$\|P_1 - P_2\| \leq \max\{\rho_{12}, \rho_{21}\}. \qquad \square$$

Let H_1 and H_2 be Hilbert spaces. An operator U from H_1 into H_2 such that $D(U) = H_1$ is called an *isometry* if $\|Uf\| = \|f\|$ for all $f \in H_1$. If U is an isometry and $R(U) = H_2$, then U is an isomorphism of H_1 onto H_2. In this case U is called a *unitary operator*. An operator U from H_1 into H_2 such that $D(U) = H_1$ is called a *partial isometry* if there exists a closed subspace M of H_1 such that

$$\|Uf\| = \|f\| \quad \text{for} \quad f \in M, \qquad Uf = 0 \quad \text{for} \quad f \in M^\perp.$$

We have $R(U) = UM$; this shows immediately that $R(U)$ is closed (if $Uf_n \to g \in H_2$, then $(P_M f_n)$ is a Cauchy sequence in M; therefore we have $P_M f_n \to f \in M$, and thus $Uf_n \to Uf = g \in R(U)$). The closed subspaces M and $R(U)$ are called the *initial and final domains of* U, respectively.

Theorem 4.34. *Let H_1 and H_2 be Hilbert spaces and let U be an operator from H_1 into H_2 such that $D(U) = H_1$.*

(a) *The following assertions are equivalent:*

(i) *U is a partial isometry with initial domain M and final domain N,*

(ii) *$R(U) = N$ and $\langle Uf, Ug\rangle = \langle P_M f, g\rangle$ for all $f, g \in H_1$,*

(iii) *$U^*U = P_M$ and $UU^* = P_N$,*

(iv) *U^* is a partial isometry with initial domain N and final domain M.*

(b) *The following assertions are equivalent*:
(i) U *is unitary*,
(ii) $R(U)=H_2$ *and* $\langle Uf, Ug\rangle = \langle f, g\rangle$ *for all* $f, g \in H_1$,
(iii) $U^*U = I_{H_1}$ *and* $UU^* = I_{H_2}$, *i.e.*, $U^* = U^{-1}$,
(iv) U^* *is unitary*.

PROOF.

(a) The equivalence of (i) and (ii) follows by (1.8) in the real case and by (1.4) in the complex case.
(i) *implies* (iv): We have $N(U^*) = R(U)^\perp = N^\perp$ and (because (i) implies (ii)) $\|U^*Uf\| = \|P_M f\| = \|Uf\|$ for all $f \in H_1$, therefore $\|U^*h\| = \|h\|$ for all $h \in R(U) = N$. Hence the operator U^* is a partial isometry with initial domain N. If we interchange the roles of U and U^* in this reasoning, then it follows that the final domain of U^* is equal to the initial domain of $U^{**} = U$, and, consequently, it is equal to M.
(iv) *implies* (i) for the same reason.
(i) *implies* (iii): As (i) implies (ii), it follows that $U^*U = P_M$. As (i) implies (iv), it follows similarly that $UU^* = P_N$.
(iii) *implies* (ii): We have $R(U) \supset R(UU^*) = R(P_N) = N$. Since $\|U^*f\|^2 = \langle UU^*f, f\rangle = \|P_N f\|^2$, we have $N(U^*) = N^\perp$ and thus $R(U) \subset N(U^*)^\perp = N$. Summing up, it follows that $R(U) = N$. Moreover, we have $\langle Uf, Ug\rangle = \langle U^*Uf, g\rangle = \langle P_M f, g\rangle$ for all $f, g \in H$.

(b) This is a special case of part (a). □

Theorem 4.35. *If P and Q are orthogonal projections on the Hilbert space H such that $\|P-Q\| < 1$, then we have*
(a) $\dim R(P) = \dim R(Q)$, $\dim R(I-P) = \dim R(I-Q)$,
(b) *P and Q are unitarily equivalent, i.e., there exists a unitary operator U in H such that $Q = UPU^{-1}$ and $P = U^{-1}QU$.*

PROOF.

(a) We have $R(P) \cap R(Q)^\perp = \{0\}$, because for $f \in R(P) \cap R(Q)^\perp, f \neq 0$ we would have $\|(P-Q)f\| = \|Pf\| = \|f\|$, consequently $\|P-Q\| \geq 1$ would hold. By Theorem 3.11 it follows from this that $\dim R(P) \leq \dim R(Q)$. The opposite inequality follows by symmetry. Hence $\dim R(P) = \dim R(Q)$. Replacing P and Q by $I-P$ and $I-Q$, respectively, we obtain that $\dim R(I-P) = \dim R(I-Q)$.

(b) By part (a) of Theorem 4.10 there exist unitary operators V and W from $R(P)$ onto $R(Q)$ and from $R(I-P)$ onto $R(I-Q)$, respectively. Then the operator $U = VP + W(I-P)$ is a unitary operator on H such that $U^{-1} = V^{-1}Q + W^{-1}(I-Q)$. We have

$$\begin{aligned} UPU^{-1} &= (VP + W(I-P))P(V^{-1}Q + W^{-1}(I-Q)) \\ &= VPV^{-1}Q = Q, \end{aligned}$$

since for $g \in R(Q)$ we have $VPV^{-1}g = g$. From this $P = U^{-1}QU$ follows immediately. □

EXERCISES

4.29. (a) Let H_1 and H_2 be Hilbert spaces. If there exists a surjective mapping $T \in B(H_1, H_2)$, then $\dim H_2 \leq \dim H_1$.
Hint: See Exercise 3.7.
(b) Give another proof for Theorem 4.35(a).
Hint: From $\|P-Q\|<1$ it follows that $R(PQ)=R(P)$ and $R(QP) = R(Q)$.

4.30. (a) If (U_m) is a sequence of isometric operators and $U_m \xrightarrow{s} U$, then U is isometric.
(b) The strong limit of a sequence of unitary operators is not necessarily unitary.
Hint: Consider the unitary operators U_m on l_2 defined by

$U_m(f_n)_{n\in\mathbb{N}} = (f_m, f_1, f_2, f_3, \ldots, f_{m-1}, f_{m+1}, f_{m+2}, \ldots)$ for $(f_n)_{n\in\mathbb{N}} \in l_2$.

We have $U_m \xrightarrow{s} U$, where

$$U(f_n)_{n\in\mathbb{N}} = (0, f_1, f_2, f_2, \ldots) \quad \text{for} \quad (f_n)_{n\in\mathbb{N}} \in l_2.$$

4.31. Let M and N be closed subspaces of the Hilbert space H such that $\dim M < \dim N$. Then we have $M^{\perp} \cap N \neq \{0\}$.
Hint: If P_N denotes the orthogonal projection onto N, then we have $M^{\perp} \cap N = N \ominus P_N M$.

4.32. Let H be a Hilbert space.
(a) If M and N are closed subspaces of H, then $M+N$ is not necessarily closed.
(b) If M is a closed subspace and P is an orthogonal projection, then PM is not necessarily closed.
Hint: Choose, for example, $H = l_2$, for M the subspace of the elements $(x_1, x_1, x_2, 2x_2, x_3, 3x_3, \ldots)$, for N the subspace of the elements $(0, y_1, 0, y_2, 0, y_3, \ldots)$, and for P the projection onto $N^{\perp}$.

5 Closed linear operators

5.1 Closed and closable operators, the closed graph theorem

In what follows H, H_1 and H_2 will always be Hilbert spaces. As long as no adjoint operators (in particular no symmetric or self-adjoint operators) are treated, we could also consider Banach spaces; the proofs may be somewhat harder, in this case. An operator T from H_1 into H_2 is said to be *closed* if its graph $G(T)$ (cf. Section 4.4) in $H_1 \times H_2$ is closed. An operator T is said to be *closable* if $\overline{G(T)}$ is a graph. From the proof of Theorem 4.15 we know that there exists then a uniquely determined operator $\bar{T}$ such that $G(\bar{T}) = \overline{G(T)}$; $\bar{T}$ is closed and is called the *closure* of T.

Let T be a closed operator. A subspace D of $D(T)$ is called a *core of T* provided that for $S = T|_D$ we have $T = \bar{S}$. One should notice that by Theorem 4.15 the operator S is surely closable, since we have $\overline{G(S)} \subset G(T)$. If T is a closable operator, then $D(T)$ is obviously a core of $\bar{T}$.

Proposition.

(a) *T is closed if and only if the following holds: If (f_n) is a sequence in $D(T)$ that is convergent in H_1 and the sequence (Tf_n) is convergent in H_2, then we have* $\lim f_n \in D(T)$ *and* $T(\lim f_n) = \lim Tf_n$.

(b) *T is closable if and only if the following holds: If (f_n) is a sequence in $D(T)$ such that $f_n \to 0$, and the sequence (Tf_n) in H_2 is convergent, then we have* $\lim Tf_n = 0$.

(c) *If T is closable, then*

$$D(\bar{T}) = \{f \in H_1: \text{there exists a sequence } (f_n) \text{ from } D(T) \text{ such that } f_n \to f \text{ and for which } (Tf_n) \text{ is also convergent}\},$$

$$\bar{T}f = \lim Tf_n \quad \text{for } f \in D(\bar{T}).$$

(d) *If T is closed, then $N(T)$ is closed.*

(e) *If T is injective, then T is closed if and only if T^{-1} is closed.*

Parts (a), (b) and (c) are reformulations of the definitions. Part (d) follows immediately from Part (a). Part (e) follows from the equality $G(T^{-1}) = VG(T)$.

Theorem 5.1. *Let T be an operator from H_1 into H_2. On $D(T)$ by*

$$\langle f, g\rangle_T = \langle f, g\rangle + \langle Tf, Tg\rangle, \quad \|f\|_T = \{\|f\|^2 + \|Tf\|^2\}^{1/2}$$

a scalar product and the corresponding norm (T-norm or graph norm) are defined. T is closed if and only if $(D(T), \langle .\,, .\rangle_T)$ is a Hilbert space.

PROOF. The properties of a semi-scalar product are obviously satisfied. Because of the inequality $\langle f, f\rangle_T \geqslant \langle f, f\rangle$ the semi-scalar product $\langle .\,, .\rangle_T$ is positive, thus it is a scalar product.

If T is closed and (f_n) is a T-Cauchy sequence in $D(T)$ (i.e., a Cauchy sequence with respect to the T-norm), then (f_n) and (Tf_n) are Cauchy sequences in H_1 and H_2, respectively; therefore there exist $f \in H_1$ and $g \in H_2$ such that $f_n \to f$, $Tf_n \to g$. By part (a) of the above proposition we have $f \in D(T)$, $Tf = g$. We also have

$$\|f_n - f\|_T = \{\|f_n - f\|^2 + \|Tf_n - g\|^2\}^{1/2} \to 0$$

as $n \to \infty$, i.e., f_n is convergent in $(D(T), \langle .\,, .\rangle_T)$.

Suppose now that $(D(T), \langle .\,, .\rangle_T)$ is complete. If (f_n) is a sequence in $D(T)$ and $f_n \to f \in H_1$, $Tf_n \to g \in H_2$, then (f_n) and (Tf_n) are Cauchy sequences. Consequently, (f_n) is a T-Cauchy sequence, i.e., there exists an $f_0 \in D(T)$ such that $\|f_n - f_0\|_T \to 0$. It follows from this that $\|f_n - f_0\| \to 0$, $\|Tf_n - Tf_0\| \to 0$, therefore $f = f_0 \in D(T)$ and $Tf = Tf_0 = g$. □

Theorem 5.2. *Every bounded operator is closable. A bounded operator T is closed if and only if $D(T)$ is closed. If T is bounded, then we have $D(\bar{T}) = \overline{D(T)}$; the closure $\bar{T}$ is the bounded extension of T onto $\overline{D(T)}$, constructed in Theorem 4.5.*

PROOF.

1. Let (f_n) be a sequence in $D(T)$ such that $f_n \to 0$ and $Tf_n \to g$. Then we have $\|Tf_n\| \leqslant \|T\|\,\|f_n\| \to 0$. Therefore $g = 0$.

2. For all $f \in D(T)$ we have

$$\|f\| \leq \|f\|_T \leq (1+\|T\|^2)^{1/2}\|f\|.$$

Therefore (f_n) is a Cauchy sequence (converging to $f \in D(T)$) if and only if (f_n) is a T-Cauchy sequence (T-converging to f). Consequently, $(D(T), \|\,.\,\|)$ is complete if and only if $(D(T), \|\,.\,\|_T)$ is complete. From this the assertion follows via Theorems 2.4 and 5.1.
3. The equality $D(\bar{T}) = \overline{D(T)}$ immediately follows from part (c) of the above proposition. Part (c) also says that $\bar{T}$ is the extension occurring in Theorem 4.5. □

Proposition. *Let T be closable and injective. The operator T^{-1} is closable if and only if $\bar{T}$ is injective. We then have $\overline{T^{-1}} = \bar{T}^{-1}$. If $\bar{T}^{-1}$ is continuous, then we have $R(\bar{T}) = \overline{R(T)}$.*

Proof. If $\bar{T}$ is injective, then $\bar{T}^{-1}$ is a closed extension of T^{-1}. If T^{-1} is closable, then $VG(\bar{T}) = \overline{VG(T)} = \overline{G(T^{-1})}$ is a graph, i.e., $\bar{T}$ is injective, and we have $\bar{T}^{-1} = \overline{T^{-1}}$. If $\bar{T}^{-1}$ is continuous, then by Theorem 5.2 we have $R(\bar{T}) = D(\bar{T}^{-1}) = D(\overline{T^{-1}}) = \overline{D(T^{-1})} = \overline{R(T)}$. □

Theorem 5.3. *Let T be a densely defined operator from H_1 into H_2.*
(a) T^* is closed.
(b) T is closable if and only if T^* is densely defined; we then have $\bar{T} = T^{**}$.
(c) If T is closable, then $(\bar{T})^* = T^*$.

Proof.
(a) By Theorem 4.16 we have $G(T^*) = (UG(T))^{\perp}$. Therefore $G(T^*)$ is closed.
(b) Since

$$\begin{aligned}\overline{G(T)} = G(T)^{\perp\perp} &= (U^{-1}G(T^*))^{\perp}\\ &= \{(f,g) \in H_1 \times H_2 : \langle f, T^*h\rangle - \langle g, h\rangle = 0 \text{ for all } h \in D(T^*)\},\end{aligned}$$

we have $(0, g) \in \overline{G(T)}$ if and only if $g \in D(T^*)^{\perp}$. Therefore $(0, g) \in \overline{G(T)}$ implies $g = 0$ if and only if $\overline{D(T^*)} = H_2$. Consequently, $\overline{G(T)}$ is a graph if and only if T^* is densely defined. If $D(T^*)$ is dense, then we have $G(T^{**}) = U^{-1}(G(T^*)^{\perp}) = U^{-1}U(G(T)^{\perp\perp}) = \overline{G(T)} = G(\bar{T})$.
(c) If T is closable, then we have $G(T^*) = U(G(T)^{\perp}) = U(\overline{G(T)}^{\perp}) = U(G(\bar{T}))^{\perp} = G((\bar{T})^*)$. Therefore $T^* = (\bar{T})^*$. □

Example 1. The operator T from Section 4.4, Example 3 is not closable, as $D(T^*) = \{0\}$.

Theorem 5.4.

(a) *An operator T from H_1 into H_2 is closable if and only if there exists a closed extension of T.*

(b) *Every symmetric operator T on the Hilbert space H is closable; $\bar{T}$ is also symmetric.*

PROOF.

(a) If T is closable, then we have $T \subset \bar{T}$. Therefore $\bar{T}$ is a closed extension of T. If S is a closed extension of T, then we have $G(T) \subset G(S) = \overline{G(S)}$, hence $\overline{G(T)} \subset G(S)$, and thus $\overline{G(T)}$ is a graph (cf. Theorem 4.15).

(b) By part (a) the operator T is closable, since $T \subset T^*$ and T^* is closed. For all $f, g \in D(\bar{T})$ there exist sequences (f_n) and (g_n) from $D(T)$ such that $f_n \to f$, $g_n \to g$, $Tf_n \to \bar{T}f$ and $Tg_n \to \bar{T}g$. As T is symmetric, we have

$$\langle \bar{T}f, g\rangle = \lim \langle Tf_n, g_n\rangle = \lim \langle f_n, Tg_n\rangle = \langle f, \bar{T}g\rangle.$$

Since $D(\bar{T})$ is dense, the operator $\bar{T}$ is symmetric. □

EXAMPLE 2. Let us consider on $L_2(M)$ the maximal operator T of multiplication by a measurable function $t: M \to \mathbb{C}$ (cf. Section 4.1, Example 1). This is the operator defined by

$$D(T) = \{f \in L_2(M) : tf \in L_2(M)\} \text{ and } Tf = tf \text{ for } f \in D(T).$$

(5.1) *T^* is the maximal multiplication operator induced by the function t^* (where $t^*(x) = t(x)^*$), in particular, we have $D(T^*) = D(T)$.*

PROOF. It is obvious that $D(T)$ is also the domain of the maximal multiplication operator induced by t^*. Since for all $f, g \in D(T)$ we have

$$\langle g, Tf\rangle = \int_M g(x)^* t(x) f(x)\,dx = \int_M (t^*(x)g(x))^* f(x)\,dx = \langle t^*g, f\rangle,$$

the maximal operators of multiplication by t and t^* are formal adjoints of each other. What remains is to prove that for $g \in D(T^*)$ we have $t^*g \in L_2(M)$. Let $g \in D(T^*)$. Then for all $f \in D(T)$ we have

$$\langle T^*g, f\rangle = \langle g, Tf\rangle = \int_M g(x)^* t(x) f(x)\,dx,$$

thus

$$\int_M (T^*g(x) - g(x)t(x)^*)^* f(x)\,dx = 0 \quad \text{for all } f \in D(T). \tag{5.2}$$

Let us define, for all $n \in \mathbb{N}$, the subsets M_n of M by $M_n = \{x \in M : |t(x)| < n\}$. Then we obviously have $M = \bigcup_{n=1}^{\infty} M_n$. For every $f \in L_2(M)$, $\chi_{M_n} f$

belongs to $D(T)$, consequently we have

$$\int_M (T^*g(x) - g(x)t(x)^*)^* \chi_{M_n}(x) f(x)\,dx = 0 \quad \text{for all} \quad f \in L_2(M).$$

Since $\chi_{M_n}(T^*g - t^*g) \in L_2(M)$, it follows that $T^*g(x) = t^*(x)g(x)$ almost everywhere in M_n. Since this holds true for all $n \in \mathbb{N}$, we have $T^*g = t^*g$, therefore $t^*g \in L_2(M)$. □

(5.3) *T is closed.*

PROOF. T is the adjoint of the maximal multiplication operator induced by t^*, hence T is closed by Theorem 5.3(a). □

(5.4) *The following assertions are equivalent:*
(a) *T is self-adjoint,*
(b) *T is symmetric,*
(c) *t is real-valued (i.e., $t(x)$ is real almost everywhere in M).*

PROOF. (b) obviously follows from (a).
(b) *implies* (c): If t is not real, then at least one of the sets $M_1 = \{x \in M : \text{Im}\, t(x) > 0\}$ or $M_2 = \{x \in M : \text{Im}\, t(x) < 0\}$ has positive measure. If M_1 is of positive measure, then for all $f \in D(T)$, different from zero and vanishing outside M_1, we have

$$\text{Im}\,\langle f, Tf\rangle = \int_{M_1} |f(x)|^2\, \text{Im}\, t(x)\,dx > 0.$$

By Theorem 4.18 the operator T is therefore *not* symmetric. We can argue similarly if M_2 has positive measure.
(c) *implies* (a): Since $t = t^*$, we have $T = T^*$ by (5.1). □

(5.5) *If M is an open subset of $\mathbb{R}^m$ and t is locally bounded on M (i.e., t is essentially bounded on any compact subset of M), then $C_0^\infty(M)$ and $L_{2,0}(M)$ are cores of T.*

PROOF. We obviously have $C_0^\infty(M) \subset L_{2,0}(M) \subset D(T)$. We prove the assertion for $C_0^\infty(M)$; the other assertion follows from this. We have to prove that for each $f \in D(T)$ and for each $\epsilon > 0$ there exists an $f_\epsilon \in C_0^\infty(M)$ such that $\|f_\epsilon - f\| + \|Tf_\epsilon - Tf\| \leq \epsilon$. If (M_n) is a sequence of open bounded subsets of M such that $\overline{M}_n \subset M_{n+1}$ and $M = \cup M_n$, then for all $n \in \mathbb{N}$ let us define

$$g_n(x) = f(x) \text{ on } M_n, \qquad g_n(x) = 0 \text{ on } M \setminus M_n.$$

We obviously have $g_n \in L_{2,0}(M)$, $g_n \to f$, and $Tg_n \to Tf$. Therefore there exists an $n_0 \in \mathbb{N}$ such that $\|g_{n_0} - f\| + \|Tg_{n_0} - Tf\| \leq \epsilon/2$. On the other hand, there exists a sequence $f_m \in C_0^\infty(M_{n_0}) \subset C_0^\infty(M)$ such that $f_m \to g_{n_0}$. Then we also have that $Tf_m \to Tg_{n_0}$. So there exists an $m_0 \in \mathbb{N}$ such that

$\|f_{m_0}-g_{n_0}\|+\|Tf_{m_0}-Tg_{n_0}\|\leqslant\epsilon/2$. The assertion follows from this by taking $f_\epsilon=f_{m_0}$. □

Let T and S be operators from H_1 into H_2 and from H_1 into H_3, respectively. The operator S is said to be *T-bounded* if $D(T)\subset D(S)$ and there exists a $C\geqslant 0$ such that $\|Sf\|\leqslant C\|f\|_T$ for all $f\in D(T)$, i.e., if S, as an operator from $(D(T),\langle .\,,.\rangle_T)$ into H_3, is bounded. Then for all $f\in D(T)$ we have

$$\|Sf\|\leqslant C(\|f\|+\|Tf\|).$$

If S is T-bounded, then the infinum of all numbers $b\geqslant 0$ for which an $a\geqslant 0$ exists such that

$$\|Sf\|\leqslant a\|f\|+b\|Tf\| \quad \text{for all} \quad f\in D(T),$$

is called the *T-bound* of S. One should notice that if c is the T-bound of S, then in general there exists no $a\geqslant 0$ such that for all $f\in D(T)$ we have $\|Sf\|\leqslant a\|f\|+c\|Tf\|$ (cf. Exercise 5.4).

Proposition. *If T is an arbitrary operator from H_1 into H_2 and $S\in B(H_1,H_3)$, then S is T-bounded with T-bound* 0.

Theorem 5.5. *Let T and S be operators from H_1 into H_2, and let S be T-bounded with T-bound less than* 1. *Then $T+S$ is closed (closable) if and only if T is closed (closable); we have $D(\overline{T+S})=D(\overline{T})$.*

PROOF. As the T-bound of S is less than 1, there exist a $b<1$ and an $a\geqslant 0$ such that $\|Sf\|\leqslant a\|f\|+b\|Tf\|$ for all $f\in D(T)$. Consequently, for all $f\in D=D(T)=D(T+S)$ we have

$$\begin{aligned}-a\|f\|+(1-b)\|Tf\|&\leqslant\|Tf\|-\|Sf\|\leqslant\|(T+S)f\|\\&\leqslant\|Tf\|+\|Sf\|\leqslant a\|f\|+(1+b)\|Tf\|.\end{aligned}$$

From this it follows with a properly chosen $C\geqslant 0$ that

$$\|Tf\|\leqslant C(\|f\|+\|(T+S)f\|) \tag{5.6}$$

$$\|(T+S)f\|\leqslant C(\|f\|+\|Tf\|) \tag{5.7}$$

for all $f\in D$. Hence there exists a $K\geqslant 0$ such that

$$\|f\|_T\leqslant K\|f\|_{T+S} \quad\text{and}\quad \|f\|_{T+S}\leqslant K\|f\|_T.$$

From this it follows that $(D,\langle .\,,.\rangle_{T+S})$ is complete if and only if $(D,\langle .\,,.\rangle_T)$ is. Let T be closable. If (f_n) is a sequence from $D(T+S)=D(T)$ such that $f_n\to 0$ and for which $((T+S)f_n)$ is convergent in H_2, then by (5.6) the sequence (Tf_n) is a Cauchy sequence. Hence $Tf_n\to 0$, because T is closable. Because of (5.7) from this it follows that $(T+S)f_n\to 0$, so $T+S$ is also closable. One can show in an analogous way that T is closable provided that $T+S$ is closable. By part (c) of the proposition preceding

Theorem 5.1 we have $f \in D(\overline{T+S})$ if and only if there exists a sequence (f_n) from $D(T+S)=D(T)$ for which $f_n \to f$ and $((T+S)f_n)$ is convergent. Since because of (5.6) and (5.7) $((T+S)f_n)$ is convergent if and only if (Tf_n) is convergent, we have $D(\overline{T+S})=D(\overline{T})$. □

Theorem 5.6 (Banach; closed graph theorem). *Let H_1 and H_2 be Hilbert spaces and let T be an operator from H_1 into H_2. Then the following statements are equivalent*:

(a) *T is closed and $D(T)$ is closed,*

(b) *T is bounded and $D(T)$ is closed,*

(c) *T is bounded and closed.*

PROOF. (a) *implies* (b): We have to show that T is bounded. Without loss of generality we may assume that $D(T)=\overline{D(T)}=H_1$ (otherwise we could consider T as an operator from the Hilbert space $D(T)$ into H_2). Consequently, T^* is defined. For all $g \in D(T^*)$ such that $\|g\| \leqslant 1$ we have

$$|\langle T^*g, f\rangle| = |\langle g, Tf\rangle| \leqslant \|Tf\| \quad \text{for all} \quad f \in H_1.$$

For the linear functionals $\{L_g : g \in D(T^*),\ \|g\| \leqslant 1\}$ on H_1, where $L_g f = \langle T^*g, f\rangle$, we therefore have

$$|L_g(f)| \leqslant \|Tf\| \quad \text{for all} \quad f \in H_1;$$

consequently they are pointwise bounded. By Theorem 4.22 there exists a $C > 0$ such that

$$\|T^*g\| = \|L_g\| \leqslant C \quad \text{for all} \quad g \in D(T^*) \text{ such that } \quad \|g\| \leqslant 1.$$

Therefore T^* is bounded and $\|T^*\| \leqslant C$. As T is closed (consequently closable), $D(T^*)$ is dense and, by Theorem 5.2, closed. We therefore have $D(T^*)=H_2$, i.e., $T^* \in B(H_2, H_1)$. Since T is closed, this implies that $T=\overline{T} = T^{**} \in B(H_1, H_2)$.

The assertions "(b) implies (c)" and "(c) implies (a)" are contained in Theorem 5.2. □

Theorem 5.7. *Let H_1 and H_2 be Hilbert spaces and let T be an operator from H_1 into H_2 such that $D(T)=H_1$ and $D(T^*)$ is dense in H_2. Then T belongs to $B(H_1, H_2)$. In particular, every symmetric operator T on the Hilbert space H such that $D(T)=H$ is bounded* (Hellinger-Toeplitz).

PROOF. By Theorem 5.3 the operator T is closable. Because $D(T)=H_1$, we have $D(T)=D(\overline{T})$, i.e., $T=\overline{T}$. Therefore T is closed and $D(T)=H_1$. Then T is bounded by Theorem 5.6. □

Theorem 5.8. *Let H_1 and H_2 be Hilbert spaces and let T be an injective operator from H_1 into H_2 such that $R(T)=H_2$. The operator T is closed if and only if $T^{-1} \in B(H_1, H_2)$.*

PROOF. By Part (e) of the proposition preceding Theorem 5.1 the operator T is closed if and only if T^{-1} is closed. The assertion follows immediately from this and from Theorem 5.6. □

Theorem 5.9. *Let H_1, H_2 and H_3 be Hilbert spaces, let T be a closed operator from H_1 into H_2, and let S be a closable operator from H_1 into H_3 such that $D(S) \supset D(T)$. Then S is T-bounded.*

PROOF. On account of Theorem 5.6 it is enough to show that the operator S_0 from $(D(T), \langle .\,, .\rangle_T)$ into H_3, defined by $D(S_0) = D(T)$ and $S_0 f = Sf$ for $f \in D(T)$, is closed. As $D(S_0) = D(T)$, it is enough to show that S_0 is closable. Let (f_n) be a sequence in $D(T)$ for which $\|f_n\|_T \to 0$ and $(S_0 f_n)$ is also convergent. Since $\|f_n\|_T \to 0$ implies $\|f_n\| \to 0$ and since S is closable, we obtain from this that $S_0 f_n = Sf_n \to 0$, thus S_0 is closable. □

EXERCISES

5.1. (a) Every Carleman operator (cf. Exercise 4.2) is closable.
(b) Any Carleman operator, defined on the whole of $L_2(M)$, is bounded.

5.2. (a) Any densely defined operator T on the Hilbert space H such that $\operatorname{Re}\langle f, Tf\rangle \geq 0$ for all $f \in D(T)$ is closable.
Hint: If $(0, g) \in \overline{G(T)}$, then $\operatorname{Re}\langle f, Tf + zg\rangle \geq 0$ for all $f \in D(T)$ and $z \in \mathbb{K}$. One infers from this that $\langle f, g\rangle = 0$ for all $f \in D(T)$.
(b) The *numerical range* of an operator T on H is defined by $W(T) = \{\langle f, Tf\rangle : f \in D(T), \|f\| = 1\}$. The set $W(T)$ is convex (cf. for example, *P. R. Halmos* [14], Problem 166). If $D(T)$ is dense and $W(T) \neq \mathbb{K}$, then T is closable.

5.3. (a) Let H be a Hilbert space, and let T be a densely defined operator from H into $\mathbb{C}^m$. The operator T is bounded if and only if it is closable (this holds true for linear functionals in particular).
(b) In the Hilbert space l_2 let the functional T be defined by $D(T) = l_{2,0}$, $Tf = \sum n f_n$ for $f \in D(T)$. Then T is not closable.

5.4. Let T and S be on $L_2(\mathbb{R})$ the maximal operators of multiplication by $t(x) = x^2$ and $s(x) = x^2 + x$, respectively. The operator S is T-bounded with T-bound 1. However, there exists no $a \geq 0$ such that $\|Sf\| \leq \|Tf\| + a\|f\|$ for all $f \in D(T)$.

5.5. Prove the results of (4.3) with the aid of the closed graph theorem, (4.1), (4.2), (4.6), and (5.3).

5.6. Let H_1, H_2 and H_3 be Hilbert spaces, let $S \in B(H_1, H_2)$, and let T be a closed operator from H_2 into H_3 such that $R(S) \subset D(T)$. Then we have $TS \in B(H_1, H_3)$.
Hint: Show that TS is closed, and $D(TS) = H_1$.

5.7. Assume that H_1, H_2 and H_3 are Hilbert spaces, T is a closable operator from H_1 into H_2, and S is a T-bounded operator from H_1 into H_3. If the sequence (f_n) from $D(T)$ is such that $f_n \xrightarrow{w} 0$ and $Tf_n \xrightarrow{w} 0$, then we also have $Sf_n \xrightarrow{w} 0$.

5.8. Let H be a Hilbert space. An operator P on H is an orthogonal projection if and only if P is symmetric, $D(P)=H$, and $P^2=P$.
Hint: Use Theorem 5.7 (Hellinger-Toeplitz).

5.9. Let H be a vector space, and let $\langle .\,,.\rangle_1$ and $\langle .\,,.\rangle_2$ be scalar products on H such that $(H,\langle .\,,.\rangle_1)$ and $(H,\langle .\,,.\rangle_2)$ are complete. If there exists a C such that $\|\,.\,\|_1 \leqslant C\|\,.\,\|_2$, then there also exists a C' such that $\|\,.\,\|_2 \leqslant C'\|\,.\,\|_1$.

5.10. Let the vector space H be endowed with two scalar products, consequently with two norms $\|\,.\,\|_1$ and $\|\,.\,\|_2$. These norms are said to be *coordinated* if we have: (k_{12}) from $\|x_n\|_1\to 0$ and $\|x_n-x\|_2\to 0$ it follows that $x=0$, or (k_{21}) from $\|x_n\|_2\to 0$ and $\|x_n-x\|_1\to 0$ it follows that $x=0$.
(a) The assumptions (k_{12}) and (k_{21}) are equivalent.
(b) If the norms $\|\,.\,\|_1$ and $\|\,.\,\|_2$ are coordinated, and $(H,\|\,.\,\|_1)$ and $(H,\|\,.\,\|_2)$ are complete, then there exist $c_1, c_2>0$ such that $\|\,.\,\|_1 \leqslant c_1\|\,.\,\|_2 \leqslant c_2\|\,.\,\|_1$.
Hint: The identity map from $(H,\|\,.\,\|_1)$ into $(H,\|\,.\,\|_2)$ is closed.

5.11. (a) Let H_1 and H_2 be Hilbert spaces. Assume that H_1 is finite dimensional. Every operator T from H_1 into H_2 is continuous.
Hint: If $\{e_1,\ldots,e_m\}$ is an ONB of $D(T)$, then $Tf=\sum_j\langle e_j,f\rangle Te_j$ for all $f\in D(T)$.
(b) If $\langle .\,,.\rangle_1$ and $\langle .\,,.\rangle_2$ are scalar products on a finite dimensional vector space H, then there exist constants $c_1, c_2 \geqslant 0$ such that $\|\,.\,\|_1 \leqslant c_1\|\,.\,\|_2 \leqslant c_2\|\,.\,\|_1$ ($\|\,.\,\|_j$ denotes the norm defined by $\langle .\,,.\rangle_j$).

5.12. Let T be an operator from H_1 into H_2, and let S be a continuously invertible operator from H_2 into H_3. If S and T are closed (closable), then ST is also closed (closable).

5.13. If S is T-bounded with T-bound b, then for any $\epsilon>0$ there exists an $a\geqslant 0$ such that $\|Sf\|^2 \leqslant a^2\|f\|^2+(b^2+\epsilon)\|Tf\|^2$ for all $f\in D(T)$. If we have $\|Sf\|^2 \leqslant a^2\|f\|^2+b^2\|Tf\|^2$, then we also have $\|Sf\| \leqslant a\|f\|+b\|Tf\|$.

5.2 The fundamentals of spectral theory

In what follows an operator T from H_1 into H_2 will be said to be *bijective* if T is injective and $R(T)=H_2$.

Theorem 5.10. *Let S and T be bijective operators from H_1 into H_2. If $D(S)\subset D(T)$, then*

$$T^{-1}-S^{-1}=T^{-1}(S-T)S^{-1}.$$

If $D(S)=D(T)$, then

$$T^{-1}-S^{-1}=T^{-1}(S-T)S^{-1}=S^{-1}(S-T)T^{-1}.$$

PROOF. It is enough to prove the first assertion. We shall prove that $T^{-1}=S^{-1}+T^{-1}(S-T)S^{-1}$. Since $T^{-1}Tf=f$ for all $f\in D(T)$ and

$SS^{-1}g = g$ for all $g \in H_2$, it follows that

$$S^{-1} + T^{-1}(S-T)S^{-1} = T^{-1}TS^{-1} + T^{-1}(S-T)S^{-1}$$
$$= T^{-1}(T+S-T)S^{-1} = T^{-1}SS^{-1} = T^{-1}. \quad \square$$

Now we show that a closed bijective operator remains bijective even after the addition of a "not too big" operator.

Theorem 5.11. *Assume that H_1 and H_2 are Hilbert spaces, T is a closed bijective operator from H_1 into H_2, S is an operator from H_1 into H_2 such that $D(S) \supset D(T)$, and $\|ST^{-1}\| < 1$. Then $T+S$ is also bijective, and we have*

$$(T+S)^{-1} = \sum_{n=0}^{\infty} (-1)^n T^{-1}(ST^{-1})^n = \sum_{n=0}^{\infty} (-1)^n (T^{-1}S)^n T^{-1}; \quad (5.8)$$

the series are convergent in the norm of $B(H_2, H_1)$.

PROOF. For all $f \in D(T)$ we have

$$\|Sf\| = \|ST^{-1}Tf\| \leq \|ST^{-1}\| \, \|Tf\|,$$

i.e., S is T-bounded with T-bound less than 1. By Theorem 5.5 the operator $T+S$ is closed, too. Moreover, $T+S$ is injective, because

$$\|(T+S)f\| \geq \|Tf\| - \|Sf\| \geq (1 - \|ST^{-1}\|)\|Tf\| > 0$$

for all $f \in D(T+S) = D(T)$, $f \neq 0$.

The two occuring series are obviously identical term by term. Let us define

$$A_p = \sum_{n=0}^{p} (-1)^n T^{-1}(ST^{-1})^n, \quad p \in \mathbb{N}.$$

Then for $q > p$ we have

$$\|A_q - A_p\| = \left\| \sum_{n=p+1}^{q} (-1)^n T^{-1}(ST^{-1})^n \right\|$$
$$\leq \|T^{-1}\| \, \|ST^{-1}\|^{p+1} \sum_{n=0}^{q-p-1} \|ST^{-1}\|^n$$
$$\leq \|T^{-1}\| \, \|ST^{-1}\|^{p+1} \sum_{n=0}^{\infty} \|ST^{-1}\|^n \leq C\|ST^{-1}\|^{p+1},$$

where $C = \|T^{-1}\| \sum_{n=0}^{\infty} \|ST^{-1}\|^n$. From this it follows that (A_p) is a Cauchy sequence in $B(H_2, H_1)$, consequently there exists an $A \in B(H_2, H_1)$ such that $A_p \to A$ as $p \to \infty$.

We have $(T+S)A = I$:

$$(T+S)A_p = \sum_{n=0}^{p} (-1)^n (T+S)T^{-1}(ST^{-1})^n = I + (-1)^p (ST^{-1})^{p+1} \to I,$$

as $p\to\infty$. Therefore for all $g\in H_2$ we have

$$A_p g\to Ag \quad\text{and}\quad (T+S)A_p g\to g \quad\text{as}\quad p\to\infty.$$

As $(T+S)$ is closed, $Ag\in D(T+S)$ and $(T+S)Ag=g$, i.e., $(T+S)A=I$. In particular, $R(T+S)=H_2$, hence $T+S$ is bijective and we have $A=(T+S)^{-1}$. □

Corollary. *The statements of Theorem* 5.11 *hold in particular if S is bounded and $\|S\|<\|T^{-1}\|^{-1}$ (then we have $\|ST^{-1}\|\leq\|S\|\,\|T^{-1}\|<1$).*

Corollary. *Let H_1 and H_2 be Hilbert spaces and let T and T_n $(n\in\mathbb{N})$ be linear operators from H_1 into H_2 such that $D(T)\subset D(T_n)$ for all $n\in\mathbb{N}$. Assume that T is closed, bijective, and $\|(T-T_n)T^{-1}\|\to 0$ as $n\to\infty$. Then there exists an $n_0\in\mathbb{N}$ such that T_n is bijective for $n\geq n_0$ and $\|T_n^{-1}-T^{-1}\|\to 0$ as $n\to\infty$. (The assumptions on T_n hold in particular if the operators $T-T_n$ are bounded and $\|T-T_n\|\to 0$ as $n\to\infty$.)*

PROOF. With $S_n=T_n-T$ we have $\|S_nT^{-1}\|\to 0$ as $n\to\infty$. Therefore there exists an $n_0\in\mathbb{N}$ such that $\|S_nT^{-1}\|<1/2$ for $n\geq n_0$. Hence by Theorem 5.11 the operator $T_n=T+S_n$ is bijective for $n\geq n_0$ and we have

$$\|T_n^{-1}-T^{-1}\|\leq\sum_{m=1}^{\infty}\|T^{-1}(S_nT^{-1})^m\|\leq\|T^{-1}\|\,\|S_nT^{-1}\|\sum_{m=1}^{\infty}2^{1-m}\to 0$$

as $n\to\infty$. □

In what follows let H be a Hilbert space over $\mathbb{K}$ and let T be an operator on H. The number $z\in\mathbb{K}$ is called an *eigenvalue* of T if there exists an $f\in D(T)$, $f\neq 0$ such that $Tf=zf$, i.e., if the operator $z-T=zI-T$ is not injective, $N(z-T)\neq\{0\}$. The subspace $N(z-T)$ is called the *eigenspace* of z, the dimension of $N(z-T)$ is called the *multiplicity* of the eigenvalue. The element f is called an *eigenelement* or *eigenvector of T belonging to the eigenvalue z*. If z is not an eigenvalue (i.e., $(z-T)$ is injective), then the operator

$$R(z,T)=(z-T)^{-1}$$

is well-defined. The set

$$\rho(T)=\{z\in\mathbb{K}: z-T \text{ is injective, and } R(z,T)\in B(H)\} \qquad (5.9)$$

is called the *resolvent set* of T. If T is not closed, then $z-T$ and $R(z,T)$ are not closed, consequently $\rho(T)=\varnothing$. This is why in most of the following cases we shall assume the closedness of T. For a closed operator T on H we have by the closed graph theorem that

$$\rho(T)=\{z\in\mathbb{K}: z-T \text{ is bijective}\}. \qquad (5.10)$$

The function

$$R(.\,, T) : \rho(T) \to B(H),\ z \quad R(z, T)$$

is called the *resolvent* of T. For any $z \in \rho(T)$ the operator $R(z, T)$ is called the *resolvent of T at the point z*. The set

$$\sigma(T) = \mathbb{K} \setminus \rho(T) = C_{\mathbb{K}}\rho(T) \tag{5.11}$$

is called the spectrum of T. The set $\sigma_p(T)$ of all the eigenvalues of T is obviously contained in $\sigma(T)$. The set $\sigma_p(T)$ is called the *point-spectrum* of T.

Theorem 5.12. *Let T be a densely defined operator on H. Then $\sigma(T^*) = \sigma(T)^*$ and $\rho(T^*) = \rho(T)^*$ (here for any subset M of the complex numbers $M^* = \{z^* : z \in M\}$).*

Proof. Because of (5.11) it is enough to prove that $\rho(T) = \rho(T^*)^*$. To prove this it is enough to show that $\rho(T) \subset \rho(T^*)^*$, since because of the equality $T^{**} = T$ we also have $\rho(T^*) \subset \rho(T)^*$, and thus $\rho(T^*)^* \subset \rho(T)$.

Let $z \in \rho(T)$. Then $z - T$ is densely defined and bijective. By Theorem 4.17(b) the operator $z^* - T^* = (z - T)^*$ is therefore injective, also, and we have $(z^* - T^*)^{-1} = ((z - T)^{-1})^* \in B(H)$. Hence $z^* - T^*$ is bijective, i.e., $z^* \in \rho(T^*)$, and thus $z \in \rho(T^*)^*$. □

Theorem 5.13. *Let S and T be closed operators on H.*

(a) *For all $z, z' \in \rho(T)$ we have the* first resolvent identity

$$R(z, T) - R(z', T) = (z' - z)R(z, T)R(z', T)$$

$$= (z' - z)R(z', T)R(z, T);$$

in particular, $R(z, T)$ and $R(z', T)$ commute.

(b) *If $D(S) \subset D(T)$, then for all $z \in \rho(S) \cap \rho(T)$ we have*

$$R(z, T) - R(z, S) = R(z, T)(T - S)R(z, S).$$

(c) *If $D(S) = D(T)$, then for all $z \in \rho(S) \cap \rho(T)$ we have* the second resolvent identity

$$R(z, T) - R(z, S) = R(z, T)(T - S)R(z, S)$$

$$= R(z, S)(T - S)R(z, T).$$

Proof. The first resolvent identity follows from Theorem 5.10 if in there we replace T by $z - T$ and S by $z - T$. The second resolvent identity follows similarly if T and S are replaced by $z - T$ and $z - S$, respectively. □

Theorem 5.14. *If T is a closed operator on the Hilbert space H, then $\rho(T)$ is open, consequently $\sigma(T)$ is closed. More precisely, if $z_0 \in \rho(T)$, then $z \in \rho(T)$*

for all $z \in \mathbb{K}$ *such that* $|z - z_0| < \|R(z_0, T)\|^{-1}$, *and for these* z *we have*

$$R(z, T) = \sum_{n=0}^{\infty} (z_0 - z)^n R(z_0, T)^{n+1}.$$

If $T \in B(H)$, *then we have* $\{z \in \mathbb{K} : |z| > \|T\|\} \subset \rho(T)$; *the spectrum* $\sigma(T)$ *is compact. Furthermore,*

$$R(z, T) = \sum_{n=0}^{\infty} z^{-n-1} T^n \quad \text{for} \quad |z| > \|T\|; \tag{5.12}$$

this series is called the von Neumann series.

PROOF. Let $z_0 \in \rho(T)$, and let $|z - z_0| < \|R(z_0, T)\|^{-1}$. If in Theorem 5.11 we replace T by $z_0 - T$ and S by $(z - z_0)I$, then it follows that $z - T = z_0 - T + (z - z_0)$ is bijective, consequently $z \in \rho(T)$. Moreover, by Theorem 5.11 we have

$$R(z, T) = ((z_0 - T) + (z - z_0))^{-1} = \sum_{n=0}^{\infty} (z_0 - z)^n R(z_0, T)^{n+1}.$$

Now let $T \in B(H)$ and let $|z| > \|T\|$. If in Theorem 5.11 we replace T by zI and S by T, then it follows that $z - T$ is bijective, therefore $z \in \rho(T)$, and

$$(z - T)^{-1} = \sum_{n=0}^{\infty} z^{-n-1} T^n.$$

Hence $\sigma(T) \subset \{z \in \mathbb{K} : |z| \leq \|T\|\}$. As $\sigma(T)$ is closed, the compactness of $\sigma(T)$ follows from this. □

Theorem 5.15. *Let* T *be a closed operator on the Hilbert space* H. *The resolvent* $R(.\,, T) : \rho(T) \to B(H)$ *is a continuous function (i.e., for any* $z_0 \in \rho(T)$ *and any sequence* (z_n) *from* $\rho(T)$ *such that* $z_n \to z_0$ *we have* $\|R(z_n, T) - R(z_0, T)\| \to 0$). *If* $\sigma(T)$ *is non-empty, then for every* $z \in \rho(T)$ *we have*

$$\|R(z, T)\| \geq d(z, \sigma(T))^{-1}.$$

For every sequence (z_n) *from* $\rho(T)$ *such that* $z_n \to z_0$, $z_0 \in \sigma(T)$ *we therefore have* $\|R(z_n, T)\| \to \infty$.

PROOF. Let $z_0, z \in \rho(T)$ such that $|z - z_0| < \|R(z_0, T)\|^{-1}$. Then by Theorem 5.14 we have

$$\|R(z, T) - R(z_0, T)\| \leq \sum_{n=1}^{\infty} |z_0 - z|^n \|R(z_0, T)\|^{n+1}.$$

As the right side is small when z is close to z_0, the continuity at z_0 follows from this for any $z_0 \in \rho(T)$. If $z \in \rho(T)$, then by Theorem 5.14 the point z'

also belongs to $\rho(T)$ for all $z' \in K$ such that $|z'-z| < \|R(z, T)\|^{-1}$. Consequently, $|z'-z| \geq \|R(z, T)\|^{-1}$ for all $z' \in \sigma(T)$, and thus

$$\|R(z, T)\|^{-1} \leq \inf \{|z'-z| : z' \in \sigma(T)\} = d(z, \sigma(T)). \qquad \square$$

Let G be an open subset of $\mathbb{C}$, and let X be a Banach space. A function $F: G \to X$ is said to be *holomorphic* if for every $z_0 \in G$ there exist an $r > 0$ and a sequence (f_n) from X such that

$$F(z) = \sum_{n=0}^{\infty} (z-z_0)^n f_n \quad \text{for} \quad |z-z_0| < r;$$

where the convergence has to be understood in the sense of the norm of X. As in function theory, one can prove that the quantity

$$r_0 = \left[\limsup_{n\to\infty} \|f_n\|^{1/n}\right]^{-1} \tag{5.13}$$

is the *radius of convergence* of the above series and that the series is uniformly convergent on each disc around z_0 of radius less than r_0, while it is divergent for all z such that $|z-z_0| > r_0$ (cf., for example, *K. Jörgens* [19], §4.4). Every holomorphic function is continuous (cf. the proof of Theorem 5.15).

Theorem 5.16. *Let T be a closed operator on the complex Hilbert space H, and let $f, g \in H$. Then the functions*

$$R(.\,, T) : \rho(T) \to B(H), \qquad z \mapsto R(z, T)$$
$$R(.\,, T)f : \rho(T) \to H, \qquad z \mapsto R(z, T)f$$
$$\langle g, R(.\,, T)f\rangle : \rho(T) \to \mathbb{C}, \quad z \mapsto \langle g, R(z, T)f\rangle$$

are holomorphic.

Proof. Let $z_0 \in \rho(T)$, and let $r = \|R(z_0, T)\|^{-1}$. Then for all $z \in \mathbb{C}$ such that $|z-z_0| < r$ we have

$$R(z, T) = \sum_{n=0}^{\infty} (z_0 - z)^n R(z_0, T)^{n+1}$$

in the sense of the norm convergence in $B(H)$,

$$R(z, T)f = \sum_{n=0}^{\infty} (z_0 - z)^n R(z_0, T)^{n+1} f,$$

in the sense of the norm convergence in H, and

$$\langle g, R(z, T)f\rangle = \sum_{n=0}^{\infty} (z_0 - z)^n \langle g, R(z_0, T)^{n+1} f\rangle.$$

According to the above definition, these three functions are therefore holomorphic. □

Theorem 5.17. *Assume that H is a Hilbert space and $T \in B(H)$, $r(T) = \limsup \|T^n\|^{1/n}$.*
(a) *We have $r(T) \leq \|T^m\|^{1/m}$ for all $m \in \mathbb{N}$, and thus $r(T) = \lim \|T^n\|^{1/n}$.*
(b) *We have $r(T) \leq \|T\|$ and $\sigma(T) \subset \{z \in \mathbb{K} : |z| \leq r(T)\}$. For all $z \in \mathbb{K}$ such that $|z| > r(T)$ the operator $R(z, T)$ is given by the Neumann series (5.12).*
(c) *If H is complex, then $\sigma(T)$ is non-empty and there exists a $z \in \sigma(T)$ such that $|z| = r(T)$, i.e., we have $r(T) = \sup \{|z| : z \in \sigma(T)\}$.*
(d) *The statement in part (c) holds for any self-adjoint T in a real Hilbert space, as well.*

$r(T)$ is called the spectral radius *of T.*

REMARK. Theorem 5.17(c) does not hold for real Hilbert spaces, as the example of the operator defined by the matrix $\begin{pmatrix} 0 & 1 \\ -1 & 0 \end{pmatrix}$ on $\mathbb{R}^2$ shows.

PROOF.
(a) Let $m \in \mathbb{N}$. Every $n \in \mathbb{N}$ can be uniquely represented in the form $n = mp_n + q_n$ with $p_n, q_n \in \mathbb{N}$ and $q_n < m$. If we denote $C = \max \{1, \|T\|, \|T^2\|, \ldots, \|T^{m-1}\|\}$, then it follows that

$$\|T^n\| \leq \|T^m\|^{p_n} \|T^{q_n}\| \leq C \|T^m\|^{p_n},$$

and thus

$$r(T) \leq \limsup_{n \to \infty} C^{1/n} \|T^m\|^{(1/m) - (1/nm)q_n} = \|T^m\|^{1/m}.$$

For every $n \in \mathbb{N}$ we therefore have $r(T) \leq \|T^n\|^{1/n}$, and thus $r(T) \leq \liminf \|T^n\|^{1/n}$, i.e., $\lim \|T^n\|^{1/n}$ exists and $r(T) = \lim \|T^n\|^{1/n}$.

(b) The radius of convergence of the series $\sum_{n=0}^{\infty} u^{n+1} T^n$, $u \in \mathbb{K}$ is obviously equal to $r(T)^{-1}$, because of (5.13). Therefore for all $z \in \mathbb{K}$ such that $|z| > r(T)$ the operator $A(z) = \sum_{n=0}^{\infty} z^{-n-1} T^n \in B(H)$ is defined; the series is *absolutely convergent* in the norm of $B(H)$ (i.e., for every $\epsilon > 0$ there exists an $n_0 \in \mathbb{N}$ such that $\sum_{n=k}^{m} |z^{-n-1}| \, \|T^n\| < \epsilon$ for $n_0 \leq k \leq m$). One can verify easily that

$$(z - T)A(z) = A(z)(z - T) = \sum_{n=0}^{\infty} z^{-n} T^n - \sum_{n=0}^{\infty} z^{-n-1} T^{n+1} = I,$$

i.e., $A(z) = (z - T)^{-1}$ and $z \in \rho(T)$. The Neumann series is therefore convergent for all $z \in \mathbb{K}$ such that $|z| > r(T)$ and it represents the operator $(z - T)^{-1}$.

(c) Let us assume that $\sigma(T) = \varnothing$. Then $\rho(T) = \mathbb{C}$, and by Theorem 5.16 the function $F_{fg} : z \mapsto \langle f, R(z, T)g \rangle$ is an entire function for all $f, g \in H$. Because of the inequality $\|(z - T)f\| \geq (\|z\| - \|T\|)\|f\|$ we have $\|R(z, T)\| \leq (\|z\| - \|T\|)^{-1}$ for $|z| > \|T\|$, and thus $F_{fg}(z) \to 0$ as $|z| \to$

∞. Consequently, F_{fg} is bounded, and by Liouville's theorem it is constant. Since $F_{fg}(z)\to 0$ as $|z|\to\infty$, we have $F_{fg}=0$ for all $f, g\in H$, therefore $R(z, T)=0$. This is a contradiction. It remains to be proved that $r(T)=\sup\{|z| : z\in\sigma(T)\}=r_0$. As $\sigma(T)$ is closed and we already know the inequality $r_0\leqslant r(T)$, it remains to show that $r(T)\leqslant r_0$. For arbitrary $f, g\in H$ the function $z\mapsto\langle f, R(z, T)g\rangle$ is holomorphic in $\{z\in\mathbb{C} : |z|>r_0\}$ and can be uniquely expanded in a Laurent series there. By (5.12) this Laurent series has the form

$$\langle f, R(z, T)g\rangle = \sum_{m=0}^{\infty} z^{-m-1}\langle f, T^m g\rangle.$$

Let $s>r_0$. Then the sequence $(s^{-m-1}\langle f, T^m g\rangle)$ is bounded for all $f, g\in H$. By applications of the Banach-Steinhaus theorem (Theorem 4.22), first to the functionals induced by $s^{-m-1}T^m g$ and then to the operators $s^{-m-1}T^m$, we obtain a $C\geqslant 0$ such that

$$|s^{-m-1}|\,\|T^m\|\leqslant C \quad\text{for all}\quad m\in\mathbb{N}.$$

This implies that $\lim\|T^m\|^{1/m}\leqslant s$. Since this holds for all $s>r_0$, it follows that $r(T)\leqslant r_0$.

(d) If T is self-adjoint, then by Theorem 4.4(b) there exists a sequence (f_n) from H such that $\|f_n\|=1$ and $\langle f_n, Tf_n\rangle\to\|T\|$ or $\langle f_n, Tf_n\rangle\to-\|T\|$. In the first case it follows that

$$\begin{aligned}\|(\|T\|-T)f_n\|^2 &= \|T\|^2\|f_n\|^2-2\|T\|\langle f_n, Tf_n\rangle+\|Tf_n\|^2\\ &\leqslant \|T\|^2-2\|T\|\langle f_n, Tf_n\rangle+\|T\|^2\to 0 \quad\text{as}\quad n\to\infty.\end{aligned}$$

Analogously, in the second case it follows that

$$\|(\|T\|+T)f_n\|\to 0 \quad\text{as}\quad n\to\infty.$$

Consequently, $\|T\|\in\sigma(T)$ in the first case, and $-\|T\|\in\sigma(T)$ in the second case. □

Example 1. Let M be a measurable subset of $\mathbb{R}^m$, let $t : M\to\mathbb{C}$ be measurable, and let T be the maximal operator of multiplication on $L_2(M)$ by t.

(5.14) *We have* $\sigma_p(T)=\{z\in\mathbb{C} : t^{-1}(z)$ *has positive measure*$\}$, *here* $t^{-1}(z)$ *denotes the set* $\{x\in M : t(x)=z\}$. (See also Exercise 5.23.)

Proof. This can be obtained immediately from the results of (4.2) if we notice that $z-T$ is the multiplication operator generated by $z-t$. □

(5.15) *We have* $z\in\sigma(T)$ *if and only if for every* $\epsilon>0$ *the set* $t^{-1}(\{w\in\mathbb{C} : |w-z|<\epsilon\})=\{x\in M : |t(x)-z|<\epsilon\}$ *is of positive measure. In particular, if M is open and t is continuous, then* $\sigma(T)$ *is the closure of the range of t.*

PROOF. Let $\Sigma = \{z \in \mathbb{C} : \{x \in M : |t(x) - z| < \epsilon\}$ have positive measure for every $\epsilon > 0\}$.

We show that $\rho(T) = \mathbb{C}\Sigma$. We have $z \in \rho(T)$ if and only if $z - T$ is bijective, thus if and only if $z - T$ is injective and $R(z - T) = L_2(M)$. By (4.3) this holds if and only if there exists a $c > 0$ such that $|z - t(x)| \geqslant c$ almost everywhere on M, i.e., if and only if $t^{-1}(\{w \in \mathbb{C} : |w - z| < c\})$ has measure zero. Consequently, $z \in \rho(T)$ if and only if $z \notin \Sigma$. □

The corresponding results hold if we consider the real Hilbert space $L_{2,\mathbb{R}}(M)$ and a real function t.

EXAMPLE 2. If U is a unitary operator on H, then $\sigma(U) \subset \{z \in \mathbb{C} : |z| = 1\}$.

PROOF. By Theorem 5.14 we have $\{z \in \mathbb{C} : |z| > 1\} \subset \rho(U)$. Since U is bijective and U^{-1} is unitary, we have $0 \in \rho(U)$, and thus by Theorem 5.14 $\{z \in \mathbb{C} : |z| < 1\} \subset \rho(U)$. The assertion follows. □

EXERCISES

5.14. Let H_1 and H_2 be Banach spaces. An operator T from H_1 into H_2 has a continuous inverse (not necessarily defined on the whole space H_2) if and only if

$$\gamma = \inf \{\|Tx\| : x \in D(T), \|x\| \geqslant 1\} > 0.$$

We have then that $\|T^{-1}\| = \gamma^{-1}$.

5.15. Let (a_n) be a sequence from $\mathbb{C}$. On l_2 by

$$D(T) = \{f = (f_n) \in l_2 : (a_n f_n) \in l_2\},$$
$$Tf = (a_n f_n) \quad \text{for} \quad f = (f_n) \in D(T)$$

a closed operator is defined. T is self-adjoint if and only if the sequence (a_n) is real. We have $\sigma_p(T) = \{a_n : n \in \mathbb{N}\}$, $\sigma(T) = \overline{\sigma_p(T)}$. Determine $R(z, T)$ for $z \notin \sigma_p(T)$.

5.16. If S and T are from $B(H)$, then $r(ST) = r(TS)$.

5.17. Assume that H_1 and H_2 are Hilbert spaces, and T is a bijective operator from H_1 onto H_2.

(a) If S is an operator from H_1 into H_2 such that $D(S) \supset D(T)$, $ST^{-1} \in B(H_2)$, $r(ST^{-1}) < 1$, and $T + S$ is closed, then $T + S$ is also bijective.

(b) If $S \in B(H_1, H_2)$ and $r(T^{-1}S) < 1$, then $T + S$ is bijective.

In both cases we have

$$(T+S)^{-1} = \sum_{n=0}^{\infty} (-1)^n T^{-1}(ST^{-1})^n = \sum_{n=0}^{\infty} (-1)^n (T^{-1}S)^n T^{-1},$$

where the series converge in the norm of $B(H_2, H_1)$.

Hint: After showing the convergence of the series, denote the sum by A, and prove that $(T + S)A = A(T + S) = I$. Compare with the proof of Theorem 5.11, as well.

5.18. Let H_1 and H_2 be Hilbert spaces, let $A, A_n \in B(H_1, H_2)$, and let A_n be bijective for all $n \in \mathbb{N}$. If $A_n \to A$ and $\|A_n^{-1}\| \leq C$ for all $n \in \mathbb{N}$, then A is also bijective and we have $A_n^{-1} \to A^{-1}$.
Hint: Use Theorem 5.11.

5.19. Assume that H_1 and H_2 are Hilbert spaces; $T_n (n \in \mathbb{N})$ and T are closed bijective operators from H_1 onto H_2; the sequence $(\|T_n^{-1}\|)$ is bounded, and for some core D of T the following holds: For every $x \in D$ there exists an $n(x) \in \mathbb{N}$ such that $x \in D(T_n)$ for $n \geq n(x)$ and also $T_n x \to Tx$ for $n \to \infty$. Then $T_n^{-1} \xrightarrow{s} T^{-1}$.
Hint: From Theorem 5.10 it follows that $T_n^{-1} g \to T^{-1} g$ for all $g \in T(D)$; $T(D)$ is dense in H_2; since $(\|T_n^{-1}\|)$ is bounded, it follows that $T_n^{-1} g \to T^{-1} g$ for all $g \in H_2$.

5.20. Prove that in Exercise 5.19 no assumptions can be removed.
(a) If T, T_n are bijective and $T_n \xrightarrow{s} T$, then we do not necessarily have $T_n^{-1} \xrightarrow{s} T^{-1}$.
Hint: Consider, on l_2, the operators $T = I$ and $T_n f = (f_1, \ldots, f_n, (1/n) f_{n+1}, (1/n) f_{n+2}, \ldots)$; (the sequence $\|T_n^{-1}\|$ is not bounded).
(b) If T is bijective and $T_n \xrightarrow{s} T$, then T_n is not necessarily bijective for large n.
Hint: Consider the operators $T = I$ and $T_n f = (f_1, \ldots, f_n, 0, 0, \ldots)$ on l_2.
(c) If the T_n are bijective, $(\|T_n^{-1}\|)$ is bounded and $T_n \xrightarrow{s} T$, then T is not necessarily bijective.
Hint: Consider the operators $T_n f = (\ldots, f_{-n-2}, f_{-n-1}, f_0, f_{-1}, f_1, f_{-2}, f_2, \ldots, f_{-n}, f_n f_{n+1}, f_{n+2}, \ldots)$ on $l_2(\mathbb{Z})$.

5.21. Assume that H is a Hilbert space over $\mathbb{K}$, Ω is an open subset of $\mathbb{K}$, and $R : \Omega \to B(H)$ satisfies the properties (i) $R(z_1) - R(z_2) = (z_2 - z_1) R(z_1) R(z_2)$ for $z_1, z_2 \in \Omega$, (ii) $N(R(z)) = \{0\}$ for all $z \in \Omega$. Then we have
(a) $R(z_1) R(z_2) = R(z_2) R(z_1)$ for all $z_1, z_2 \in \Omega$,
(b) With $K(z) = R(z)^{-1} (z \in \Omega)$ we have $K(z_1) - z_1 = K(z_2) - z_2$ for all $z_1, z_2 \in \Omega$; i.e., we can define $T = z - K(z)$.
(c) T is closed; we have $\Omega \subset \rho(T)$ and $R(z) = (z - T)^{-1}$ for $z \in \Omega$.

5.22. Assume that H_1 and H_2 are isomorphic Hilbert spaces, M is a dense subspace of H_2, and $M \neq H_2$. Then $\overline{B(H_1, M)} \neq B(H_1, H_2)$. (From this it follows that $B(H_1, \hat{H})$ is in general not a completion of $B(H_1, H)$ if $\hat{H}$ is a completion of the pre-Hilbert space H.)
Hint: Let U be an isomorphism from H_1 onto H_2. Assume that there exists a sequence (U_n) from $B(H_1, M)$ such that $U_n \to U$. Then $R(U_n) = H_2$ for large n.

5.23. Let T be the operator of multiplication by a measurable function t on $L_2(M)$ (cf. Example 1). Every eigenvalue of T has infinite multiplicity.

5.24. Let $T \in B(H)$, let $N_T = \{f \in H : T^n f \to 0 \text{ as } n \to \infty\}$, and let $B_T = \{f \in H :$ the sequence $(T^n f)$ is bounded$\}$.
(a) If $r(T) < 1$, then $N_T = H$.
(b) If $B_T = H$, then $r(T) \leq 1$.

Hint: Use Theorem 4.22.

(c) From $N_T = H$ it does not follow in general that $r(T) < 1$.
Hint: Consider the operator $Tf(x) = xf(x)$ on $L_2(0, 1)$.

5.25. If $T \in B(H)$, $\lambda \in \mathbb{K}$ such that $|\lambda| = \|T\|$, and $u \in H$ is such that $Tu = \lambda u$, then $T^*u = \lambda^* u$.
Hint: Show that $\|T^*u - \lambda^* u\|^2 \leq 0$!

5.26. Let $A \in B(H_1, H_2)$ and $B \in B(H_2, H_1)$.

(a) For $\lambda \neq 0$ we have $\lambda \in \rho(AB)$ if and only if $\lambda \in \rho(BA)$. In this case we have $(\lambda - BA)^{-1} = \frac{1}{\lambda}(I + B(\lambda - AB)^{-1}A)$.

(b) Assertion (a) does not hold for $\lambda = 0$.
Hint: Let A be an isometric operator such that $R(A) \neq H_2$, and let $B = A^*$.

(c) Assertion (a) holds for $\lambda = 0$, as well, provided that at least one of the operators is bijective.
Hint: $AB = A(BA)A^{-1}$.

5.27. Let T be an operator on H with non-empty resolvent set, and let $\lambda_0 \in \rho(T)$. We have $\lambda \in \sigma(T)$ if and only if $(\lambda - \lambda_0)^{-1} \in \sigma(R(\lambda_0, T))$; for all $\lambda \in \rho(T)$ we have $R(\lambda, T) = R(\lambda_0, T)(I - (\lambda_0 - \lambda)R(\lambda_0, T))^{-1} = (I - (\lambda_0 - \lambda)R(\lambda_0, T))^{-1}R(\lambda_0, T)$. In particular, $\sigma(R(\lambda_0, T)) = \{(\lambda - \lambda_0)^{-1} : \lambda \in \sigma(T)\}$.
Hint: Use the first resolvent identity.

5.28. An operator $T \in B(H)$ is said to be *nilpotent* if there exists an $n \in \mathbb{N}$ such that $T^n = 0$. The operator T is said to be *quasi-nilpotent* if $\|T^n\|^{1/n} \to 0$, i.e., $r(T) = 0$.

(a) Every nilpotent operator is quasi-nilpotent.

(b) If T is a quasi-nilpotent operator on a complex Hilbert space, then we have $\sigma(T) = \{0\}$.

(c) If $k(x, y)$ is continuous on $0 \leq y \leq x \leq 1$, then the *Volterra integral operator* K defined on $L_2(0, 1)$ by $D(K) = L_2(0, 1)$ and $(Kf)(x) = \int_0^x k(x, y)f(y)\, dy$ is quasi-nilpotent.
Hint: Put $k(x, y) = 0$ for $x < y$ and $M = \max\{|k(x, y)| : 0 \leq y \leq x \leq 1\}$; then $|k^{(n)}(x, y)| \leq (M^n/(n-1)!)|x - y|^{n-1}$ for the kernel $k^{(n)}$ of K^n, consequently $\|K^n\| \leq (M^n/(n-1)!)$.

5.29. Let H be a complex Hilbert space, let $T \in B(H)$, and let $p(t) = \sum_{j=0}^n a_j t^j$ be a polynomial with $p(T) = \sum_{j=0}^n a_j T^j = 0$ $(T^0 = I)$.

(a) If $p(z) \neq 0$, then $z \in \rho(T)$ and $R(z, T) = [p(z)]^{-1}q(z, T)$ with $q(z, T) = \sum_{j=0}^{n-1} z^j \sum_{k=j+1}^n a_k T^{k-j-1}$.

(b) If p is a polynomial of minimal degree, then $\sigma(T)$ is the set of zeros of p.

5.30. Let T be a closed operator on the Hilbert space H. Then the function $z \mapsto R(z, T)$ is holomorphic on $\rho(T)$ as a function with values in $B(H, D(T))$, where $D(T)$ is equipped with the T-norm.
Hint: Use the inequality $\|[R(z_0, T)]^{n+1}\|_{B(H, D(T))} \leq \|R(z_0, T)\|_{B(H, D(T))} \times \|R(z_0, T)\|^n$ and Theorem 5.14.

5.3 Symmetric and self-adjoint operators

Theorem 5.18. *Let T be a Hermitian operator on the pre-Hilbert space H. Every eigenvalue of T is real; eigenvectors belonging to different eigenvalues are orthogonal. If H is complex, then for any $z \in \mathbb{C} \setminus \mathbb{R}$ the operator $z - T$ is continuously invertible, and we have $\|R(z, T)\| \leq |\operatorname{Im} z|^{-1}$ (this holds true in particular for symmetric and self-adjoint operators).*

PROOF. Let z be an eigenvalue of T and let $f \in N(z-T)$, $f \neq 0$. Then $z^*\|f\|^2 = \langle Tf, f\rangle = \langle f, Tf\rangle = z\|f\|^2$, thus $z = z^*$, i.e., $z \in \mathbb{R}$. If z_1, z_2 are two distinct eigenvalues and f_1, f_2 are corresponding eigenelements of T, then, as the z_j are real, we have $(z_1 - z_2)\langle f_1, f_2\rangle = \langle Tf_1, f_2\rangle - \langle f_1, Tf_2\rangle = 0$. Therefore $\langle f_1, f_2\rangle = 0$. With $z = x + iy$ $(x, y \in \mathbb{R})$ we have for all $f \in D(T)$ that

$$\|(z-T)f\|^2 = \|(x-T)f + iyf\|^2$$

$$= \|(x-T)f\|^2 + |y|^2\|f\|^2 \geq |\operatorname{Im} z|^2\|f\|^2.$$

For $z \in \mathbb{C} \setminus \mathbb{R}$ it follows from this that $(z - T)$ is injective and that for $g = (z-T)f \in D(R(z, T))$ we have

$$\|R(z, T)g\| = \|f\| \leq |\operatorname{Im} z|^{-1}\|(z-T)f\| = |\operatorname{Im} z|^{-1}\|g\|,$$

therefore

$$\|R(z, T)\| \leq |\operatorname{Im} z|^{-1}. \qquad \square$$

Now we prove a simple criterion for the self-adjointness of a symmetric operator.

Theorem 5.19. *Let T be a symmetric operator on the Hilbert space H. If $H = N(s-T) + R(s-T)$ for some $s \in \mathbb{R}$, then T is self-adjoint and $H = N(s-T) \oplus R(s-T)$. Special case: If $R(s-T) = H$, then T is self-adjoint and $N(s-T) = \{0\}$.*

PROOF. From $\bar{T} = T^{**}$ and $T \subset T^*$ it follows that $N(s-T) \subset N(s-\bar{T}) = R(s-T^*)^\perp \subset R(s-T)^\perp$, thus that $N(s-T) \perp R(s-T)$ and $H = N(s-T) \oplus R(s-T)$. Therefore we have

$$N(s-T) = R(s-T)^\perp$$

and

$$R(s-T) = N(s-T)^\perp \supset N(s-\bar{T})^\perp \supset R(s-T^*).$$

We show that $D(T^*) \subset D(T)$. This together with the inclusion $T \subset T^*$ implies $T = T^*$. Let $f \in D(T^*)$, $g = (s - T^*)f$. Because of the inclusion $R(s-T^*) \subset R(s-T)$ there exists an $f_0 \in D(T) \subset D(T^*)$ such that $(s - T^*)f_0 = (s-T)f_0 = g = (s-T^*)f$. We therefore have $f - f_0 \in N(s-T^*) = R(s-T)^\perp = N(s-T) \subset D(T)$ and thus $f \in D(T)$. $\square$

A symmetric operator T on a Hilbert space is said to be *essentially self-adjoint* provided that $\bar{T}$ is self-adjoint.

Theorem 5.20. *A symmetric operator T on a Hilbert space is essentially self-adjoint if and only if T^* is symmetric. We then have $\bar{T}=T^*$.*

PROOF. If T is essentially self-adjoint, then $T^*=(\bar{T})^*=\bar{T}=T^{**}$, consequently T^* is self-adjoint (therefore symmetric) and we have $\bar{T}=T^*$. If T^* is symmetric, then since $\bar{T}$ is symmetric by Theorem 5.4(b) and $(\bar{T})^*=T^*$ holds, we have $\bar{T}\subset(\bar{T})^*=T^*\subset T^{**}=\bar{T}$, hence $\bar{T}=(\bar{T})^*$. □

Theorem 5.21. *Let T be a symmetric operator on a complex Hilbert space H. The operator T is self-adjoint (essentially self-adjoint) if and only if $R(z_\pm - T)=H$ ($\overline{R(z_\pm - T)}=H$) for some z_+ with $\operatorname{Im} z_+>0$ and some z_- with $\operatorname{Im} z_-<0$ (this then holds for all z_+ with $\operatorname{Im} z_+>0$ and z_- with $\operatorname{Im} z_-<0$).*

PROOF. Let $R(z_\pm - T)=H$ for some z_+ such that $\operatorname{Im} z_+>0$ and some z_- such that $\operatorname{Im} z_-<0$. As T is symmetric, the operators $z_\pm - T$ are injective, therefore bijective, and by Theorem 5.18 we have $\|(z_+ - T)^{-1}\|\leqslant|\operatorname{Im} z_\pm|^{-1}$. By Theorem 5.11 the operator $z-T$ is also bijective for all z such that $|z-z_+|<|\operatorname{Im} z_+|$ or $|z-z_-|<|\operatorname{Im} z_-|$. Since $\|(z-T)^{-1}\|\leqslant|\operatorname{Im} z|^{-1}$ for all z such that $\operatorname{Im} z\neq 0$, we can iterate this procedure and obtain that $(z-T)$ is bijective for all $z\in\mathbb{C}\setminus\mathbb{R}$, in particular for $z=\pm i$ (cf. also Exercise 5.33).

As T is symmetric, we have $T\subset T^*$. To prove that $T=T^*$, it is enough to prove that $D(T^*)\subset D(T)$. To this end, let $f\in D(T^*)$ and let $f_0=(i-T)^{-1}(i-T^*)f$. Then we have $f_0\in D(T)\subset D(T^*)$, $Tf_0=T^*f_0$, and $(i-T^*)(f-f_0)=(i-T^*)f-(i-T^*)f_0=(i-T)f_0-(i-T)f_0=0$. Therefore

$$f-f_0\in N(i-T^*)=R(-i-T)^\perp=H^\perp=\{0\},$$

thus $f=f_0\in D(T)$.

If $\overline{R(z_\pm - T)}=H$, then $R(z_\pm - \bar{T})=H$ by the proposition following Theorem 5.2. As $\bar{T}$ is also symmetric by Theorem 5.4(b), the self-adjointness of $\bar{T}$ follows from the part already proved.

Let T now be self-adjoint and let $z\in\mathbb{C}\setminus\mathbb{R}$. Then $R(z,T)$ is closed and bounded, therefore $D(R(z,T))=R(z-T)$ is closed. Consequently, it is enough to show that $R(z-T)^\perp=\{0\}$. To this end, let $h\perp R(z-T)$. Then by Theorem 4.13(b) we have $h\in N((z-T)^*)=N(z^*-T)$. Since T is self-adjoint, T has no non-real eigenvalue, thus $h=0$.

If T is essentially self-adjoint, i.e., $\bar{T}$ is self-adjoint, then by the proposition following Theorem 5.2 we have

$$\overline{R(z_\pm - T)}=R\left(z_+ - \bar{T}\right)=H.$$ □

Theorem 5.22. *If T is a symmetric operator on the complex Hilbert space H and for some $n\in\mathbb{N}$, $n\geqslant 2$ we have $\overline{R(i-T^n)}=H$ or $\overline{R(-i-T^n)}=H$ (respec-*

tively $R(i-T^n)=H$ or $R(-i-T^n)=H$), then T is essentially self-adjoint (respectively self-adjoint).

PROOF.
(a) Let $\overline{R(i-T^n)}=\overline{R(T^n-i)}=H$. There are numbers $\gamma_\pm \in \mathbb{C}$ such that $\operatorname{Im}\gamma_+>0$, $\operatorname{Im}\gamma_-<0$ and $\gamma_\pm^n=i$. We then have
$$(T^n-i)=(T-\gamma_\pm)(T^{n-1}+T^{n-2}\gamma_\pm+\cdots+T\gamma_\pm^{n-2}+\gamma_\pm^{n-1}),$$
hence $\overline{R(T-\gamma_\pm)}\supset\overline{R(T^n-i)}=H$. Therefore T is essentially self-adjoint.
(b) If $\overline{R(T^n+i)}=H$, then we choose $\gamma_\pm\in\mathbb{C}$ so that $\gamma_\pm^n=-i$. The proof of self-adjointness is similar. □

Theorem 5.23.
(a) *The symmetric operator T on the complex Hilbert space H is self-adjoint if and only if $\sigma(T)\subset\mathbb{R}$.*
(b) *If T is self-adjoint on the (real or complex) Hilbert space H, then $s\in\sigma_p(T)$ if and only if $\overline{R(s-T)}\neq H$. For $z\notin\sigma_p(T)$ we have $R(z,T)^*=R(z^*,T)$.*

PROOF.
(a) By Theorems 5.18 and 5.21 the operator T is self-adjoint if and only if $z-T$ is surjective and continuously invertible for all $z\in\mathbb{C}\setminus\mathbb{R}$, i.e., if and only if $\mathbb{C}\setminus\mathbb{R}\subset\rho(T)$, or, equivalently, $\sigma(T)\subset\mathbb{R}$.
(b) Assume that T is self-adjoint and $z\notin\sigma_p(T)$. As $\sigma_p(T)\subset\mathbb{R}$, we also have $z^*\notin\sigma_p(T)$. Therefore $R(z-T)^\perp=N(z^*-T^*)=N(z^*-T)=\{0\}$, i.e., $\overline{R(z-T)}=H$. Now let $\overline{R(z-T)}=H$. Since $R(z-T)=H$ for all $z\in\mathbb{C}\setminus\mathbb{R}$, we also have $\overline{R(z^*-T)}=H$, and thus $N(z-T)=N(z-T^*)=R(z^*-T)^\perp=\{0\}$, i.e., $z\notin\sigma_p(T)$. If $z\notin\sigma_p(T)$, then $z-T^*=z-T$ is densely defined, injective and $\overline{R(z-T)}=H$. Therefore $R(z,T)^*=((z-T)^{-1})^*=((z-T)^*)^{-1}=(z^*-T)^{-1}=R(z^*,T)$ by Theorem 4.17(b). □

Now we obtain an especially simple characterization of the spectrum (or the resolvent set) of a self-adjoint operator.

Theorem 5.24. *If T is self-adjoint, then the following statements are equivalent:*
(i) *$z\in\rho(T)$,*
(ii) *there exists a $c>0$ such that $\|(z-T)f\|\geqslant c\|f\|$ for all $f\in D(T)$ (i.e., $(z-T)$ is injective and $\|R(z,T)\|\leqslant c^{-1}$),*
(iii) *$R(z-T)=H$.*
(This theorem is in general false for symmetric operators.)

PROOF. If $z\in\rho(T)$, then $(z-T)$ is injective and $R(z,T)$ is continuous. If $(z-T)$ is injective and $R(z,T)$ is continuous, then $z\notin\sigma_p(T)$ and thus by Theorem 5.23(b) the set $D(R(z,T))=R(z-T)$ is dense in H; as $R(z,T)$ is closed, we have $R(z-T)=D(R(z,T))=H$. If $R(z,T)=H$ and $z\in\mathbb{R}$, then $N(z-T)=N(z^*-T^*)=R(z-T)^\perp=\{0\}$; therefore $z-T$ is bijective, i.e., $z\in\rho(T)$. If $\operatorname{Im} z\neq 0$, then $z\in\rho(T)$ by Theorem 5.23(a). □

For the proof of any further criteria for self-adjointness we need some auxiliary results. In complex Hilbert spaces the theorem of Rellich-Kato (Theorem 5.28) can be proved directly somewhat more rapidly (cf. Exercise 5.35). Here we obtain it (also for real Hilbert spaces) as a special case of more general results. The auxiliary results gathered here will be used at other places, as well.

Theorem 5.25. *Assume that H_1 and H_2 are Hilbert spaces, A and B are operators from H_1 into H_2 such that*

$$D(A)\subset D(B) \quad \text{and} \quad \|Bf\|\leqslant C\|Af\| \quad \text{for} \quad f\in D(A)$$

with some $C\geqslant 0$. For every $\kappa\in\mathbb{K}$ let P_κ denote the orthogonal projection (from H_2) onto $\overline{R(A+\kappa B)}$. Then $\|P_\kappa-P_0\|\to 0$ as $\kappa\to 0$.

Proof. For $|\kappa|<(1/2C)$ and for all $f\in D(A)$ we have

$$\|Bf\|\leqslant C\|Af\|\leqslant C\{\|(A+\kappa B)f\|+|\kappa|\,\|Bf\|\}\leqslant C\|(A+\kappa B)f\|+\tfrac{1}{2}\|Bf\|,$$

therefore

$$\|Bf\|\leqslant 2C\|(A+\kappa B)f\|.$$

For $h\in R(P_0)^\perp=\overline{R(A)}^\perp=R(A)^\perp$ we thus have

$$\begin{aligned}\|P_\kappa h\| &= \sup\{|\langle h,g\rangle| : g\in R(A+\kappa B),\ \|g\|\leqslant 1\}\\ &= \sup\{|\langle h,(A+\kappa B)f\rangle| : f\in D(A),\ \|(A+\kappa B)f\|\leqslant 1\}\\ &= \sup\{|\langle h,\kappa Bf\rangle| : f\in D(A),\ \|(A+\kappa B)f\|\leqslant 1\}\\ &\leqslant |\kappa|\,\|h\|\sup\{\|Bf\| : f\in D(A),\ \|(A+\kappa B)f\|\leqslant 1\}\\ &\leqslant 2C|\kappa|\,\|h\|.\end{aligned}$$

We can prove in a completely analogous way that for all $h\in R(P_\kappa)^\perp=R(A+\kappa B)^\perp$

$$\|P_0 h\|\leqslant C|\kappa|\,\|h\|.$$

Via Theorem 4.33 we obtain that $\|P_\kappa-P_0\|\leqslant 2C|\kappa|$. This proves the theorem. □

Theorem 5.26. *Assume that H_1 and H_2 are Hilbert spaces, T and S are operators from H_1 into H_2; T is closed, and S is T-bounded. Furthermore, denote by Ω the set*

$$\Omega=\{z\in\mathbb{K} : T+zS \text{ is closed}\}$$

and for every $z\in\Omega$ let Q_z denote the orthogonal projection from $H_1\times H_2$ onto $G(T+zS)$. Then Ω is open and the function $z\mapsto Q_z$ is continuous on Ω (with respect to the norm topology of $B(H_1\times H_2)$).

PROOF. If $z_0 \in \Omega$, then $T+z_0S$ is closed and $D(T+z_0S)=D(T)$. The operator S is therefore $(T+z_0S)$-bounded by Theorem 5.9. Hence $T+zS$ is closed for z sufficiently near z_0. Thus Ω is open. Let us define the operators A and B from H_1 into $H_1 \times H_2$ by

$$D(A)=D(T), \qquad Af=(f, (T+z_0S)f),$$
$$D(B)=D(T), \qquad Bf=(0, Sf).$$

Then the assumptions of Theorem 5.25 are obviously satisfied and the equalities $R(A)=G(T+z_0S)$ and $R(A+\kappa B)=G(T+(z_0+\kappa)S)$ imply the continuity of $z \mapsto Q_z$ at the point z_0. □

With this we can now obtain a general theorem on the perturbation of a closed operator and its adjoint.

Theorem 5.27. *Assume that H_1 and H_2 are Hilbert spaces, T and S are operators from H_1 into H_2. Let T be densely defined, let S be T-bounded, let S^* be T^*-bounded, and let*

$$\Omega = \{z \in \mathbb{K} : T+zS \text{ and } T^*+z^*S^* \text{ are closed}\},$$
$$\Omega_0 = \text{the connected component of } \Omega \text{ that contains zero.}$$

Then $(T+zS)^*=T^*+z^*S^*$ for all $z \in \Omega_0$.

PROOF. Let Q_z be the orthogonal projection (in $H_1 \times H_2$) onto $G(T+zS)$ and let Q_z' be the orthogonal projection onto $U^{-1}G(T^*+z^*S^*)$, where U is defined as in Section 4.4. By Theorem 5.26 the operators Q_z and Q_z' depend continuously on z for $z \in \Omega$ and $Q_0+Q_0'=I_{H_1\times H_2}=I$ (as we have $G(T)\oplus U^{-1}G(T^*)=H_1\times H_2$). For every $z \in \mathbb{K}$ we have $T^*+z^*S^* \subset (T+zS)^*$, i.e.,

$$U^{-1}G(T^*+z^*S^*) \subset U^{-1}G(T+zS)^* = G(T+zS)^{\perp}.$$

By Theorem 4.30(a) we therefore have $Q_zQ_z'=Q_z'Q_z=0$, i.e., $I-Q_z-Q_z'$ is an orthogonal projection for any $z \in \Omega$. Consequently, $\|I-Q_z-Q_z'\|$ assumes only the values 0 and 1. Since, on the other hand, $\|I-Q_z-Q_z'\|$ depends continuously on $z \in \Omega$, it follows that $\|I-Q_z-Q_z'\|=\|I-Q_0-Q_0'\|=0$ for all $z \in \Omega_0$, therefore $G(T^*+z^*S^*)=UG(T+zS)^{\perp}=G((T+zS)^*)$, and thus $T^*+z^*S^*=(T+zS)^*$ for all $z \in \Omega_0$. □

If the relative bounds of S with respect to T and of S^* with respect to T^* are less than 1, then it follows that $(T+S)^*=T^*+S^*$ (*Hess-Kato* [42]), since in this case $\{z \in \mathbb{K} : |z| \leq 1\} \subset \Omega_0$ by Theorem 5.5. If we specialize this result for self-adjoint operators T and symmetric operators S, then we obtain the following important result.

Theorem 5.28 (Rellich-Kato). *If T is self-adjoint (essentially self-adjoint) on the Hilbert space H, the operator S is symmetric and T-bounded with*

T-bound less than 1, then $T+S$ is self-adjoint (essentially self-adjoint with $\overline{T+S}=\bar{T}+\bar{S}$ and $D(\overline{T+S})=D(\bar{T})$).

PROOF.

(a) If T is self-adjoint ($T=T^*$), then because of the inclusion $S\subset S^*$ the operator S^* is T^*-bounded with T^*-bound less than 1. In Theorem 5.27 we therefore have $\{z\in\mathbb{K}:|z|\leqslant 1\}\subset\Omega_0$, and thus $(T+S)^*=T^*+S^*=T+S$.

(b) Let T now be essentially self-adjoint, i.e., let $\bar{T}$ be self-adjoint. First we show that $\bar{S}$ is $\bar{T}$-bounded with $\bar{T}$-bound less than 1 (more precisely, equal to the T-bound of S). To this end, let $f\in D(\bar{T})$. Then there exists a sequence (f_n) from $D(T)$ such that $f_n\to f$, $Tf_n\to\bar{T}f$. From the T-boundedness of S it follows that (Sf_n) is a Cauchy sequence, therefore $f\in D(\bar{S})$, $Sf_n\to\bar{S}f$ and

$$\|\bar{S}f\|=\lim\|Sf_n\|\leqslant\lim(a\|f_n\|+b\|Tf_n\|)=a\|f\|+b\|\bar{T}f\|.$$

By part (a) the operator $\bar{T}+\bar{S}$ is therefore self-adjoint. From the inclusion $\overline{T+S}\subset\bar{T}+\bar{S}$ and from the closedness of $\bar{T}+\bar{S}$ it follows that $\overline{T+S}\subset\bar{T}+\bar{S}$. By Theorem 5.5 and by the equality $D(\bar{T}+\bar{S})=D(\bar{T})$ it follows that $D(\overline{T+S})=D(\bar{T}+\bar{S})$, and therefore $\overline{T+S}=\bar{T}+\bar{S}$. □

Theorem 5.29 (Wüst [58]). *Let T be a self-adjoint operator on the Hilbert space H, let S be symmetric and T-bounded, and let*

$$\Omega=\{z\in\mathbb{K}: T+zS \text{ and } T+z^*S \text{ are closed}\},$$
$$\Omega_0=\text{the connected component of } \Omega \text{ containing zero}.$$

For every $z\in\mathbb{R}\cap\Omega_0$ the operator $T+zS$ is self-adjoint. Special case: If $T+tS$ is closed for all $t\in[0,1]$, then $T+S$ is self-adjoint.

PROOF. We have $(T+zS)^*=T+zS^*=T+zS$ for every $z\in\mathbb{R}\cap\Omega_0$ by Theorem 5.27. □

Theorem 5.30 (Wüst [57]). *Let T be essentially self-adjoint on the complex Hilbert space H, let S be symmetric with $D(T)\subset D(S)$ and let $\|Sf\|\leqslant a\|f\|+\|Tf\|$ for all $f\in D(T)$ with some $a\geqslant 0$. Then $T+S$ is essentially self-adjoint.*

PROOF. Let $A=T+S$. We show that $R(\mu-A)^\perp=\{0\}$ for $\mu\in\{i,-i\}$. For this, let (t_n) be a sequence from $(0,1)$ such that $t_n\to 1$. By Theorem 5.28 the operator $A_n=T+t_nS$ is essentially self-adjoint for any $n\in\mathbb{N}$, and we have

$$\|(A-A_n)f\|=(1-t_n)\|Sf\|\leqslant a\|f\|+\|Tf\|-t_n\|Sf\|$$
$$\leqslant a\|f\|+\|(T+t_nS)f\|=a\|f\|+\|A_nf\|. \qquad (5.16)$$

Now let $h \in R(\mu - A)^{\perp}$. As A_n is essentially self-adjoint, there exists an $f_n \in D(A_n) = D(A)$ such that

$$\|(\mu - A_n) f_n - h\| \leqslant \frac{1}{n}, \quad n \in \mathbb{N}.$$

Therefore

$$h = \lim (\mu - A_n) f_n. \tag{5.17}$$

Because of the inequality $\|(\mu - A_n)^{-1}\| \leqslant 1$ we have $\|f_n\| \leqslant \|(\mu - A_n) f_n\|$. Consequently,

$$\limsup \|f_n\| \leqslant \|h\|.$$

By (5.17) it follows from this that

$$\limsup \|A_n f_n\| \leqslant 2\|h\|$$

and thus because of (5.16)

$$\limsup \|(A_n - A) f_n\| \leqslant \limsup \left[a\|f_n\| + \|A_n f_n\| \right] \leqslant c\|h\|,$$

with $c = a + 2$. As $D(T)$ is dense, for every $\epsilon > 0$ there exists an $h_\epsilon \in D(T)$ such that $\|h - h_\epsilon\| < \epsilon$. Because of the relation $h \in R(\mu - A)^{\perp}$, it follows that

$$\begin{aligned} \|h\|^2 &= \lim \langle h, (\mu - A_n) f_n \rangle = \lim \langle h, (A - A_n) f_n \rangle \\ &= \lim \left[\langle h - h_\epsilon, (A - A_n) f_n \rangle + \langle h_\epsilon, (A - A_n) f_n \rangle \right] \\ &\leqslant \|h - h_\epsilon\| \limsup \|(A - A_n) f_n\| + \limsup \|(A - A_n) h_\epsilon\| \|f_n\| \\ &\leqslant c\|h\|\epsilon + \limsup (1 - t_n) \|S h_\epsilon\| \|f_n\| = c\|h\|\epsilon. \end{aligned}$$

Since this holds true for all $\epsilon > 0$, we have $h = 0$. □

Exercises

5.31. Let T be a symmetric operator on the Hilbert space H.

(a) If H is complex, then T is essentially self-adjoint if and only if $\sigma_p(T^*) \subset \mathbb{R}$.

(b) If $H = N(s - T) + R(s - T)$ for some $s \in \mathbb{R}$, then T is essentially self-adjoint and we have $N(s - \bar{T}) = \overline{N(s - T)}$ and $R(s - \bar{T}) = \overline{R(s - T)}$

(c) If $s - T$ is continuously invertible and $\overline{R(s - T)} = H$ for some $s \in \mathbb{R}$, then T is essentially self-adjoint.

5.32. Let H be a real Hilbert space, and let T be an operator on H.

(a) The space $H_{\mathbb{C}} = H \times H$, with the *addition* $(f_1, g_1) + (f_2 g_2) = (f_1 + f_2, g_1 + g_2)$, *multiplication by a scalar* $(a + ib)(f, g) = (af - bg, ag + bf)$ and *scalar product* $\langle (f_1, g_1), (f_2, g_2) \rangle = \langle f_1, f_2 \rangle + i\langle f_1, g_2 \rangle - i\langle g_1, f_2 \rangle + \langle g_1, g_2 \rangle$, is a complex Hilbert space, the *complexification* of H.

(b) By setting $D(T_{\mathbb{C}}) = \{(f, g) \in H_{\mathbb{C}} : f, g \in D(T)\}$ and $T_{\mathbb{C}}(f_1, f_2) = (Tf_1, Tf_2)$ a linear operator $T_{\mathbb{C}}$ is defined on $H_{\mathbb{C}}$. We have:

(i) $T_{\mathbb{C}}$ is bounded (respectively belongs to $B(H_{\mathbb{C}})$) if and only if T is bounded (respectively belongs to $B(H)$). We have $\|T_{\mathbb{C}}\| = \|T\|$.

(ii) $D(T_C)$ is dense if and only if $D(T)$ is dense.
(iii) T_C is symmetric if and only if T is symmetric.
(iv) T_C is (essentially) self-adjoint if and only if T is (essentially) self-adjoint.

T_C is called the *complexification* of T.

5.33. Let T be a closed symmetric operator on the complex Hilbert space H, and let P_z denote the orthogonal projection onto $R(z-T)$ for $z \in \mathbb{C}\setminus\mathbb{R}$.
(a) The mapping $z \mapsto P_z$ is continuous on $\mathbb{C}\setminus\mathbb{R}$ with respect to the norm convergence in $B(H)$.
(b) If $R(z_0 - T) = H$ for some $z_0 \in \mathbb{C}$ such that $\operatorname{Im} z_0 > 0$ (respectively $\operatorname{Im} z_0 < 0$), then $R(z-T) = H$ for all $z \in \mathbb{C}$ such that $\operatorname{Im} z > 0$ (respectively $\operatorname{Im} z < 0$).
Hint: Use Theorem 5.25.

5.34. Let T be (essentially) self-adjoint on the Hilbert space H, and let S be symmetric such that $D(T) \subset D(S)$ and $\operatorname{Re}\langle Tf, Sf\rangle \geq -(a\|f\|^2 + b\|Tf\|\,\|Sf\|)$ for all $f \in D(T)$ with some $b < 1$. Then $T + tS$ is (essentially) self-adjoint for all $t \geq 0$.
Hint: For $t \geq 0$ we have $\|\bar{T}f\| + \|f\| \leq C_1(\|(\bar{T} + t\bar{S})f\| + \|f\|) \leq C_2(\|\bar{T}f\| + \|f\|)$.

5.35. (a) Let T be a self-adjoint operator on a complex Hilbert space, and let S be T-bounded with T-bound < 1. If $c > 0$ is large enough, then $\|Sf\| \leq b\|(\pm ic - T)f\|$ for all $f \in D(T)$ with some $b < 1$. Using this, prove the theorem of Rellich-Kato (Theorem 5.28) for complex Hilbert spaces.
(b) Using part (a) and Exercise 5.32, prove the theorem of Rellich-Kato for real Hilbert spaces.

5.4 Self-adjoint extensions of symmetric operators

If S is a symmetric operator, then $S \subset S^*$. For every symmetric extension T of S we have (cf. the proposition preceding Theorem 4.19) $S \subset T \subset T^* \subset S^*$. It seems plausible that there exists a sufficiently "large" extension T of S that is self-adjoint. For this we would have

$$S \subset T = T^* \subset S^*. \tag{5.18}$$

In this section we begin with some simple investigations concerning the problem of the existence of a self-adjoint extension of a symmetric operator (cf. also Sections 5.5 and 8.1 to 8.3).

Theorem 5.31.
(a) *If $T_1 \subset T_2$ are self-adjoint operators, then $T_1 = T_2$.*
(b) *If S is a symmetric operator and T_1 and T_2 are self-adjoint extensions of S such that $D(T_1) \subset D(T_2)$, then $T_1 = T_2$.*
(c) *If S is essentially self-adjoint, then $\bar{S}$ is the only self-adjoint extension of S.*

PROOF.
(a) It follows that $T_1 \subset T_2 = T_2^* \subset T_1^* = T_1$, thus $T_1 = T_2$.
(b) For all $f \in D(T_1) \subset D(T_2)$ and for all $g \in D(S) \subset D(T_1) \subset D(T_2)$ we have

$$\langle T_2 f, g\rangle = \langle f, T_2 g\rangle = \langle f, Sg\rangle = \langle f, T_1 g\rangle = \langle T_1 f, g\rangle.$$

Since $D(S)$ is dense, it follows that $T_2 f = T_1 f$ for all $f \in D(T_1)$, i.e., $T_1 \subset T_2$. From part (a) it follows that $T_1 = T_2$.
(c) If T is a self-adjoint extension of S, then we have $\bar{S} \subset T$, since T is closed. The equality $\bar{S} = T$ now follows from part (a). □

The following theorem ensures the existence of self-adjoint extensions for two large classes of symmetric operators. A symmetric operator S on the Hilbert space H is said to be *bounded from below* if there exists a $\gamma \in \mathbb{R}$ such that $\langle f, Sf\rangle \geq \gamma\|f\|^2$ for all $f \in D(S)$. Every γ of this kind is called a *lower bound* of S. The least upper bound of all lower bounds is also a lower bound. If 0 is a lower bound of S, then S is said to be *non-negative*. The concepts *bounded from above*, *upper bound*, and *non-positive* are defined similarly. If an operator is bounded from either below or above, then it is said to be *semi-bounded*. Besides the semi-bounded operators, the following simple theorem also treats the continuously invertible symmetric operators, i.e., those symmetric operators S, for which $\|Sf\| \geq \gamma\|f\|$ with some $\gamma > 0$ (cf. Exercise 5.14).

Theorem 5.32. *Let S be a symmetric operator on the (real or complex) Hilbert space H, and assume that $\langle f, Sf\rangle \geq \gamma\|f\|^2$ with some $\gamma \in \mathbb{R}$ (respectively $\|Sf\| \geq \gamma\|f\|$ with some $\gamma > 0$) for all $f \in D(S)$. Then for each $\kappa \in (-\infty, \gamma)$ (respectively $\kappa \in (-\gamma, \gamma)$) there exists a self-adjoint extension T_κ of S such that $\langle f, T_\kappa f\rangle \geq \kappa\|f\|^2$ (respectively $\|T_\kappa f\| \geq |\kappa|\,\|f\|$) for all $f \in D(T_\kappa)$. We have $N(\kappa - T_\kappa) = N(\kappa - S^*) = R(\kappa - T_\kappa)^\perp$.*

PROOF. The operator S is closable and $\bar{S}$ obviously satisfies the same assumptions, also. Therefore we can assume without loss of generality that S is closed. In the first case we have for all $\kappa \in (-\infty, \gamma)$ and $f \in D(S)$, $f \neq 0$ that

$$\begin{aligned}\|(S-\kappa)f\| &\geq |\langle (S-\kappa)f, f\rangle|\,\|f\|^{-1}\\ &\geq \langle Sf, f\rangle\,\|f\|^{-1} - \kappa\|f\| \geq (\gamma-\kappa)\|f\|.\end{aligned}$$

In the second case we have for all $\kappa \in (-\gamma, \gamma)$ and $f \in D(S)$ that

$$\|(S-\kappa)f\| \geq \|Sf\| - |\kappa|\,\|f\| \geq (\gamma - |\kappa|)\|f\|.$$

Consequently, in both cases $S - \kappa$ is continuously invertible. The range $R(S-\kappa) = D((S-\kappa)^{-1})$ is therefore closed. From this it follows that

$$R(S-\kappa) + N(S^* - \kappa) = R(S-\kappa) + R(S-\kappa)^\perp = H. \qquad (5.19)$$

Because of the equality $N(S^* - \kappa) \cap D(S) = N(S - \kappa) = \{0\}$, the sum $D(S) + N(S^* - \kappa)$ is a direct sum. Hence we can define

$$D(T_\kappa) = D(S) + N(S^* - \kappa),$$
$$T_\kappa(f_1 + f_2) = Sf_1 + \kappa f_2 \quad \text{for} \quad f_1 \in D(S), f_2 \in N(S^* - \kappa).$$

We obviously have $N(\kappa - T_\kappa) = N(\kappa - S^*)$. The operator T_κ is symmetric, because $D(T_\kappa)$ is dense (since $D(T_\kappa) \supset D(S)$) and for all $f_1, g_1 \in D(S)$, $f_2, g_2 \in N(S^* - \kappa) = R(S - \kappa)^\perp$ we have (observe that $(T_\kappa - \kappa)f_2 = (T_\kappa - \kappa)g_2 = 0$)

$$\begin{aligned} \langle f_1 + f_2, (T_\kappa - \kappa)(g_1 + g_2)\rangle &= \langle f_1 + f_2, (S - \kappa)g_1\rangle \\ &= \langle f_1, (S - \kappa)g_1\rangle = \langle (S - \kappa)f_1, g_1\rangle \\ &= \langle (S - \kappa)f_1, g_1 + g_2\rangle \\ &= \langle (T_\kappa - \kappa)(f_1 + f_2), g_1 + g_2\rangle. \end{aligned}$$

By Theorem 5.19 the operator T_κ is self-adjoint, since because of (5.19) we have $H = R(S - \kappa) + N(S^* - \kappa) = R(T_\kappa - \kappa) + N(T_\kappa - \kappa)$. Besides, for all $f_1 \in D(S)$, $f_2 \in N(S^* - \kappa)$ we have

$$\begin{aligned} \langle f_1 + f_2, T_\kappa(f_1 + f_2)\rangle &= \langle f_1, Sf_1\rangle + \langle S^* f_2, f_1\rangle + \kappa\left[\langle f_1, f_2\rangle + \|f_2\|^2\right] \\ &\geqslant \gamma\|f_1\|^2 + \kappa\left[\langle f_2, f_1\rangle + \langle f_1, f_2\rangle + \|f_2\|^2\right] \geqslant \kappa\|f_1 + f_2\|^2 \end{aligned}$$

in the first case, and

$$\begin{aligned} \|T_\kappa(f_1 + f_2)\|^2 &= \langle Sf_1 + \kappa f_2, Sf_1 + \kappa f_2\rangle \\ &= \|Sf_1\|^2 + \kappa\langle f_1, S^* f_2\rangle + \kappa\langle S^* f_2, f_1\rangle + \kappa^2\|f_2\|^2 \\ &\geqslant \gamma^2\|f_1\|^2 + \kappa^2\left[\langle f_1, f_2\rangle + \langle f_2, f_1\rangle + \|f_2\|^2\right] \geqslant \kappa^2\|f_1 + f_2\|^2 \end{aligned}$$

in the second case. □

In order to sharpen the results of Theorem 5.32, we first prove an extension theorem for bounded Hermitian operators.

Theorem 5.33. *If A is a bounded Hermitian operator on the Hilbert space H, then there exists a self-adjoint extension $B \in B(H)$ of A such that $\|B\| = \|A\|$. If $R(A)$ is dense, then every self-adjoint extension of A is injective.*

Proof. If $\|A\| = 0$, then $B = 0$ is the required extension. Therefore let $\|A\| \neq 0$. Without loss of generality we may assume that $\|A\| = 1$. Since along with A its closure $\bar{A}$ is also Hermitian and $\|A\| = \|\bar{A}\|$, we may also assume that A is closed, i.e., $D(A)$ is closed. Let P be the orthogonal projection onto $D(A)$. Then we have

$$A = A_1 + A_2 \quad \text{with} \quad A_1 = PA, \quad A_2 = (I - P)A.$$

We consider A_1 as an operator from H into the Hilbert space $D(A)$ with $D(A_1)=D(A)$ and A_2 as an operator from H into the Hilbert space $D(A)^\perp$ with $D(A_2)=D(A)$. First we show: There exist extensions $\tilde{A}_1$ and $\tilde{A}_2$ of A_1 and A_2, respectively, such that $D(\tilde{A}_1)=D(\tilde{A}_2)=H$, $R(\tilde{A}_1)\subset D(A)$, $R(\tilde{A}_2)\subset D(A)^\perp$ and

$$\|\tilde{A}_1 f\|^2+\|\tilde{A}_2 f\|^2 \leqslant \|f\|^2 \quad \text{for all} \quad f\in H.$$

We define the operator $\tilde{A}_1$ by

$$\tilde{A}_1 = (AP)^*.$$

Then we have $\|\tilde{A}_1\|=\|AP\|\leqslant\|A\|$ and $R(\tilde{A}_1)\subset N(AP)^\perp\subset D(A)$. Moreover, for all $f\in H$ and $g\in D(A)$ we have

$$\langle f,\tilde{A}_1 g\rangle = \langle APf, g\rangle = \langle Pf, Ag\rangle = \langle f, A_1 g\rangle.$$

Therefore $\tilde{A}_1 g=A_1 g$, and thus $A_1\subset\tilde{A}_1$. Because of the relation $\|\tilde{A}_1 f\|\leqslant\|\tilde{A}_1\|\,\|f\|\leqslant\|A\|\,\|f\|=\|f\|$ the equality

$$[f,g] = \langle f,g\rangle - \langle \tilde{A}_1 f, \tilde{A}_1 g\rangle$$

defines a semiscalar product on H. The set

$$N=\{f\in H: [f,f]=0\}$$

is a closed subspace of H. (If $f,g\in N$, $a\in\mathbb{K}$, then $af\in N$ and $[f+g,f+g]=2\,\mathrm{Re}\,[f,g]\leqslant 2\{[f,f][g,g]\}^{1/2}=0$, therefore $f+g\in N$. If (f_n) is a sequence from N such that $f_n\to f\in H$, then $[f,f]=\langle f,f\rangle-\langle\tilde{A}_1 f,\tilde{A}_1 f\rangle=\lim_{n\to\infty}\{\langle f_n,f_n\rangle-\langle\tilde{A}_1 f_n,\tilde{A}_1 f_n\rangle=0$, therefore $f\in N$). Let $H_0=N^\perp$ and let P_0 be the orthogonal projection onto H_0. By construction, we have $[f,f]\neq 0$ for all non-zero $f\in H_0$, i.e., $[.\,,.]$ is a scalar product on H_0. For every $f\in H$ we have

$$\begin{aligned}[P_0 f, P_0 f] &= [f-(I-P_0)f, f-(I-P_0)f] \\ &= [f,f]-2\,\mathrm{Re}\,[f,(I-P_0)f]+[(I-P_0)f,(I-P_0)f] = [f,f].\end{aligned}$$

For $f\in D(A)\cap N$ we have

$$\begin{aligned}\|A_2 f\|^2 &= \|(I-P)Af\|^2 = \|Af\|^2-\|PAf\|^2 \\ &= \|Af\|^2-\|A_1 f\|^2 \leqslant \|f\|^2-\|A_1 f\|^2 = [f,f] = 0.\end{aligned}$$

Therefore $A_2 f=A_2 g$ for $f-g\in D(A)\cap N$. Consequently, the equalities

$$\begin{aligned}D(\hat{A}_2) &= P_0 D(A),\\ \hat{A}_2 g &= A_2 f \quad \text{for} \quad g=P_0 f\in D(\hat{A}_2)\end{aligned}$$

define a linear operator from H_0 into $D(A)^\perp$, and for all $g = P_0 f \in D(\hat{A}_2)$ we have

$$\|\hat{A}_2 g\|^2 = \|A_2 f\|^2 = \langle (I-P)Af, Af\rangle = \langle Af, Af\rangle - \langle A_1 f, A_1 f\rangle$$
$$\leqslant \langle f, f\rangle - \langle A_1 f, A_1 f\rangle = [f, f] = [P_0 f, P_0 f] = [g, g].$$

From this it follows that $\hat{A}_2$ can be extended to an operator C from H_0 into $D(A)^\perp$ such that $D(C) = H_0$ and

$$\|Cf\|^2 \leqslant [f, f] \quad \text{for all} \quad f \in H_0$$

(cf. also Exercise 4.16). Let us now define $\tilde{A}_2$ by

$$\tilde{A}_2 = CP_0.$$

Then for all $f \in H$

$$\|\tilde{A}_2 f\|^2 = \|CP_0 f\|^2 \leqslant [P_0 f, P_0 f] = [f, f],$$

consequently

$$\|\tilde{A}_1 f\|^2 + \|\tilde{A}_2 f\|^2 \leqslant \|f\|^2.$$

Since for all $f \in D(A)$ we have

$$\tilde{A}_2 f = CP_0 f = A_2 P_0 f = A_2 f,$$

the operator $\tilde{A}_2$ is an extension of A_2. Therefore $\tilde{A} = \tilde{A}_1 + \tilde{A}_2$ is an extension of $A = A_1 + A_2$ and $\|\tilde{A}\| \leqslant 1$. Since for all $f \in D(A)$ and all $g \in H$ we have

$$\langle \tilde{A}^* f, g\rangle = \langle f, (\tilde{A}_1 + \tilde{A}_2) g\rangle = \langle f, \tilde{A}_1 g\rangle + \langle f, \tilde{A}_2 g\rangle$$
$$= \langle f, \tilde{A}_1 g\rangle = \langle f, (AP)^* g\rangle = \langle Af, g\rangle,$$

the operator $\tilde{A}^*$ is also an extension of A. Hence

$$B = \tfrac{1}{2}(\tilde{A} + \tilde{A}^*)$$

is a self-adjoint extension of A such that $\|B\| = 1$.

Now let $R(A)$ be dense and let B be a self-adjoint extension of A. Then $R(B)$ is also dense, consequently $N(B) = N(B^*) = R(B)^\perp = \{0\}$. □

Theorem 5.34. *Let S be a symmetric operator on the (real or complex) Hilbert space* H.

(a) *If $\|Sf\| \geqslant \gamma \|f\|$ for all $f \in D(S)$ with some $\gamma > 0$, then there exists a self-adjoint extension T of S such that $\|Tf\| \geqslant \gamma\|f\|$ for all $f \in D(T)$.*
(b) *If S is bounded from below, then there exists a self-adjoint extension T of S with the same lower bound (cf. also Theorem 5.38, Friedrichs' extension).*

PROOF.
(a) The operator $A = S^{-1}$ is Hermitian ($\langle ASf, Sg\rangle = \langle f, Sg\rangle = \langle Sf, g\rangle = \langle Sf, ASg\rangle$ for all $Sf, Sg \in D(A) = R(S)$) and bounded, $\|A\| \leqslant \gamma^{-1}$; A is

injective and $R(A) = D(S)$ is dense. Therefore by Theorem 5.33 there exists an injective self-adjoint extension B of A such that $\|B\| = \|A\| \leq \gamma^{-1}$. Then $T = B^{-1}$ is a self-adjoint extension of S and $\|Tf\| \geq \gamma\|f\|$ for all $f \in D(T)$.

(b) Without loss of generality we may assume that $\gamma = 0$. As in the proof of Theorem 5.32, we can show that $I + S$ is continuously invertible. Let us define A by

$$A = (I - S)(I + S)^{-1}.$$

Then A is Hermitian, because for all $f = (I + S)f_0, g = (I + S)g_0 \in D(A) = R(I + S)$ we have

$$\begin{aligned}\langle Af, g\rangle &= \langle (I - S)f_0, (I + S)g_0\rangle \\ &= \langle f_0, g_0\rangle - \langle Sf_0, g_0\rangle + \langle f_0, Sg_0\rangle - \langle Sf_0, Sg_0\rangle \\ &= \langle f_0, g_0\rangle - \langle f_0, Sg_0\rangle + \langle Sf_0, g\rangle - \langle Sf_0, Sg_0\rangle \\ &= \langle (I + S)f_0, (I - S)g_0\rangle = \langle f, Ag\rangle.\end{aligned}$$

It follows from the definition of A that

$$I - A = (I + S)(I + S)^{-1} - (I - S)(I + S)^{-1} = 2S(I + S)^{-1},$$
$$I + A = 2(I + S)^{-1}.$$

Consequently, $I + A$ is injective and

$$S = (I - A)(I + A)^{-1}.$$

A is bounded with norm $\|A\| \leq 1$, since for all $f \in D(A)$ we have

$$\begin{aligned}\|f\|^2 - \|Af\|^2 &= \langle (I - A)f, (I + A)f\rangle \\ &= \langle (I - A)(I + A)^{-1}(I + A)f, (I + A)f\rangle \\ &= \langle S(I + A)f, (I + A)f\rangle \geq 0.\end{aligned}$$

Therefore by Theorem 5.33 there exists a self-adjoint extension B of A such that $\|B\| = \|A\|$. By the same theorem $I + B$ is injective, since $R(I + A) = D(S)$ is dense. The operator

$$T = (I - B)(I + B)^{-1}$$

is therefore an extension of S. For all $f = (I + B)f_0$, $g = (I + B)g_0 \in D(T) = R(I + B)$ we have

$$\langle Tf, g\rangle = \langle (I - B)f_0, (I + B)g_0\rangle = \langle (I + B)f_0, (I - B)g_0\rangle = \langle f, Tg\rangle,$$
$$\langle f, Tf\rangle = \langle (I + B)f_0, (I - B)f_0\rangle = \|f_0\|^2 - \|Bf_0\|^2 \geq 0,$$

i.e., T is symmetric and bounded from below with lower bound 0. We have $I + T = 2(I + B)^{-1}$ and $I - T = 2B(I + B)^{-1}$, hence $B = (I - T)(I + T)^{-1}$. From this it follows that $R(I + T) = D(B) = H$, thus T is self-adjoint (cf. Theorem 5.19). □

Exercise

5.36. Let A be symmetric and semi-bounded.
(a) If A has only one semi-bounded self-adjoint extension, then A is essentially self-adjoint.
(b) If the lower bound of A is positive and A has only one positive extension, then A is essentially self-adjoint.

5.5 Operators defined by sesquilinear forms (Fiedrichs' extension)

In what follows H will always be a Hilbert space. A sesquilinear form s on H is said to be *bounded* if there exists a $C \geqslant 0$ such that $|s(f, g)| \leqslant C\|f\|\,\|g\|$ for all $f, g \in H$. The smallest such C is called the *norm* of s. It will be denoted by $\|s\|$. If $T \in B(H)$, then the equality $t(f, g) = \langle Tf, g\rangle$ defines a bounded sesquilinear form on H. We obviously have $\|t\| = \|T\|$. Conversely, every bounded sesquilinear form induces an operator on $B(H)$.

Theorem 5.35. *If t is a bounded sesquilinear form on H, then there exists exactly one $T \in B(H)$ such that $t(f, g) = \langle Tf, g\rangle$ for all $f, g \in H$. We then have $\|T\| = \|t\|$.*

Proof. For every $f \in H$ the function $g \mapsto t(f, g)$ is a continuous linear functional on H, since we have $|t(f, g)| \leqslant \|t\|\,\|f\|\,\|g\|$. Therefore for each $f \in H$ there exists exactly one $\tilde{f} \in H$ such that $t(f, g) = \langle \tilde{f}, g\rangle$. The mapping $f \mapsto \tilde{f}$ is obviously linear. Let us define T by the equality $Tf = \tilde{f}$ for all $f \in H$. By Theorem 4.3(b) the operator T is bounded with norm

$$\begin{aligned}\|T\| &= \sup\{|\langle Tf, g\rangle| : f, g \in H, \|f\| = \|g\| = 1\}\\ &= \sup\{|t(f, g)| : f, g \in H, \|f\| = \|g\| = 1\} = \|t\|.\end{aligned}$$

If T_1 and T_2 are from $B(H)$ and $\langle T_1 f, g\rangle = t(f, g) = \langle T_2 f, g\rangle$ for all $f, g \in H$, then it follows that $T_1 = T_2$, i.e., T is uniquely determined. □

For unbounded sesquilinear forms the situation is much more complicated. We consider only a special case.

Theorem 5.36. *Let $(H, \langle .\,, .\rangle)$ be a Hilbert space and let H_1 be a dense subspace of H. Assume that a scalar product $\langle .\,, .\rangle_1$ is defined on H_1 in such a way that $(H_1, \langle .\,, .\rangle_1)$ is a Hilbert space and with some $\kappa > 0$ we have $\kappa\|f\|^2 \leqslant \|f\|_1^2$ for all $f \in H_1$. Then there exists exactly one self-adjoint operator T on H such that*

$$D(T) \subset H_1 \text{ and } \langle Tf, g\rangle = \langle f, g\rangle_1 \text{ for } f \in D(T), g \in H_1. \quad (5.20)$$

T is bounded from below with lower bound κ. The operator T can be defined by the equalities

$$D(T) = \{ f \in H_1 : \text{there exists an } \hat{f} \in H \text{ such that } \langle f, g\rangle_1 = \langle \hat{f}, g\rangle \text{ for all } g \in H_1 \}, \tag{5.21}$$
$$Tf = \hat{f}.$$

$D(T)$ is dense in H_1 with respect to the norm $\|.\|_1$.

PROOF. *Existence*: The element $\hat{f}$ in (5.21) is uniquely determined, since H_1 is dense. Since the mapping $f \mapsto \hat{f}$ is also linear, (5.21) defines a linear operator. We can also consider T as an operator from $H_1 = (H_1, \langle .,. \rangle_1)$ into H. This operator will be denoted by T_0. If J denotes the operator from H into H_1 defined by

$$D(J) = H_1 \subset H, \quad Jf = f \quad \text{for} \quad f \in D(J),$$

then by (5.21) we obviously have

$$T_0 = J^*.$$

J is closed, since for any sequence (f_n) from $D(J) = H_1$ such that $f_n \to f$ [in H] and $f_n \to h$ [in H_1] we have $f_n \to h$ [in H] because of the inequality $\|.\|^2 \leqslant \kappa^{-1}\|.\|_1^2$, i.e., $f = h \in D(J)$. Therefore $T_0 = J^*$ is densely defined. Thus $D(T) = D(T_0)$ is also dense in H_1 (with respect to $\|.\|_1$) and, consequently, in H (with respect to $\|.\|$), as well. By (5.21) we have for all $f, g \in D(T)$ that

$$\langle Tf, g\rangle = \langle f, g\rangle_1 = [\langle g, f\rangle_1]^* = [\langle Tg, f\rangle]^* = \langle f, Tg\rangle,$$

i.e., T is symmetric. The self-adjointness of T will follow from Theorem 5.19 if we prove that $R(T) = H$. For this let $f \in H$ be arbitrary. Then $g \mapsto \langle f, g\rangle$ is a continuous linear functional on H_1, since we have

$$|\langle f, g\rangle| \leqslant \|f\| \, \|g\| \leqslant \kappa^{-1/2}\|f\| \, \|g\|_1.$$

Consequently, there exists an $\tilde{f} \in H_1$ such that

$$\langle f, g\rangle = \langle \tilde{f}, g\rangle_1 \quad \text{for all} \quad g \in H_1.$$

By (5.21) this means, however, that $\tilde{f} \in D(T)$ and $f = T\tilde{f}$. The semi-boundedness follows from the inequality

$$\langle Tf, f\rangle = \langle f, f\rangle_1 \geqslant \kappa\|f\|^2, \quad f \in D(T).$$

T obviously satisfies (5.20), as well.

Uniqueness: Every operator S that satisfies (5.20) is obviously a restriction of the operator T defined by (5.21). Since T is self-adjoint, it follows that $S \subset T \subset S^*$. If S is self-adjoint, then we necessarily have $S = T$. □

In what follows let D be a dense subspace of H and let s be a *semi-bounded* sesquilinear form on D, more precisely, let the inequality

$s(f,f) \geq \gamma\|f\|^2$ be satisfied by some $\gamma \in \mathbb{R}$ for all $f \in D$. Then the equality $\langle f, g\rangle_s = (1-\gamma)\langle f, g\rangle + s(f,g)$ defines a scalar product on D such that $\|f\|_s \geq \|f\|$ for all $f \in D$. Moreover, we assume that $\|\,.\,\|_s$ is *compatible* with $\|\,.\,\|$ in the following sense: If (f_n) is a $\|\,.\,\|_s$-Cauchy sequence from D and $\|f_n\| \to 0$, then we also have $\|f_n\|_s \to 0$ (cf. Exercise 5.37; in the theory of sesquilinear forms such a sesquilinear form s is said to be *closable*). Let H_s now be a $\|\,.\,\|_s$-completion of D (for example the one that was constructed in Section 4.3). It follows from the compatibility assumption that H_s may be considered as a subspace of H if the embedding of H_s into H is defined as follows: Let (f_n) be a $\|\,.\,\|_s$-Cauchy sequence in D. Then (f_n) is a Cauchy sequence in H. Let the element $\lim f_n$ from H correspond to the element $[(f_n)]$ of H_s. On the basis of the compatibility assumption, this correspondence is injective and the embedding is continuous with norm ≤ 1. The spaces H and H_s are related the same way as H and H_1 were in Theorem 5.36 (with $\kappa = 1$). Let

$$\bar{s}(f,g) = \langle f, g\rangle_s - (1-\gamma)\langle f, g\rangle \quad \text{for } f, g \in H_s.$$

Therefore $\bar{s}(f,g) = s(f,g)$ for $f, g \in D$. The sesquilinear form $\bar{s}$ is called the *closure* of s.

Theorem 5.37. *Assume that H is a Hilbert space, D is a dense subspace of H and s is a semi-bounded symmetric sesquilinear form on D with lower bound γ. Let $\|\,.\,\|_s$ be compatible with $\|\,.\,\|$. There exists exactly one semi-bounded self-adjoint operator T with lower bound γ such that*

$$D(T) \subset H_s \text{ and } \langle Tf, g\rangle = s(f,g) \text{ for all } f \in D \cap D(T), g \in D. \quad (5.22)$$

We have

$$D(T) = \{f \in H_s : \text{there exists an } \hat{f} \in H \text{ such that } s(f,g) = \langle \hat{f}, g\rangle \text{ for all } g \in D\}, \quad (5.23)$$
$$Tf = \hat{f} \quad \text{for} \quad f \in D(T).$$

Proof. If we replace $(H_1, \langle\,.\,,\,.\,\rangle_1)$ by $(H_s, \langle\,.\,,\,.\,\rangle_s)$ in Theorem 5.36, then we obtain exactly one self-adjoint operator T_0 such that $D(T_0) \subset H_s$ and

$$\langle T_0 f, g\rangle = \langle f, g\rangle_s \quad \text{for all} \quad f \in D(T_0), g \in H_s.$$

T_0 is semi-bounded with lower bound 1. The operator $T = T_0 - (1-\gamma)$ obviously possesses the required properties. The uniqueness follows from the uniqueness of T_0. Formula (5.22) implies (5.23), since D is dense (in H_s and in H). □

If S is a semi-bounded symmetric operator with lower bound γ, then the equality

$$s(f,g) = \langle Sf, g\rangle, \quad f, g \in D(S)$$

defines a semi-bounded sesquilinear form s on $D(S)$ with lower bound γ. In this case

$$\langle f, g\rangle_s = \langle Sf, g\rangle + (1-\gamma)\langle f, g\rangle \text{ and } \|f\|_s^2 = \langle Sf, f\rangle + (1-\gamma)\|f\|^2$$

for $f, g \in D(S)$. The norm $\|\,.\,\|_s$ is compatible with $\|\,.\,\|$: Let (f_n) be a $\|\,.\,\|_s$-Cauchy sequence from $D(S)$ such that $f_n \to 0$. Then for all $n, m \in \mathbb{N}$ we have

$$\begin{aligned}\|f_n\|_s^2 &= \langle f_n, f_n\rangle_s = |\langle f_n, f_n - f_m\rangle_s + \langle f_n, f_m\rangle_s| \\ &\leq \|f_n\|_s \|f_n - f_m\|_s + \|(S+1-\gamma)f_n\|\,\|f_m\|.\end{aligned}$$

The sequence $(\|f_n\|_s)$ is bounded, $\|f_n - f_m\|_s$ is small for large n and m and for any fixed n we have $\|(S+1-\gamma)f_n\|\,\|f_m\| \to 0$ as $m \to \infty$. Consequently it follows that $\|f_n\|_s \to 0$ as $n \to \infty$. This fact makes the following construction of a self-adjoint extension (*Friedrichs' extension*) of a semi-bounded symmetric operator possible, where the lower (upper) bound remains unchanged.

Theorem 5.38. *Let S be a semi-bounded symmetric operator with lower bound γ. Then there exists a semi-bounded self-adjoint extension of S with lower bound γ. If we define $s(f, g) = \langle Sf, g\rangle$ for $f, g \in D(S)$, and H_s as above, then we have: The operator T defined by*

$$D(T) = D(S^*) \cap H_s \quad \text{and} \quad Tf = S^*f \quad \text{for} \quad f \in D(T)$$

is a self-adjoint extension of S with lower bound γ. The operator T is the only self-adjoint extension of S having the property $D(T) \subset H_s$.

PROOF. By Theorem 5.37 there exists exactly one self-adjoint operator T with $D(T) \subset H_s$ and

$$\langle Tf, g\rangle = s(f, g) = \langle Sf, g\rangle \quad \text{for} \quad f \in D(S) \cap D(T), g \in D(S).$$

γ is a lower bound for T. By (5.23) we have

$$\begin{aligned} D(T) = \{f \in H_s : \text{There exists an } \hat{f} \in H \text{ with } \bar{s}(f, g) \\ = \langle \hat{f}, g\rangle \quad \text{for all} \quad g \in D(S)\} \end{aligned} \tag{5.24}$$

$$Tf = \hat{f} \quad \text{for} \quad f \in D(T)$$

We can replace $\bar{s}(f, g)$ by $\langle f, Sg\rangle$ in (5.24): If we choose a sequence (f_n) from $D(S)$ such that $\|f_n - f\|_s \to 0$, then we obtain

$$\bar{s}(f, g) = \lim \bar{s}(f_n, g) = \lim \langle f_n, Sg\rangle = \langle f, Sg\rangle.$$

Consequently, it follows that $D(T) = D(S^*) \cap H_s$ and $T = S^*|_{D(T)}$. Because of the inclusions $S \subset S^*$ and $D(S) \subset H_s$ it follows from this that T is an extension of S. Let A be an arbitrary self-adjoint extension of S such that $D(A) \subset H_s$. Then $A \subset S^*$ and $D(T) = D(S^*) \cap H_s$ imply that $A \subset T$, consequently $A = T$. □

Our arguments so far enable us to study the operator product A^*A, as well. If $A\in B(H_1, H_2)$, then we already know from Exercise 4.19 that A^*A is self-adjoint.

Theorem 5.39. *Let $(H_1, \langle . , . \rangle_1)$ and $(H_2, \langle . , . \rangle_2)$ be Hilbert spaces and let A be a densely defined closed operator from H_1 into H_2. Then A^*A is a self-adjoint operator on H_1 with lower bound 0 (A^*A is non-negative). $D(A^*A)$ is a core of A. We have $N(A^*A)=N(A)$.*

Proof. As A is closed, $D(A)$ is a Hilbert space with the scalar product $\langle f,g\rangle_A=\langle Af,Ag\rangle_2+\langle f,g\rangle_1$, and $\|f\|_A \geqslant \|f\|_1$ for all $f\in D(A)$. Therefore by Theorem 5.36 there exists a self-adjoint operator T with lower bound 1 for which

$$D(T) = \{f\in D(A) : \text{there exists an } \hat{f}\in H_1 \text{ such that } \langle f,g\rangle_A=\langle \hat{f},g\rangle_1 \quad \text{for all} \quad g\in D(A)\},$$

$$Tf = \hat{f} \quad \text{for} \quad f\in D(T).$$

On account of the equality $\langle f,g\rangle_A=\langle Af,Ag\rangle_2+\langle f,g\rangle_1$, this definition says that $f\in D(T)$ if and only if $Af\in D(A^*)$ (i.e., $f\in D(A^*A)$) and $Tf=\hat{f}=A^*Af+f$. Hence it follows that $T=A^*A+1$, $A^*A=T-1$, i.e., A^*A is self-adjoint and non-negative. From Theorem 5.36 it follows that $D(A^*A)$ is dense in $D(A)$ with respect to $\|\,.\,\|_A$, i.e., $D(A^*A)$ is a core of A. If $f\in N(A)$, then $Af=0\in D(A^*)$ and $A^*Af=0$. Therefore $N(A)\subset N(A^*A)$. If $f\in N(A^*A)$, then $\|Af\|^2=\langle A^*Af,f\rangle=0$. Therefore $N(A^*A)\subset N(A)$, and thus $N(A^*A)=N(A)$. □

Theorem 5.40. *Let A_1 and A_2 be densely defined closed operators from H into H_1 and from H into H_2, respectively. Then $A_1^*A_1=A_2^*A_2$ if and only if $D(A_1)=D(A_2)$ and $\|A_1f\|=\|A_2f\|$ for all $f\in D(A_1)=D(A_2)$.*

Proof. Assume that $D(A_1)=D(A_2)$ and $\|A_1f\|=\|A_2f\|$ for all $f\in D(A_1)$. It follows from (1.4) in the complex case and from (1.8) in the real case that

$$\langle A_1f, A_1g\rangle = \langle A_2f, A_2g\rangle \quad \text{for all} \quad f,g\in D(A_1)=D(A_2).$$

Then the construction of Theorem 5.39 provides the same operator for $A=A_1$ and $A=A_2$, consequently $A_1^*A_1=A_2^*A_2$. If this equality holds, then for all $f\in D(A_1^*A_1)=D(A_2^*A_2)$ we have

$$\|A_1f\|^2=\langle A_1^*A_1f,f\rangle=\langle A_2^*A_2f,f\rangle=\|A_2f\|^2$$

(here we have used the inclusions $D(A_1^*A_1)\subset D(A_1)$ and $D(A_2^*A_2)\subset D(A_2)$). By Theorem 5.39 the subspace $D(A_1^*A_1)=D(A_2^*A_2)$ is a core of A_1 and A_2. As the A_1-norm and the A_2-norm coincide on $D(A_1^*A_1)=D(A_2^*A_2)$, it follows finally that $D(A_1)=D(A_2)$ and $\|A_1f\|=\|A_2f\|$ for all $f\in D(A_1)=D(A_2)$. □

EXERCISE

5.37. Let $H = L_2(0, 1)$. Then $D = C^1[0, 1]$ (the space of continuously differentiable functions on $[0, 1]$) is a dense subspace of H. The equality $s(f, g) = \langle f, g\rangle + f'(\frac{1}{2})^* g'(\frac{1}{2})$ defines a sesquilinear form on D such that $\|f\|_s \geq \|f\|$. The norm $\|\,.\,\|_s$ is not compatible with $\|\,.\,\|$.

5.6 Normal operators

A densely defined operator T on a Hilbert space H is said to be *normal* if

$$D(T) = D(T^*) \quad \text{and} \quad \|Tf\| = \|T^*f\| \quad \text{for all} \quad f \in D(T).$$

Every self-adjoint operator is obviously normal.

EXAMPLE 1. Let $M \subset \mathbb{R}^m$ be measurable and let $t : M \to \mathbb{C}$ be a measurable function. The maximal multiplication operator T induced by t (cf. Section 4.1, Example 1 and Section 5.1, Example 2) is normal. By (5.1) we have $D(T) = D(T^*)$ and for $f \in D(T)$ we obviously have

$$\|Tf\|^2 = \int_M |t(x)f(x)|^2 \, dx = \int_M |t(x)^* f(x)|^2 \, dx = \|T^*f\|^2.$$

EXAMPLE 2. According to Section 4.6 an isomorphism U of a Hilbert space H onto itself is called a unitary operator on H. We have $D(U) = D(U^*) = H$ and $\|Uf\| = \|U^*f\| = \|f\|$ for all $f \in H$, i.e., every unitary operator on a Hilbert space H is normal.

Proposition.
(1) *Every normal operator T is closed and maximal normal (i.e., for every normal operator N the inclusion $T \subset N$ implies $T = N$).*
(2) *Let T be densely defined and closed. Then the following assertions are equivalent:*
  (i) *T is normal,*
  (ii) *T^* is normal,*
  (iii) *$T^*T = TT^*$.*
(3) *If T is normal, then $z + T$ is also normal for every $z \in \mathbb{K}$.*

PROOF.
(1) The T-norm and the T^*-norm coincide on $D(T)$. Since T^* is closed, that T is closed follows from Theorem 5.1. If N is normal and $T \subset N$, then $D(T) \subset D(N) = D(N^*) \subset D(T^*) = D(T)$. Therefore $D(T) = D(N)$, and thus $T = N$.
(2) The equivalences (i)$\Leftrightarrow$(iii) and (ii)$\Leftrightarrow$(iii) immediately follow from Theorem 5.40 with $A_1 = T$, $A_2 = T^*$ and from Proposition (1).

(3) We have $D((z+T)^*) = D(z^* + T^*) = D(T^*) = D(T) = D(z+T)$ and for all $f \in D(T)$

$$\begin{aligned} \|(z+T)f\|^2 &= |z|^2\|f\|^2 + 2\,\mathrm{Re}\,\langle zf, Tf\rangle + \|Tf\|^2 \\ &= |z^*|^2\|f\|^2 + 2\,\mathrm{Re}\,\langle z^*f, T^*f\rangle + \|T^*f\|^2 \\ &= \|(z^*+T^*)f\|^2. \end{aligned}$$

□

Theorem 5.41. *Let T be a normal operator.*

(a) *For every $z \in \mathbb{K}$ we have $N(z-T) = N(z^* - T^*)$.*

(b) *If z_1, z_2 are distinct eigenvalues of T and f_1, f_2 are corresponding eigenvectors, then $f_1 \perp f_2$.*

PROOF.

(a) The statement is evident, as $z - T$ is normal.

(b) By part (a) we have

$$\begin{aligned} (z_1 - z_2)\langle f_1, f_2\rangle &= \langle z_1^* f_1, f_2\rangle - \langle f_1, z_2 f_2\rangle \\ &= \langle T^* f_1, f_2\rangle - \langle f_1, Tf_2\rangle = 0. \end{aligned}$$

Consequently $\langle f_1, f_2\rangle = 0$. □

Theorem 5.42. *Let T be normal and injective. Then we have:*

(i) *$R(T)$ is dense,*

(ii) *T^* is injective,*

(iii) *T^{-1} is normal,*

(iv) *$R(T) = R(T^*)$.*

PROOF. $R(T)$ is dense because of the equalities $R(T)^\perp = N(T^*) = N(T) = \{0\}$. Consequently, T^* is injective, and we have $(T^*)^{-1} = (T^{-1})^*$. Therefore it follows that

$$\begin{aligned} (T^{-1})^* T^{-1} &= (T^*)^{-1} T^{-1} = (TT^*)^{-1} = (T^*T)^{-1} \\ &= T^{-1}(T^*)^{-1} = T^{-1}(T^{-1})^*. \end{aligned}$$

i.e., T^{-1} is normal. It also follows that

$$R(T) = D(T^{-1}) = D((T^{-1})^*) = D((T^*)^{-1}) = R(T^*).$$

Corollary. *If T is a normal operator, then $R(z, T) = (z-T)^{-1}$ is normal for all $z \notin \sigma_p(T)$. This holds in particular for a self-adjoint T.*

Theorem 5.43. *If T is normal, then*

$$\begin{aligned} \rho(T) &= \{z \in \mathbb{K} : (z-T) \text{ is continuously invertible}\} \\ &= \{z \in \mathbb{K} : R(z-T) = H\}, \\ \sigma_p(T) &= \{z \in \mathbb{K} : \overline{R(z-T)} \neq H\}. \end{aligned}$$

(Compare with Theorems 5.23 and 5.24 for self-adjoint operators.)

PROOF. If $z \in \rho(T)$, then $(z-T)^{-1} \in B(H)$ (in particular, $z-T$ is continuously invertible). If $z-T$ is continuously invertible, then (as T is closed)

$$R(z-T) = D((z-T)^{-1}) = \overline{D((z-T)^{-1})} = \overline{R(z-T)},$$

and

$$R(z-T)^{\perp} = N(z^*-T^*) = N(z-T) = \{0\}.$$

Consequently $R(z-T) = H$. If $R(z-T) = H$, then

$$N(z-T) = N(z^*-T^*) = R(z-T)^{\perp} = \{0\},$$

i.e., $z-T$ is injective. As T is closed and $R(z-T) = H$, it follows that $(z-T)^{-1} \in B(H)$, i.e., $z \in \rho(T)$.

If $z \in \sigma_p(T)$, then $R(z-T)^{\perp} = N(z^*-T^*) = N(z-T) \neq \{0\}$. Therefore $\overline{R(z-T)} \neq H$. If $\overline{R(z-T)} \neq H$, then $N(z-T) = R(z-T)^{\perp} \neq \{0\}$. Therefore $z \in \sigma_p(T)$. □

Theorem 5.44. *If $T \in B(H)$ is normal, then the spectral radius $r(T)$ equals $\|T\|$.*

PROOF. By Theorem 4.46 we have for all $T \in B(H)$

$$\begin{aligned}\|T^*T\| &= \sup\{|\langle T^*Tf, f\rangle| : f \in H, \|f\| \leq 1\}\\ &= \sup\{\|Tf\|^2 : f \in H, \|f\| \leq 1\} = \|T\|^2.\end{aligned}$$

In particular, $\|T^2\| = \|T\|^2$ for a self-adjoint T. We obtain by induction that for a self-adjoint T

$$\|T^{2^n}\| = \|T\|^{2^n} \quad \text{for all} \quad n \in \mathbb{N}.$$

Let T now be normal. Since $r(T) \leq \|T\|$ always holds, we only have to prove that $r(T) \geq \|T\|$. Because of the equality $r(T) = r(T^*)$ and the self-adjointness of TT^* we have

$$\begin{aligned}r(T)^2 &= r(T)r(T^*) = \lim_{n\to\infty} \{\|T^{2^n}\|\,\|(T^*)^{2^n}\|\}^{1/2^n}\\ &\geq \lim_{n\to\infty} \|T^{2^n}(T^*)^{2^n}\|^{1/2^n} = \lim_{n\to\infty} \|(TT^*)^{2^n}\|^{1/2^n}\\ &= \|TT^*\| = \|T\|^2.\end{aligned}$$

□

EXERCISES

5.38. If $T \in B(H)$, then there exist uniquely determined self-adjoint operators $T_1, T_2 \in B(H)$ such that $T = T_1 + iT_2$. We have $T_1 = \frac{1}{2}(T+T^*)$, $T_2 = (1/2i)(T-T^*)$. T_1 is called the *real part* of T and T_2 is called the *imaginary part* of T. The operator T is normal if and only if T_1 and T_2 *commute*, i.e., if and only if $T_1T_2 = T_2T_1$.

5.39. Let T be an operator on the Hilbert space H and let M be a closed subspace of H. The subspace M is said to be *invariant under* T if $T(M \cap D(T)) \subset M$. If M and $M^{\perp}$ are invariant under T and $D(T) = [M \cap D(T)] + [M^{\perp} \cap D(T)]$, then M is called a *reducing subspace* of T.

(a) If M is a reducing subspace, then $M^{\perp}$ is a reducing subspace, also.

(b) If $T \in B(H)$, then M is a reducing subspace of T if and only if M is invariant under T and T^*.
(c) M is a reducing subspace of T if and only if $TP \supset PT$ for the orthogonal projection P onto M.
(d) If T is densely defined, $D(T) = D(T^*)$, and M is a reducing subspace of T, then M is a reducing subspace of T^*.

5.40. Let T be a normal operator on H.
(a) If M is a reducing subspace of T, then M is also a reducing subspace of T^*.
(b) We have $R(T) = R(T^*)$. The restriction of T onto $N(T)^\perp \cap D(T)$ is an injective normal operator on $N(T)^\perp$.
Hint: Theorem 5.42.

5.41. Let T be a bounded normal operator on H.
(a) If $\lambda \in \sigma(T)$, then there is a sequence (f_n) from H such that $\|f_n\| = 1$ and $(\lambda - T)f_n \to 0$.
(b) In complex case we have $\|T\| = \sup\{|\langle f\, Tf\rangle| : f \in D(T),\ \|f\| = 1\}$.

5.42. If T_n and T are bounded normal operators such that $T_n \overset{s}{\to} T$, then we also have $T_n^* \overset{s}{\to} T^*$.
Hint: $T_n^* \overset{w}{\to} T^*$ and $\|T_n^* f\| \to \|T^* f\|$ for all f.

5.43. Assume that $H = \oplus_{n \in \mathbb{N}} H_n$, P_n is the orthogonal projection onto H_n, and T_n is an operator on H_n. The *orthogonal sum* $T = \oplus_{n \in \mathbb{N}} T_n$ of the operators T_n is defined by

$$D(T) = \left\{ f \in H : P_n f \in D(T_n),\ \sum_{n \in \mathbb{N}} \|T_n P_n f\|^2 < \infty \right\},$$

$$Tf = \sum_{n \in \mathbb{N}} T_n P_n f \quad \text{for} \quad f \in D(T).$$

(a) If all T_n are self-adjoint (normal, closed), then T is also self-adjoint (normal, closed).
(b) If D_n is a core of $T_n (n \in \mathbb{N})$, then $L\{D_n : n \in \mathbb{N}\}$ is a core of T.
(c) If each T_n is bounded and $\sup\{\|T_n\| : n \in \mathbb{N}\} < \infty$, then T is also bounded and $\|T\| = \sup\{\|T_n\| : n \in \mathbb{N}\}$.
(d) If all T_n are non-negative, then T is also non-negative. The Friedrichs extension of T is the orthogonal sum of the Friedrichs extensions of T_n.

5.44. (a) Let A be a bounded self-adjoint operator and P an orthogonal projection in a Hilbert space H. Then PAP is self-adjoint. (See also exercise 6.13 for unbounded A.)
(b) A corresponding result for normal operators does not hold in general. Counterexample is

$$H = \mathbb{C}^3 : N = \begin{pmatrix} 0 & 0 & 1 \\ 1 & 0 & 0 \\ 0 & 1 & 0 \end{pmatrix}, P = \begin{pmatrix} 0 & 0 & 0 \\ 0 & 1 & 0 \\ 0 & 0 & 1 \end{pmatrix};$$

N is unitary with $Ne_1 = e_2$, $Ne_2 = e_3$ and $Ne_3 = e_1$; P is the projection onto $L(e_2, e_3)$. For dim $H = 2$ such an example cannot exist.
(c) If N is normal and P is an orthogonal projection with $PN \subset NP$, then PNP is also normal.

Special classes of linear operators 6

6.1 Finite rank and compact operators

Let H_1 and H_2 be Hilbert spaces. An operator T from H_1 into H_2 is said to be of *finite rank* (of *rank m*) if $R(T)$ is finite-dimensional (m-dimensional).

Theorem 6.1. *Let T be an operator from H_1 into H_2 such that $D(T) = H_1$. The operator T is a bounded operator of rank m if and only if there are linearly independent elements $f_1, \ldots, f_m$ from H_1 and linearly independent elements $g_1, \ldots, g_m$ from H_2 such that*

$$Tf = \sum_{j=1}^{m} \langle f_j, f\rangle g_j \quad \text{for all} \quad f \in H_1. \tag{6.1}$$

Then

$$T^*g = \sum_{j=1}^{m} \langle g_j, g\rangle f_j \quad \text{for all} \quad g \in H_2, \tag{6.2}$$

and $\|T\| \leq \sum_{j=1}^{m} \|f_j\| \, \|g_j\|$. The operator T is of rank m if and only if T^* is of rank m. There is no loss of generality in assuming that either $\{f_1, \ldots, f_m\}$ or $\{g_1, \ldots, g_m\}$ is an orthonormal system.[1]

PROOF. If T has the form (6.1), then $R(T) \subset L(g_1, \ldots, g_m)$. For every $j_0 \in \{1, \ldots, m\}$ there exists an $h_{j_0} \in L(f_1, \ldots, f_m)$ such that $h_{j_0} \neq 0$, and $h_{j_0} \perp f_j$ for $j \neq j_0$. Since $\langle f_{j_0}, h_{j_0}\rangle \neq 0$ and $Th_{j_0} = \langle f_{j_0}, h_{j_0}\rangle g_{j_0}$, it follows that all g_j are contained in $R(T)$. This implies $R(T) = L(g_1, \ldots, g_m)$; hence $R(T)$

[1] From Theorem 7.6 it follows that both $\{f_1, \ldots, f_m\}$ and $\{g_1, \ldots, g_m\}$ may be chosen to be orthonormal if suitable scalar factors are added, i.e., $Tf = \sum_{j=1}^{m} s_j \langle f_j, f\rangle g_j$.

is m-dimensional. Because of the inequality

$$\|Tf\| \leq \sum_{j=1}^{m} |\langle f_j, f\rangle| \, \|g_j\| \leq \|f\| \sum_{j=1}^{m} \|f_j\| \, \|g_j\|,$$

T is bounded and $\|T\| \leq \sum_{j=1}^{m} \|f_j\| \, \|g_j\|$.

Assume that T is bounded, $\dim R(T) = m$, $\{g_1, \ldots, g_m\}$ is an ONB of $R(T)$, and $f_j = T^* g_j$ for $j = 1, \ldots, m$. Then for all $f \in H_1$ we have

$$Tf = \sum_{j=1}^{m} \langle g_j, Tf\rangle g_j = \sum_{j=1}^{m} \langle f_j, f\rangle g_j.$$

It remains to show that the elements $f_1, \ldots, f_m$ are linearly independent. Let us assume that this does not hold. There is no loss of generality in assuming that $f_1 = \sum_{j=2}^{m} a_j f_j$. Then

$$Tf = \sum_{j=1}^{m} \langle f_j, f\rangle g_j = \sum_{j=2}^{m} \langle f_j, f\rangle (a_j^* g_1 + g_j), \quad f \in H_1.$$

It would follow from this that $R(T)$ is at most $(m-1)$-dimensional, which contradicts the assumption. If T has the form (6.1), then for all $f \in H_1$, $g \in H_2$

$$\langle g, Tf\rangle = \sum_{j=1}^{m} \langle f_j, f\rangle \langle g, g_j\rangle = \Big\langle \sum_{j=1}^{m} \langle g_j, g\rangle f_j, f\Big\rangle.$$

Hence (6.2) holds for T^*. If T is of rank m, then (6.2) implies that T^* is of rank m, too. The opposite direction follows the same way. Our construction shows that $\{g_1, \ldots, g_m\}$ can be chosen to be an ONS. If the same reasoning is applied to T^*, then we obtain that $\{f_1, \ldots, f_m\}$ can be chosen to be an ONS. □

EXAMPLE 1. If $\{f_1, \ldots, f_m\}$ and $\{g_1, \ldots, g_m\}$ are orthonormal systems in H_1 and H_2, respectively, and $\mu_1, \ldots, \mu_m \in \mathbb{K}$, then

$$Tf = \sum_{j=1}^{m} \mu_j \langle f_j, f\rangle g_j, \quad f \in H_1$$

defines an operator $T \in B(H_1, H_2)$ of rank m with $\|T\| = \max\{|\mu_j| : j = 1, \ldots, m\}$. We only have to prove the last statement. For all $f \in H$

$$\|Tf\|^2 = \sum_{j=1}^{m} |\mu_j|^2 |\langle f_j, f\rangle|^2 \leq \|f\|^2 \max\left\{|\mu_j|^2 : j = 1, \ldots, m\right\}.$$

Therefore $\|T\| \leq \max\{|\mu_j| : j = 1, \ldots, m\}$. If $j_0 \in \{1, \ldots, m\}$ is chosen such that $|\mu_{j_0}| = \max\{|\mu_j| : j = 1, \ldots, m\}$, then $\|Tf_{j_0}\| = |\mu_{j_0}|$; thus $\|T\| \geq |\mu_{j_0}|$.

Let H_1 and H_2 be Hilbert spaces. An operator T from H_1 into H_2 is said to be *compact* if every bounded sequence (f_n) from $D(T)$ contains a subsequence (f_{n_k}) for which (Tf_{n_k}) is convergent (i.e., T maps bounded sets onto relatively compact sets; cf. Exercise 6.1).

Theorem 6.2. *Every compact operator is bounded. If T is compact, then $\overline{T}$ is also compact.*

PROOF. Assume that T is not bounded. Then there exists a sequence (f_n) from $D(T)$ with the properties $\|f_n\| \leq 1$ and $\|Tf_n\| \geq n$ for all $n \in \mathbb{N}$. Therefore, no subsequence (f_{n_k}) has the property that (Tf_{n_k}) is convergent; thus T is not compact. Let T be compact. If (f_n) is a bounded sequence from $D(\overline{T}) = \overline{D(T)}$, then there exists a sequence (g_n) from $D(T)$ such that $\|g_n - f_n\| \leq n^{-1}$. Since T is compact, there exists a subsequence (g_{n_k}) of (g_n), for which (Tg_{n_k}) is convergent. Then $(\overline{T}f_{n_k})$ is also convergent because of the inequality $\|\overline{T}f_{n_k} - Tg_{n_k}\| \leq n^{-1}\|T\|$. □

On the basis of Theorem 6.2, together with Exercise 4.7, there is no loss of generality in assuming that compact operators from H_1 into H_2 always belong to $B(H_1, H_2)$.

Theorem 6.3. *Let H_1 and H_2 be Hilbert spaces. An operator $T \in B(H_1, H_2)$ is compact if and only if $Tf_n \to 0$ for every weak null-sequence (f_n) from H_1.*

PROOF. Let T be compact. It is sufficient to prove that every weak null-sequence (f_n) from H_1 has a subsequence (f_{n_k}) such that $Tf_{n_k} \to 0$. Let (f_n) be a weak null-sequence from H_1. As T is compact, there exists a subsequence (f_{n_k}) such that (Tf_{n_k}) is convergent; say $g = \lim Tf_{n_k}$. As $f_{n_k} \xrightarrow{w} 0$, by Theorem 4.27 (ii) and Theorem 6.2 it follows that $Tf_{n_k} \xrightarrow{w} 0$, and thus $g = \lim Tf_{n_k} = w - \lim Tf_{n_k} = 0$.

Let T now send every weak null-sequence from H_1 to a null-sequence. Consider a bounded sequence (f_n) from H_1. By Theorem 4.25 there exists a weakly convergent subsequence (f_{n_k}) of (f_n); say $f_{n_k} \xrightarrow{w} f$. Then $f_{n_k} - f \xrightarrow{w} 0$ and $T(f_{n_k} - f) \to 0$, by assumption. Therefore (Tf_{n_k}) is convergent. Consequently, T is compact. □

Proposition. *Let H be an infinite dimensional Hilbert space. If $T \in B(H)$ is compact, then $0 \in \sigma(T)$.*

PROOF. If (e_n) is an arbitrary orthornormal sequence in H, then $Te_n \to 0$. Consequently, T is not continuously invertible. □

Theorem 6.4. *Let H_1, H_2 and H_3 be Hilbert spaces.*

(a) *If $S \in B(H_2, H_3)$ and $T \in B(H_1, H_2)$, and one of these operators is compact, then ST is compact.*
(b) *If $T_1, T_2 \in B(H_1, H_2)$ are compact and $a, b \in \mathbb{K}$, then $aT_1 + bT_2$ is compact.*
(c) *$T \in B(H_1, H_2)$ is compact if and only if T^*T is compact.*
(d) *$T \in B(H_1, H_2)$ is compact if and only if T^* is compact.*

(e) *If (T_n) is a sequence of compact operators from $B(H_1, H_2)$ and $\|T_n - T\| \to 0$ for some $T \in B(H_1, H_2)$, then T is compact.*

PROOF.
(a) First let S be compact. If (f_n) is a weak null-sequence in H_1, then by Theorem 4.27 (ii) the sequence (Tf_n) is also a weak null-sequence. As S is compact, then $STf_n \to 0$; hence ST is compact. Now let T be compact. Then $Tf_n \to 0$ for every weak null-sequence (f_n) from H_1. Since S is continuous, we also have $STf_n \to 0$. Therefore ST is compact in this case, also.
(b) If $f_n \xrightarrow{w} 0$ in H_1, then $T_1 f_n \to 0$ and $T_2 f_n \to 0$. Hence $(aT_1 + bT_2)f_n \to 0$.
(c) If T is compact, then the operator T^*T is also compact by part (a). Let T^*T be compact. If (f_n) is a weak null-sequence in H_1, then $T^*Tf_n \to 0$, therefore

$$\|Tf_n\|^2 = \langle Tf_n, Tf_n\rangle = \langle T^*Tf_n, f_n\rangle \to 0.$$

Consequently, T is compact.
(d) If T is compact, then $(T^*)^*T^* = TT^*$ is also compact by part(a). Hence T^* is also compact by part(c).
(e) Let (f_n) be a weak null-sequence from H_1. The sequence (f_n) is then bounded, say $\|f_n\| \leq C$ for all $n \in \mathbb{N}$. We show that $Tf_n \to 0$, i.e., for every $\epsilon > 0$ there exists an $n_0 \in \mathbb{N}$ such that $\|Tf_n\| \leq \epsilon$ for $n \geq n_0$. Let $\epsilon > 0$ be given. Since $\|T_n - T\| \to 0$, there exists an $m_0 \in \mathbb{N}$ such that

$$\|T_{m_0} - T\| \leq \tfrac{1}{2}\epsilon C^{-1}.$$

Since T_{m_0} is compact, there exists an $n_0 \in \mathbb{N}$ for which

$$\|T_{m_0} f_n\| \leq \tfrac{1}{2}\epsilon \quad \text{for all} \quad n \geq n_0.$$

It follows from this that for all $n \geq n_0$

$$\|Tf_n\| \leq \|(T - T_{m_0})f_n\| + \|T_{m_0} f_n\| \leq \epsilon. \qquad \square$$

We denote by $B_\infty(H_1, H_2)$ the set of compact operators from $B(H_1, H_2)$ (the index ∞ will be justified in Section 7.1). If $H = H_1 = H_2$, then we briefly write $B_\infty(H)$.

Proposition. *It follows from Theorem 6.4 that $B_\infty(H)$ is a closed two-sided ideal of $B(H)$.*

Every rank 1 bounded operator from H_1 into H_2 is compact (because T has the form $Tf = \langle g, f\rangle h$ with $g \in H_1$, $h \in H_2$; for every weak null-sequence (f_n) we therefore have $Tf_n = \langle g, f_n\rangle h \to 0$). By Theorem 6.4(b) then every finite rank operator is also compact, and by Theorem 6.4(e) so is every operator that is the limit, with respect to the norm of $B(H_1, H_2)$, of a sequence of finite rank operators. Actually, these are all the compact operators.

Theorem 6.5. *An operator $T \in B(H_1, H_2)$ is compact if and only if there exists a sequence (T_n) of finite rank operators from $B(H_1, H_2)$ for which $\|T_n - T\| \to 0$. For every compact operator T the subspaces $N(T)^{\perp}$ and $R(T)$ are separable.*

PROOF. One direction has already been proved. Let T now be compact. First we show that $N(T)^{\perp}$ is separable. Let $\{e_\alpha : \alpha \in A\}$ be an ONB of $N(T)^{\perp}$. As T is compact, $Te_{\alpha_n} \to 0$ for every sequence (α_n) from A such that $\alpha_n \neq \alpha_m$ for $n \neq m$. It follows from this that for every $\epsilon > 0$ there exist only finitely many $\alpha \in A$ such that $\|Te_\alpha\| \geq \epsilon$. Consequently, the set A is at most countable, i.e., $N(T)^{\perp}$ is separable.

Let $\{e_n : n \in \mathbb{N}\}$ be an ONB of $N(T)^{\perp}$ (if $N(T)^{\perp}$ is finite dimensional, then the following reasoning is simpler), and let P_m be the orthogonal projection onto $L(e_1, \ldots, e_m)$. Then $P_m \xrightarrow{s} P$, the orthogonal projection onto $N(T)^{\perp}$. The operators $T_m = TP_m$ are of rank at most m ($T_m f = \sum_{n=1}^{m} \langle e_n, f\rangle Te_n$); thus they are compact. For every m there exists an $f_m \in H$ such that $\|f_m\| = 1$ and $\|(T - T_m)f_m\| \geq \|T - T_m\|/2$. As $\langle (P - P_m)f_m, g\rangle = \langle f_m, (P - P_m)g\rangle \to 0$, it follows that $(P - P_m)f_m \xrightarrow{w} 0$. Therefore $(T - T_m)f_m = T(P - P_m)f_m \to 0$, since T is compact. Hence $\|T - T_m\| \leq 2\|(T - T_m)f_m\| \to 0$, i.e., $T_m \to T$. If $\{g_n : n \in \mathbb{N}\}$ is a countable dense subset of $N(T)^{\perp}$, then $\{Tg_n : n \in \mathbb{N}\}$ is dense in $R(T)$. Therefore $R(T)$ is also separable. □

Now we want to study the spectrum of a compact operator. For this we need the following theorem.

Theorem 6.6. *Assume that H is a Hilbert space, $T \in B_\infty(H)$, $\lambda \in \mathbb{K}$, $\lambda \neq 0$. Then $R(\lambda - T)$ is closed.*

PROOF. Let $g \in \overline{R(\lambda - T)}$. Then there exists a sequence (f_n') from H such that $g_n = (\lambda - T)f_n' \to g$. If f_n is the orthogonal projection of f_n' onto $N(\lambda - T)^{\perp}$, then $g_n = (\lambda - T)f_n$. If the sequence (f_n) is not bounded, then we can assume that $\|f_n\| \to \infty$ (since this holds for some subsequence of (f_n)). For $h_n = \|f_n\|^{-1}f_n$ we then have $\|h_n\| = 1$ and $(\lambda - T)h_n \to 0$. As T is compact, there exists a subsequence (h_{n_k}) of (h_n), for which (Th_{n_k}) is convergent. Then the sequence $h_{n_k} = \lambda^{-1}(\|f_{n_k}\|^{-1}g_{n_k} + Th_{n_k})$ tends to an element $h \in H$. Because of the relations $h_n \in N(\lambda - T)^{\perp}$ and $\|h_n\| = 1$ we have $h \in N(\lambda - T)^{\perp}$ and $\|h\| = 1$. On the other hand, $(\lambda - T)h = \lim_{k\to\infty} (\lambda - T)h_{n_k} = 0$; thus $h \in N(\lambda - T)$. This is a contradiction.

The sequence (f_n) is therefore bounded. Since T is compact, there exists a subsequence (f_{n_k}) of (f_n) for which (Tf_{n_k}) is convergent. Then the sequence $f_{n_k} = \lambda^{-1}(g_{n_k} + Tf_{n_k})$ also tends to an element $f \in H$, and we have

$$g = \lim_{k\to\infty} (\lambda - T)f_{n_k} = (\lambda - T)f \in R(\lambda - T).$$ □

Theorem 6.7. *Let H be a Hilbert space over $\mathbb{K}$, and let $T \in B_\infty(H)$. We have $\sigma(T) \cap (\mathbb{K}\setminus\{0\}) = \sigma_p(T) \cap (\mathbb{K}\setminus\{0\})$. If H is infinite dimensional, then $\sigma(T) = \sigma_p(T) \cup \{0\}$. The operator T has at most countably many eigenvalues that can* *cluster only at* 0. *Every non-zero eigenvalue has finite multiplicity. The number* $\lambda \neq 0$ *is an eigenvalue of* T *if and only if* λ^* *is an eigenvalue of* T^*.

PROOF. $0 \in \sigma(T)$ by the proposition following Theorem 6.2, if H is infinite dimensional. To prove $\sigma(T) \cap (\mathbb{K} \setminus \{0\}) = \sigma_p \cap (\mathbb{K} \setminus \{0\})$, it is sufficient to show that if $\lambda \neq 0$ is not an eigenvalue of T, then $\lambda \notin \sigma(T)$. For this, let us assume that $\lambda \neq 0$, $N(\lambda - T) = 0$. Since $R(\lambda - T)$ is closed, $\lambda - T$ is a bijective mapping from H onto the Hilbert space $R(\lambda - T)$. In what follows we show that $R(\lambda - T) = H$. From this we can infer that $\lambda \in \rho(T)$, i.e., $\lambda \notin \sigma(T)$.

Let us now assume that $R(\lambda - T) \neq H$. Define H_0, and H_n for $n \geq 1$ by the equalities $H_0 = H$ and $H_n = R((\lambda - T)^n)$. Then for $n \in \mathbb{N}$ the subspace H_{n+1} is a closed subspace, strictly smaller than H_n (the subspace H_{n+1} is closed, because $(\lambda - T)^{-1}$ is continuous and H_n is closed; from $H_{n_0} = H_{n_0+1}$ it would follow that $H = (\lambda - T)^{-n_0} H_{n_0} = (\lambda - T)^{-n_0} H_{n_0+1} = H_1$). If for every $n \in \mathbb{N}$ we choose an element $f_n \in H_{n-1} \ominus H_n$ such that $\|f_n\| = 1$, then (f_n) is an orthonormal sequence. Since T is compact, $Tf_n \to 0$. On the other hand, for all $n \in \mathbb{N}$

$$Tf_n = \lambda f_n - (\lambda - T) f_n.$$

Here $(\lambda - T)f_n$ belongs to H_n; consequently it is orthogonal to f_n. Therefore,

$$\|Tf_n\| \geq |\lambda| \, \|f_n\| = |\lambda|.$$

This contradicts the fact that $Tf_n \to 0$. Hence $R(\lambda - T) = H$.

Now we show that the multiplicity of every non-zero eigenvalue λ is finite. If we had $\dim N(\lambda - T) = \infty$, then there would exist an orthonormal sequence (f_n) from $N(\lambda - T)$. Since T is compact, we would then have $Tf_n \to 0$, which contradicts the equalities $\|Tf_n\| = |\lambda| \, \|f_n\| = |\lambda|$.

In the next step we show that the eigenvalues can cluster only at 0. It also follows from this that there are at most countably many eigenvalues. Let us assume that there exists a sequence (λ_n) of pairwise different eigenvalues of T such that $\lambda_n \to \lambda \neq 0$. Then there exists a sequence (f_n) from H such that $\|f_n\| = 1$ and $(\lambda_n - T)f_n = 0$. We know from linear algebra that the family $\{f_n : n \in \mathbb{N}\}$ is linearly independent. Let $H_n = L(f_1, \ldots, f_n)$ ($H_0 = \{0\}$), $g_n \in H_n \ominus H_{n-1}$ such that $\|g_n\| = 1$ for $n \in \mathbb{N}$. Then $g_n \xrightarrow{w} 0$; hence $Tg_n \to 0$. On the other hand, for all $n \in \mathbb{N}$

$$Tg_n = \lambda_n g_n - (\lambda_n - T) g_n,$$

where $(\lambda_n - T) g_n \in H_{n-1}$, since it follows from the equality $g_n = \sum_{j=1}^{n} a_{nj} f_j$ that

$$(\lambda_n - T) g_n = \sum_{j=1}^{n} a_{nj} (\lambda_n - T) f_j = \sum_{j=1}^{n-1} a_{nj} (\lambda_n - \lambda_j) f_j \in H_{n-1}.$$

Because of the relation $g_n \perp H_{n-1}$ we therefore have $\|Tg_n\| \geq |\lambda_n|$, which contradicts the fact that $Tg_n \to 0$.

The last assertion follows from the equalities $\sigma(T^*) = \sigma(T)^*$ and $\sigma(T^*) \cap (\mathbb{K} \setminus \{0\}) = \sigma_p(T^*) \cap (\mathbb{K} \setminus \{0\})$ (the last equality holds, as T^* is also compact). □

Theorem 6.8. *Let H be a Hilbert space and let $T \in B_\infty(H)$. If $\lambda \neq 0$ is an eigenvalue of T (hence λ^* is an eigenvalue of T^*) then $N(\lambda - T)$ and $N(\lambda^* - T^*) = R(\lambda - T)^\perp$ have the same dimension.*

Proof. We have $\dim N(\lambda - T) \leq \dim N(\lambda^* - T^*)$ or $\dim N(\lambda^* - T^*) \leq \dim N(\lambda - T)$. We treat the first case. There exists then an isometric mapping V of $N(\lambda - T)$ into $N(\lambda^* - T^*) = R(\lambda - T)^\perp$. Let P denote the orthogonal projection onto $N(\lambda - T)$. The operator P is of finite rank, hence compact. For the compact operator $T_1 = T + VP$ we then have $N(\lambda - T_1) = \{0\}$, i.e., λ is not an eigenvalue for T_1. By Theorem 6.7, λ^* is then not an eigenvalue for T_1^*, i.e., $\{0\} = N(\lambda^* - T_1^*) = R(\lambda - T_1)^\perp = (R(\lambda - T) \oplus R(V))^\perp$. Consequently, $R(V) = R(\lambda - T)^\perp$, and thus $\dim N(\lambda - T) = \dim R(V) = \dim R(\lambda - T)^\perp = \dim N(\lambda^* - T^*)$. The other case can be treated similarly. □

Exercises

6.1. Let H be a Hilbert space. A subset A of H is said to be *compact* provided that every sequence (f_n) from A has a subsequence that is convergent in A. The subset A is said to be *relatively compact* if $\bar{A}$ is compact.

(a) A is compact if and only if every open cover of A contains a finite cover of A.

(b) An operator T from H_1 into H_2 is compact if and only if the set $TA = \{Tf : f \in A\}$ is relatively compact for every bounded subset A of $D(T)$.

(c) A subset A of l_2 is relatively compact if and only if it is bounded and for every $\varepsilon > 0$ there exists an $n_0 \in \mathbb{N}$ such that for all $f = (f_n) \in A$ we have $\sum_{n=n_0}^{\infty} |f_n|^2 \leq \varepsilon$.

(d) Let A be a set of continuous functions defined on $\mathbb{R}^m$, having the following properties: A is bounded in the sense of $L_2(\mathbb{R}^m)$, A is equicontinuous on every compact subset of $\mathbb{R}^m$, and for every $\epsilon > 0$ there exists an $r \geq 0$ such that $\int_{|x| \geq r} |f(x)|^2 \, dx \leq \epsilon$ for all $f \in A$. Then A is a relatively compact subset of $L_2(\mathbb{R}^m)$.

6.2. (a) Let (t_n) be a null-sequence from $\mathbb{C}$. The equalities $D(T) = l_2$ and $Tf = (t_n f_n)$ for all $f = (f_n) \in l_2$ define a compact operator $T \in B(l_2)$.

(b) Let $t : [0, 1] \times [0, 1] \to \mathbb{C}$ be continuous. The equalities $D(T) = L_2(0, 1)$ and $Tf(x) = \int_0^1 t(x, y) f(y) \, dy$ for all $f \in L_2(0, 1)$ define a compact operator $T \in B(L_2(0, 1))$.

6.3. If T is a compact operator on the Hilbert space H, then the set $T\{f \in H : \|f\| \leq 1\}$ is compact.

6.4. A sesquilinear form $s(.,.)$ defined on a Hilbert space is said to be compact if $f_n \xrightarrow{w} 0$ and $g_n \xrightarrow{w} 0$ imply $s(f_n, g_n) \to 0$.
(a) Every compact sesquilinear form is bounded.
(b) A sesquilinear form is compact if and only if the operator induced by it (cf. Theorem 5.35) is compact.

6.5. (a) Let T be a compact operator on H. For every $\lambda \in \mathbb{K}$, $\lambda \neq 0$ we have the *Fredholm alternative*: Either the equations $(\lambda - T)f = g$ and $(\lambda^* - T^*)h = k$ are uniquely solvable for all $g, k \in H$ or the homogeneous equations $(\lambda - T)f = 0$ and $(\lambda^* - T^*)h = 0$ have nontrivial solutions.
(b) The spaces of solutions of the two homogeneous equations have the same dimension, and $(\lambda - T)f = g$ is solvable if and only if g is orthogonal to every solution h of the equation $(\lambda^* - T^*)h = 0$.

6.6. Let H be a separable Hilbert space.
(a) If (P_n) is an increasing sequence of finite rank projections on H such that $P_n \xrightarrow{s} I$, then $P_n T P_n \to T$ for every compact operator T on H.
(b) $B_\infty(H)$ is a separable subspace of $B(H)$ (cf. Exercise 4.8).

6.6 . Let $A_n, A \in B(H_2, H_3)$ and $B \in B(H_1, H_2)$. If $A_n \xrightarrow{s} A$ and B is compact, then $A_n B \to AB$.

6.2 Hilbert-Schmidt operators and Carleman operators

We begin by studying one of the most important classes of compact operators, the class of Hilbert-Schmidt operators. Let H_1 and H_2 be Hilbert spaces. An operator $T \in B(H_1, H_2)$ is called a *Hilbert-Schmidt operator* if there exists an orthonormal basis $\{e_\alpha : \alpha \in A\}$ of H_1 such that

$$\sum_{\alpha \in A} \|Te_\alpha\|^2.$$

Theorem 6.9. *An operator $T \in B(H_1, H_2)$ is a Hilbert-Schmidt operator if and only if T^* is a Hilbert-Schmidt operator. Then*

$$\|T\| < \left(\sum_{\alpha \in A} \|Tf_\alpha\|^2 \right)^{1/2} = \left(\sum_{\beta \in B} \|T^* e'_\beta\|^2 \right)^{1/2} < \infty \tag{6.3}$$

for arbitrary orthonormal bases $\{f_\alpha : \alpha \in A\}$ of H_1 and $\{e'_\beta : \beta \in B\}$ of H_2.

The common value of the square roots in (6.3) is called the *Hilbert-Schmidt norm* of T and is denoted by $|||T|||$. We have $\|T\| \leq |||T||| = |||T^*|||$, because of (6.3). The set of Hilbert-Schmidt operators is denoted by $B_2(H_1, H_2)$ or $B_2(H)$ (for the justification of the index 2, cf. Section 7.1).

It is not hard to show that $B_2(H_1, H_2)$, equipped with the Hilbert–Schmidt norm, is a Hilbert space (cf. Exercise 6.11).

PROOF. Let $T \in B(H_1, H_2)$ be a Hilbert-Schmidt operator and let $\{e_\alpha : \alpha \in A\}$ be an ONB of H_1 such that $\Sigma \|Te_\alpha\|^2 < \infty$. If $\{e'_\beta : \beta \in B\}$ is an arbitrary ONB of H_2, then

$$\sum_{\beta \in B} \|T^* e'_\beta\|^2 = \sum_{\beta \in B} \sum_{\alpha \in A} |\langle T^* e'_\beta, e_\alpha \rangle|^2 = \sum_{\alpha \in A} \sum_{\beta \in B} |\langle e'_\beta, Te_\alpha \rangle|^2$$
$$= \sum_{\alpha \in A} \|Te_\alpha\|^2,$$

i.e., T^* is a Hilbert-Schmidt operator. One can prove the converse in the same way and obtain the equality sign in (6.3) at the same time. If $\{e'_\beta : \beta \in B\}$ is an ONB of H_2, then for all $f \in H_1$

$$\|Tf\|^2 = \sum_\beta |\langle e'_\beta, Tf \rangle|^2 \leqslant \|f\|^2 \sum_\beta \|T^* e'_\beta\|^2.$$

Consequently $\|T\|^2 \leqslant \Sigma_\beta \|T^* e'_\beta\|^2$. Hence, (6.3) is completely proved. □

Theorem 6.10. *An operator S from H_1 into H_2 is the restriction of a Hilbert-Schmidt operator T if and only if S is closable and $D(\bar{S})$ contains an orthonormal basis $\{e_\gamma : \gamma \in \Gamma\}$ such that $\Sigma_{\gamma \in \Gamma} \|\bar{S} e_\gamma\|^2 < \infty$. Every Hilbert-Schmidt operator is compact.*

PROOF. Let $T \in B(H_1, H_2)$ be a Hilbert-Schmidt operator and let $S \subset T$. Then S is bounded, hence $D(\bar{S}) = \overline{D(S)}$. Consequently, $D(\bar{S})$ is a Hilbert space, so it contains an ONB $\{e_\gamma : \gamma \in \Gamma\}$. Since $\{e_\gamma : \gamma \in \Gamma\}$ is contained in an ONB of H_1, it follows by Theorem 6.9 that $\Sigma \|\bar{S} e_\gamma\|^2 = \Sigma \|Te_\gamma\|^2 < \infty$.

Let S now have the given property. We show that $\bar{S}$ is compact (the compactness of an arbitrary Hilbert-Schmidt operator follows from this). Let (f_n) be in $D(\bar{S})$ and assume that $f_n \xrightarrow{w} 0$, $\epsilon > 0$. The subset $\{e_\gamma : \langle e_\gamma, f_n \rangle \neq 0$ for some $n \in \mathbb{N}\}$ of $\{e_\gamma : \gamma \in \Gamma\}$ is at most countable; we denote it simply by $\{e_1, e_2, \ldots\}$. We have

$$f_n = \sum_j \langle e_j, f_n \rangle e_j \quad \text{and} \quad \sum_j \|\bar{S} e_j\|^2 < \infty.$$

Because of the inequality

$$\sum_j |\langle e_j, f_n \rangle| \|\bar{S} e_j\| \leqslant \left\{ \sum_j |\langle e_j, f_n \rangle|^2 \sum_j \|\bar{S} e_j\|^2 \right\}^{1/2} = \|f_n\| \left\{ \sum_j \|\bar{S} e_j\|^2 \right\}^{1/2},$$

the series $\Sigma_j \langle e_j, f_n \rangle \bar{S} e_j$ is convergent. Since $\bar{S}$ is closed, it follows that

$$\bar{S} f_n = \sum_j \langle e_j, f_n \rangle \bar{S} e_j.$$

It follows from this that for all $N \in \mathbb{N}$ and $n \in \mathbb{N}$,

$$\begin{aligned}\|\bar{S}f_n\| &\leq \sum_j |\langle e_j, f_n\rangle| \|\bar{S}e_j\| \\ &\leq \sum_{j \leq N} |\langle e_j, f_n\rangle| \|\bar{S}e_j\| + \Big\{\sum_{j>N} |\langle e_j, f_n\rangle|^2 \sum_{j>N} \|\bar{S}e_j\|^2\Big\}^{1/2} \\ &\leq \sum_{j \leq N} |\langle e_j, f_n\rangle| \|\bar{S}e_j\| + \|f_n\| \Big\{\sum_{j>N} \|\bar{S}e_j\|^2\Big\}^{1/2}.\end{aligned}$$

Because of the inequality $\sum_j \|\bar{S}e_j\|^2 < \infty$ and the boundedness of $(\|f_n\|)$ there exists an $N_0 \in \mathbb{N}$ such that

$$\|f_n\| \Big\{\sum_{j>N_0} \|\bar{S}e_j\|^2\Big\}^{1/2} < \tfrac{1}{2}\epsilon \quad \text{for all} \quad n \in \mathbb{N}.$$

Since $f_n \xrightarrow{w} 0$, there exists an $n_0 \in \mathbb{N}$ such that

$$\sum_{j \leq N_0} |\langle e_j, f_n\rangle| \|\bar{S}e_j\| < \tfrac{1}{2}\epsilon \quad \text{for all} \quad n \geq n_0.$$

It follows from this that $\|\bar{S}f_n\| < \epsilon$ for $n \geq n_0$, hence that $\bar{S}f_n \to 0$, i.e., that $\bar{S}$ is compact.

Since the compact operator $\bar{S}$ is continuous, we have $D(\bar{S}) = \overline{D(S)}$. We define an extension T of S by the equalities $D(T) = H_1$ and

$$T(f+g) = \bar{S}f \quad \text{for} \quad f \in D(\bar{S}),\, g \in D(\bar{S})^{\perp}.$$

T is a Hilbert-Schmidt operator, since $T \in B(H_1, H_2)$ with $\|T\| = \|S\|$ (cf. Exercise 4.7) and if $\{g_\beta : \beta \in B\}$ is an ONB of $D(\bar{S})^{\perp}$, then $\{e_\gamma : \gamma \in \Gamma\} \cup \{g_\beta : \beta \in B\}$ is an ONB of H_1 such that

$$\sum_\gamma \|Te_\gamma\|^2 + \sum_\beta \|Tg_\beta\|^2 = \sum_\gamma \|\bar{S}e_\gamma\|^2 < \infty.$$

□

Corollary. *Let $T \in B(H_1, H_2)$, $S \in B(H_2, H_3)$, and let one of these operators be a Hilbert-Schmidt operator. Then ST is a Hilbert-Schmidt-operator. $B(H_2)$ is therefore a two-sided ideal of $B(H)$.*

PROOF. Let T be a Hilbert-Schmidt operator and let $\{e_\alpha : \alpha \in A\}$ be an ONB of H_1. Then

$$\sum_{\alpha \in A} \|STe_\alpha\|^2 \leq \|S\|^2 \sum_{\alpha \in A} \|Te_\alpha\|^2 < \infty.$$

If S is a Hilbert-Schmidt operator, then T^*S^* is a Hilbert-Schmidt operator using Theorem 6.9 and what we have just proved. Hence $ST = (T^*S^*)^*$ is also a Hilbert-Schmidt operator. □

Now we return to our earlier definition of a Hilbert-Schmidt operator (cf. Section 4.1, Example 3).

Theorem 6.11. *Let M_1 and M_2 be measurable subsets of $\mathbb{R}^p$ and $\mathbb{R}^q$, respectively*[2]. *The operator $T \in B(L_2(M_1), L_2(M_2))$ is a Hilbert-Schmidt operator if and only if there exists a kernel $K \in L_2(M_2 \times M_1)$ such that*

$$Tf(x) = \int_{M_1} K(x, y) f(y) \, dy \text{ almost everywhere in } M_2, f \in L_2(M_1). \tag{6.4}$$

The adjoint operator T^* is then induced by the adjoint kernel $K^+(x, y) = K(y, x)^*$.

PROOF. If T is of this form, then $T \in B(L_2(M_1), L_2(M_2))$, by Section 4.1, Example 3. If $\{e_n : n \in \mathbb{N}\}$ and $\{f_m : m \in \mathbb{N}\}$ are orthonormal bases of $L_2(M_1)$ and $L_2(M_2)$, respectively, then $\{g_{nm} : (n, m) \in \mathbb{N} \times \mathbb{N}\}$ is an ONB of $L_2(M_2 \times M_1)$, where $g_{nm}(x, y) = f_m(x) e_n(y)^*$ (cf. Theorem 3.8). Hence

$$\sum_n \|Te_n\|^2 = \sum_{n,m} |\langle Te_n, f_m\rangle|^2 = \sum_{n,m} \left| \int_{M_2} \int_{M_1} K(x, y)^* e_n(y)^* \, dy f_m(x) \, dx \right|^2$$

$$= \sum_{n,m} |\langle K, g_{nm}\rangle|^2 = \|K\|^2 < \infty,$$

i.e., T is a Hilbert-Schmidt operator.

Let T now be a Hilbert-Schmidt operator, and let e_n, f_m and g_{nm} be defined as above. Let us define c_{nm} by

$$c_{nm} = \langle f_m, Te_n\rangle \quad \text{for} \quad (n, m) \in \mathbb{N} \times \mathbb{N}.$$

Then

$$\sum_{n,m} |c_{nm}|^2 = \sum_{n,m} |\langle f_m, Te_n\rangle|^2 = \sum_n \|Te_n\|^2 = |||T|||^2 < \infty;$$

so there exists a $K \in L_2(M_2 \times M_1)$ such that

$$\langle g_{nm}, K\rangle = c_{nm} \quad \text{for all} \quad (n, m) \in \mathbb{N} \times \mathbb{N}.$$

If T_0 denotes the operator induced by K in the sense of (6.4), then with $h(x, y) = g(x) f^*(y)$ we have for all $f \in L_2(M_1)$, $g \in L_2(M_2)$ that

$$\langle g, T_0 f\rangle = \int_{M_2} g(x)^* \int_{M_1} K(x, y) f(y) \, dy \, dx$$

$$= \langle h, K\rangle = \sum_{n,m} c_{nm} \langle h, g_{nm}\rangle = \sum_{n,m} \langle f_m, Te_n\rangle \langle g, f_m\rangle \langle e_n, f\rangle$$

$$= \left\langle \sum_m \langle f_m, g\rangle f_m, \sum_n \langle e_n, f\rangle Te_n \right\rangle = \langle g, Tf\rangle.$$

Hence $T = T_0$. □

[2] M_1 and M_2 can be replaced by arbitrary measure spaces in most cases. Sometimes we have to assume that the M_j are σ-finite.

Theorem 6.12. *Let T be an operator from a Hilbert space H into $L_2(M)$. Then the following assertions are equivalent:*

(i) *T is the restriction of a Hilbert-Schmidt operator,*

(ii) *there exists a function $k : M \to H$ such that $\|k(\cdot)\| \in L_2(M)$ and*

$$Tf(x) = \langle k(x), f\rangle \text{ almost everywhere in } M,\ f \in D(T),$$

(iii) *there exists a function $\kappa \in L_2(M)$ such that*

$$|Tf(x)| \leqslant \|f\|\kappa(x) \text{ almost everywhere in } M,\ f \in D(T)$$

(*of course, the sets of exceptional points in* (ii) *and* (iii) *depend on f and on the choice of the representative of Tf.*)

Proof. (i) *implies* (ii): Let $T \subset S$ and let $S \in B_2(H, L_2(M))$. We show (ii) for the operator S. Since S is compact, $N(S)^\perp$ is separable (Theorem 6.5). Let $\{e_1, e_2, \ldots\}$ be an ONS of $N(S)^\perp$. Then

$$\sum_n \int_M |Se_n(x)|^2 \, dx = \sum_n \|Se_n\|^2 < \infty;$$

consequently, by B. Levi's theorem, $\sum_n |Se_n(x)|^2 < \infty$ almost everywhere in M and $\int_M \sum_n |Se_n(x)|^2 \, dx < \infty$. Hence we can define the function $k : M \to H$ by the formula

$$k(x) = \begin{cases} \sum_n (Se_n(x))^* e_n & \text{if } \sum_n |Se_n(x)|^2 < \infty, \\ 0 & \text{otherwise.} \end{cases}$$

With this function k we have for all $f \in H$ that

$$\langle k(x), f\rangle = \sum_n \langle e_n, f\rangle Se_n(x) = Sf(x)$$

almost everywhere in M. Because of the equality $\|k(x)\|^2 = \sum_n |Se_n(x)|^2$, the function $\|k(\cdot)\|$ belongs to $L_2(M)$.

(ii) *implies* (iii) by taking $\kappa(x) = \|k(x)\|$.

(iii) *implies* (i): It follows from (iii) that

$$\|Tf\|^2 \leqslant \|f\|^2 \int_M |\kappa(x)|^2 \, dx$$

for all $f \in D(T)$. So T is bounded. $\overline{T}$ also satisfies (iii). This can be seen in the following way. For every $f \in D(\overline{T}) = \overline{D(T)}$ there exists a sequence (f_n) from $D(T)$ such that $f_n \to f$; therefore also $Tf_n \to \overline{T}f$. By Theorem 2.1 there exists a subsequence (f_{n_k}) such that $Tf_{n_k}(x) \to \overline{T}f(x)$ almost everywhere. Hence

$$|\overline{T}f(x)| = \lim_{k\to\infty} |Tf_{n_k}(x)| \leqslant \lim_{k\to\infty} \|f_{n_k}\|\kappa(x) = \|f\|\kappa(x).$$

almost everywhere. This holds also for the operator $S \in B(H, L_2(M))$ defined by

$$S(f+g) = \bar{T}f \quad \text{for} \quad f \in D(\bar{T}), g \in D(\bar{T})^{\perp}.$$

We show that S is a Hilbert-Schmidt operator and $|||S|||^2 \leq \int_M |\kappa(x)|^2 \, dx$. It is obviously sufficient to show that

$$\sum_{j=1}^{n} \|Se_j\|^2 \leq \int_M |\kappa(x)|^2 \, dx \quad \text{for every finite ONS } \{e_1, \ldots, e_n\}$$

in H. Let $Se_j(\cdot)$ be an arbitrary (however, in what follows fixed) representative of Se_j. Let us define $A : L(e_1, \ldots, e_n) \to L_2(M)$ by the equality

$$\left(A\left(\sum a_j e_j\right)\right)(x) = \sum a_j Se_j(x).$$

Then $A = S|_{L(e_1, \ldots, e_n)}$ in the sense of $L_2(M)$. Since the set $L_r(e_1, \ldots, e_n)$ of linear combinations of $e_1, \ldots, e_n$ with rational coefficients is countable, there exists a subset N of M of measure 0 for which

$$|Af(x)| \leq \|f\| \kappa(x) \quad \text{for all} \quad x \in M \setminus N, f \in L_r(e_1, \ldots, e_n).$$

If $f = \sum_{j=1}^{n} a_j e_j \in L(e_1, \ldots, e_n)$, and we choose rational sequences $(a_{jk})_{k \in \mathbb{N}}$ so that $a_{jk} \to a_j$ as $k \to \infty$, then with $f_k = \sum_{j=1}^{n} a_{jk} e_j$ it follows that

$$|Af(x)| = \lim_{k \to \infty} |Af_k(x)| \leq \lim_{k \to \infty} \|f_k\| \kappa(x) = \|f\| \kappa(x)$$

for all $x \in M \setminus N$. Consequently,

$$|Af(x)| \leq \|f\| \kappa(x) \quad \text{for all} \quad x \in M \setminus N, f \in L(e_1, \ldots, e_n).$$

For every $x \in M \setminus N$ the mapping $f \mapsto Af(x)$ is a continuous linear functional on $L(e_1, \ldots, e_n)$ whose norm is not greater than $\kappa(x)$. Hence there exists a $k(x) \in L(e_1, \ldots, e_n) \subset H$ such that $\|k(x)\| \leq \kappa(x)$ and

$$Af(x) = \langle k(x), f \rangle \quad \text{for} \quad x \in M \setminus N, f \in L(e_1, \ldots, e_n).$$

It follows that

$$\sum_{j=1}^{n} \|Ae_j\|^2 = \int_M \sum_{j=1}^{n} |\langle k(x), e_j \rangle|^2 \, dx = \int_M \|k(x)\|^2 \, dx \leq \int_M |\kappa(x)|^2 \, dx. \qquad \square$$

A linear operator T from a Hilbert space H into $L_2(M)$ is called a *Carleman operator* if there exists a function $k : M \to H$ such that for all $f \in D(T)$

$$Tf(x) = \langle k(x), f \rangle \quad \text{almost everywhere in } M. \tag{6.5}$$

By Theorem 6.12 every Hilbert-Schmidt operator from H into $L_2(M)$ is a Carleman operator.

A function $k : M \to H$ is said to be *measurable* if the function $\langle k(\cdot), f \rangle : M \to \mathbb{K}, x \mapsto \langle k(x), f \rangle$ is measurable for every $f \in H$. If T is a

Carleman operator and k is an *inducing* function of T (in the sense of (6.5)), then the function $\langle k(\cdot), f\rangle$ is measurable for all $f \in D(T)$. This obviously also holds for all $f \in \overline{D(T)}$. If P denotes the orthogonal projection onto $\overline{D(T)}$, then $\langle Pk(\cdot), f\rangle = \langle k(\cdot), Pf\rangle$ is therefore measurable for every $f \in H$, i.e., $Pk(\cdot)$ is measurable, and for all $f \in D(T)$

$$Tf(x) = \langle Pk(x), f\rangle \quad \text{almost everywhere in } M.$$

Consequently, there is no loss of generality in assuming that k is measurable.

If $k : M \rightarrow H$ is measurable, then the equalities

$$\begin{aligned} &D(T_k) = \{f \in H : \langle k(\cdot), f\rangle \in L_2(M)\}, \\ &T_k f(x) = \langle k(x), f\rangle \quad \text{almost everywhere in } M, f \in D(T_k), \end{aligned} \tag{6.6}$$

define an operator T_k from H into $L_2(M)$. The operator T_k is called the *maximal Carleman operator* induced by k. An operator is a Carleman operator if and only if it is the restriction of a maximal Carleman operator.

Theorem 6.13.

(a) *Every Carleman operator is closable. The closure of a Carleman operator is a Carleman operator. Every maximal Carleman operator is closed.*

(b) *If T_1, T_2 are Carleman operators (with inducing functions k_1, k_2) and $a, b \in \mathbb{C}$, then $aT_1 + bT_2$ is a Carleman operator (with inducing function $ak_1 + bk_2$).*

(c) *If T is a (maximal) Carleman operator from H_1 into $L_2(M)$ (induced by k) and $S \in B(H_2, H_1)$, then TS is a (maximal) Carleman operator from H_2 into $L_2(M)$ (induced by S^*k).*

(d) *Let T be an operator from H into $L_2(M)$, and let P be the orthogonal projection onto $\overline{D(T)}$. The operator T is a Carleman operator if and only if TP is a Carleman operator.*

(e) *If H is separable and k_1, k_2 are inducing functions of a Carleman operator T, then $Pk_1(x) = Pk_2(x)$ almost everywhere in M, where P denotes the orthogonal projection onto $\overline{D(T)}$.*

PROOF.

(a) Since an operator is a Carleman operator if and only if it is a restriction of a maximal Carleman operator, it is sufficient to show that every maximal Carleman operator is closed. Let $k : M \rightarrow H$ be measurable, and let T_k be defined by (6.6). Take a sequence (f_n) from $D(T_k)$ for which $f_n \rightarrow f \in H$, $T_k f_n \rightarrow g \in L_2(M)$. Since $\langle k(x), f_n\rangle \rightarrow \langle k(x), f\rangle$ (for all $x \in M$), we have

$$\langle k(x), f\rangle = g(x) \quad \text{almost everywhere in } M.$$

Hence $f \in D(T_k)$ and $T_k f = g$, i.e., T_k is closed.

(b) This assertion is obvious.

(c) Again, it is sufficient to show the statement for maximal Carleman operators. Let T_k be the maximal Carleman operator induced by k.

Then

$$D(T_kS) = \{f\in H_2 : Sf\in D(T_k)\} = \{f\in H_2 : \langle k(\cdot), Sf\rangle\in L_2(M)\}$$
$$= \{f\in H_2 : \langle S^*k(\cdot), f\rangle\in L_2(M)\},$$
$$T_kSf(x) = \langle k(x), Sf\rangle = \langle S^*k(x), f\rangle \text{ almost everywhere in } M,$$

i.e., T_kS is the maximal Carleman operator induced by S^*k.

(d) By part (c), along with T the operator TP is also a Carleman operator. Since T is a restriction of TP (we have $D(TP)=D(T)+D(T)^{\perp}$ and $T=(TP)|_{D(T)}$), along with TP the operator T is also a Carleman operator.

(e) If $\{e_1, e_2, \ldots\}$ is an ONB of $\overline{D(T)}$ that is contained in $D(T)$, then it follows that

$$\langle k_1(x) - k_2(x), e_n\rangle = 0 \text{ almost everywhere, for all } n\in\mathbb{N},$$

and thus

$$\|P(k_1(x)-k_2(x))\|^2 = \sum_n |\langle k_1(x) - k_2(x), e_n\rangle|^2$$
$$= 0 \quad \text{almost everywhere.}$$ □

Theorem 6.14 (Korotkov [46]). *An operator T from H into $L_2(M)$ is a Carleman operator if and only if there exists a measurable function $\kappa : M\to\mathbb{R}$ such that for all $f\in D(T)$*

$$|Tf(x)| \leqslant \|f\|\kappa(x) \quad \text{almost everywhere in } M. \tag{6.7}$$

PROOF. If T is a Carleman operator induced by k, then (6.7) holds with $\kappa(x)=\|k(x)\|$. Let (6.7) now be satisfied. Then there exists a bounded function $g : M\to(0, \infty)$ for which $g\kappa\in L_2(M)$; for example we can choose the function

$$g(x) = \left[(1+|x|)^m(1+\kappa(x))\right]^{-1}, \quad x\in M \text{ (for } M\subset\mathbb{R}^m).$$

If $G\in B(L_2(M))$ is the operator of multiplication by g, then for all $f\in D(T)$,

$$|GTf(x)| \leqslant \|f\|\, g(x)\kappa(x) \quad \text{almost everywhere in } M.$$

By Theorem 6.12 the operator GT is therefore the restriction of a Hilbert-Schmidt operator and there exists a function $k' : M\to H$ such that $GTf(x) =\langle k'(x), f\rangle$ almost everywhere in M. With $k(x)=g(x)^{-1}k'(x)$ we therefore have for all $f\in D(T)$ that

$$Tf(x) = g(x)^{-1}\langle k'(x), f\rangle = \langle k(x), f\rangle \quad \text{almost everywhere in } M.$$ □

Theorem 6.15. *Let H be a separable Hilbert space, and let T be an operator from H into $L_2(M)$. The operator T is a Carleman operator if and only if the series $\sum_n |Te_n(x)|^2$ converges almost everywhere for every orthonormal system $\{e_1, e_2, \ldots\}$ in $D(T)$.*

Proof. If T is a Carleman operator, then, as in the proof of Theorem 6.14, we can find a $g: M \to (0, \infty)$ for which GT is the restriction of a Hilbert-Schmidt operator. Then for every ONS $\{e_1, e_2, \ldots\}$ from $D(T)$

$$\sum_n \int_M |GTe_n(x)|^2 \, dx = \sum_n \|GTe_n\|^2 < \infty.$$

Consequently, by B. Levi's theorem,

$$g(x)^2 \sum_n |Te_n(x)|^2 = \sum_n |GTe_n(x)|^2 < \infty \quad \text{almost everywhere in } M.$$

Division by g^2 gives the assertion.

Now suppose the series $\Sigma_n |Te_n(x)|^2$ is almost everywhere convergent for every ONS $\{e_1, e_2, \ldots\}$ in $D(T)$. First we show: If $\{e_1, e_2, \ldots\}$ is an arbitrary ONS in $D(T)$, then $T|_{L(e_1, e_2, \ldots)}$ is a Carleman operator. For this, let $M_0 = \{x \in M : \Sigma_n |Te_n(x)|^2 = \infty\}$. M_0 is of measure zero by our assumption. The function

$$k(x) = \begin{cases} \sum_n (Te_n(x))^* e_n & \text{for} \quad x \in M \setminus M_0, \\ 0 & \text{otherwise}, \end{cases}$$

induces $T|_{L(e_1, e_2, \ldots)}$, since for $f \in L(e_1, e_2, \ldots)$

$$Tf(x) = \sum_n \langle e_n, f \rangle Te_n(x) = \langle \sum_n (Te_n(x))^* e_n, f \rangle$$
$$= \langle k(x), f \rangle \quad \text{almost everywhere in } M.$$

Let $\{e_1, e_2, \ldots\}$ now be an ONB of $D(T)$, let k be an inducing function of $T_0 = T|_{L(e_1, e_2, \ldots)}$ and let T_k be the maximal Carleman operator induced by k. We show that $T \subset T_k$. For this, let $f \in D(T)$, let $\{f_1, f_2, \ldots\}$ be the ONS that arises from $\{f, e_1, e_2, \ldots\}$ by orthogonalization, and let k' be an inducing function of $T_1' = T|_{L(f_1, f_2, \ldots)}$. Because of the inclusion $L(e_1, e_2, \ldots) \subset L(f_1, f_2, \ldots)$ we have $T_0 \subset T_1$; hence k and k' are inducing functions of T_0. If P is the orthogonal projection onto $\overline{D(T)} = \overline{D(T_0)} = \overline{L(e_1, e_2, \ldots)}$, then by Theorem 6.13(e) we have $Pk'(x) = Pk(x)$ almost everywhere. Since $f \in \overline{D(T)}$, it follows from this that

$$\langle k(x), f \rangle = \langle Pk(x), f \rangle = \langle Pk'(x), f \rangle = \langle k'(x), f \rangle = T_1 f(x);$$

so $f \in D(T_k)$ and $Tf = T_1 f = T_k f$. □

Theorem 6.16. *An operator T from a separable Hilbert space H into $L_2(M)$ is a Carleman operator if and only if $Tf_n(x) \to 0$ almost everywhere in M for every null-sequence (f_n) from $D(T)$.*

Proof. It is evident from the definition that every Carleman operator has this property. It remains to prove the reverse direction. By Theorem 6.15 it is sufficient to show that the series $\Sigma_n |Te_n(x)|^2$ is almost everywhere convergent for every ONS $\{e_1, e_2, \ldots\}$ from $D(T)$. Let $\{e_1, e_2, \ldots\}$ be an ONS in $D(T)$. Assume that there exists a measurable subset $N \subset M$ such that $\lambda(N) > 0$ and $\Sigma_n |Te_n(x)|^2 = \infty$ for $x \in N$ (λ stands for Lebesgue measure). For all $m, l \in \mathbb{N}$ let us define $N_{m,l}$ by the equality

$$N_{m,l} = \left\{ x \in N : \sum_{n=1}^{l} |Te_n(x)|^2 \geq m^2 \right\}.$$

Then $N = \bigcup_{l \in \mathbb{N}} N_{m,l}$ for every $m \in \mathbb{N}$, and there exists an $l(m) \in \mathbb{N}$ such that

$$\lambda(N_{m,l(m)}) \geq (1 - 3^{-m})\lambda(N).$$

Consequently, for $N_0 = \bigcap_{m \in \mathbb{N}} N_{m,l(m)}$ we have

$$\lambda(N_0) \geq \left(1 - \sum_{m=1}^{\infty} 3^{-m}\right)\lambda(N) > 0.$$

For all $m \in \mathbb{N}$ we have

$$\sum_{n=1}^{l(m)} |Te_n(x)|^2 \geq m^2 \text{ for } x \in N_0.$$

By Exercise 6.9, for every $m \in \mathbb{N}$ there exist finitely many elements $(x_{m,j}) = (\xi_{m,j,1}, \ldots, \xi_{m,j,l(m)}) \in \mathbb{C}^{l(m)}$, $j = 1, 2, \ldots, p(m)$ for which we have: $|x_{m,j}|^2 = \Sigma_{n=1}^{l(m)} |\xi_{m,j,n}|^2 \leq 2m^{-2}$ and for every $x = (\xi_1, \ldots, \xi_{l(m)}) \in \mathbb{C}^{l(m)}$ with $|x|^2 \geq m^2$ there exists a $j \in \{1, \ldots, p(m)\}$ for which

$$\left| \sum_{n=1}^{l(m)} \xi_{m,j,n} \xi_n \right| \geq 1.$$

Let us set

$$g_{m,j} = \sum_{n=1}^{l(m)} \xi_{m,j,n} e_n.$$

Then for every $m \in \mathbb{N}$ and for every $x \in N_0$ there exists a $j \in \{1, \ldots, p(m)\}$ such that

$$|Tg_{m,j}(x)| = \left| \sum_{n=1}^{l(m)} \xi_{m,j,n} Te_n(x) \right| \geq 1.$$

Thus, for the sequence

$$(g_n) = (g_{1,1}, g_{1,2}, \ldots, g_{1,p(1)}, g_{2,1}, \ldots, g_{2,p(2)}, g_{3,1}, \ldots)$$

we have: $g_n \to 0$ and for every $x \in N_0$ there exists an arbitrarily large $n \in \mathbb{N}$ such that $Tg_n(x) \geq 1$. This contradicts the assumption. □

Theorem 6.17. *An operator T from $L_2(M_1)$ into $L_2(M_2)$ is a Carleman operator if and only if there exists a measurable function $K: M_2 \times M_1 \to \mathbb{C}$ such that $K(x, \cdot) \in L_2(M_1)$ almost everywhere in M_2 and*

$$Tf(x) = \int_{M_1} K(x,y)f(y)\,dy \quad \textit{almost everywhere in } M_2, f \in D(T). \tag{6.8}$$

Such a kernel K is called a Carleman kernel.

PROOF. If T is induced by a Carleman kernel K in the sense of (6.8), then the assumption of Theorem 6.14 (Korotkov) is fulfilled with $\kappa(x) = \|K(x,.)\|$; so T is a Carleman operator. If T is a Carleman operator, then we proceed as in the proof of Theorem 6.14. GT is then a Hilbert-Schmidt operator from $L_2(M_1)$ into $L_2(M_2)$; therefore, by Theorem 6.11, it is induced by a kernel $K' \in L_2(M_2 \times M_1)$. The kernel $K(x,y) = g(x)^{-1}K'(x,y)$ is then a Carleman kernel and it induces T. □

Let $K: M \to H$ be a measurable function, and let $M_n = \{x \in M : \|k(x)\| \leq n \text{ and } |x| \leq n\}$. Let

$$D(T_{k,0}) = \{g \in L_2(M)\text{: there exists an } n \in \mathbb{N} \text{ such that}$$
$$g(x) = 0 \text{ almost everywhere in } M \setminus M_n\}.$$

For every $g \in D(T_{k,0})$ the equality

$$\langle T_{k,0}g, f\rangle = \int_{M_n} g^*(x)\langle k(x), f\rangle\,dx \quad \text{for all} \quad f \in H$$

uniquely defines an element $T_{k,0}g$, since because of the inequalities

$$\left|\int_{M_n} g^*(x)\langle k(x), f\rangle\,dx\right| \leq n\|f\|\int_{M_n}|g(x)|\,dx \leq n\lambda(M_n)^{1/2}\|g\|\,\|f\|$$

the function $f \mapsto \int_{M_n} g^*(x)\langle k(x), f\rangle\,dx$ is a continuous linear functional on H. The mapping $g \mapsto T_{k,0}g$ is obviously linear. $T_{k,0}$ is therefore an operator from $L_2(M)$ into H; $D(T_{k,0})$ is dense in $L_2(M)$. The operator $T_{k,0}$ is called the *semi-Carleman operator* induced by k.

Theorem 6.18. *We have $(T_{k,0})^* = T_k$. (In what follows we write $T^*_{k,0}$ for $(T_{k,0})^*$.)*

PROOF. By the definition of $T_{k,0}$ we have for all $f \in D(T_k)$ and $g \in D(T_{k,0})$

$$\langle g, T_k f\rangle = \langle g, \langle k(\cdot), f\rangle\rangle = \int_M g(x)^*\langle k(x), f\rangle\,dx = \langle T_{k,0}g, f\rangle,$$

i.e., the operators T_k and $T_{k,0}$ are formal adjoints of each other; therefore $T_k \subset T^*_{k,0}$. It remains to prove that $D(T^*_{k,0}) \subset D(T_k)$. Let $f \in D(T^*_{k,0})$. Then for every $g \in L_2(M_n)$ and all $n \in \mathbb{N}$ we have

$$\int_{M_n} g^*(x)\langle k(x), f\rangle\, dx = \langle T_{k,0}g, f\rangle = \langle g, T_{k,0}^* f\rangle = \int_{M_n} g^*(x) T_{k,0}^* f(x)\, dx.$$

Consequently,

$$\int_{M_n} g^*(x)\{\langle k(x), f\rangle - T_{k,0}^* f(x)\}\, dx = 0 \quad \text{for all} \quad g \in L_2(M_n).$$

Because of the relation $\{\langle k(\cdot), f\rangle - T_{k,0}^* f(\cdot)\}|_{M_n} \in L_2(M_n)$ it follows from this that

$$T_{k,0}^* f(x) = \langle k(x), f\rangle \quad \text{almost everywhere in } M_n.$$

As this holds for all n, it follows that $\langle k(\cdot), f\rangle = T_{k,0}^* f \in L_2(M)$, i.e., $f \in D(T_k)$. □

If $K: M_2 \times M_1 \to \mathbb{C}$ is a Carleman kernel and k denotes the mapping $k: M_2 \to L_2(M_1)$, $k(x) = K(x, \cdot)$, then we write $T_K = T_k$ and $T_{K,0} = T_{k,0}$. It follows from the definition of $T_{k,0}$ (by Fubini's theorem) that for all $f \in D(T_{K,0})$ we have

$$T_{K,0} f(x) = \int_{M_2} K(y, x)^* f(y)\, dy \quad \text{almost everywhere in } M_1.$$

Theorem 6.19. *Let T be a densely defined Carleman operator from $L_2(M_1)$ into $L_2(M_2)$ that is induced by the Carleman kernel K. The adjoint T^* is a Carleman operator if and only if K^+ is a Carleman kernel ($K^+(x, y) = K(y, x)^*$ for $(x, y) \in M_1 \times M_2$) and $\bar{T} \supset T_{K^+,0}$. Then T^* is induced by K^+.*

Proof. By assumption, $T \subset T_K$. As T is closable, $D(T^*)$ is dense. If T^* is defined by the Carleman kernel $H: M_1 \times M_2 \to \mathbb{C}$, then $T^* \subset T_H$. Consequently, $\bar{T} = T^{**} \supset T_H^* = \overline{T_{H,0}} \supset T_{H,0}$. It remains to prove that $H(x, y) = K^+(x, y)$ almost everywhere in $M_1 \times M_2$. Let

$$M_{1,n} = \left\{ x \in M_1 : \int_{M_2} |H(x, y)|^2\, dy \leq n \text{ and } |x| \leq n \right\}.$$

Then $H|_{M_{1,n} \times M_2} \in L_2(M_{1,n} \times M_2)$ and $H^+|_{M_2 \times M_{1,n}} \in L_2(M_2 \times M_{1,n})$. Hence the function H^+, as a kernel on $M_2 \times M_{1,n}$, is a Carleman kernel. Therefore, for every $n \in \mathbb{N}$, $T_{H,0}|_{L_2(M_{1,n})}$ is a Carleman operator induced by H^+ and K. Consequently, by Theorem 6.13(e) we have $H^+(x, y) = K(x, y)$ almost everywhere in $M_2 \times M_{1,n}$ for every $n \in \mathbb{N}$. Hence $H^+(x, y) = K(x, y)$ almost everywhere. In particular, K^+ is a Carleman kernel and $\bar{T} \supset T_{K^+,0}$. If K^+ is a Carleman kernel and $\bar{T} \supset T_{K^+,0}$, then $T^* = \bar{T}^* \subset (T_{K^+,0})^* = T_{K^+}$; so T^* is a Carleman operator induced by K^+. □

Corollary. *If T is a symmetric Carleman operator on $L_2(M)$ with inducing kernel K, and T^* is also a Carleman operator, then the kernel K is Hermitian, i.e., $K(x, y) = K^+(x, y)$ almost everywhere. In particular, the kernel of a self-adjoint Carleman operator is Hermitian.*

EXERCISES

6.7. Let H_1 and H_2 be Hilbert spaces and let H_1 be separable. A closable operator T from H_1 into H_2 is the restriction of a Hilbert-Schmidt operator if and only if there exists an orthonormal basis $\{e_1, e_2, \dots\}$ of $D(T)$ such that $\Sigma_n \|Te_n\|^2 < \infty$ (cf. Theorem 6.10).

6.8. Let H_1 be a separable Hilbert space and let T be an operator from H_1 into $L_2(M)$. Assume that
(i) If (f_n) is a null-sequence from $D(T)$ and $(Tf_n(x))$ is convergent almost everywhere in M, then $Tf_n(x) \to 0$ almost everywhere,
(ii) there exists an orthonormal basis of $D(T)$ such that $\Sigma_n |Te_n(x)|^2 < \infty$ almost everywhere in M.
Then T is a Carleman operator.
Hint: cf. Theorem 6.15. Remark: (i) cannot be replaced by the assumption of the closability of T.

6.9. For every $m \in \mathbb{N}$ and $C > 0$ there are finitely many elements $x_j = (\xi_{j,1}, \dots, \xi_{j,m}) \in \mathbb{C}^m, j = 1, \dots, p = p(m, C)$, for which $|x_j| \leq 2C^{-1}$ and for which we have: for every $x = (\xi_1, \dots, \xi_m) \in \mathbb{C}^m$ with $|x| \geq C$ there exists a $j \in \{1, \dots, p\}$ such that

$$|\langle x, x_j \rangle| = \left| \sum_{k=1}^{m} \xi_k^* \xi_{j,k} \right| \geq 1.$$

6.10. If K is a Hermitian Carleman kernel over $M \times M$ and $T_{K,0}$ is bounded, then T_K is from $B(L_2(M))$ and T_K is self-adjoint.

6.11. Let H_1 and H_2 be Hilbert spaces.
(a) The equality $\langle T, S \rangle = \Sigma_{\alpha \in A} \langle Te_\alpha, Se_\alpha \rangle$ defines a scalar product on the space $B_2(H_1, H_2)$ (here let $\{e_\alpha : \alpha \in A\}$ be an arbitrary orthonormal basis of H_1). The corresponding norm is the Hilbert-Schmidt norm.
(b) $B_2(H_1, H_2)$ is a Hilbert space with this scalar product.
(c) With this norm, $B_2(H)$ is a Banach algebra (without identity element in the case dim H is infinite).
(d) We have $\langle S, T \rangle = \langle T^*, S^* \rangle$ for all $S, T \in B_2(H_1, H_2)$.

6.12. Let T be an operator on $L_2(M)$, and let $(z_0 - T)^{-1}$ be a Carleman operator for some $z_0 \in \rho(T)$.
(a) $(z - T)^{-1}$ is a Carleman operator for all $z \in \rho(T)$.
(b) If T is self-adjoint and $k_z(x, y)$ is the kernel of $(z - T)^{-1}$ for some $z \in \rho(T)$, then $k_{z^*}(x, y) = k_z(y, x)^*$.
Hint: Notice that $[(z - T)^{-1}]^* = (z^* - T)^{-1}$, and use Theorem 6.19.

6.3 Matrix operators and integral operators

Let H_1 and H_2 be infinite dimensional Hilbert spaces over $\mathbb{K}$, let $\{e_n : n \in \mathbb{N}\}$ and $\{e'_n : n \in \mathbb{N}\}$ be orthonormal bases of H_1 and H_2, respectively, and let $(a_{jk})_{j,k\in\mathbb{N}}$ be an (infinite) matrix with $a_{jk} \in \mathbb{K}$. (If H_1 or H_2 is finite dimensional, then some simplifications arise.) First we show that the formulae

$$D(A) = \left\{ f \in H_1 : \lim_{m\to\infty} \sum_{k=1}^{m} a_{jk}\langle e_k, f\rangle \text{ exists for all } j \in \mathbb{N}, \text{ and } \sum_{j=1}^{\infty} \left| \sum_{k=1}^{\infty} a_{jk}\langle e_k, f\rangle \right|^2 < \infty \right\}, \tag{6.9}$$

$$Af = \sum_{j=1}^{\infty} \left(\sum_{k=1}^{\infty} a_{jk}\langle e_k, f\rangle \right) e'_j \quad \text{for} \quad f \in D(A)$$

define a linear operator from H_1 into H_2: If $f, g \in D(A)$ and $a, b \in \mathbb{K}$, then the limit

$$\lim_{m\to\infty} \sum_{k=1}^{m} a_{jk}\langle e_k, af+bg\rangle = \lim_{m\to\infty} \left\{ a \sum_{k=1}^{m} a_{jk}\langle e_k, f\rangle + b \sum_{k=1}^{m} a_{jk}\langle e_k, g\rangle \right\}$$
$$= a \sum_{k=1}^{\infty} a_{jk}\langle e_k, f\rangle + b \sum_{k=1}^{\infty} a_{jk}\langle e_k, g\rangle$$

obviously exists, and we have

$$\sum_{j=1}^{\infty} \left| \sum_{k=1}^{\infty} a_{jk}\langle e_k, af+bg\rangle \right|^2 = \sum_{j=1}^{\infty} \left| a \sum_{k=1}^{\infty} a_{jk}\langle e_k, f\rangle + b \sum_{k=1}^{\infty} a_{jk}\langle e_k, g\rangle \right|^2$$
$$\leq 2\left\{ |a|^2 \sum_{j=1}^{\infty} \left| \sum_{k=1}^{\infty} a_{jk}\langle e_k, f\rangle \right|^2 + |b|^2 \sum_{j=1}^{\infty} \left| \sum_{k=1}^{\infty} a_{jk}\langle e_k, g\rangle \right|^2 \right\} < \infty.$$

Therefore, it follows that $af + bg \in D(A)$ and $A(af+bg) = aAf + bAg$.

Theorem 6.20. *Let H_1, H_2, $\{e_n : n \in \mathbb{N}\}$, and $\{e'_n : n \in \mathbb{N}\}$ be as above. If (a_{jk}) is a matrix such that $\sum_{j=1}^{\infty}|a_{jk}|^2 < \infty$ for all $k \in \mathbb{N}$, and A is the operator from H_1 into H_2 defined by (6.9), then the following holds: $D(A)$ is dense in H_1, A^* is a restriction of the operator A^+ from H_2 into H_1 induced by the* *adjoint matrix* $(a_{jk}^+) = (a_{kj}^*)$ *analogously to* (6.9):

$$D(A^+) = \left\{ g \in H_2 : \lim_{m\to\infty} \sum_{k=1}^{m} a_{jk}^+ \langle e_k', g\rangle \text{ exists for all } j \in \mathbb{N}, \text{ and } \sum_{j=1}^{\infty} \left| \sum_{k=1}^{\infty} a_{jk}^+ \langle e_k', g\rangle \right|^2 < \infty \right\}.$$

$$A^+ g = \sum_{j=1}^{\infty} \left(\sum_{k=1}^{\infty} a_{jk}^+ \langle e_k', g\rangle \right) e_j \quad \text{for} \quad g \in D(A^+).$$

If we also have $\sum_{k=1}^{\infty} |a_{jk}|^2 < \infty$ *for all* $j \in \mathbb{N}$, *then* A *is closed and* A^* *is densely defined.*

Proof. For all $n \in \mathbb{N}$ we have

$$\sum_{k=1}^{\infty} a_{jk} \langle e_k, e_n\rangle = a_{jn} \quad \text{and} \quad \sum_{j=1}^{\infty} \left| \sum_{k=1}^{\infty} a_{jk} \langle e_k, e_n\rangle \right|^2 = \sum_{j=1}^{\infty} |a_{jn}|^2 < \infty.$$

Consequently, $L(\{e_n : n \in \mathbb{N}\}) \subset D(A)$; so $D(A)$ is dense. Let $A_0 = A|_{L(\{e_n : n \in \mathbb{N}\})}$. We show that $A_0^* = A^+$; this then implies that $A^* \subset A^+$. It is easy to see that A_0 and A^+ are formal adjoints of each other, i.e., that we have $A^+ \subset A_0^*$. What remains is to prove that $D(A_0^*) \subset D(A^+)$. Let $g \in D(A_0^*)$. Then, because of the relation $e_k \in D(A_0)$, we have for every $k \in \mathbb{N}$ that

$$\langle e_k, A_0^* g\rangle = \langle A_0 e_k, g\rangle = \left\langle \sum_{j=1}^{\infty} a_{jk} e_j', g \right\rangle = \sum_{j=1}^{\infty} a_{jk}^* \langle e_j', g\rangle.$$

Consequently,

$$\sum_{k=1}^{\infty} \left| \sum_{j=1}^{\infty} a_{jk}^* \langle e_j', g\rangle \right|^2 = \|A_0^* g\|^2 < \infty,$$

i.e., $g \in D(A^+)$.

If we also have $\sum_{k=1}^{\infty} |a_{jk}|^2 < \infty$, then $A = (A_0^+)^*$, where $A_0^+ = A^+|_{L(\{e_n' : n \in \mathbb{N}\})}$. Hence A is closed, and A^* is densely defined. □

Theorem 6.21. *Let* H_1 *and* H_2 *be separable Hilbert spaces, and let* T *be a densely defined operator from* H_1 *into* H_2. *The operator* T *is closable if and only if there exist orthonormal bases* $\{e_n : n \in \mathbb{N}\}$ *of* H_1 *and* $\{e_n' : n \in \mathbb{N}\}$ *of* H_2, *and a matrix* (a_{jk}) *with the properties:* $\sum_{j=1}^{\infty} |a_{jk}|^2 < \infty$ *for all* $k \in \mathbb{N}$, $\sum_{k=1}^{\infty} |a_{jk}|^2 < \infty$ *for all* $j \in \mathbb{N}$, *and* T *is a restriction of the* *operator A defined by (6.9). The orthonormal bases can be chosen from $D(T)$ and $D(T^*)$, respectively.*

PROOF. If T has this form, then it is closable by Theorem 6.20. If T is closable, then there are orthonormal bases $\{e_n : n \in \mathbb{N}\}$ of H_1 in $D(T)$ and (since $D(T^*)$ is dense) $\{e'_n : n \in \mathbb{N}\}$ of H_2 in $D(T^*)$. With

$$a_{jk} = \langle e'_j, Te_k \rangle = \langle T^*e'_j, e_j \rangle, \quad (j, k) \in \mathbb{N} \times \mathbb{N}$$

we then have

$$\sum_{j=1}^{\infty} |a_{jk}|^2 = \|Te_k\|^2 < \infty, \quad \sum_{k=1}^{\infty} |a_{jk}|^2 = \|T^*e'_j\|^2 < \infty,$$

for all $k \in \mathbb{N}$ and for all $j \in \mathbb{N}$, respectively. For every $f \in D(T)$

$$\begin{aligned}\sum_{j=1}^{\infty} \left| \sum_{k=1}^{\infty} a_{jk} \langle e_k, f \rangle \right|^2 &= \sum_{j=1}^{\infty} \left| \sum_{k=1}^{\infty} \langle T^*e'_j, e_k \rangle \langle e_k, f \rangle \right|^2 \\ &= \sum_{j=1}^{\infty} |\langle T^*e'_j, f \rangle|^2 = \sum_{j=1}^{\infty} |\langle e'_j, Tf \rangle|^2 \\ &= \|Tf\|^2 < \infty.\end{aligned}$$

Therefore $f \in D(A)$. Moreover, we have

$$\begin{aligned}Tf &= \sum_{j=1}^{\infty} \langle e'_j, Tf \rangle e'_j = \sum_{j=1}^{\infty} \langle T^*e'_j, f \rangle e'_j = \sum_{j=1}^{\infty} \left(\sum_{k=1}^{\infty} \langle T^*e'_j, e_k \rangle \langle e_k, f \rangle \right) e'_j \\ &= \sum_{j=1}^{\infty} \left(\sum_{k=1}^{\infty} a_{jk} \langle e_k, f \rangle \right) e'_j = Af.\end{aligned}$$

□

Proposition. *If T is a symmetric operator on a separable Hilbert space H, then there exists an orthonormal basis $\{e_n : n \in \mathbb{N}\}$ and a Hermitian matrix (a_{jk}) such that $\sum_{k=1}^{\infty} |a_{jk}|^2 < \infty$ for all $j \in \mathbb{N}$, and*

$$Tf = \sum_{j=1}^{\infty} \left(\sum_{k=1}^{\infty} a_{jk} \langle e_k, f \rangle \right) e_j \quad \text{for all} \quad f \in D(T).$$

PROOF. In the proof of the preceding theorem choose for $\{e_n : n \in \mathbb{N}\}$ and $\{e'_n : n \in \mathbb{N}\}$ the same orthonormal basis in $D(T) \subset D(T^*)$. The series of equalities

$$a_{jk} = \langle e_j, Te_k \rangle = \langle Te_j, e_k \rangle = \langle e_k, Te_j \rangle^* = a^*_{kj}$$

shows this matrix is Hermitian. □

Corollary. *If T is a symmetric operator on l_2 and $l_{2,0} \subset D(T)$, then there exists a Hermitian matrix (a_{jk}) such that $\Sigma_{k=1}^{\infty}|a_{jk}|^2 < \infty$ for all $j \in \mathbb{N}$ and*

$$Tf = T(f_n) = \left(\sum_{k=1}^{\infty} a_{nk} f_k\right)_{n \in \mathbb{N}} \quad \text{for all} \quad f \in D(T).$$

This can be deduced from the previous proposition if we choose the basis $\{e_n : n \in \mathbb{N}\}$ so that $e_n = (\delta_{nj})_{j \in \mathbb{N}}$.

Now we prove some simple criteria for the boundedness of the operators that are induced by matrices in the sense of (6.9).

Theorem 6.22. *If $\Sigma_{j,k}|a_{jk}|^2 = C^2 < \infty$, then the operator A defined by* (6.9) *is a Hilbert-Schmidt operator, and $|||A||| = C$.*

PROOF. By Theorem 6.20, A is densely defined and closed. For the basis $\{e_n : n \in \mathbb{N}\}$ we have

$$\sum_n \|Ae_n\|^2 = \sum_n \left\| \sum_j \left(\sum_k a_{jk} \langle e_k, e_n \rangle \right) e_j' \right\|^2 = \sum_{n,j} |a_{jn}|^2 = C^2.$$

Therefore, by Theorem 6.10, A is a restriction of a Hilbert-Schmidt operator. Since A is densely defined and bounded, we have $D(A) = \overline{D(A)} = H_1$, and thus $A \in B(H_1, H_2)$. Hence A is a Hilbert-Schmidt operator, and $|||A||| = C$. □

Theorem 6.23. *Let $a_{jk} = b_{jk} c_{jk}$; furthermore, let*

$$\sum_k |b_{jk}|^2 \leqslant C_1^2 \quad \text{for all} \quad j \in \mathbb{N},$$

and

$$\sum_j |c_{jk}|^2 \leqslant C_2^2 \quad \text{for all} \quad k \in \mathbb{N},$$

then the operator A defined by (6.9) *is from $B(H_1, H_2)$, and $\|A\| \leqslant C_1 C_2$.*

PROOF. For every $f \in H_1$ we have

$$\begin{aligned}
\sum_j \left| \sum_k a_{jk} \langle e_k, f \rangle \right|^2 &= \sum_j \left| \sum_k b_{jk} c_{jk} \langle e_k, f \rangle \right|^2 \\
&\leqslant \sum_j \left\{ \sum_k |b_{jk}|^2 \sum_k |c_{jk} \langle e_k, f \rangle|^2 \right\} \\
&\leqslant C_1^2 \sum_{j,k} |c_{jk}|^2 |\langle e_k, f \rangle|^2 \\
&\leqslant C_1^2 C_2^2 \sum_k |\langle e_k, f \rangle|^2 \\
&= C_1^2 C_2^2 \|f\|^2.
\end{aligned}$$

It follows from this that $H_1 = D(A)$ and $\|Af\| \leqslant C_1C_2\|f\|$ for all $f \in D(A) = H_1$. □

Proposition. *The assumptions of Theorem 6.23 are satisfied in particular if*

$$\sum_k |a_{jk}| \leqslant C_1^2 \quad \textit{for all} \quad j \in \mathbb{N}$$

and

$$\sum_j |a_{jk}| \leqslant C_2^2 \quad \textit{for all} \quad k \in \mathbb{N}.$$

For the *proof*, in Theorem 6.23 let us take $b_{jk} = |a_{jk}|^{1/2}$ and $c_{jk} = |a_{jk}|^{1/2} \operatorname{sgn} a_{jk}$, where $\operatorname{sgn} a = 0$ for $a = 0$ and $\operatorname{sgn} a = |a|^{-1}a$ for $a \neq 0$.

EXAMPLE 1. The matrix

$$a_{jk} = \begin{cases} j^{-1} & \text{for } k \leqslant j, \\ 0 & \text{for } k > j \end{cases}$$

defines a bounded operator A by (6.9) and we have $\|A\| \leqslant \sqrt{6}$. For the proof, let us define $b_{jk} = c_{jk} = 0$ for $j < k$ and

$$b_{jk} = j^{-1/4}k^{-1/4}, \; c_{jk} = j^{-3/4}k^{1/4} \quad \text{for} \quad k \leqslant j.$$

Then we have

$$\sum_k |b_{jk}|^2 = j^{-1/2} \sum_{k=1}^{j} k^{-1/2} \leqslant j^{-1/2} \int_0^j x^{-1/2}\, \mathrm{d}x = 2,$$

and

$$\begin{aligned}\sum_j |c_{jk}|^2 &= k^{1/2} \sum_{j=k}^{\infty} j^{-3/2} \leqslant k^{1/2}\left\{k^{-3/2} + \int_k^\infty x^{-3/2}\, \mathrm{d}x\right\} \\ &= k^{-1} + 2 \leqslant 3,\end{aligned}$$

for all $j \in \mathbb{N}$ and $k \in \mathbb{N}$, respectively. The assertion follows from this with the aid of Theorem 6.23.

Similar arguments can be made for operators T from $L_2(M_1)$ into $L_2(M_2)$, that are induced by a measurable function (*kernel*) $K: M_2 \times M_1 \rightarrow \mathbb{C}$. Let K be such a kernel. The equalities

$$D(T_K) = \left\{ f \in L_2(M_1) : K(x, .)f \text{ is integrable over } M_1 \text{ almost everywhere in } M_2, \text{ and } \int_{M_1} K(.\,, y) f(y)\, \mathrm{d}y \text{ is in } L_2(M_2) \right\},$$

$$T_K f(x) = \int_{M_1} K(x, y) f(y)\, \mathrm{d}y \text{ almost everywhere in } M_2 \tag{6.10}$$

define a linear operator from $L_2(M_1)$ into $L_2(M_2)$ (notice that the notation T_K is compatible with that of Section 6.2). Special cases of such operators (Hilbert-Schmidt operators and Carleman operators) have already been studied in Section 6.2. Now we want to give a further criterion for the boundedness of an integral operator (formally this is an exact analogue of Theorem 6.23 for matrix operators).

Theorem 6.24. *Let $k : M_2 \times M_1 \to \mathbb{C}$ be measurable, and let K_1 and K_2 be measurable functions defined on $M_2 \times M_1$ such that $K(x, y) = K_1(x, y)K_2(x, y)$ and*

$$\int_{M_1} |K_1(x, y)|^2 \, dy \leq C_1^2 \quad \textit{almost everywhere in } M_2,$$

$$\int_{M_2} |K_2(x, y)|^2 \, dx \leq C_2^2 \quad \textit{almost everywhere in } M_1.$$

Then the operator T_K is in $B(L_2(M_1), L_2(M_2))$, and $\|T_K\| \leq C_1C_2$. The adjoint T_K^* is equal to the operator T_{K^+} induced by the adjoint kernel K^+.

Proof. For every $r > 0$ let $M_2(r) = \{x \in M_2 : |x| \leq r\}$. The Lebesque measure of $M_2(r)$ is finite. For every $f \in L_2(M_1)$ Fubini's theorem implies that

$$\int_{M_2(r)\times M_1} |K(x, y)f(y)| dx \, dy \leq \int_{M_2(r)} \left\{ 1 + \left[\int_{M_1} |K(x, y)f(y)| \, dy \right]^2 \right\} dx$$

$$= \lambda(M_2(r)) + \int_{M_2(r)} \left[\int_{M_1} |K_1(x, y)| \, |K_2(x, y)f(y)| \, dy \right]^2 dx$$

$$\leq \lambda(M_2(r)) + \int_{M_2(r)} \left[\int_{M_1} |K_1(x, y)|^2 \, dy \int_{M_1} |K_2(x, y)f(y)|^2 \, dy \right] dx$$

$$\leq \lambda(M_2(r)) + C_1^2 \int_{M_1} |f(y)|^2 \left[\int_{M_2} |K_2(x, y)|^2 \, dx \right] dy$$

$$\leq \lambda(M_2(r)) + C_1^2 C_2^2 \|f\|^2 < \infty.$$

By Fubini's theorem again, the integral $\int_{M_1} K(x, y)f(y) \, dy$ exists almost everywhere in $M_2(r)$ and defines a measurable function there. Since this holds for all $r > 0$, the equality

$$T_K f(x) = \int_{M_1} K(x, y)f(y) \, dy \quad \text{almost everywhere in } M_2$$

defines a measurable function on M_2 for every $f \in L_2(M_1)$. For all those $x \in M_2$, for which this integral exists, we have

$$|T_K f(x)| \leq \int_{M_1} |K_1(x, y)| \, |K_2(x, y)f(y)| \, dy$$

$$\leq C_1 \left\{ \int_{M_1} |K_2(x, y)f(y)|^2 \, dy \right\}^{1/2},$$

and thus

$$\int_{M_2} |T_K f(x)|^2 \, dx \leq C_1^2 \int_{M_2} \int_{M_1} |K_2(x, y) f(y)|^2 \, dy \, dx$$
$$= C_1^2 \int_{M_1} |f(y)|^2 \int_{M_2} |K_2(x, y)|^2 \, dx \, dy \leq C_1^2 C_2^2 \|f\|^2.$$

Hence, for every $f \in L_2(M_1)$ the function $T_K f$ belongs to $L_2(M_2)$, and $\|T_K f\| \leq C_1 C_2 \|f\|$. Since the mapping $f \mapsto T_K f$ is obviously linear, we then have $T_K \in B(L_2(M_1), L_2(M_2))$ with $\|T_K\| \leq C_1 C_2$. The corresponding arguments show that K^+ induces an operator $T_{K^+} \in B(L_2(M_2), L_2(M_1))$ and that for all $f \in L_2(M_1)$ and $g \in L_2(M_2)$ the integral

$$\int_{M_2} |g(x)| \int_{M_1} |K(x, y) f(y)| \, dy \, dx$$

exists. Therefore, it follows from Fubini's theorem that

$$\langle g, T_K f \rangle = \int_{M_2} g(x)^* \int_{M_1} K(x, y) f(y) \, dy \, dx$$
$$= \int_{M_1} f(y) \left\{ \int_{M_2} K^+(y, x) g(x) \, dx \right\}^* dy = \langle T_{K^+} g, f \rangle.$$

Hence, $T_K^* = T_{K^+}$. □

Corollary. *If $K : M_2 \times M_1 \to \mathbb{C}$ is measurable, and*

$$\int_{M_1} |K(x, y)| \, dy \leq C_1^2 \quad \textit{almost everywhere in } M_2,$$

$$\int_{M_2} |K(x, y)| \, dx \leq C_2^2 \quad \textit{almost everywhere in } M_1,$$

then the operator T_K defined by (6.10) *is from $B(L_2(M_1), L_2(M_2))$, and $\|T_K\| \leq C_1 C_2$.*

This follows from Theorem 6.24 if we take $K_1(x, y) = |K(x, y)|^{1/2}$ and $K_2(x, y) = |K(x, y)|^{1/2} \operatorname{sgn} K(x, y)$.

It is important to observe that the operators occurring in Theorem 6.24 are not necessarily compact (cf. the following example).

EXAMPLE 2. Let $M_1 = M_2 = \mathbb{R}$. The kernel $K(x, y) = \exp(-|x - y|)$ satisfies the assumption of the above corollary. Assume that $f \in L_2(M_1)$, $f \neq 0$, and $f(x) \geq 0$ almost everywhere. Then $T_K f \neq 0$. Let us set $f_n(x) = f(x - n)$. Then $f_n \xrightarrow{w} 0$, and

$$T_K f_n(x) = \int \exp(-|x - y|) f(y - n) \, dy$$
$$= \int \exp(-|(x - n) - y|) f(y) \, dy = (T_K f)(x - n).$$

Consequently, $\|T_K f_n\| = \|T_K f\|$ for all n. So we do not have $T_K f_n \to 0$, i.e., T_K is not compact.

Theorem 6.25. *Let $K : M_2 \times M_1 \to \mathbb{C}$ be measurable, and let K_1, K_2 be measurable such that $K(x,y) = K_1(x,y)K_2(x,y)$. Let (M_{1j}) and (M_{2j}) be increasing sequences of measurable subsets of M_1 and M_2 such that $M_1 = \bigcup_{j=1}^{\infty} M_{1j}$ and $M_2 = \bigcup_{j=1}^{\infty} M_{2j}$, respectively. Assume that for all $j \in \mathbb{N}$*

$$\int_{M_{2j} \times M_{1j}} |K(x,y)|^2 \,dx\,dy < \infty,$$

and for every $\epsilon > 0$ there exists a $j_0 = j_0(\epsilon) \in \mathbb{N}$ for which

$$\int_{M_1} |K_1(x,y)|^2 \,dy \leqslant \epsilon \quad \textit{almost everywhere in } M_2 \setminus M_{2j_0},$$

$$\int_{M_2} |K_2(x,y)|^2 \,dx \leqslant \epsilon \quad \textit{almost everywhere in } M_1 \setminus M_{1j_0},$$

$$\int_{M_1 \setminus M_{1j_0}} |K_1(x,y)|^2 \,dy \leqslant \epsilon \quad \textit{almost everywhere in } M_2,$$

$$\int_{M_2 \setminus M_{2j_0}} |K_2(x,y)|^2 \,dx \leqslant \epsilon \quad \textit{almost everywhere in } M_1.$$

Then $T_K \in B(L_2(M_1), L_2(M_2))$, and T_K is compact.

Proof. Let us set

$$H_j(x,y) = \begin{cases} K(x,y) & \text{for} \quad (x,y) \in M_{2j} \times M_{1j}, \\ 0 & \text{otherwise} \end{cases}$$

and $L_j = K - H_j$. Then T_{H_j} is a Hilbert-Schmidt (hence compact) operator, and $\|T_{L_j}\| \leqslant \epsilon$ provided that $j \geqslant j_0(\epsilon)$. The operator T_K is therefore the limit of a sequence of Hilbert-Schmidt operators, so that it is compact (Theorem 6.4(e)). □

Example 3. Let $M \subset \mathbb{R}^m$ be measurable and let

$$K(x,y) = f_1(x)f_2(y)f_3(x-y), \quad (x,y) \in M \times M;$$

where f_1 and f_2 are bounded measurable functions defined on $\mathbb{R}^m$ and $f_j(x) \to 0$ as $|x| \to \infty (j = 1, 2)$; f_3 is measurable and $\int |f_3(x)|\,dx < \infty$. With $M_{1j} = M_{2j} = \{x \in M : |x| \leqslant j\}$ and

$$K_1(x,y) = f_1(x)|f_2(x-y)|^{1/2}$$

$$K_2(x,y) = f_2(y)|f_3(x-y)|^{1/2} \operatorname{sgn} f_3(x-y),$$

the above theorem is applicable; hence T_K is compact.

EXAMPLE 4. Let M be a bounded measurable subset of $\mathbb{R}^m$, and let

$$K(x,y) = \begin{cases} |x-y|^{\alpha-m}H(x,y) & \text{for } (x,y)\in M\times M,\ x\neq y,\\ 0 \ \text{for} \ x=y\in M; \end{cases}$$

where $\alpha>0$, and H is a bounded measurable function on $M\times M$. If we set

$$K_n(x,y) = \begin{cases} K(x,y) & \text{for } |x-y|\geq \frac{1}{n},\\ 0 \ \text{for} \ |x-y|<\frac{1}{n} \end{cases}$$

and $L_n = K - K_n$, then the K_n induce Hilbert-Schmidt operators, and the operator induced by L_n converges to 0 as $n\to\infty$. Hence T_K is compact.

6.4 Differential operators on $L_2(a, b)$ with constant coefficients

In the following let (a, b) be an arbitrary (non-empty) open interval in $\mathbb{R}$, i.e., let $-\infty\leq a<b\leq\infty$. Furthermore, let

$$A_n(a,b) = \{f:(a,b)\to\mathbb{C} : f, f', \ldots, f^{(n-2)} \text{ are continuously differentiable on } (a,b) \text{ and } f^{(n-1)} \text{ is absolutely continuous on } (a,b)\}.$$

Hence, for $f\in A_n(a,b)$ there exists an "nth derivative" $f^{(n)}$ for which the following holds: $f^{(n)}$ is integrable over every compact subinterval of (a, b), and for $a<\alpha<\beta<b$ and every function g, absolutely continuous on (a, b), we have (cf. Appendix A5)

$$\int_\alpha^\beta g(x)f^{(j)}(x)\,dx = g(\beta)f^{(j-1)}(\beta) - g(\alpha)f^{(j-1)}(\alpha) - \int_\alpha^\beta g'(x)f^{(j-1)}(x)\,dx$$

for $j\in\{1,\ldots,n\}$.

Theorem 6.26. *For every $n\in\mathbb{N}$, every interval (a, b), and every $\epsilon>0$ there exists a $C>0$ such that for all $j\in\{0, 1,\ldots,n-1\}$ and all $f\in A_n(a,b)$*

$$\int_a^b|f^{(j)}(x)|^2\,dx \leq \epsilon\int_a^b|f^{(n)}(x)|^2\,dx + C\int_a^b|f(x)|^2\,dx.$$

Here we have to consider an integral to be equal to ∞ in case the integrand is not integrable. The relations $f\in A_n(a,b)\cap L_2(a,b)$ and $f^{(n)}\in L_2(a,b)$ therefore imply that $f^{(j)}\in L_2(a,b)$ for $j\in\{1,\ldots,n-1\}$.

Proof. We prove this by induction on n. For $n=1$ we only have to treat the case $j=0$. Then for every $\epsilon \geqslant 0$ we may choose $C=1$. The assertion is just as obvious for $n=2$ and $j=0$. Therefore, let $n=2, j=1$.

Let (α, β) be an arbitrary bounded subinterval of (a, b), let $L=\beta-\alpha>0$, and let $J_1=(\alpha, \alpha+\frac{1}{3}L)$, $J_2=[\alpha+\frac{1}{3}L, \beta-\frac{1}{3}L]$, $J_3=(\beta-\frac{1}{3}L, \beta)$. For arbitrary $s \in J_1$ and $t \in J_3$, by the mean value theorem there exists an $x_0=x_0(s, t) \in (\alpha, \beta)$ such that

$$f'(x_0) = (t-s)^{-1}[f(t)-f(s)].$$

With this x_0 it follows for all $x \in (\alpha, \beta)$ that

$$\begin{aligned} |f'(x)| &= \left| f'(x_0) + \int_{x_0}^{x} f''(y)\, dy \right| \\ &\leqslant 3L^{-1}(|f(s)|+|f(t)|) + \int_\alpha^\beta |f''(y)|\, dy. \end{aligned}$$

This holds for all $s \in J_1$ and $t \in J_3$. The inequality can be integrated over J_1 with respect to s and over J_3 with respect to t. We obtain that

$$\begin{aligned} 3^{-2}L^2|f'(x)| &\leqslant \int_{J_1} |f(s)|\, ds + \int_{J_3} |f(t)|\, dt + 3^{-2}L^2 \int_\alpha^\beta |f''(y)|\, dy \\ &\leqslant \int_\alpha^\beta |f(y)|\, dy + 3^{-2}L^2 \int_\alpha^\beta |f''(y)|\, dy \\ &\leqslant \left\{ L \int_\alpha^\beta |f(y)|^2\, dy \right\}^{1/2} + 3^{-2}L^2 \left\{ L \int_\alpha^\beta |f''(y)|^2\, dy \right\}^{1/2}. \end{aligned}$$

It follows from this that

$$|f'(x)|^2 \leqslant 2L\left\{ 3^4 L^{-4} \int_\alpha^\beta |f(y)|^2\, dy + \int_\alpha^\beta |f''(y)|^2\, dy \right\},$$

and thus by integration over (α, β) that

$$\int_\alpha^\beta |f'(x)|^2\, dx \leqslant 162L^{-2} \int_\alpha^\beta |f(y)|^2\, dy + 2L^2 \int_\alpha^\beta |f''(y)|^2\, dy.$$

If we divide (a, b) into (finitely or infinitely many) disjoint intervals of length L, then we obtain

$$\int_a^b |f'(x)|^2\, dx \leqslant 162L^{-2} \int_a^b |f(x)|^2\, dx + 2L^2 \int_a^b |f''(x)|^2\, dx.$$

As L can be chosen arbitrarily small, the assertion follows for $n=2$ and $j=1$.

Let us now assume that the assertion holds for $n \leqslant k$ $(k \geqslant 2)$. Let $f \in A_{k+1}(a, b)$. Since the theorem holds for $n=2$ and $j=1$, for every $\eta>0$ there exists a $C_1 \geqslant 0$ such that

$$\int_a^b |f^{(k)}(x)|^2\, dx \leqslant C_1 \int_a^b |f^{(k-1)}(x)|^2\, dx + \eta \int_a^b |f^{(k+1)}(x)|^2\, dx.$$

By the induction hypothesis there exists a $C_2 > 0$ for which

$$\int_a^b |f^{(k-1)}(x)|^2\, dx \leq C_2 \int_a^b |f(x)|^2\, dx + \tfrac{1}{2} C_1^{-1} \int_a^b |f^{(k)}(x)|^2\, dx;$$

consequently,

$$\int_a^b |f^{(k)}(x)|^2\, dx \leq \tfrac{1}{2} \int_a^b |f^{(k)}(x)|^2\, dx + C_1 C_2 \int_a^b |f(x)|^2\, dx + \eta \int_a^b |f^{(k+1)}(x)|^2\, dx,$$

and thus

$$\int_a^b |f^{(k)}(x)|^2\, dx \leq 2C_1 C_2 \int_a^b |f(x)|^2\, dx + 2\eta \int_a^b |f^{(k+1)}(x)|^2\, dx.$$

This is the assertion for $n = k+1$ and $j = k$. For $j < k$ the assertion now follows easily with the aid of the induction hypothesis. □

In what follows we shall use the notation

$$W_{2,n}(a, b) = \{ f \in A_n(a, b) \cap L_2(a, b) : f^{(n)} \in L_2(a, b) \}.$$

$W_{2,n}(a, b)$ is called the *Sobolev space* of order n over (a, b).

Theorem 6.27. *Let (a, b) be an arbitrary open interval in $\mathbb{R}$, and let $f \in W_{2,n}(a, b)$. If $-\infty < a$, then $f^{(j)}$ can be extended continuously to a for all $j \in \{0, 1, \ldots, n-1\}$; if $a = -\infty$, then $\lim_{x \to -\infty} f^{(j)}(x) = 0$. The corresponding assertion holds for the point b.*

PROOF. Let $c \in (a, b)$. If $a > -\infty$, then

$$\int_a^c |f^{(j+1)}(x)|\, dx \leq \left\{ (c-a) \int_a^c |f^{(j+1)}(x)|^2\, dx \right\}^{1/2} < \infty,$$

because $f^{(j+1)} \in L_2(a, b)$ (Theorem 6.26). Hence the limit

$$\lim_{x \to a} f^{(j)}(x) = \lim_{x \to a} \left\{ f^{(j)}(c) - \int_x^c f^{(j+1)}(s)\, ds \right\} = f^{(j)}(c) - \int_a^c f^{(j+1)}(s)\, ds$$

exists.

Now let $a = -\infty$. For all $x \in (-\infty, c)$ we have

$$\int_x^c f^{(j)}(s) f^{(j+1)}(s)\, dx = \tfrac{1}{2}\{ f^{(j)}(c)^2 - f^{(j)}(x)^2 \}.$$

The integral here converges as $x \to -\infty$, therefore the limit $\lim_{x \to -\infty} f^{(j)}(x)^2$ also exists. If this limit were different from zero, then $f^{(j)}$ could not be in $L_2(a, c)$; which would contradict Theorem 6.26. □

Now we shall study differential operators on $L_2(a, b)$ that are induced by the differential form $\tau_n = (-i)^n(d^n/dx^n)$. The *minimal operator* $T_{n,0}$ *induced by* τ_n *is defined by*

$$D(T_{n,0}) = C_0^\infty(a, b) \quad \text{and} \quad T_{n,0}f = \tau_n f \quad \text{for} \quad f \in D(T_{n,0}).$$

The *maximal operator* T_n *induced by* τ_n is defined by

$$D(T_n) = W_{2,n}(a, b) \quad \text{and} \quad T_n f = \tau_n f \quad \text{for} \quad f \in D(T_n).$$

The maximal operator is obviously defined on the largest possible subspace on which τ_n can act meaningfully. Below we shall show that $T_{n,0}^* = T_n$; if $A \subset T_{n,0}$ and $\bar{A} \neq \overline{T_{n,0}}$, then $A^* \supsetneqq T_{n,0}^* = T_n$, i.e., A^* is not a differential operator induced by τ_n; this is why $T_{n,0}$ is called the minimal operator (it is also usual to call $\overline{T_{n,0}}$ the minimal operator). For $n=0$ we obviously have $\overline{T_{0,0}} = T_0 = I$.

Theorem 6.28. *We have $k \in R(T_{n,0})$ if and only if $k \in C_0^\infty(a, b)$ and*

$$\int_a^b x^j k(x)\,dx = 0 \quad \text{for all} \quad j \in \{0, 1, \ldots, n-1\}.$$

PROOF. If $k = T_{n,0}g$ for some $g \in D(T_{n,0}) = C_0^\infty(a, b)$, then we obviously have $k \in C_0^\infty(a, b)$, and for $j \in \{0, 1, \ldots, n-1\}$

$$\int_a^b x^j k(x)\,dx = \int_a^b x^j(\tau_n g)(x)\,dx = (-1)^n \int_a^b (\tau_n x^j) g(x)\,dx = 0.$$

Conversely, assume now that $k \in C_0^\infty(a, b)$ has this property, and $[\alpha, \beta]$ is a compact subinterval of (a, b) that contains the support of k. Set

$$g(x) = (i)^n \int_a^x \int_a^{x_n} \cdots \int_a^{x_2} k(x_1)\,dx_1 \ldots dx_n.$$

Then we obviously have $g \in C^\infty(a, b)$ and $g(x) = 0$ for $x \in (a, \alpha]$. For $x \in [\beta, b)$ we have

$$\begin{aligned} g(x) &= (i)^n \int_a^x \int_{x_1}^x \cdots \int_{x_{n-1}}^x k(x_1)\,dx_n \ldots dx_1 \\ &= (i)^n \int_a^x k(x_1)\left\{\int_{x_1}^x \cdots \int_{x_{n-1}}^x dx_n \ldots dx_2\right\} dx_1 \\ &= (i)^n \int_a^b k(x_1)p(x_1)\,dx_1 = 0, \end{aligned}$$

because $p(x_1)$ is a polynomial of degree $n-1$. Hence, $g \in C_0^\infty(a, b)$ and $k = T_{n,0}g$. □

Theorem 6.29. *We have $T_{n,0}^* = T_n$.*

PROOF. Let $f \in D(T_n)$, $g \in D(T_{n,0}) = C_0^\infty(a, g)$, and let $[\alpha, \beta]$ be a compact subinterval of (a, b) that contains the support of g. It follows via integration by parts that

$$\begin{aligned}\langle T_n f, g\rangle &= \int_a^b (T_n f)^*(x) g(x)\,dx = \int_a^b (\tau_n f)^*(x) g(x)\,dx \\ &= \int_\alpha^\beta (\tau_n f)^*(x) g(x)\,dx = \int_\alpha^\beta f^*(x)(\tau_n g)(x)\,dx \\ &= \int_a^b f^*(x)(\tau_n g)(x)\,dx = \langle f, T_{n,0} g\rangle\end{aligned}$$

(the integrated terms vanish, since g, together with all of its derivatives, vanishes at α and β). Consequently, $T_{n,0}$ and T_n are formal adjoints of each other. What remains is to prove the inclusion $T_{n,0}^* \subset T_n$. Let $f \in D(T_{n,0}^*)$ and $h \in A_n(a, b)$ such that $\tau_n h = T_{n,0}^* f$. Such an h exists; for example we can define h in the following way:

$$h(x) = (\mathrm{i})^n \int_c^x \int_c^{x_n} \dots \int_c^{x_2} (T_{n,0}^* f)(x_1)\,dx_1 \dots dx_n$$

with some $c \in (a, b)$. Then it follows by integration by parts (let $[\alpha, \beta]$ be chosen as above) that for all $g \in D(T_{n,0})$

$$\begin{aligned}\langle f, T_{n,0} g\rangle = \langle T_{n,0}^* f, g\rangle &= \int_\alpha^\beta (T_{n,0}^* f)^*(x) g(x)\,dx \\ &= \int_\alpha^\beta h^*(x)\tau_n g(x)\,dx = \int_a^b h^*(x) T_{n,0} g(x)\,dx.\end{aligned}$$

Hence, for all $k \in R(T_{n,0})$ we have

$$\int_a^b (f(x) - h(x))^* k(x)\,dx = 0.$$

The null space of the linear functional $F : C_0^\infty(a, b) \to \mathbb{C}$ defined as

$$F(k) = \int_a^b (f(x) - h(x))^* k(x)\,dx \quad \text{for} \quad k \in C_0^\infty(a, b)$$

contains therefore $R(T_{n,0})$. By Theorem 6.28 we have

$$R(T_{n,0}) = \bigcap_{j=0}^{n-1} N(F_j);$$

where the F_j are the linear functionals

$$F_j : C_0^\infty(a, b) \to \mathbb{C}, \quad F_j(k) = \int_a^b x^j k(x)\,dx.$$

By Theorem 4.1 there exist complex numbers $c_0, c_1, \dots, c_{n-1}$ such that

$$F = \sum_{j=0}^{n-1} c_j F_j.$$

Hence, with $p(x) = \sum_{j=0}^{n-1} c_j^* x^j$ we have

$$\int_a^b (f(x) - h(x) - p(x))^* k(x)\,dx = 0 \quad \text{for all} \quad k \in C_0^\infty(a, b).$$

For every compact subinterval $[\alpha, \beta]$ we have $(f-h-p)|_{[\alpha,\beta]} \in L_2(\alpha, \beta)$ and $(f-h-p)|_{[\alpha,\beta]} \perp C_0^\infty(\alpha, \beta)$, and thus $(f-h-p)|_{[\alpha,\beta]}=0$. Since this holds for every compact subinterval $[\alpha, \beta]$, it follows that

$$f(x) = h(x) + p(x) \quad \text{almost everywhere in } (a, b).$$

It follows from this that $f \in A_n(a, b) \cap L_2(a, b)$ and $\tau_n f = T_{n,0}^* f \in L_2(a, b)$. Hence $f \in D(T_n)$. □

Theorem 6.30. *In the case $(a, b) = \mathbb{R}$ we have $\overline{T_{n,0}} = T_n = T_n^*$; thus $T_{n,0}$ is essentially self-adjoint and T_n is self-adjoint.*

PROOF. The relations $T_{n,0}^* = T_n$ and $T_{n,0} \subset T_n$ imply that $T_n^* \subset T_{n,0}^* = T_n$. We show that T_n is symmetric; then it will follow that $T_n^* \subset T_n \subset T_n^*$; hence $T_n = T_n^* = T_{n,0}^{**} = \overline{T_{n,0}}$. For $f, g \in D(T_n)$ we have

$$\begin{aligned}\langle f, T_n g\rangle &= \lim_{c\to\infty} \int_{-c}^{c} f^*(x)(\tau_n g)(x)\,dx \\ &= \lim_{c\to\infty} \left\{R(c) + \int_{-c}^{c} (\tau_n f)^*(x) g(x)\,dx\right\} \\ &= \lim_{c\to\infty} \int_{-c}^{c} (\tau_n f)^*(x) g(x)\,dx = \langle T_n f, g\rangle,\end{aligned}$$

because $R(c)$ is a linear combination of terms of the form $f^{(j)}(\pm c)^* g^{(k)}(\pm c)$ with $j+k=n-1$, and so $R(c)\to 0$ as $c\to\infty$, by Theorem 6.27. □

Theorem 6.31. *In case $(a, b) \neq \mathbb{R}$ we have*

$$D(\overline{T_{n,0}}) = \left\{ f \in W_{2,n}(a, b): \begin{matrix} f(a) = f'(a) = \cdots = f^{(n-1)}(a) = 0 \text{ if } a > -\infty, \\ f(b) = f'(b) = \cdots = f^{(n-1)}(b) = 0 \text{ if } b < \infty \end{matrix} \right\}.$$

In this case $\overline{T_{n,0}} \neq T_n$ (for $n > 0$), and none of these operators are self-adjoint.

PROOF. We write $W_{2,n}^0(a, b)$ for the subspace given in the theorem. Let S_n be the operator induced on $W_{2,n}^0(a, b)$ by τ_n. Then one verifies easily that S_n and T_n are formal adjoints of each other. Therefore $S_n \subset T_n^* = \overline{T_{n,0}}$. Let $f \in D(T_n^*)$, and let $a > -\infty$. For every $j \in \{0, 1, \ldots, n-1\}$ there exists a $g_j \in D(T_n)$ such that $g_j^{(k)}(a) = \delta_{jk}$ for $k \in \{0, 1, \ldots, n-1\}$ and $g_j(x)$ vanishes identically in some neighborhood of b (we choose an arbitrary smooth function $\tilde{g}_j$ such that $\tilde{g}_j^{(k)}(a) = \delta_{jk}$, and set $g_j = \varphi \tilde{g}_j|_{(a,b)}$, where $\varphi \in C_0^\infty(\mathbb{R})$ and $\varphi(x) = 1$ in some neighborhood of a, $\varphi(x) = 0$ in some neighborhood of b). With these g_j we have

$$\begin{aligned}0 = \langle f, T_n g_j\rangle - \langle T_n^* f, g_j\rangle &= (-i)^n \sum_{k=0}^{n-1} (-1)^{n-k-1} f^{(n-k-1)}(a)^* g_j^{(k)}(a) \\ &= (-i)^n (-1)^{n-j-1} f^{(n-j-1)}(a)^*.\end{aligned}$$

As this holds for all $j \in \{0, 1, \ldots, n-1\}$, it follows that

$$f(a) = f'(a) = \cdots = f^{(n-1)}(a) = 0.$$

It follows analogously that

$$f(b) = f'(b) = \cdots = f^{(n-1)}(b) = 0,$$

in case $b < \infty$. Consequently $D(\overline{T_{n,0}}) = D(T_n^*) \subset W_{2,n}^0(a, b)$, and thus $S_n = \overline{T_{n,0}} = T_n^*$.

On the basis of our reasoning so far, it is clear that (for $n > 0$) $D(\overline{T_{n,0}}) \neq D(T_n)$, thus $\overline{T_{n,0}} \neq T_n$. Because of the equality $\overline{T_{n,0}}^* = T_n$, none of these operators are self-adjoint. □

The question of whether $T_{n,0}$ has self-adjoint extensions in the case $(a, b) \neq \mathbb{R}$, will in general be answered in Chapter 8. For even n we can now give the following theorem.

Theorem 6.32. *The operator $T_{2n,0}$ is non-negative. The Friedrichs extension $T_{2n,F}$ of $T_{2n,0}$ is given by the formulae*

$$D(T_{2n,F}) = \left\{ f \in D(T_{2n}) : \begin{array}{l} f(a) = f'(a) = \cdots = f^{(n-1)}(a) = 0, \text{ if } a > -\infty, \\ f(b) = f'(b) = \cdots = f^{(n-1)}(b) = 0, \text{ if } b < \infty \end{array} \right\}$$

$$= W_{2,2n}(a, b) \cap W_{2,n}^0(a, b),$$

$$T_{2n,F} f = \tau_{2n} f \quad \text{for} \quad f \in D(T_{2n,F}).$$

Proof. For all $f \in D(T_{2n,0}) = C_0^\infty(a, b)$ we obviously have

$$\langle f, T_{2n,0} f \rangle = (-1)^n \int_a^b f(x)^* f^{(2n)}(x)\, dx = \int_a^b |f^{(n)}(x)|^2\, dx \geq 0.$$

In this case the sesquilinear form s used in Theorem 5.38 is

$$s(f, g) = \langle f^{(n)}, g^{(n)} \rangle, \quad f, g \in C_0^\infty(a, b).$$

Therefore the s-norm $\|\cdot\|_s$ is equal to the $T_{n,0}$-norm $\|\cdot\|_{T_{n,0}}$ (cf. Theorem 5.1). Hence, the completion of $C_0^\infty(a, b)$ with respect to $\|\cdot\|_s$ is equal to the completion of $D(T_{n,0}) = C_0^\infty(a, b)$ with respect to $\|\cdot\|_{T_{n,0}}$, and thus it is equal to $D(\overline{T_{n,0}})$. Consequently, by Theorem 6.31

$$H_s = D(\overline{T_{n,0}}) = W_{2,n}^0(a, b)$$

$$= \left\{ f \in W_{2,n}(a, b) : \begin{array}{l} f(a) = f'(a) = \cdots = f^{(n-1)}(a) = 0 \text{ if } a > -\infty, \\ f(b) = f'(b) = \cdots = f^{(n-1)}(b) = 0 \text{ if } b < \infty \end{array} \right\}.$$

The assertion follows from this, because of the equalities $D(T_{2n,F}) = D(T^*_{2n,0}) \cap H_s = D(T_{2n}) \cap H_s$. □

In some cases we can explicitly calculate the spectrum of the self-adjoint operators induced by τ_n.

Theorem 6.33.
(a) *If* $(a, b) = \mathbb{R}$, *then*

$$\sigma(T_{2n}) = [0, \infty) \quad \textit{and} \quad \sigma(T_{2n-1}) = \mathbb{R} \quad \textit{for all} \quad n \in \mathbb{N}.$$

(b) *If* (a, b) *is a half-line* ((a, ∞) *or* $(-\infty, b)$), *then for every self-adjoint extension* A *of* $T_{2n,0}$ *we have*

$$\sigma(A) \supset [0, \infty).$$

For the Friedrichs extension $T_{2n,F}$ *of* $T_{2n,0}$ *we have*

$$\sigma(T_{2n,F}) = [0, \infty).$$

PROOF.
(a) $T_{2n} = \overline{T_{2n,0}}$ is non-negative. Hence for every $s<0$ we have

$$\|(s - T_{2n})f\|^2 = |s|^2\|f\|^2 - 2 \operatorname{Re} s\langle f, T_{2n}f\rangle + \|T_{2n}f\|^2 \geqslant |s|^2\|f\|^2,$$

i.e., $s - T_{2n}$ is continuously invertible. Therefore, $(-\infty, 0) \subset \rho(T_{2n})$. Now we show that every $s \geqslant 0$ lies in $\sigma(T_{2n})$. To this end, let $\varphi : \mathbb{R} \to \mathbb{R}$ be infinitely often differentiable and let

$$\varphi(x) = \begin{cases} 1 & \text{for} \quad x \leqslant 0, \\ 0 & \text{for} \quad x \geqslant 1, \end{cases} \qquad 0 \leqslant \varphi(x) \leqslant 1 \quad \text{for} \quad 0 < x < 1.$$

Furthermore, for all $m \in \mathbb{N}$ let

$$\varphi_m(x) = \varphi(|x| - m), \quad x \in \mathbb{R}.$$

Let us set (with $s^{1/2n} \geqslant 0$)

$$f_m(x) = (2m)^{-1/2}\varphi_m(x)\, e^{is^{1/2n}x}, \quad x \in \mathbb{R}.$$

Then $f_m \in C_0^\infty(\mathbb{R})$, $\|f_m\| \to 1$ as $m \to \infty$ and $(s - T_{2n})f_m = (s - \tau_{2n})f_m \to 0$, as a simple calculation shows. Consequently, $s - T_{2n}$ is not continuously invertible, and thus $[0, \infty) \subset \sigma(T_{2n})$.

We can show analogously that $\mathbb{R} \subset \sigma(T_{2n-1})$ if for every $s \in \mathbb{R}$ we define the sequence (f_m) of functions by the equalities

$$f_m(x) = (2m)^{-1/2}\varphi_m(x)\, e^{is^{1/(2n-1)}x}, \quad x \in \mathbb{R}$$

(here the root $s^{1/(2n-1)}$ has to be taken in such a way that $s^{1/(2n-1)}<0$ for $s<0$ and $s^{1/(2n-1)}\geq 0$ for $s\geq 0$).

(b) Without loss of generality we may assume that $(a, b)=(0, \infty)$. Then the inclusion $[0, \infty)\subset\sigma(A)$ follows as in part (a) if φ_m is replaced by

$$\psi_m(x) = \varphi_m(x-m-2)$$

(notice that $\psi_m\in C_0^\infty(0, \infty)$). Since the Friedrichs extension $T_{2n,F}$ is non-negative, it follows that $\sigma(T_{2n,F})\subset[0, \infty)$ (cf. the beginning of the proof of part (a)); consequently, $\sigma(T_{2n,F})=[0, \infty)$. □

Theorem 6.34. *Let T be a self-adjoint operator on $L_2(a, b)$ induced by τ_n (i.e., $T_{n,0}\subset T=T^*\subset T_n$). Let*

$$(\sigma f)(x) = \sum_{j=0}^{n} a_j(x)\frac{d^jf(x)}{dx^j}$$

be a differential expression such that $\sup\{|a_n(x)| : x\in(a, b)\}<1$ *and* $\sup\{|a_j(x)| : x\in(a, b), j=0, 1, \ldots, n-1\}<\infty$. *Assume that the operator S defined by the equalities*

$$D(S)=D(T) \quad \text{and} \quad Sf=\sigma f \quad \text{for} \quad f\in D(S)$$

is symmetric. Then $T+S$ is self-adjoint.

Proof. If $c=\sup\{|a_n(x)| : x\in(a, b)\}$, then by Theorem 6.26, the operator S is obviously T-bounded with T-bound $c<1$. The assertion follows from this by Theorem 5.28. □

Exercise

6.13 Let A be an (unbounded) self-adjoint operator in a Hilbert space H and P an orthogonal projection with $PD(A)\subset D(A)$. Then the operator PAP is in general not self-adjoint. Counterexamples may be constructed along the following lines: Let $H=H_0\oplus H_0$, B self-adjoint in H_0, C symmetric with B-bound <1 and $P(f_1, f_2)=(f_1, 0)$ for $(f_1, f_2)\in H=H_0\oplus H_0$.

(a) The operator $A_0\begin{pmatrix} 0 & B \\ B & 0\end{pmatrix}$ with $D(A_0)=D(B)\oplus D(B)$, $A_0(f_1, f_2)=(Bf_2, Bf_1)$ is self-adjoint.

(b) The operator $A=\begin{pmatrix} C & B \\ B & C\end{pmatrix}=A_0+\begin{pmatrix} C & 0 \\ 0 & C\end{pmatrix}$ is self-adjoint.

(c) If C is not (essentially) self-adjoint on $D(B)$, then PAP is not (essentially) self-adjoint. If C has no self-adjoint extension, then PAP has no self-adjoint extension.

(d) Possible examples for B and C are as follows:

(α) C essentially self-adjoint, but not self-adjoint: $B=-d^2/dx^2$, $C=id/dx$ in $L_2(\mathbb{R})$ with $D(B)=D(C)=W_{2,2}(\mathbb{R})$,

(β) C not essentially self-adjoint: $B=-d^2/dx^2$, $C=id/dx$ in $L_2(0, 1)$ with $D(B)=D(C)=\{f\in W_{2,2}(0, 1) : f(0)=f(1)=0\}$,

(γ) C has no self-adjoint extension: $B=-d^2/dx^2$, $C=id/dx$ in $L_2(0, \infty)$ with $D(B)=D(C)=\{f\in W_{2,2}(0, \infty) : f(0)=0\}$ (see also Section 8.2, Example 1).

7 The spectral theory of self-adjoint and normal operators

7.1 The spectral theorem for compact operators, the spaces $B_p(H_1, H_2)$

We studied the spectrum of compact operators thoroughly in Section 6.1. For compact normal operators the results obtained there may be sharpened.

Theorem 7.1. (The spectral theorem for compact normal operators.)

(a) *Let T be a compact normal operator on a complex Hilbert space H and let $\{\lambda_1, \lambda_2, \ldots\}$ be the non-zero eigenvalues of T; furthermore let $\{P_1, P_2, \ldots\}$ be the orthogonal (finite rank) projections onto the corresponding eigenspaces (cf. Theorems 6.7 and 5.41), then*

$$T = \sum_j \lambda_j P_j; \tag{7.1}$$

this series converges in the norm of $B(H)$. If T is self-adjoint, then this holds in real Hilbert spaces, as well.

(b) *If (λ_j) is a null-sequence (or a finite sequence) from $\mathbb{K}\setminus\{0\}$ such that $\lambda_j \neq \lambda_k$ for $j \neq k$, and the P_j are non-zero orthogonal projections of finite rank such that $P_j P_k = 0$ for $j \neq k$, then the series (7.1) is convergent in $B(H)$, and $T = \sum_j \lambda_j P_j$ is compact and normal. Furthermore, $\{\lambda_1, \lambda_2, \ldots\}$ is the set of non-zero eigenvalues of T and the $R(P_j)$ are the corresponding eigenspaces. The representation (7.1) is therefore unique in this sense. If the λ_j are real, then T is self-adjoint.*

PROOF.

(a) Let M be the closed linear hull of $\{R(P_j) : j = 1, 2, \ldots\}$, and let P be the orthogonal projection onto $M^\perp$.

If $M^\perp \neq \{0\}$, then for all $f \in M^\perp$, $g \in H$, and $j = 1, 2, \ldots$ we have

$$\langle Tf, P_j g\rangle = \langle f, T^* P_j g\rangle = \lambda_j^* \langle f, P_j g\rangle = 0,$$

i.e., $TM^\perp \subset M^\perp$; we can prove in just the same way that $T^* M^\perp \subset M^\perp$. Hence it follows that the restriction S of T to $M^\perp$ is a normal operator on the Hilbert space $M^\perp$. Every eigenvalue of S is an eigenvalue of T, every corresponding eigenvector of S is an eigenvector of T contained in $M^\perp$; therefore it follows that S can only have the eigenvalue 0 (since M is the closed linear hull of all eigenvectors of T that belong to non-zero eigenvalues). Therefore (cf. Theorem 6.7), $\sigma(S) = \{0\}$, and thus $r(S) = 0$ by Theorem 5.17(c) and (d). Hence $S = 0$ by Theorem 5.44, i.e., T vanishes on $M^\perp$.

Consequently, it follows for $f \in H$ (whether $M^\perp = \{0\}$ or $M^\perp \neq \{0\}$) that

$$Tf = TPf + T\sum_j P_j f = \sum_j TP_j f = \sum_j \lambda_j P_j f.$$

If the sequence (λ_j) is infinite, then for every $f \in H$ and $m \in \mathbb{N}$

$$\left\| \left(T - \sum_{j=1}^{m} \lambda_j P_j \right) f \right\|^2 = \sum_{j=m+1}^{\infty} |\lambda_j|^2 \|P_j f\|^2 \leqslant \sup\{|\lambda_j|^2 : j \geqslant m+1\} \|f\|^2.$$

Since the sequence (λ_j) is a null-sequence (cf. Theorem 6.7), this implies the norm convergence of (7.1).

(b) We can show the convergence of the series just as in part (a). For every $m \in \mathbb{N}$ the operator $\sum_{j=1}^{m} \lambda_j P_j$ is of finite rank, consequently compact. The compactness of T follows by Theorem 6.4(e). It is easy to verify that T is normal. All the λ_j are obviously eigenvalues of T, and every $f \in R(P_j)$ is an eigenelement of T belonging to the eigenvalue λ_j. If $\lambda \neq 0$ is an eigenvalue of T and $f \neq 0$ is a corresponding eigenvector, then

$$0 = \|(\lambda - T)f\|^2 = \sum_j |\lambda - \lambda_j|^2 \|P_j f\|^2 + |\lambda|^2 \left\| f - \sum_j P_j f \right\|^2,$$

i.e., $|\lambda - \lambda_j| \, \|P_j f\| = 0$ for all j and $f = \sum_j P_j f$. Since $f \neq 0$, there is a j_0 such that $P_{j_0} f \neq 0$. Hence $\lambda = \lambda_{j_0}$. Consequently, $\lambda \neq \lambda_j$ for all $j \neq j_0$, and thus $P_j f = 0$ for $j \neq j_0$. It follows from this that $f \in R(P_{j_0})$. □

Theorem 7.2. (The expansion theorem for compact normal operators). *If T is a compact normal operator on a complex Hilbert space, then there exists a zero-sequence (or a finite sequence) (μ_j) from $\mathbb{C}$ and an orthonormal sequence (f_j) from H such that*

$$Tf = \sum_j \mu_j \langle f_j, f\rangle f_j \quad \textit{for all} \quad f \in H. \tag{7.2}$$

Conversely, every operator defined by (7.2) *is compact and normal; the numbers μ_j are eigenvalues of T; the elements f_j are corresponding eigenvectors. If T is self-adjoint, then this is true in real Hilbert spaces, as well.*

PROOF. Let $T=\sum_j\lambda_jP_j$ be the representation from Theorem 7.1. For every j let $\{g_{j,1},g_{j,2},\dots,g_{j,k_j}\}$ be an orthonormal basis of $R(P_j)$; furthermore, let $\mu_{j,k}=\lambda_j$ for $k=1,2,\dots,k_j$. Then by Theorem 7.1 and Theorem 3.7

$$Tf=\sum_j\lambda_jP_jf=\sum_j\lambda_j\sum_k\langle g_{j,k},P_jf\rangle g_{j,k}=\sum_j\sum_k\mu_{j,k}\langle g_{j,k},f\rangle g_{j,k}.$$

(7.2) follows from this by changing the indices. We leave the rest of the proof to the reader. □

Theorem 7.3. *Let T be a compact normal operator on a complex Hilbert space, and let (λ_j) be the sequence of non-zero eigenvalues of T; every eigenvalue counted according to its multiplicity, and $|\lambda_n|\geqslant|\lambda_{n+1}|$ for all n. Then*

$$|\lambda_1|=\|T\|$$
$$|\lambda_{n+1}|=\inf_{g_1,\dots,g_n\in H}\sup\{\|Tf\|:f\in H, f\perp g_1,\dots,g_n,\|f\|=1\}\qquad(7.3)$$

for $n\in\mathbb{N}$ (cf. also Exercise 7.2). If T is self-adjoint, then this holds in real Hilbert spaces, as well.

PROOF. It follows from the relations $r(T)=\|T\|$ and $\sigma(T)\setminus\{0\}\subset\sigma_p(T)$ that $|\lambda_1|=\|T\|$ (cf. Theorem 5.17(c) and (d)). Let (f_j) be an orthonormal sequence such that $Tf_j=\lambda_jf_j$. If we choose $g_j=f_j$ for $j=1,\dots,n$, then for every $f\perp g_1,\dots,g_n$ we have

$$\|Tf\|^2=\Big\|\sum_{j>n}\lambda_j\langle f_j,f\rangle f_j\Big\|^2=\sum_{j>n}|\lambda_j|^2|\langle f_j,f\rangle|^2$$
$$\leqslant|\lambda_{n+1}|^2\sum_{j>n}|\langle f_j,f\rangle|^2\leqslant|\lambda_{n+1}|^2\|f\|^2.$$

It follows from this that $|\lambda_{n+1}|\geqslant\inf\sup\{\dots\}$. If $g_1,\dots,g_n$ are arbitrary, then there exists an $f\in L(f_1,\dots,f_{n+1})$ for which $\|f\|=1$ and $f\perp g_1,\dots,g_n$. We have for this f that

$$\|Tf\|^2=\Big\|\sum_{j=1}^{n+1}\lambda_j\langle f_j,f\rangle f_j\Big\|^2=\sum_{j=1}^{n+1}|\lambda_j|^2|\langle f_j,f\rangle|^2$$
$$=|\lambda_{n+1}|^2\|f\|^2\geqslant|\lambda_{n+1}|^2.$$

It follows from this that $|\lambda_{n+1}|\leqslant\inf\sup\{\dots\}$. Consequently, the theorem is proved. □

If $T = \Sigma_j \lambda_j P_j$ is the representation of the compact normal operator T, given in Theorem 7.1, then for all $n \in \mathbb{N}$ the equalities

$$T^n = \sum_j \lambda_j^n P_j.$$

obviously hold. If $\lambda_j^{1/n}$ is chosen in some way, then the operator

$$A_n = \sum_j \lambda_j^{1/n} P_j$$

has the property $(A_n)^n = T$. The roots $\lambda_j^{1/n}$ can be chosen in a unique way if we require that, for example, $0 \leqslant \arg \lambda_j^{1/n} < 2\pi/n$. Therefore, we have the following theorem.

Theorem 7.4. *Let $n \in \mathbb{N}$, $n \geqslant 2$. Every normal compact operator T on a complex Hilbert space has exactly one normal compact*[1] *nth root whose eigenvalues all lie in $\{z \in \mathbb{C} : 0 \leqslant \arg z < 2\pi/n\}$. Every non-negative self-adjoint compact operator has exactly one non-negative compact nth root.*

PROOF. The operator $A_n = \Sigma_j \lambda_j^{1/n} P_j$ with $0 \leqslant \arg \lambda_j^{1/n} < 2\pi/n$ has the required property. Let $B = \Sigma_k \mu_k Q_k$ be an operator having the same property. Then we have in particular that

$$\sum_k \mu_k^n Q_k = B^n = T = \sum_j \lambda_j P_j.$$

The uniqueness statement of Theorem 7.1 assures that $\mu_j^n = \lambda_j$ and $Q_j = P_j$ (this is true perhaps only after an appropriate reindexing). The inequalities $0 \leqslant \arg \mu_j < 2\pi/n$ imply that $\mu_j = \lambda_j^{1/n}$; consequently, $B = A_n$. If T is non-negative, then $\lambda_j \geqslant 0$ for all j; the condition $0 \leqslant \arg \lambda_j^{1/n} < 2\pi/n$ then implies that $\lambda_j^{1/n} \geqslant 0$. □

If T is a compact operator from H_1 into H_2, then T^*T is compact, self-adjoint, and non-negative. Hence we can define the absolute value of T by the equality $|T| = (T^*T)^{1/2}$, where $(T^*T)^{1/2}$ is the uniquely determined non-negative square root of T^*T (a definition for arbitrary densely defined and closed operators will be given in Section 7.3). $|T|$ is obviously compact. The term "absolute value" is justified by the following theorem.

[1] Actually, the compactness of the nth root does not have to be assumed. Every normal nth root A_n of a normal compact operator T is compact: If $(A_n)^n = T$, then $(A_n^* A_n)^n = T^*T$; consequently, $(A_n^* A_n)^n$ is compact. The compactness of $A_n^* A_n$ follows from this by Theorem 7.20 (for $n = 2$ this follows from Theorem 6.4(c) because of the equality $(A_2^* A_2)^2 = (A_2^* A_2)^*(A_2^* A_2)$). The compactness of A_n follows from Theorem 6.4(c).

Theorem 7.5. *Let T be compact. Then $\| |T|f\| = \|Tf\|$ for all $f \in H$. There exists an isometric operator U from $\overline{R(|T|)}$ onto $\overline{R(T)}$ such that $T = U|T|$, and $|T| = U^{-1}T$. The representation $T = U|T|$ is called the* polar decomposition *of T.*

Proof. For all $f \in H$ we have

$$\| |T|f\|^2 = \langle |T|f, |T|f\rangle = \langle |T|^2 f, f\rangle = \langle T^* Tf, f\rangle$$
$$= \langle Tf, Tf\rangle = \|Tf\|^2.$$

If for every $f \in H$ we set

$$V(|T|f) = Tf,$$

then V is obviously a linear isometric mapping of $R(|T|)$ onto $R(T)$. Then $U = \overline{V}$ is an isometry from $\overline{R(|T|)}$ onto $\overline{R(T)}$ (cf. the proof of Theorem 4.11), and we have $T = U|T|$. □

If T is a compact operator, then the non-zero eigenvalues of $|T|$ are called the *singular numbers* or *singular values* or *s-numbers* of T. In the following let $(s_j(T))$ denote the (possibly finite) non-increasing sequence of the singular numbers of T; every number counted according to its multiplicity as an eigenvalue of $|T|$. For $0 < p < \infty$ we denote by $B_p(H_1, H_2)$ the set of all compact operators T out of $B(H_1, H_2)$ for which

$$\sum_j [s_j(T)]^p < \infty.$$

We write $B_\infty(H_1, H_2)$ for the set of compact operators belonging to $B(H_1, H_2)$ (cf. Section 6.1). For $B_p(H, H)$ we briefly write $B_p(H)$.

Theorem 7.6. *Let $T \in B_\infty(H_1, H_2)$, $s_j = s_j(T)$. Then there exist orthonormal sequences (f_j) from H_1 and (g_j) from H_2 (these sequences can be finite) for which*

$$Tf = \sum_j s_j \langle f_j, f\rangle g_j \quad \textit{for all} \quad f \in H_1,$$

$$T^* g = \sum_j s_j \langle g_j, g\rangle f_j \quad \textit{for all} \quad g \in H_2,$$

$$|T| f = \sum_j s_j \langle f_j, f\rangle f_j \quad \textit{for all} \quad f \in H_1,$$

$$|T^*| g = \sum_j s_j \langle g_j, g\rangle g_j \quad \textit{for all} \quad g \in H_2.$$

The elements f_j and g_j are eigenelements of $|T|$ and $|T^*|$, respectively. In particular, T, $|T|$, T^* and $|T^*|$ have the same singular values, and the following assertions are equivalent:

$$T \in B_p(H_1, H_2),\ |T| \in B_p(H_1),\ T^* \in B_p(H_2, H_1),\ |T^*| \in B_p(H_2).$$

Proof. By Theorem 7.2 the compact operator $|T|$ has a representation of the above form. Since $|T|$ is non-negative, all the s_j are positive. Consequently, with the operator U from Theorem 7.5 it follows for all $f \in H_1$ that

$$Tf = U|T|f = U\Big(\sum_j s_j\langle f_j, f\rangle f_j\Big) = \sum_j s_j\langle f_j, f\rangle Uf_j.$$

Along with (f_j), the sequence (Uf_j) is also an orthonormal sequence; therefore, we have the required representation of T. For all $f \in H_1$ and $g \in H_2$ we have

$$\begin{aligned}\langle f, T^*g\rangle = \langle Tf, g\rangle &= \Big\langle \sum_j s_j\langle f_j, f\rangle g_j, g\Big\rangle \\ &= \Big\langle f, \sum_j s_j\langle g_j, g\rangle f_j\Big\rangle;\end{aligned}$$

the required representation of T^* follows from this. We can deduce from these identities that

$$T^{**}T^*g = TT^*g = \sum_k s_k\Big\langle f_k, \sum_j s_j\langle g_j, g\rangle f_j\Big\rangle g_k = \sum_j s_j^2\langle g_j, g\rangle g_j$$

for all $g \in H_2$. Consequently, we have the required representation of $|T^*| = (T^{**}T^*)^{1/2}$. The remaining assertions are now clear. □

With the aid of Theorem 7.3 we have the opportunity of determining the singular numbers.

Theorem 7.7. *Let S and T be from $B_\infty(H, H_1)$. Then*

$$s_1(T) = \|T\|,$$

$$s_{j+1}(T) = \inf_{g_1, \ldots, g_j \in H} \sup\{\|Tf\| : f \in H, f \perp g_1, \ldots, g_j, \|f\| = 1\} \quad (7.4)$$

for all $j \in \mathbb{N}$, and

$$s_{j+k+1}(S+T) \leq s_{j+1}(S) + s_{k+1}(T) \quad \text{for all } j, k \in \mathbb{N}_0. \quad (7.5)$$

If $T \in B_\infty(H, H_1)$ and $S \in B_\infty(H_1, H_2)$, then for all $j, k \in N_0$

$$s_{j+k+1}(ST) \leq s_{j+1}(S)s_{k+1}(T). \quad (7.6)$$

If $T \in B_\infty(H, H_1)$ and $S \in B(H_1, H_2)$, then

$$s_j(ST) = s_j(T^*S^*) \leq \|S\|s_j(T) = \|S^*\|s_j(T^*). \quad (7.7)$$

PROOF. The formulae (7.4) follow from (7.3), since the $s_j(T)$ are the eigenvalues of $|T|$ and since $\|Tf\| = \| \, |T| f \|$ for all $f \in H$. For $S, T \in B_\infty(H, H_1)$ and $j, k \in \mathbb{N}_0$

$$\begin{aligned}
s_{j+k+1}(S+T) &= \inf_{g_1, \ldots, g_{j+k}} \sup \{\|(S+T)f\| : f \in H, f \perp g_1, \ldots, g_{j+k}, \|f\| = 1\} \\
&\leqslant \inf_{g_1, \ldots, g_{j+k}} \sup \{\|Sf\| + \|Tf\| : f \in H, f \perp g_1, \ldots, g_{j+k}, \|f\| = 1\} \\
&\leqslant \inf_{g_1, \ldots, g_{j+k}} \Big[\sup \{\|Sf\| : f \in H, f \perp g_1, \ldots, g_j, \|f\| = 1\} \\
&\qquad + \sup \{\|Tf\| : f \in H, f \perp g_{j+1}, \ldots, g_{j+k}, \|f\| = 1\}\Big] \\
&= \inf_{g_1, \ldots, g_j} \sup \{\|Sf\| : f \in H, f \perp g_1, \ldots, g_j, \|f\| = 1\} \\
&\quad + \inf_{g_{j+1}, \ldots, g_{j+k}} \sup \{\|Tf\| : f \in H, f \perp g_{j+1}, \ldots, g_{j+k}, \|f\| = 1\} \\
&= s_{j+1}(S) + s_{k+1}(T).
\end{aligned}$$

For $T \in B_\infty(H, H_1)$, $S \in B_\infty(H_1, H_2)$ and $j, k \in \mathbb{N}_0$ we have (set $\|Tf\|^{-1}\|STf\| = 0$ if $Tf = 0$)

$$\begin{aligned}
s_{j+k+1}(ST) &= \inf_{g_1, \ldots, g_{j+k} \in H} \sup \{\|STf\| : f \in H, f \perp g_1, \ldots, g_{j+k}, \|f\| = 1\} \\
&\leqslant \inf_{\substack{g_1, \ldots, g_k \in H \\ g_{k+1}, \ldots, g_{k+j} \in H_1}} \sup \{\|STf\| : f \in H, f \perp g_1, \ldots, g_k, T^* g_{k+1}, \\
&\qquad \ldots, T^* g_{k+j}, \|f\| = 1\} \\
&= \inf_{\substack{g_1, \ldots, g_k \in H \\ g_{k+1}, \ldots, g_{k+j} \in H_1}} \sup \left\{ \frac{\|STf\|}{\|Tf\|} \|Tf\| : f \in H, f \perp g_1, \ldots, g_k; \right. \\
&\qquad \left. Tf \perp g_{k+1}, \ldots, g_{k+j}; \|f\| = 1 \right\} \\
&\leqslant \inf_{\substack{g_1, \ldots, g_k \in H \\ g_{k+1}, \ldots, g_{k+j} \in H_1}} \Big[\sup \{\|Sh\| : h \in H_1, h \perp g_{k+1}, \ldots, g_{k+j}, \|h\| = 1\} \\
&\qquad \times \sup \{\|Tf\| : f \in H, f \perp g_1, \ldots, g_k, \|f\| = 1\}\Big] \\
&= s_{j+1}(S) s_{k+1}(T).
\end{aligned}$$

If $T \in B_\infty(H, H_1)$ and $S \in B(H_1, H_2)$, then for all $j \in \mathbb{N}_0$

$$\begin{aligned}
s_{j+1}(ST) &= \inf_{g_j, \ldots g_j \in H} \sup \{\|STf\| : f \in H, f \perp g_1, \ldots, g_j, \|f\| = 1\} \\
&\leqslant \|S\| \inf_{g_1, \ldots, g_j \in H} \sup \{\|Tf\| : \ldots\} = \|S\| s_{j+1}(T).
\end{aligned}$$

The remaining equalities follow from the equalities $s_j(A) = s_j(A^*)$. □

In what follows we write

$$\|T\|_p = \left\{ \sum_j |s_j(T)|^p \right\}^{1/p}$$

for $T \in B_p(H, H_1)$ $(0<p<\infty)$.

Theorem 7.8.

(a) *If $S, T \in B_p(H, H_1)$ $(0<p<\infty)$, then $S+T$ also belongs to $B_p(H, H_1)$, and*

$$\|S+T\|_p \leqslant 2^{1/p}(\|S\|_p + \|T\|_p) \quad \textit{for} \quad p \geqslant 1,$$ [2]

$$\|S+T\|_p^p \leqslant 2(\|S\|_p^p + \|T\|_p^p) \quad \textit{for} \quad p \leqslant 1.$$

The sets $B_p(H, H_1)$ are therefore vector spaces.

(b) *If $T \in B_p(H, H_1)$, $S \in B_q(H_1, H_2)$ $(0<p, q<\infty)$ and $(1/r) = (1/p) + (1/q)$, then $ST \in B_r(H, H_2)$ and*

$$\|ST\|_r \leqslant 2^{1/r}\|S\|_q \|T\|_p.$$ [2]

(c) *If $T \in B_p(H, H_1)$ and $S \in B(H_1, H_2)$, then $ST \in B_p(H, H_2)$, and we have*

$$\|ST\|_p \leqslant \|S\| \, \|T\|_p.$$

The corresponding assertions hold for $T \in B(H, H_1)$ and $S \in B_p(H_1, H_2)$.

REMARK. Theorem 7.8(a) and (c) imply that the sets $B_p(H)$ are two-sided ideals of $B(H)$ and we have $\|ST\|_p \leqslant \|S\| \, \|T\|_p$ and $\|TS\|_p \leqslant \|S\| \, \|T\|_p$ for $S \in B(H)$ and $T \in B_p(H)$.

PROOF.

(a) By virtue of (7.5) we have

$$\begin{aligned} \sum_j s_j(S+T)^p &= \sum_j \left\{ s_{2j-1}(S+T)^p + s_{2j}(S+T)^p \right\} \\ &\leqslant \sum_j \left\{ (s_j(S)+s_j(T))^p + (s_j(S)+s_{j+1}(T))^p \right\}. \end{aligned}$$

If $p \geqslant 1$, then it follows by the Minkowski inequality for the l_p-norm that

$$\begin{aligned} \|S+T\|_p^p &\leqslant \left[\left(\sum_j s_j(S)^p \right)^{1/p} + \left(\sum_j s_j(T)^p \right)^{1/p} \right]^p \\ &\quad + \left[\left(\sum_j s_j(S)^p \right)^{1/p} + \left(\sum_j s_{j+1}(T)^p \right)^{1/p} \right]^p \\ &\leqslant 2\left[\|S\|_p + \|T\|_p \right]^p. \end{aligned}$$

[2] More accurate studies show that the $\|\ .\ \|_p$ are norms for $p \geqslant 1$ (cf. Theorem 7.12 for $p = 1$), and that $\|ST\|_1 \leqslant \|S\|_q \|T\|_p$ for $(1/p) + (1/q) = 1$; cf. [5], Lemma XI. 9.14 or [10], Theorem III, 7.1.

If $p \leqslant 1$, then we can use the elementary inequality $|\alpha|^p + |\beta|^p \geqslant |\alpha + \beta|^p$ (proof: it is sufficient to prove the case $|\alpha| + |\beta| = 1$). With the aid of this inequality we obtain that

$$\|S+T\|_p^p \leqslant \sum_j \{2s_j(S)^p + s_j(T)^p + s_{j+1}(T)^p\}$$
$$\leqslant 2\sum_j (s_j(s)^p + s_j(T)^p) = 2(\|S\|_p^p + \|T\|_p^p).$$

(b) As $(r/p)+(r/q)=1$, it follows from (7.6) with the aid of Hölder's inequality that

$$\|ST\|_r = \Big(\sum_j s_j(ST)^r\Big)^{1/r}$$
$$\leqslant \Big\{\sum_j s_j(S)^r s_j(T)^r + \sum_j s_j(S)^r s_{j+1}(T)^r\Big\}^{1/r}$$
$$\leqslant \Big\{\Big(\sum_j s_j(S)^q\Big)^{r/q}\Big(\sum_j s_j(T)^p\Big)^{r/p} + \Big(\sum_j s_j(S)^q\Big)^{r/q}\Big(\sum_j s_{j+1}(T)^p\Big)^{r/p}\Big\}^{1/r}$$
$$\leqslant 2^{1/r}\|S\|_q\|T\|_p.$$

Assertion (c) follows from (7.7). □

Theorem 7.9. *Let $p, q, r > 0$ with $(1/p)+(1/q)=(1/r)$. We have $T \in B_r(H, H_1)$ if and only if there exist operators $T_1 \in B_p(H, H_2)$ and $T_2 \in B_q(H_2, H_1)$ (with an arbitrary Hilbert space H_2) for which $T = T_2T_1$; the operators T_1 and T_2 can be chosen such that $\|T\|_r = \|T_1\|_p\|T_2\|_q$.*

Proof. By Theorem 7.8(b) we have $T_2T_1 \in B_r(H, H_1)$ for $T_1 \in B_p(H, H_2)$ and $T_2 \in B_q(H_2, H_1)$. Now let $T \in B_r(H, H_1)$ and let (cf. Theorem 7.6)

$$Tf = \sum_j s_j(T)\langle f_j, f\rangle g_j, \quad f \in H,$$

where (f_j) and (g_j) are orthonormal sequences in H and H_1, respectively. If $\{h_1, h_2, \ldots\}$ is an ONS in a Hilbert space H_2 and we define T_1 and T_2 by the equalities

$$T_1 f = \sum_j s_j(T)^{r/p}\langle f_j, f\rangle h_j, \quad f \in H,$$

$$T_2 h = \sum_j s_j(T)^{r/q}\langle h_j, h\rangle g_j, \quad h \in H_2,$$

then obviously $T = T_2T_1$. The numbers $s_j(T)^{r/p}$ and $s_j(T)^{r/q}$ are the singular numbers of T_1 and T_2, respectively. Therefore, $T_1 \in B_p(H, H_2)$ and

$T_2 \in B_q(H_2, H_1)$. Moreover,

$$\|T_1\|_p = \Big(\sum_j s_j(T)^{(r/p)p}\Big)^{1/p} = \|T\|_r^{r/p} \quad \text{and} \quad \|T_2\|_q = \|T\|_r^{r/q};$$

consequently, $\|T\|_r = \|T_1\|_p \|T_2\|_q$. □

Theorem 7.10.

(a) *The set $B_2(H, H_1)$ coincides with the set of Hilbert-Schmidt operators. For $T \in B_2(H, H_1)$ we have $\|T\|_2 = |||T|||$.*

(b) *We have $T \in B_1(H, H_1)$ if and only if there exist operators $T_1 \in B_2(H, H_2)$ and $T_2 \in B_2(H_2, H_1)$ (with an arbitrary Hilbert space H_2) such that $T = T_2 T_1$; the operators T_1 and T_2 can be chosen such that $\|T\|_1 = \|T_1\|_2 \|T_2\|_2$.*

PROOF.

(a) Let $T \in B_2(H, H_1)$. If $f_1, f_2, \ldots$ are the orthonormalized eigenelements of $|T|$ that belong to the non-zero eigenvalues $s_j(T)$ and if $\{g_\alpha : \alpha \in A\}$ is an ONB of $N(|T|) = N(T)$, then $\{f_1, f_2, \ldots\} \cup \{g_\alpha : \alpha \in A\}$ is an ONB of H, and we have

$$\sum_j \|Tf_j\|^2 + \sum_\alpha \|Tg_\alpha\|^2 = \sum_j \||T|f_j\|^2 = \sum_j s_j(T)^2 = \|T\|_2^2 < \infty.$$

Consequently, T is a Hilbert-Schmidt operator with $|||T||| = \|T\|_2$. If T is a Hilbert-Schmidt (therefore compact) operator from H into H_1 and $f_1, f_2, \ldots$ are chosen as above, then

$$\sum_j s_j(T)^2 = \sum_j \|Tf_j\|^2 \leq |||T|||^2 < \infty,$$

i.e., $T \in B_2(H, H_1)$.

(b) This follows from Theorem 7.9 for $r = 1$ and $p = q = 2$. □

The set $B_1(H, H_1)$ is also called the *trace class* of operators from H into H_1. This term originates from the fact that for $T \in B_1(H)$ a *trace* can be defined by

$$\operatorname{tr}(T) = \sum_\alpha \langle e_\alpha, Te_\alpha \rangle, \tag{7.8}$$

where $\{e_\alpha : \alpha \in A\}$ is an ONB of H. This is so, because then $T = T_2 T_1$ with appropriately chosen Hilbert-Schmidt operators T_1, T_2 and

$$\sum_\alpha |\langle e_\alpha, Te_\alpha \rangle| = \sum_\alpha |\langle T_2^* e_\alpha, T_1 e_\alpha \rangle| \leq \Big(\sum_\alpha \|T_2^* e_\alpha\|^2 \sum_\alpha \|T_1 e_\alpha\|^2\Big)^{1/2} < \infty.$$

For matrices the trace does not depend on the choice of the basis with respect to which the matrix is determined. Analogously, the definition of the trace does not depend on the choice of the ONB in the above case.

Theorem 7.11.
(a) *Definition* (7.8) *of the trace of an operator* $T \in B_1(H)$ *is independent of the choice of the orthonormal basis.*
(b) *If* $T_1 \in B_2(H_1, H_2)$ *and* $T_2 \in B_2(H_2, H_1)$ *or* $T_1 \in B_1(H_1, H_2)$ *and* $T_2 \in B(H_2, H_1)$, *then* $\operatorname{tr}(T_1T_2) = \operatorname{tr}(T_2T_1)$. (*This also holds for* $T_1 \in B_p(H_1, H_2)$ *and* $T_2 \in B_q(H_2, H_1)$ *with* $(1/p)+(1/q)=1$; *cf. Exercise* 7.6(c).)

PROOF.
(a) By Theorem 7.10(b) there exist operators $T_1 \in B_2(H, H_2)$ and $T_2 \in B_2(H_2, H)$ such that $T = T_2T_1$. If $\{e_\alpha : \alpha \in A\}$ and $\{f_\beta : \beta \in B\}$ are arbitrary orthonormal bases of H and H_2, respectively, then

$$\sum_\alpha \langle e_\alpha, T_2T_1e_\alpha\rangle = \sum_\alpha \langle T_2^*e_\alpha, T_1e_\alpha\rangle = \sum_\alpha \sum_\beta \langle T_2^*e_\alpha, f_\beta\rangle\langle f_\beta, T_1e_\alpha\rangle$$

$$= \sum_\beta \sum_\alpha \langle T_1^*f_\beta, e_\alpha\rangle\langle e_\alpha, T_2f_\beta\rangle$$

$$= \sum_\beta \langle f_\beta, T_1T_2f_\beta\rangle, \tag{7.9}$$

where all sums have at most countably many summands and the sums are absolutely convergent. If we choose another ONB $\{e'_\alpha : \alpha \in A\}$ of H, then it follows from this that

$$\sum_\alpha \langle e_\alpha, T_2T_1e_\alpha\rangle = \sum_\beta \langle f_\beta, T_1T_2f_\beta\rangle = \sum_\alpha \langle e'_\alpha, T_2T_1e'_\alpha\rangle.$$

This is the required independence.
(b) The first assertion has been proved in part (a). If $T_1 \in B_1(H_1, H_2)$ and $T_2 \in B(H_2, H_1)$, then there exist operators $B \in B_2(H_1, H_3)$ and $A \in B_2(H_3, H_2)$ such that $T_1 = AB$. Thus it follows from (7.9) that

$$\begin{aligned}\operatorname{tr}(T_2T_1) &= \operatorname{tr}((T_2A)B) = \operatorname{tr}(B(T_2A)) \\ &= \operatorname{tr}((BT_2)A) = \operatorname{tr}(A(BT_2)) = \operatorname{tr}(T_1T_2),\end{aligned}$$

since T_2A, B, BT_2, and A are Hilbert-Schmidt operators. □

Theorem 7.12. *An operator T from H_1 into H_2 such that $D(T) = H_1$ is in $B_1(H_1, H_2)$ if and only if there are sequences (φ_n) from H_1 and (ψ_m) from H_2 such that $\|\varphi_n\| = \|\psi_n\| = 1$, and there is a sequence (z_n) from $\mathbb{K}$ for which $\Sigma|z_n| < \infty$ and*

$$Tf = \Sigma z_n\langle\varphi_n, f\rangle\psi_n, \quad f \in H_1. \tag{7.10}$$

The norm $\|T\|_1$ is the infimum of those sums $\Sigma|z_n|$ for which there are normed sequences (φ_n) from H_1 and (ψ_n) from H_2 such that (7.10) *holds. $\|\,.\,\|_1$ is a norm on $B_1(H_1, H_2)$, the so-called* trace norm.

PROOF. If $T \in B_1(H_1, H_2)$, then Theorem 7.6 gives the required representation (7.10); moreover, it follows that $\|T\|_1$ is greater than or equal to the given infimum. If (7.10) holds, then T is compact, since T is the limit of the finite rank operators T_m, $T_m f = \sum_{n=1}^m z_n \langle \varphi_n, f\rangle \psi_n$ in the sense of the norm of $B(H_1, H_2)$ (because $\|T - T_m\| \leq \sum_{n>m} |z_n|$). Hence, we obtain with the notation of Theorem 7.6 that

$$\sum s_j(T) = \sum \langle Tf_j, g_j\rangle = \sum_j \langle \sum_n z_n \langle \varphi_n, f_j\rangle \psi_n, g_j\rangle$$
$$\leq \sum_j \sum_n |z_n|\, |\langle \varphi_n, f_j\rangle \langle \psi_n, g_j\rangle|$$
$$\leq \sum_n |z_n| \left\{ \sum_j |\langle \varphi_n, f_j\rangle|^2 \sum_j |\langle \psi_n, g_j\rangle|^2 \right\}^{1/2}$$
$$\leq \sum_n |z_n|\, \|\varphi_n\|\, \|\psi_n\| = \sum_n |z_n|;$$

thus, $T \in B_1(H_1, H_2)$. As this holds for every representation of the form (7.10), the equality given for $\|T\|_1$ follows. The fact that $\|\,.\,\|_1$ is a norm follows immediately from this. □

EXERCISES

7.1. If T is a self-adjoint compact operator and $n \in \mathbb{N}$ is odd, then there is exactly one (compact) self-adjoint operator A such that $A^n = T$.

7.2. In (7.3) and (7.4) we can replace "inf sup" by "min max".

7.3. Theorems 7.1 and 7.2 do not hold for normal operators on real Hilbert spaces. As an example, one can consider the operator induced by the matrix $\begin{pmatrix} 0 & 1 \\ -1 & 0 \end{pmatrix}$ on $\mathbb{R}^2$; this operator has no eigenvalue.

7.4. If T is a normal operator on H and there exists a $z_0 \in \rho(T)$ such that $R(z_0, T)$ is compact, then $R(z, T)$ is compact for every $z \in \rho(T)$, and there exist a sequence (λ_j) from $\mathbb{K}$ such that $|\lambda_j| \to \infty$ as $j \to \infty$ and a sequence (P_j) of finite rank orthogonal projections such that $P_j P_k = \delta_{jk} P_j$, $D(T) = \{f \in H : \sum_j |\lambda_j|^2 \|P_j f\|^2 < \infty\}$, and $Tf = \sum_j \lambda_j P_j f$ for $f \in D(T)$.

7.5. Let $\{\lambda_\alpha : \alpha \in A\}$ be a family in $\mathbb{K}\setminus\{0\}$ for which $\lambda_\alpha \neq \lambda_\beta$ for $\alpha \neq \beta$, and let $\{P_\alpha : \alpha \in A\}$ be a family of orthogonal projections on the Hilbert space H such that $P_\alpha P_\beta = \delta_{\alpha\beta} P_\alpha$ for $\alpha, \beta \in A$.
(a) The equalities

$$D(T) = \left\{ f \in H : \sum_{\alpha \in A} |\lambda_\alpha|^2 \|P_\alpha f\|^2 < \infty \right\} \quad \text{and}$$
$$Tf = \sum_{\alpha \in A} \lambda_\alpha P_\alpha f \quad \text{for} \quad f \in D(T)$$

define a normal operator on H.
(b) Every λ_α such that $P_\alpha \neq 0$ is an eigenvalue of T, and $N(\lambda_\alpha - T) = R(P_\alpha)$.

(c) T is compact if and only if we have:
(i) $\dim R(P_\alpha) < \infty$ for all $\alpha \in A$,
(ii) for every $\epsilon > 0$ there are only finitely many $\alpha \in A$ such that $|\lambda_\alpha| \geq \epsilon$ and $P_\alpha \neq 0$.

7.6. (a) Let $0 < p < \infty$. For every $T \in B_p(H_1, H_2)$ there exists a sequence (T_n) of finite rank operators such that $\|T_n\|_p \leq \|T\|_p$ and $\|T - T_n\|_p \to 0$ as $n \to \infty$.
Hint: Theorem 7.6.
(b) For every $T \in B_1(H)$ we have $|\mathrm{tr}\,(T)| \leq \|T\|_1$.
Hint: If $Tf = \Sigma_j \lambda_j \langle f_j, f\rangle g_j$, then define the trace of T by means of an ONB that contains $\{f_1, f_2, \ldots\}$.
(c) For $T \in B_p(H_1, H_2)$, $S \in B_q(H_2, H_1)$, where $(1/p) + (1/q) = 1$, we have $\mathrm{tr}\,(ST) = \mathrm{tr}\,(TS)$.
Hint: Without loss of generality we may assume that $q < 2$. Let (T_n) be chosen as in part (a). Then $T_n \in B_2(H_1, H_2)$, $S \in B_2(H_2, H_1)$; consequently, by Theorem 7.11(b) we have $\mathrm{tr}\,(ST_n) = \mathrm{tr}\,(T_nS)$. Moreover, $|\mathrm{tr}\,(ST) - \mathrm{tr}\,(ST_n)| \leq 2\|S\|_q\|T - T_n\|_p$ and $|\mathrm{tr}\,(TS) - \mathrm{tr}\,(T_nS)| \leq 2\|S\|_q\|T - T_n\|_p \to 0$.

7.7. Prove Theorem 7.2 without reference to Theorem 6.7:
(a) There is an eigenvalue μ_1 of T such that $|\mu_1| = \|T\|$.
Hint: By Theorem 5.17(c) and (d) and Theorem 5.43 there exist a $\mu_1 \in \mathbb{K}$ and a sequence (g_n) from H for which $|\mu_1| = \|T\|$, $\|g_n\| = 1$, and $(\mu_1 - T)g_n \to 0$. The sequence (g_n) has a convergent subsequence (g_{n_k}); the element $f_1 = \lim g_{n_k}$ is an eigenvector of T belonging to the eigenvalue μ_1.
(b) Let the eigenvalues $\mu_1, \ldots, \mu_n$ $(|\mu_1| \geq |\mu_2| \geq \ldots \geq |\mu_n|)$ and the eigenelements $f_1, \ldots, f_n$ be determined. The restriction T_n of T to $L(f_1, \ldots, f_n)^\perp$ is a normal operator on $L(f_1, \ldots, f_n)^\perp$. μ_{n+1} and f_{n+1} are obtained by using (a) for T_n.
(c) Prove (7.2).

7.8. Let T and S be operators on H such that $\rho(T) \cap \rho(S) \neq \emptyset$, and let $\lambda_0 \in \rho(T) \cap \rho(S)$, $0 < p \leq \infty$. If $R(\lambda_0, T) - R(\lambda_0, S) \in B_p(H)$, then $R(\lambda, T) - R(\lambda, S) \in B_p(H)$ for all $\lambda \in \rho(A) \cap \rho(B)$.
Hint: We have

$$\begin{aligned}&(\lambda - T_1)^{-1} - (\lambda - T_2)^{-1}\\ &= (\lambda_0 - T_1)(\lambda - T_1)^{-1}\left[(\lambda_0 - T_1)^{-1} - (\lambda_0 - T_2)^{-1}\right](\lambda_0 - T_2)(\lambda - T_2)^{-1}.\end{aligned}$$

7.9. Give a proof, independent of Theorem 7.12, that $\|\cdot\|_1$ is a norm on $B_1(H_1, H_2)$.
Hint: Let $Af = \Sigma s_j'\langle f_j', f\rangle g_j'$, $Bf = \Sigma s_j''\langle f_j'', f\rangle g_j''$, and $(A + B)f = \Sigma s_j \langle f_j, f\rangle g_j$ as in Theorem 7.6. Then

$$\begin{aligned}\|A + B\|_1 &= \sum \langle (A + B)f_j, g_j\rangle\\ &= \sum_j \sum_k \langle f_j, A^* g_k'\rangle\langle g_k', g_j\rangle + \sum_j \sum_k \langle f_j, B^* g_k''\rangle\langle g_k'', g_1\rangle\\ &\leq \|A\|_1 + \|B\|_1.\end{aligned}$$

7.10. (a) If $T_n \in B_\infty(H_1, H_2)$, $T \in B(H_1, H_2)$, and $\|T_n - T\| \to 0$ as $n \to \infty$, then for every $k \in \mathbb{N}$ we have $s_k(T_n) \to s_k(T)$ as $n \to \infty$.
Hint: Use (7.5).
(b) If $T_n \in B_p(H_1, H_2)$, $T \in B(H_1, H_2)$, $\|T - T_n\| \to 0$ as $n \to \infty$, and $\liminf_{n\to\infty} \|T_n\|_p < \infty$, then $T \in B_p(H_1, H_2)$, and $\|T\|_p < \liminf_{n\to\infty} \|T_n\|_p$.
(c) If (T_n) is a sequence from $B_p(H_1, H_2)$ such that $\|T_n - T_m\|_p \to 0$ as $n, m \to \infty$, then there exists a $T \in B_p(H_1, H_2)$ such that $\|T - T_n\|_p \to 0$ as $n \to \infty$. (Consequently, $B_1(H_1, H_2)$ is a Banach space.)

7.11. Let $S, T \in B(H_1, H_2)$ be bijective. Then $S - T$ is in $B_p(H_1, H_2)$ if and only if $S^{-1} - T^{-1}$ is in $B_p(H_2, H_1)$.
Hint: $S^{-1} - T^{-1} = S^{-1}(T - S)T^{-1}$.

7.12. Let $S, T \in B(H)$, $S - T \in B_p(H)$.
(a) We have $S^n - T^n \in B_p(H)$ for all $n \in \mathbb{N}$.
Hint: $S^n - T^n = \Sigma_{j=0}^{n-1} T^j(S - T)S^{n-j-1}$.
(b) We have $p(S) - p(T) \in B_p(H)$ for every polynomial p.

7.13. If $T \in B_p(H_1, H_2)$ for some $p < \infty$, then $T^*T \in B_{p/2}(H_1)$ and $(T^*T)^n \in B_{p/2n}(H) \subset B_1(H)$ for $n > p/2$. We have $\|T\| = \lim_{n\to\infty} [\operatorname{tr}(T^*T)^n]^{1/2n}$.
Hint: $\{\Sigma_{j=1}^k |a_j|^n\}^{1/n} \to \max\{|a_j| : j = 1, \ldots, k\}$ as $n \to \infty$.

7.14. Let H_1 and H_2 be Hilbert spaces, and let $T \in B_\infty(H_1, H_2)$. For every $\epsilon > 0$ there exists a finite-dimensional subspace M_ϵ of H_1 such that $\|Tf\| \leq \epsilon\|f\|$ for all $f \in M_\epsilon^\perp$.
Hint: Use the representation in Theorem 7.6.

7.15. Let H_1 and H_2 be Hilbert spaces. Let $H_1' = B(H_1, \mathbb{K})$ be the Hilbert space of continuous linear functionals on H_1 (cf. Exercise 4.3(a)).
(a) $H_1' \otimes H_2$ is isomorphic to the space of bounded finite rank operators from H_1 into H_2; we can make the element $\Sigma_{j=1}^m c_j L_j \otimes g_j$ from $H_1' \otimes H_2$ correspond to the operator T for which $D(T) = H_1$ and $Tf = \Sigma_{j=1}^m c_j L_j(f) g_j$.
(b) We have $\|\Sigma_{j=1}^m c_j L_j \otimes g_j\| = \|T\|_2$; the space $H_1' \hat{\otimes} H_2$ can be identified with the space $B_2(H_1, H_2)$ of Hilbert-Schmidt operators from H_1 into H_2.
(c) A norm $\|\cdot\|_\tau$ on $H_1' \otimes H_2$ is called a "cross"-norm if $\|L \otimes g\|_\tau = \|L\|\,\|g\|$. The completion $H_1' \hat{\otimes}_\tau H_2$ of $H_1' \otimes H_2$ with respect to $\|.\|_\tau$ can be identified with a subspace $B_\tau(H_1, H_2)$ of $B_\infty(H_1, H_2)$. The spaces $B_\tau(H_1, H_2)$ are two-sided ideals of $B(H_1, H_2)$.
(d) For every "cross"-norm $\|.\|_\tau$ we have $B_1(H_1, H_2) \subset B_\tau(H_1, H_2)$. Furthermore, $B_\tau(H_1, H_2) = B_1(H_1, H_2)$, if we choose

$$\left\|\sum_{j=1}^n L_j \otimes g_j\right\|_\tau = \inf\left\{\sum_{j=1}^l \|M_j\|\,\|k_j\| : \sum_{j=1}^l M_j \otimes k_j = \sum_{j=1}^n L_j \otimes g_j\right\}.$$

We have $B_\tau(H_1, H_2) = B_2(H_1, H_2)$ if we choose

$$\left\|\sum_{j=1}^n L_j \otimes g_j\right\|_\tau = \left\|\sum_{j=1}^n L_j \otimes g_j\right\|.$$

7.2 Integration with respect to a spectral family

(7.11) *A spectral family on a Hilbert space H is a function $E:\mathbb{R}\to B(H)$ having the following properties*:

(a) *$E(t)$ is an orthogonal projection for every $t\in\mathbb{R}$*,
(b) *$E(s)\leq E(t)$ for $s\leq t$* (monotonicity),
(c) *$E(t+\epsilon)\overset{s}{\to}E(t)$ for all $t\in\mathbb{R}$, as $\epsilon\to 0+$* (continuity from the right),
(d) *$E(t)\overset{s}{\to}0$ as $t\to-\infty$, $E(t)\overset{s}{\to}I$ as $t\to\infty$.*

REMARK. Property (7.11(c)) is not essential, it is needed only in the uniqueness statement of Theorem 7.17. Right continuity can be replaced by left continuity; then we have to replace $\lim_{\delta\to 0+}$ by $\lim_{\delta\to 0-}$ in Theorem 7.17.

EXAMPLE 1. Let M be a measurable subset of $\mathbb{R}^m$ and let $g:M\to\mathbb{R}$ be a measurable function. For every $t\in\mathbb{R}$ let

$$M(t)=\{x\in M: g(x)\leq t\}.$$

$M(t)$ is obviously a measurable subset of M.

(7.12) *The equality $E(t)f=\chi_{M(t)}f$ for $f\in L_2(M)$ and $t\in\mathbb{R}$ defines a spectral family on $L_2(M)$.*

PROOF. Properties (7.11(a)) and (7.11(b)) are evidently satisfied. We show (7.11(c)): Let $t\in\mathbb{R}$, let $f\in L_2(M)$, and let (ϵ_n) be a null-sequence of positive numbers. Then

$$\|(E(t+\epsilon_n)-E(t))f\|^2=\int_M(\chi_{M(t+\epsilon_n)}(x)-\chi_{M(t)}(x))|f(x)|^2\,dx.$$

Because of the relations $0\leq\chi_{M(t+\epsilon_n)}(x)-\chi_{M(t)}(x)\leq 1$ and $\chi_{M(t+\epsilon_n)}(x)-\chi_{M(t)}(x)\to 0$ as $n\to\infty$ for all $x\in M$, it follows by Lebesgue's theorem that

$$(E(t+\epsilon_n)-E(t))f\to 0\quad\text{as}\quad n\to\infty.$$

As this holds for every zero-sequence (ϵ_n), property (7.11(c)) follows. We can show (7.11(d)) similarly, since we have

$$\chi_{M(t)}(x)\to 1\quad\text{for all}\quad x\in M,\text{ as } t\to\infty.$$
$$\chi_{M(t)}(x)\to 0\quad\text{for all}\quad x\in M,\text{ as } t\to-\infty.$$

□

EXAMPLE 2. Let $\{\rho_\alpha:\alpha\in A\}$ be a family of right continuous non-decreasing functions defined on $\mathbb{R}$.

(7.13) *The equalities*

$$E(t)(f_\alpha)=\chi_{(-\infty,t]}(f_\alpha)$$
$$=(\chi_{(-\infty,t]}f_\alpha)\quad\text{for}\quad (f_\alpha)\in\bigoplus_{\alpha\in A}L_2(\mathbb{R},\rho_\alpha),\ t\in\mathbb{R}$$

define a spectral family on $\oplus_{\alpha\in A}L_2(\mathbb{R},\rho_\alpha)$.

Proof. It is clear that the $E(t)$ are orthogonal projections. The increasing character follows from the equality

$$\langle (f_\alpha), (E(t)-E(s))(f_\alpha)\rangle = \sum_\alpha \int_{(s,t]} |f_\alpha(x)|^2 \, d\rho_\alpha(x) \geq 0$$

for $t > s$ and $(f_\alpha) \in \oplus_{\alpha\in A} L_2(\mathbb{R}, \rho_\alpha)$. For $t \to s_+$ we have

$$\int_{(s,t]} |f_\alpha(x)|^2 \, d\rho_\alpha(x) \to 0 \quad \text{for all} \quad \alpha \in A.$$

Because

$$\int_{(s,t]} |f_\alpha(x)|^2 \, d\rho_\alpha(x) \leq \int_{\mathbb{R}} |f_\alpha(x)|^2 \, d\rho_\alpha(x) \quad \text{for all} \quad s, t \text{ and all } \alpha$$

and

$$\sum_{\alpha\in A} \int_{\mathbb{R}} |f_\alpha(x)|^2 \, d\rho_\alpha(x) = \|(f_\alpha)_{\alpha\in A}\|^2 < \infty$$

it follows from this that

$$\|(E(t)-E(s))(f_\alpha)\|^2 = \sum_\alpha \int_{(s,t]} |f_\alpha(x)|^2 \, d\rho_\alpha(x) \to 0, \quad t \to s_+,$$

i.e., E is right continuous. The remaining assertions $E(t) \to I$ as $t \to +\infty$ and $E(t) \to 0$ as $t \to -\infty$ are clear. □

Example 3. Let (λ_j) be a sequence of pairwise different real numbers, and let (P_j) be a sequence of orthogonal projections on H such that $P_j P_k = 0$ for $j \neq k$ and $\oplus_j R(P_j) = H$.

(7.14) *The equalities*

$$E(t)f = \sum_{\{j:\lambda_j \leq t\}} P_j f \quad \text{for} \quad f \in H \text{ and } t \in \mathbb{R}$$

define a spectral family on H.

Proof. Properties (7.11(a)), (7.11(b)) and (7.11(d)) are clear; we leave their proof to the reader, and only prove the right continuity here. For every $t \in \mathbb{R}$ and $\epsilon > 0$

$$\|(E(t+\epsilon)-E(t))f\|^2 = \sum_{\{j:t<\lambda_j\leq t+\epsilon\}} \|P_j f\|^2.$$

The sum converges to zero as $\epsilon \to 0+$ (since the series $\sum_j \|P_j f\|^2$ is convergent and for every $n_0 \in \mathbb{N}$ there exists an $\epsilon > 0$ such that $\lambda_j \notin (t, t+\epsilon]$ for $j \leq n_0$). □

Let E be a spectral family on the Hilbert space H. For every $f \in H$ define

$$\rho_f(t) = \langle f, E(t)f\rangle = \|E(t)f\|^2, \quad t \in \mathbb{R}. \tag{7.15}$$

The function $\rho_f : \mathbb{R} \to \mathbb{R}$ is obviously bounded, non-decreasing, and right continuous; $\lim_{t\to-\infty} \rho_f(t) = 0$, $\lim_{t\to\infty} \rho_f(t) = \|f\|^2$.

A function $u : \mathbb{R} \to \mathbb{K}$ is said to be *E-measurable* if it is ρ_f-measurable for every $f \in H$ (cf. Appendix A). Non-trivial examples of E-measurable functions (for every spectral family E) are all continuous functions, all step functions, and all functions that are pointwise limits of step functions; all Borel measurable functions are E-measurable. The function u_0 with $u_0(t) = 1$ for all $t \in \mathbb{R}$ is in $L_2(\mathbb{R}, \rho_f)$ for every $f \in H$. Consequently, by Theorem A 14, every bounded E-measurable function $u : \mathbb{R} \to \mathbb{K}$ belongs to $L_2(\mathbb{R}, \rho_f)$ for all $f \in H$.

For a step function u we can define the integral $\int u(t)\, \mathrm{d}E(t)$ by the equality

$$\int \sum_{j=1}^{n} c_j \chi_{J_j}(t)\, \mathrm{d}E(t) = \sum_{j=1}^{n} c_j E(J_j),$$

where

$$E((a,b]) = E(b) - E(a), \quad E((a,b)) = E(b-) - E(a),$$
$$E([a,b]) = E(b) - E(a-), \quad E([a,b)) = E(b-) - E(a-)$$

(here let $E(t-) = s-\lim_{\epsilon\to 0+} E(t-\epsilon)$; this limit exists by Theorem 4.32 and is an orthogonal projection). For any step function $u : \mathbb{R} \to \mathbb{K}$ and for every $f \in H$ we obviously have

$$\left\| \int u(t)\, \mathrm{d}E(t) f \right\|^2 = \int |u(t)|^2\, \mathrm{d}\rho_f(t), \tag{7.16}$$

as a simple calculation shows. If $f \in H$ and $u \in L_2(\mathbb{R}, \rho_f)$, then by Section 2.2, Examples 7 and 13, there exists a sequence (u_n) of step functions for which $u_n \to u$ in $L_2(\mathbb{R}, \rho_f)$.[3] Then

$$\left\| \int u_n(t)\, \mathrm{d}E(t) f - \int u_m(t)\, \mathrm{d}E(t) f \right\|^2 = \int |u_n(t) - u_m(t)|^2\, \mathrm{d}\rho_f(t)$$
$$\to 0 \text{ as } n, m \to \infty,$$

i.e., the sequence $(\int u_n\, \mathrm{d}E(t) f)$ is a Cauchy sequence in H. Therefore, we can make the definition

$$\int u(t)\, \mathrm{d}E(t) f = \lim_{n\to\infty} \int u_n(t)\, \mathrm{d}E(t) f. \tag{7.17}$$

This definition is obviously independent of the choice of the sequence (u_n), and we have

$$\left\| \int u(t)\, \mathrm{d}E(t) f \right\|^2 = \int |u(t)|^2\, \mathrm{d}\rho_f(t). \tag{7.18}$$

[3] Here and in the sequel $L_2(\mathbb{R}, \rho_f)$ is meant to be the real or the complex L_2-space according as H is real or complex.

For $u, v \in L_2(\mathbb{R}, \rho_f)$ and $a, b \in \mathbb{K}$ it follows from (7.17) immediately that

$$\int (au(t) + bv(t))\, \mathrm{d}E(t)f = a\int u(t)\, \mathrm{d}E(t)f + b\int v(t)\, \mathrm{d}E(t)f. \quad (7.19)$$

The *integral* just defined *is* therefore *linear*. □

In our further studies we shall use the following auxiliary theorem; for functions that are pointwise limits of step functions (all functions explicitly occurring in the following are of this kind) this auxiliary theorem is not needed.

Auxiliary Theorem 7.13. *Let E be a spectral family on H and let $u : \mathbb{R} \to \mathbb{K}$ be an E-measurable function. If $\{f_1, \ldots, f_p\}$ is a finite set in H, then there exists a sequence (u_n) of step functions that converges to u almost everywhere with respect to ρ_{f_j} for $j = 1, \ldots, p$. If u is bounded, then the sequence can be chosen to be bounded.*

Proof. It is enough to show that there exists an $h \in H$ for which every ρ_h-null set is also a ρ_{f_j}-null set for $j = 1, \ldots, p$ (then we choose a sequence of step functions that converges to u ρ_h-almost everywhere). In order to prove this, it is enough to find, for any two elements $f_1 = f$ and $f_2 = g$, an element $h \in H$ for which every set of ρ_h-measure zero is of ρ_f- and ρ_g-measure zero; the rest is simple induction.

For this, set $M = \overline{L\{E(t)f : t \in \mathbb{R}\}}$. Let us introduce the notations: P is the orthogonal projection onto M, $g_1 = Pg$, $g_2 = (I - P)g = g - g_1$, and $h = f + g_2$. Then for arbitrary intervals I_1 and I_2 in $\mathbb{R}$ we obviously have $E(I_1)f \perp E(I_2)g_2$ and $E(I_1)g_1 \perp E(I_2)g_2$ (since $E(I_1)f \in M$, $E(I_1)g_1 \in M$, and $E(I_2)g_2 \in M^\perp$).

Let N be a set of ρ_h-measure zero. Then there exists a sequence (S_n) for which $N \subset S_n$, and $S_n = \cup_m J_{nm}$, where the J_{nm} are at most countably many mutually disjoint intervals for fixed $n \in \mathbb{N}$, and

$$\sum_m \rho_h(J_{nm}) \to 0 \quad \text{as} \quad n \to \infty.$$

Because of the equalities

$$\rho_h(J_{nm}) = \|E(J_{nm})h\|^2 = \|E(J_{nm})f\|^2 + \|E(J_{nm})g_2\|^2 = \rho_f(J_{nm}) + \rho_{g_2}(J_{nm})$$

for all n, m, it follows from this that

$$\sum_m \rho_f(J_{nm}) \to 0 \quad \text{and} \quad \sum_m \rho_{g_2}(J_{nm}) \to 0 \quad \text{as} \quad n \to \infty.$$

Consequently, N is a set of ρ_f-measure zero.

For all $t \in \mathbb{R}$ we have

$$\Big\|\sum_m E(J_{nm})E(t)f\Big\|^2 = \Big\|E(t)\sum_m E(J_{nm})f\Big\|^2 \leq \Big\|\sum_m E(J_{nm})f\Big\|^2$$
$$= \sum_m \|E(J_{nm})f\|^2 = \sum_m \rho_f(J_{nm}) \to 0 \quad \text{as} \quad n \to \infty.$$

Since the set $\{E(t)f : t \in \mathbb{R}\}$ is total in M and the norms of the operators $k \mapsto \Sigma_m E(J_{nm})k$ are less than or equal to 1, Theorem 4.23 implies that $\Sigma_m E(J_{nm})k \to 0$ for all $k \in M$. As $g_1 \in M$, it follows from this that

$$\sum_m \rho_{g_1}(J_{nm}) = \sum_m \|E(J_{nm})g_1\|^2 = \Big\|\sum_m E(J_{nm})g_1\Big\|^2 \to 0,$$

consequently,

$$\sum_m \rho_g(J_{nm}) = \sum_m \big(\rho_{g_1}(J_{nm}) + \rho_{g_2}(J_{nm})\big) \to 0,$$

as $n \to \infty$. Therefore, the ρ_g-measure of N equals zero. □

For $f, g \in H$ and a bounded E-measurable function $u : \mathbb{R} \to \mathbb{K}$, according to the polarization identities (1.4) and (1.8), respectively, we define

$$\int u(t)\,\mathrm{d}\langle g, E(t)f\rangle$$
$$= \tfrac{1}{4}\Big\{\int u(t)\,\mathrm{d}\rho_{g+f}(t) - \int u(t)\,\mathrm{d}\rho_{g-f}(t) + \mathrm{i}\int u(t)\,\mathrm{d}\rho_{g-\mathrm{i}f}(t)$$
$$-\mathrm{i}\int u(t)\,\mathrm{d}\rho_{g+\mathrm{i}f}(t)\Big\} \quad \text{if} \quad \mathbb{K} = \mathbb{C},$$
$$= \tfrac{1}{4}\Big\{\int u(t)\,\mathrm{d}\rho_{g+f}(t) - \int u(t)\,\mathrm{d}\rho_{g-f}(t)\Big\} \quad \text{if} \quad \mathbb{K} = \mathbb{R}.$$

With this definition we obtain for any bounded E-measurable functions u and v that

$$\int v(t)^* u(t)\,\mathrm{d}\langle g, E(t)f\rangle = \Big\langle \int v(t)\,\mathrm{d}E(t)g, \int u(t)\,\mathrm{d}E(t)f\Big\rangle. \tag{7.20}$$

For step functions u and v this is evident. In the general case this follows by Lebesgue's theorem if, according to Auxiliary theorem 7.13, we choose bounded sequences (u_n) and (v_n) of step functions that converge $\bar{\rho_f}, \bar{\rho_g}, \bar{\rho_{g+f}}, \bar{\rho_{g-f}}, \bar{\rho_{g+\mathrm{i}f}}, \bar{\rho_{g-\mathrm{i}f}}$ almost everywhere to u and v, respectively.

Now we are in a position to generate linear operators by means of *integrals with respect to a spectral family.*

Theorem 7.14. *Let E be a spectral family on the Hilbert space H, and let $u : \mathbb{R} \to \mathbb{K}$ be an E-measurable function. Then the formulae*

$$D(\hat{E}(u)) = \{f \in H : u \in L_2(\mathbb{R}, \rho_f)\}$$
$$\hat{E}(u)f = \int u(t)\,\mathrm{d}E(t)f \quad \text{for} \quad f \in D(\hat{E}(u)) \tag{7.21}$$

define a normal operator $\hat{E}(u)$ on H. For (7.21) we briefly write

$$\hat{E}(u) = \int u(t)\,\mathrm{d}E(t).$$

If $u, v : \mathbb{R} \to \mathbb{K}$ are arbitrary E-measurable functions, $a, b \in \mathbb{K}$, and

$$\varphi_n(x) = \begin{cases} 1 & \text{if } |u(x)| \leq n, \\ 0 & \text{otherwise}, \end{cases} \qquad \psi_n(x) = \begin{cases} 1 & \text{if } |v(x)| \leq n, \\ 0 & \text{otherwise}, \end{cases}$$

then it follows:

(a) *For all $f \in D(\hat{E}(u))$ and $g \in D(\hat{E}(v))$ we have*

$$\langle \hat{E}(v)g, \hat{E}(u)f \rangle = \lim_{n\to\infty} \int \psi_n(t)v(t)^*\varphi_n(t)u(t)\,\mathrm{d}\langle g, E(t)f\rangle;$$

for the latter we briefly write $\int v(t)^*u(t)\,\mathrm{d}\langle g, E(t)f\rangle$.

(b) *For all $f \in D(\hat{E}(u))$ we have*

$$\|\hat{E}(u)f\|^2 = \int |u(t)|^2\,\mathrm{d}\rho_f(t).$$

(c) *If u is bounded, then $\hat{E}(u) \in B(H)$ and*

$$\|\hat{E}(u)\| \leq \sup\{|u(t)| : t \in \mathbb{R}\}.$$

(d) *If $u(t) = 1$ for all $t \in \mathbb{R}$, then $\hat{E}(u) = I$.*

(e) *For every $f \in D(\hat{E}(u))$ and all $g \in H$ we have*

$$\langle g, \hat{E}(u)f \rangle = \int u(t)\,\mathrm{d}\langle g, E(t)f\rangle.$$

(f) *If $u(t) \geq c$ for all $t \in \mathbb{R}$, then*

$$\langle f, \hat{E}(u)f \rangle \geq c\|f\|^2 \quad \text{for all} \quad f \in D(\hat{E}(u)).$$

(g) $\hat{E}(au + bv) \supset a\hat{E}(u) + b\hat{E}(v)$, $D(\hat{E}(u) + \hat{E}(v)) = D(\hat{E}(|u| + |v|))$.

(h) $\hat{E}(uv) \supset \hat{E}(u)\hat{E}(v)$, $D(\hat{E}(u)\hat{E}(v)) = D(\hat{E}(v)) \cap D(\hat{E}(uv))$.

(i) $\hat{E}(u^*) = \hat{E}(u)^*$, $D(\hat{E}(u^*)) = D(\hat{E}(u))$.

PROOF. The mapping $\hat{E}(u) : D(\hat{E}(u)) \to H$ is well-defined, because of (7.17)–(7.21). We show that this mapping is linear. It is clear that $f \in D(\hat{E}(u))$ and $a \in \mathbb{K}$ imply that $af \in D(\hat{E}(u))$ and $\hat{E}(u)(af) = a\hat{E}(u)f$.

First assume that u is bounded. Then $D(\hat{E}(u)) = H$. By Auxiliary theorem 7.13, for arbitrary $f, g \in H$ there is a bounded sequence (u_n) of step functions that converges to u almost everywhere with respect to ρ_f, ρ_g and ρ_{f+g}. Then $u_n \to u$ in $L_2(\mathbb{R}, \rho_f)$, $L_2(\mathbb{R}, \rho_g)$ and $L_2(\mathbb{R}, \rho_{f+g})$; therefore, $\hat{E}(u_n)f \to \hat{E}(u)f$, $\hat{E}(u_n)g \to \hat{E}(u)g$ and $\hat{E}(u_n)(f+g) \to \hat{E}(u)(f+g)$. Since $\hat{E}(u_n)$ is obviously linear, it follows that

$$\begin{aligned} \hat{E}(u)(f+g) &= \lim_{n\to\infty} \hat{E}(u_n)(f+g) \\ &= \lim_{n\to\infty} (\hat{E}(u_n)f + \hat{E}(u_n)g) = \hat{E}(u)f + \hat{E}(u)g. \end{aligned}$$

Now let u be an arbitrary E-measurable function, and let $f, g \in D(\hat{E}(u))$. Then $(|\varphi_n(t)u(t)|)$ converges to $|u(t)|$ monotonically for all $t \in \mathbb{R}$. Using the identity we have just proved and on the basis of (7.18) we obtain that

$$\begin{aligned}\left\{\int|\varphi_n(t)u(t)|^2\,\mathrm{d}\rho_{f+g}(t)\right\}^{1/2} &= \|\hat{E}(\varphi_n u)(f+g)\| \\ &\leq \|\hat{E}(\varphi_n u)f\| + \|\hat{E}(\varphi_n u)g\| \\ &= \left\{\int|\varphi_n(t)u(t)|^2\,\mathrm{d}\rho_f(t)\right\}^{1/2} + \left\{\int|\varphi_n(t)u(t)|^2\,\mathrm{d}\rho_g(t)\right\}^{1/2} \\ &\leq \left\{\int|u(t)|^2\,\mathrm{d}\rho_f(t)\right\}^{1/2} + \left\{\int|u(t)|^2\,\mathrm{d}\rho_g(t)\right\}^{1/2} \\ &= \|\hat{E}(u)f\| + \|\hat{E}(u)g\| < \infty.\end{aligned}$$

It follows from this by B. Levi's theorem that $u \in L_2(\mathbb{R}, \rho_{f+g})$; consequently, $f+g \in D(\hat{E}(u))$ and

$$\begin{aligned}\hat{E}(u)(f+g) &= \lim_{n\to\infty} \hat{E}(\varphi_n u)(f+g) \\ &= \lim_{n\to\infty} (\hat{E}(\varphi_n u)f + \hat{E}(\varphi_n u)g) = \hat{E}(u)f + \hat{E}(u)g.\end{aligned}$$

So $\hat{E}(u)$ is linear. We obtain from the proof of (i) that $\hat{E}(u)$ is normal.

(a) This equality is clear for step functions u and v. For bounded E-measurable functions u and v the equality follows by means of Auxiliary theorem 7.13. In both cases the passage to the limit does not actually take place, as $\varphi_n = \psi_n = 1$ for large n. If u and v are arbitrary E-measurable functions, then we have for $f \in D(\hat{E}(u))$ and $g \in D(\hat{E}(v))$ that

$$\begin{aligned}\langle \hat{E}(v)g, \hat{E}(u)f\rangle &= \lim_{n\to\infty} \langle \hat{E}(\psi_n v)g, \hat{E}(\varphi_n u)f\rangle \\ &= \lim_{n\to\infty} \int \psi_n(t)v(t)^*\varphi_n(t)u(t)\,\mathrm{d}\langle g, E(t)f\rangle.\end{aligned}$$

(b) follows from (7.18) with $v = u$, $g = f$.

(c) Since u is bounded, we have $u \in L_2(\mathbb{R}, \rho_f)$ for all $f \in H$. Consequently, $D(\hat{E}(u)) = H$. The estimate of the norm immediately follows from (b).

(d) By (c) we have $\hat{E}(u) \in B(H)$. Furthermore, $\chi_{(-n,n]}(t) \to u(t) = 1$ for all $t \in \mathbb{R}$; therefore,

$$\hat{E}(u)f = \lim_{n\to\infty} \hat{E}(\chi_{(-n,n]})f = \lim_{n\to\infty} (E(n)f - E(-n)f) = f.$$

(e) follows from (a) with $v = 1$, by taking (d) into account.

(f) immediately follows from (e).

(g)[4] If $f \in D(\hat{E}(u) + \hat{E}(v)) = D(\hat{E}(u)) \cap D(\hat{E}(v))$, then $u, v \in L_2(\mathbb{R}, \rho_f)$; consequently, $au + bv \in L_2(\mathbb{R}, \rho_f)$, i.e., $f \in D(\hat{E}(au+bv))$. The equality $\hat{E}(au+bv)f = a\hat{E}(u)f + b\hat{E}(v)f$ therefore follows from (7.19). Since for E-measurable functions u, v we have $u, v \in L_2(\mathbb{R}, \rho_f)$ if and only if $|u| + |v| \in L_2(\mathbb{R}, \rho_f)$, we have $D(\hat{E}(u) + \hat{E}(v)) = D(\hat{E}(|u| + |v|))$.

[4] Properties (g), (h), and (i) follow more easily from Theorem 7.16.

(h) By (a) and (d) we have for all bounded E-measurable functions φ, ψ and all $f, g \in H$ that

$$\langle g, \hat{E}(\varphi)f\rangle = \langle \hat{E}(1)g, \hat{E}(\varphi)f\rangle = \langle \hat{E}(\varphi^*)g, \hat{E}(1)f\rangle = \langle \hat{E}(\varphi^*)g, f\rangle;$$

consequently,

$$\begin{aligned}\langle g, \hat{E}(\varphi)\hat{E}(\psi)f\rangle &= \langle \hat{E}(\varphi^*)g, \hat{E}(\psi)f\rangle = \int \varphi(t)\psi(t)\,\mathrm{d}\langle g, E(t)f\rangle \\ &= \langle g, \hat{E}(\varphi\psi)f\rangle.\end{aligned}$$

For bounded E-measurable functions φ, ψ we therefore have $\hat{E}(\varphi)\hat{E}(\psi) = \hat{E}(\varphi\psi)$.

Let $f \in D(\hat{E}(u)\hat{E}(v))$, i.e., let $f \in D(\hat{E}(v))$ and $\hat{E}(v)f \in D(\hat{E}(u))$. As the function $\varphi_n u$ is bounded for fixed $n \in \mathbb{N}$, it follows that $\varphi_n u\psi_m v \to \varphi_n uv$ in $L_2(\mathbb{R}, \rho_f)$ as $m \to \infty$. Consequently,

$$\begin{aligned}\hat{E}(u)\hat{E}(v)f &= \lim_{n\to\infty} \hat{E}(\varphi_n u)\{\lim_{m\to\infty} \hat{E}(\psi_m v)f\} \\ &= \lim_{n\to\infty} \lim_{m\to\infty} \hat{E}(\varphi_n u)\hat{E}(\psi_m v)f \\ &= \lim_{n\to\infty} \lim_{m\to\infty} \hat{E}(\varphi_n u\psi_m v)f = \lim_{n\to\infty} \hat{E}(\varphi_n uv)f.\end{aligned}$$

The existence of this limit means that the sequence $(\varphi_n uv)$ is a Cauchy sequence in $L_2(\mathbb{R}, \rho_f)$. Since, moreover, $\varphi_n(t)u(t)v(t) \to u(t)v(t)$ for all $t \in \mathbb{R}$, it follows that uv belongs to $L_2(\mathbb{R}, \rho_f)$; consequently, $f \in D(\hat{E}(uv))$ and $\hat{E}(u)\hat{E}(v)f = \hat{E}(uv)f$. Therefore, $D(\hat{E}(u)\hat{E}(v)) \subset D(\hat{E}(v)) \cap D(\hat{E}(uv))$ and $\hat{E}(u)\hat{E}(v) \subset \hat{E}(uv)$. If $f \in D(\hat{E}(v)) \cap D(\hat{E}(uv))$, then

$$\begin{aligned}\hat{E}(uv)f &= \lim_{n\to\infty} \lim_{m\to\infty} \hat{E}(\varphi_n u\psi_m v)f = \lim_{n\to\infty} \lim_{m\to\infty} \hat{E}(\varphi_n u)\hat{E}(\psi_m v)f \\ &= \lim_{n\to\infty} \hat{E}(\varphi_n u)\hat{E}(v)f.\end{aligned}$$

The existence of this limit means that $u \in L_2(\mathbb{R}, \rho_{\hat{E}(v)f})$; consequently, $\hat{E}(v)f \in D(\hat{E}(u))$, and thus $f \in D(\hat{E}(u)\hat{E}(v))$.

(i) We first show that $D(\hat{E}(u))$ is dense. For this we prove that for every $f \in H$ and every $m \in \mathbb{N}$ we have $\hat{E}(\varphi_m)f \in D(\hat{E}(u))$; because of the limit relation $f = \lim_{m\to\infty} \hat{E}(\varphi_m)f$ it follows from this that $D(\hat{E}(u))$ is dense. Let $f \in H$, $m \in \mathbb{N}$ and $g = \hat{E}(\varphi_m)f$. Then by (h) we have for all $n > m$ that

$$\int |\varphi_n(t)u(t)|^2\,\mathrm{d}\rho_g(t) = \|\hat{E}(\varphi_n u)\hat{E}(\varphi_m)f\|^2 = \|\hat{E}(\varphi_m u)f\|^2 < \infty.$$

Therefore, $u \in L_2(\mathbb{R}, \rho_g)$, i.e., $\hat{E}(\varphi_m)f = g \in D(\hat{E}(u))$ for all $m \in \mathbb{N}$.

We obviously have $D(\hat{E}(u^*)) = D(\hat{E}(u))$. By (e) we have for $f, g \in D(\hat{E}(u)) = D(\hat{E}(u^*))$ that

$$\begin{aligned}\langle g, \hat{E}(u)f\rangle &= \int u(t)\,\mathrm{d}\langle g, E(t)f\rangle = \left\{\int u(t)^*\,\mathrm{d}\langle f, E(t)g\rangle\right\}^* \\ &= \langle f, \hat{E}(u^*)g\rangle^* = \langle \hat{E}(u^*)g, f\rangle,\end{aligned}$$

i.e., $\hat{E}(u)$ and $\hat{E}(u^*)$ are formal adjoints of each other. It remains to prove that $D(\hat{E}(u)^*) \subset D(\hat{E}(u^*))$. Let $g \in D(\hat{E}(u)^*)$. Then for all $f \in D(\hat{E}(u))$

$$\langle \hat{E}(u)^* g, f\rangle = \langle g, \hat{E}(u)f\rangle = \lim_{n\to\infty} \langle g, \hat{E}(\varphi_n u)f\rangle = \lim_{n\to\infty} \langle \hat{E}(\varphi_n u^*)g, f\rangle.$$

In particular, for every $f \in H$ and $m \in \mathbb{N}$

$$\begin{aligned}\langle \hat{E}(\varphi_m)\hat{E}(u)^* g, f\rangle &= \langle \hat{E}(u)^* g, \hat{E}(\varphi_m)f\rangle \\ &= \lim_{n\to\infty} \langle \hat{E}(\varphi_n u^*)g, \hat{E}(\varphi_m)f\rangle = \lim_{n\to\infty} \langle \hat{E}(\varphi_m\varphi_n u^*)g, f\rangle \\ &= \langle \hat{E}(\varphi_m u^*)g, f\rangle.\end{aligned}$$

Consequently, for all $g \in D(\hat{E}(u)^*)$ and all $m \in \mathbb{N}$

$$\hat{E}(\varphi_m)\hat{E}(u)^* g = \hat{E}(\varphi_m u^*)g,$$

and thus

$$\hat{E}(u)^* g = \lim_{m\to\infty} \hat{E}(\varphi_m u^*)g = \lim_{m\to\infty} \int \varphi_m(t)u(t)^* \, \mathrm{d}E(t)g.$$

The existence of this limit means that the sequence $(\varphi_m u^*)$ converges in $L_2(\mathbb{R}, \rho_g)$, i.e., $u^* \in L_2(\mathbb{R}, \rho_g)$, and thus $g \in D(\hat{E}(u^*))$.

We have in particular $D(\hat{E}(u)^*) = D(\hat{E}(u))$, and (by (b)) $\|\hat{E}(u)f\| = \|\hat{E}(u)^* f\|$ for all $f \in D(\hat{E}(u))$, i.e., $\hat{E}(u)$ is normal. □

EXAMPLE 1 (*continued*). Let E be the spectral family of Example 1. If $u : \mathbb{R} \to \mathbb{C}$ is a step function, $u(t) = \sum_{j=1}^n c_j \chi_{J_j}(t)$, then

$$\begin{aligned}\left(\int u(t) \, \mathrm{d}E(t)f\right)(x) &= \sum_{j=1}^n c_j \chi_{J_j}(g(x))f(x) \\ &= u(g(x))f(x), \quad f \in L_2(M).\end{aligned}$$

Therefore, it follows for every E-measurable function u that

$$D(\hat{E}(u)) = \{f \in L_2(M) : (u \circ g)f \in L_2(M)\}$$

and

$$(\hat{E}(u)f)(x) = u(g(x))f(x) \quad \text{for} \quad f \in D(\hat{E}(u)),$$

i.e., $\hat{E}(u)$ is the maximal operator of multiplication by $u \circ g$. For $u = \mathrm{id}$ we obtain the operator of multiplication by g.

EXAMPLE 2 (*continued*). If E is the spectral family of Example 2 on $\oplus_{\alpha\in A} L_2(\mathbb{R}, \rho_\alpha)$ and $u : \mathbb{R} \to \mathbb{C}$ is an E-measurable function (cf. Exercise 7.18), then

$$D(\hat{E}(u)) = \left\{(f_\alpha) \in \bigoplus_{\alpha\in A} L_2(\mathbb{R}, \rho_\alpha) : (uf_\alpha) \in \bigoplus_{\alpha\in A} L_2(\mathbb{R}, \rho_\alpha)\right\}$$

and

$$\hat{E}(u)(f_\alpha) = (uf_\alpha) \quad \text{for} \quad f \in D(\hat{E}(u)).$$

We call this operator the *maximal multiplication operator induced by u on* $\oplus_{\alpha\in A} L_2(\mathbb{R}, \rho_\alpha)$. The proof is along the same lines as in Example 1.

EXAMPLE 3 (*continued*). Let E be the spectral family of Example 3. Then every function $u: \mathbb{R}\to\mathbb{K}$ is E-measurable, since for all $t\in\{\lambda_j : j=1, 2, \ldots\}$

$$u(t) = \sum_n u(\lambda_n)\chi_{\{\lambda_n\}}(t) = \lim_{m\to\infty} \sum_{n=1}^{m} u(\lambda_n)\chi_{\{\lambda_n\}}(t);$$

as $\mathbb{R}\setminus\{\lambda_j : j=1, 2, \ldots\}$ is a set of ρ_f-measure zero for all $f\in H$, we have found a sequence of step functions that converges to u ρ_f-almost everywhere for all $f\in H$. It is now easy to see that for all $u: \mathbb{R}\to\mathbb{K}$

$$D(\hat{E}(u)) = \left\{ f\in H : \sum_j |u(\lambda_j)|^2 \|P_j f\|^2 < \infty \right\},$$

$$\hat{E}(u)f = \sum_j u(\lambda_j)P_j f \quad \text{for} \quad f\in D(\hat{E}(u)).$$

For $u=\mathrm{id}$ we obtain in particular that

$$D(\hat{E}(\mathrm{id})) = \left\{ f\in H : \sum_j |\lambda_j|^2 \|P_j f\|^2 < \infty \right\}$$

and

$$\hat{E}(\mathrm{id})f = \sum_j \lambda_j P_j f \quad \text{for} \quad f\in D(\hat{E}(\mathrm{id})).$$

Theorem 7.15. *Let H_1 and H_2 be Hilbert spaces, let U be a unitary operator from H_1 onto H_2, and let E be a spectral family on H_1. Then by the formula*

$$F(t) = UE(t)U^{-1}, \quad t\in\mathbb{R}$$

a spectral family is defined on H_2. A function $u: \mathbb{R}\to\mathbb{K}$ is F-measurable if and only if it is E-measurable. If $\hat{F}$ is defined analogously to $\hat{E}$, and u is E-measurable, then

$$\hat{F}(u) = U\hat{E}(u)U^{-1}.$$

PROOF. It is clear that F is a spectral family on H_2. If $\rho_f(t) = \|E(t)f\|^2$ and $\sigma_g(t) = \|F(t)g\|^2$, then $\rho_f(t) = \sigma_{Uf}(t)$ obviously holds. Consequently, u is E-measurable if and only if it is F-measurable, and $L_2(\mathbb{R}, \rho_f) = L_2(\mathbb{R}, \sigma_{Uf})$. The equality $\hat{F}(u) = U\hat{E}(u)U^{-1}$ is evident for any step function u. The assertion follows from this fact immediately. □

Theorem 7.16. *Let E be a spectral family on the Hilbert space H. Then there exists a family $\{\rho_\alpha : \alpha\in A\}$ of right continuous non-decreasing functions (the cardinality of A is at most the dimension of H) and a unitary operator*

$U: H \to \oplus_{\alpha\in A} L_2(\mathbb{R}, \rho_\alpha)$ for which

$$E(t) = U^{-1}F(t)U \quad \text{for all} \quad t \in \mathbb{R}$$

with the spectral family F from Example 2. For every E-measurable function u

$$\hat{E}(u) = U^{-1}\hat{F}(u)U,$$

where $\hat{F}(u)$ is the maximal operator of multiplication by u on

$$\oplus_{\alpha\in A} L_2(\mathbb{R}, \rho_\alpha).$$

PROOF. For any $f \in H$, $f \neq 0$ let $H_f = \overline{L\{E(t)f : t \in \mathbb{R}\}}$ and let $\rho_f(t) = \|E(t)f\|^2$. Then the formula

$$U_{f,0}\left(\sum_{j=1}^{n} c_j E(t_j)f\right) = \sum_{j=1}^{n} c_j \chi_{(-\infty, t_j]}$$

defines an isometric mapping of $L\{E(t)f : t \in \mathbb{R}\}$ into $L_2(\mathbb{R}, \rho_f)$, as can be verified easily. For all $g \in L\{E(t)f : t \in \mathbb{R}\}$

$$U_{f,0}(E(t)g) = \chi_{(-\infty, t]} U_{f,0} g.$$

The range of $U_{f,0}$ contains the space of left continuous step functions, thus it is dense in $L_2(\mathbb{R}, \rho_f)$ (observe that the left continuous step functions are dense in the space of step functions). The closure $U_f = \overline{U_{f,0}}$ is therefore a unitary operator from H_f onto $L_2(\mathbb{R}, \rho_f)$, and for all $g \in H_f$

$$U_f(E(t)g) = \chi_{(-\infty, t]} U_f g.$$

With the aid of Zorn's lemma we see immediately that there exists a maximal system $\{H_{f_\alpha} : \alpha \in A\}$ such that $H_{f_\alpha} \perp H_{f_\beta}$ for $\alpha \neq \beta$ (partial ordering = inclusion, upper bound = union). We write H_α for H_{f_α}, and show that $H = \oplus_{\alpha\in A} H_\alpha$. If we had $H \neq \oplus_{\alpha\in A} H_\alpha$, then there would be a $g \in H$, $g \neq 0$ such that $g \perp H_\alpha$ for all $\alpha \in A$. Then we would also have $E(t)g \perp H_\alpha$ for all $\alpha \in A$, and thus $H_g \perp \oplus_{\alpha\in A} H_\alpha$; this contradicts maximality.

Let $\rho_\alpha = \rho_{f_\alpha}$, let $U_\alpha : H_\alpha \to L_2(\mathbb{R}, \rho_\alpha)$ be the corresponding unitary operators, and let P_α be the orthogonal projections onto H_α, then

$$Ug = (U_\alpha P_\alpha g), \quad g \in H$$

is a unitary operator from H onto $\oplus_{\alpha\in A} L_2(\mathbb{R}, \rho_\alpha)$. For all $g \in H$ we have

$$UE(t)g = (U_\alpha P_\alpha E(t)g) = (U_\alpha E(t)P_\alpha g)$$
$$= (\chi_{(-\infty, t]} U_\alpha P_\alpha g) = F(t)Ug.$$

$\hat{E}(u) = U^{-1}\hat{F}(u)U$ follows by Theorem 7.13. The rest of the assertion follows from Example 2. □

Exercises

7.16. Let E be a spectral family, and let $u : \mathbb{R} \to \mathbb{K}$ be a continuous function. For $-\infty < a < b < \infty$ we can define the integral $\int_a^b u(t)\, dE(t)$ as a Riemann-Stieltjes integral, i.e., the integral is the limit in $B(H)$ of the sums

$$\sum_{j=1}^{n} u(t_j) \neq (E(t_j) - E(t_{j-1}))$$

with $a = t_0 < t_1 < \cdots < t_n = b$, provided that the maximal length of the intervals tends to 0.

7.17. Let E be a spectral family on H.
(a) For every sequence (f_n) from H there is an $h \in H$ for which we have: Every set of ρ_h-measure zero is of ρ_{f_j}-measure zero for all $j \in \mathbb{N}$.
(b) If H is separable, then there exists an $h \in H$ for which we have: Every set of ρ_h-measure zero is of ρ_f-measure zero for all $f \in H$.
(c) If H is separable, and $u : \mathbb{R} \to \mathbb{K}$ is an E-measurable function, then there exists a sequence (u_n) of step functions that converges to u ρ_f-almost everywhere for all $f \in H$.

7.18. Let E be the spectral family of Example 2. A function $u : \mathbb{R} \to \mathbb{C}$ is E-measurable if and only if it is ρ_α-measurable for every $\alpha \in A$.

7.3 The spectral theorem for self-adjoint operators

If u is a real-valued E-measurable function on $\mathbb{R}$, then the operator $\hat{E}(u)$ is self-adjoint by Theorem 7.14(i). We show in this section that every self-adjoint operator can be represented in this way and there exists exactly one such representation with $u = \mathrm{id}$.

Theorem 7.17 (Spectral theorem). *For every self-adjoint operator T on the Hilbert space H there exists exactly one spectral family E for which $T = \hat{E}(\mathrm{id})$, or in another notation, $T = \int t\, dE(t)$ (cf. Theorem 7.14). In the complex case the spectral family E is given by*

$$\langle g, (E(b) - E(a))f\rangle = \lim_{\delta \to 0+} \lim_{\epsilon \to 0+} \frac{1}{2\pi i} \int_{a+\delta}^{b+\delta} \langle g, (R(t - i\epsilon, T) - R(t + i\epsilon, T))f\rangle\, dt \tag{7.22}$$

for all $f, g \in H$ and $-\infty < a \leq b < \infty$. We say that E is the spectral family of T.

Proof. First we assume that H is complex.

Uniqueness: If $T = \hat{E}(\mathrm{id})$, then $z - T = \hat{E}(z - \mathrm{id})$ by Theorem 7.14(g). Then for all $z \in \mathbb{C}$ such that $\mathrm{Im}\, z \neq 0$ we have by Theorem 7.14(h), with the notation $u_z(t) = (z - t)^{-1}$, that

$$(z - T)\hat{E}(u_z)f = \hat{E}((z - \mathrm{id})u_z)f = f \quad \text{for all } f \in H,$$
$$\hat{E}(u_z)(z - T)f = \hat{E}(u_z(z - \mathrm{id}))f = f \quad \text{for all } f \in D(T).$$

Consequently, $\hat{E}(u_z) = R(z, T)$ for all $z \in \mathbb{C}$ such that $\operatorname{Im} z \neq 0$. This implies via Theorem 7.14(e) that

$$\langle f, R(z, T)f\rangle = \int (z-t)^{-1}\, d\rho_f(t) \quad \text{for all} \quad f \in H.$$

The functions $\rho_f(t) = \|E(t)f\|^2$ and $F(z) = \langle f, R(z, T)f\rangle$ therefore satisfy the assumption of Theorem B1 of the Appendix, and thus for all $t \in \mathbb{R}$

$$\begin{aligned}\|E(t)f\|^2 = \langle f, E(t)f\rangle &= \lim_{\delta\to 0+}\lim_{\epsilon\to 0+}\frac{-1}{\pi}\int_{-\infty}^{t+\delta}\operatorname{Im}\langle f, R(s+i\epsilon, T)f\rangle\, ds\\ &= \lim_{\delta\to 0+}\lim_{\epsilon\to 0+}\frac{1}{2\pi i}\int_{-\infty}^{t+\delta}\langle f, (R(s-i\epsilon, T)-R(s+i\epsilon, T))f\rangle\, ds.\end{aligned}$$

(7.22) follows from this with the aid of the polarization formula (1.4). Since (7.22) holds for all $f, g \in H$, the uniqueness has been proven.

Existence: If there exists a spectral family E such that $T = \hat{E}(\mathrm{id})$, then (7.22) must hold. Therefore we study whether (7.22) defines a spectral family E with the property $\hat{E}(\mathrm{id}) = T$. For every $f \in H$ the function F_f defined by the equality $F_f(z) = \langle f, R(z, T)f\rangle$ satisfies the assumptions of Theorem B3 (Appendix), since F_f is holomorphic for $\operatorname{Im} z > 0$ by Theorem 5.16 and we have

$$\begin{aligned}\operatorname{Im} F_f(z) &= \operatorname{Im}\langle f, R(z, T)f\rangle = \operatorname{Im}\langle (z-T)R(z, T)f, R(z, T)f\rangle\\ &= \|R(z, T)f\|^2 \operatorname{Im} z^* < 0 \quad \text{for} \quad \operatorname{Im} z > 0\end{aligned}$$

and (by Theorem 5.18)

$$|F_f(z)\operatorname{Im} z| \leq |\operatorname{Im} z|^{-1}\|f\|^2|\operatorname{Im} z| = \|f\|^2.$$

Consequently,

$$\langle f, R(z, T)f\rangle = F_f(z) = \int (z-t)^{-1}\, dw(f, t), \tag{7.23}$$

where

$$w(f, t) = \lim_{\delta\to 0+}\lim_{\epsilon\to 0+}\frac{1}{2\pi i}\int_{-\infty}^{t+\delta}\langle f, (R(s-i\epsilon, T)-R(s+i\epsilon, T))f\rangle\, ds.$$

$w(f, t)$ is a non-decreasing and right continuous function of t, and $w(f, t) \to 0$ as $t \to -\infty$, $w(f, t) \leq \|f\|^2$ for all $t \in \mathbb{R}$. Equation (7.23) holds for all $z \in \mathbb{C}\setminus\mathbb{R}$ since $\langle f, R(z^*, T)f\rangle = \langle f, R(z, T)f\rangle^*$. Furthermore, we define

$$w(g, f, t) = \lim_{\delta\to 0+}\lim_{\epsilon\to 0+}\frac{1}{2\pi i}\int_{-\infty}^{t+\delta}\langle g, (R(s-i\epsilon, T)-R(s+i\epsilon, T))f\rangle\, ds;$$

the existence of this limit follows by means of the polarization identity for the sesquilinear form $(g, f) \mapsto \langle g, (R(s-i\epsilon, T)-R(s+i\epsilon, T))f\rangle$.

The mapping $(g, f) \mapsto w(g, f, t)$ is a bounded non-negative sesquilinear form on H for every $t \in \mathbb{R}$. The sesquilinearity is clear from the definition; moreover, $w(f, f, t) = w(f, t) \geq 0$ for all $t \in \mathbb{R}$. The Schwarz inequality and the inequality $w(f, t) \leq \|f\|^2$ imply for all $f, g \in H$ and $t \in \mathbb{R}$ that

$$|w(g, f, t)|^2 \leq w(g, t)w(f, t) \leq \|g\|^2\|f\|^2.$$

Therefore, by Theorem 5.35 there exists, for every $t \in \mathbb{R}$, an operator $E(t) \in B(H)$ for which $\|E(t)\| \leq 1$ and

$$\langle g, E(t)f \rangle = w(g, f, t) \quad \text{for all} \quad f, g \in H.$$

It is obvious that $E(t)$ is self-adjoint and $E(t) \geq 0$.

Now we show that E is a spectral family. For this we first show that $E(s)E(t) = E(\min(s, t))$ for all $s, t \in \mathbb{R}$. For all $z \in \mathbb{C} \setminus \mathbb{R}$ and for all $f \in H$

$$\langle g, R(z, T)f \rangle = \int (z-t)^{-1} \, \mathrm{d}w(g, f, t) = \int (z-t)^{-1} \, \mathrm{d}\langle g, E(t)f \rangle. \tag{7.24}$$

This follows from (7.23) using the polarization identity. Consequently, the first resolvent identity implies for all $z, z' \in \mathbb{C} \setminus \mathbb{R}$ with $z \neq z'$

$$\begin{aligned}
\int (z-t)^{-1} \, \mathrm{d}\langle R(z'^*, T)g, E(t)f \rangle &= \langle R(z'^*, T)g, R(z, T)f \rangle \\
&= \langle g, R(z', T)R(z, T)f \rangle \\
&= (z'-z)^{-1}\{\langle g, R(z, T)f \rangle - \langle g, R(z', T)f \rangle\} \\
&= (z'-z)^{-1} \int [(z-t)^{-1} - (z'-t)^{-1}] \, \mathrm{d}\langle g, E(t)f \rangle \\
&= \int (z-t)^{-1}(z'-t)^{-1} \, \mathrm{d}\langle g, E(t)f \rangle \\
&= \int (z-t)^{-1} \, \mathrm{d}_t \int_{-\infty}^{t} (z'-s)^{-1} \, \mathrm{d}\langle g, E(s)f \rangle.
\end{aligned}$$

It follows from this by Theorem B2 (Appendix) that

$$\begin{aligned}
\int (z'-s)^{-1} \, \mathrm{d}_s\langle g, E(s)E(t)f \rangle &= \langle g, R(z', T)E(t)f \rangle \\
&= \langle R(z'^*, T)g, E(t)f \rangle \\
&= \int_{-\infty}^{t} (z'-s)^{-1} \, \mathrm{d}\langle g, E(s)f \rangle.
\end{aligned}$$

Therefore it follows for all $f, g \in H$ and $s, t \in \mathbb{R}$ again by Theorem B2 (Appendix) that

$$\langle g, E(s)E(t)f \rangle = \begin{cases} \langle g, E(s)f \rangle & \text{for} \quad s \leq t, \\ \langle g, E(t)f \rangle & \text{for} \quad t \leq s. \end{cases}$$

This means that for all $s, t \in \mathbb{R}$

$$E(s)E(t) = E(\min(s, t)).$$

In particular, $E(t)^2 = E(t)$. Therefore, the $E(t)$ are orthogonal projections for all $t \in \mathbb{R}$, and $E(s) \leqslant E(t)$ for $s \leqslant t$ (cf. Theorems 4.29 and 4.31). Thus (7.11(a)) and (7.11(b)) are satisfied. The right continuity (7.11(c)) follows from the formula

$$\|E(t+\epsilon)f - E(t)f\|^2 = \|E(t+\epsilon)f\|^2 - \|E(t)f\|^2$$
$$= w(f, t+\epsilon) - w(f, t) \to 0 \quad \text{as} \quad \epsilon \to 0+,$$

since $w(f, \cdot)$ is right continuous. Moreover, $\|E(t)f\|^2 = w(f, t) \to 0$ as $t \to -\infty$. Therefore, $E(t) \xrightarrow{s} 0$ as $t \to -\infty$. It only remains to prove that $E(t) \xrightarrow{s} I$ as $t \to \infty$. As $E(\cdot)$ is monotone, $E(t)$ strongly converges to an orthogonal projection $E(\infty)$ as $t \to \infty$ (cf. Theorem 4.32). We have

$$\langle f, E(\infty)f \rangle = \lim_{s \to \infty} \langle f, E(s)f \rangle \geqslant \langle f, E(t)f \rangle.$$

Consequently, $E(\infty) \geqslant E(t)$ for all $t \in \mathbb{R}$. Let $F = I - E(\infty)$. Then

$$E(t)F = E(t)(I - E(\infty)) = E(t) - E(t) = 0, \quad t \in \mathbb{R}.$$

It follows from this for all $f, g \in H$, Im $z \neq 0$ that

$$\langle g, R(z, T)Ff \rangle = \int (z-t)^{-1} \, d\langle g, E(t)Ff \rangle = 0.$$

Hence $R(z, T)Ff = 0$ for all $f \in H$, and thus $F = 0$, i.e., $E(\infty) = I$. Consequently, we have proved that E is a spectral family.

$R(z, T) = \hat{E}(u_z)$ by (7.24) and Theorem 7.14. This implies that $\hat{E}(z - \text{id}) = z - T$ and $\hat{E}(\text{id}) = T$ (Theorem 7.14(h) and (g) respectively). Hence, the theorem is proved in the complex case.

If T is a self-adjoint operator on the real Hilbert space H, then we consider the self-adjoint operator T_C on the complex Hilbert space H_C (cf. Exercise 5.32). By what we have just proved, T_C has exactly one spectral family E_C for which $\hat{E}_C(\text{id}) = T$. The restriction E of E_C to H is a spectral family on H such that $\hat{E}(\text{id}) = T$. If F were another spectral family on H with the property that $\hat{F}(\text{id}) = T$, then the complexification of F would be another spectral family F_C on H_C such that $\hat{F}_C(\text{id}) = T_C$. Therefore, $F_C = E_C$, and thus $F = E$. The details are left to the reader (cf. Exercise 7.25). $\square$

EXAMPLE 1. If T is the operator of multiplication by a real function g on $\bigoplus_{\alpha \in A} L_2(\mathbb{R}, \rho_\alpha)$ (cf. Section 7.2, Example 2 (continued)), then $E(t)$ is the operator of multiplication by the characteristic function of $\{s \in \mathbb{R} : g(s) \leqslant t\}$. We also write briefly that $E(t) = \chi_{\{s \in \mathbb{R} \,:\, g(s) \leqslant t\}}$. The corresponding result holds for multiplication operators on $L_2(M)$. The proofs are contained in Examples 1 and 2 of Section 7.2.

EXAMPLE 2. Let P be an orthogonal projection on H. Then

$$E_P(t) = \begin{cases} 0 & \text{for} \quad t<0, \\ I-P & \text{for} \quad 0 \leqslant t < 1, \\ I & \text{for} \quad t \geqslant 1 \end{cases}$$

is the spectral family of P, since for this spectral family we have

$$\int u(t)\, \mathrm{d}E(t) = u(0)(I-P) + u(1)P;$$

consequently,

$$\int t\, \mathrm{d}E(t) = P.$$

In particular, the spectral family of the zero operator is given by

$$E_0(t) = \begin{cases} 0 & \text{for} \quad t<0, \\ I & \text{for} \quad t \geqslant 0, \end{cases}$$

and that of the identity operator by

$$E_I(t) = \begin{cases} 0 & \text{for} \quad t<1, \\ I & \text{for} \quad t \geqslant 1. \end{cases}$$

EXAMPLE 3. Assume that T is a compact self-adjoint operator on H, (λ_j) is the sequence of non-zero eigenvalues of T, (P_j) is the sequence of the orthogonal projections onto the eigenspaces $N(\lambda_j - T)$ and P_0 is the orthogonal projection onto $N(T)$. Then the equality

$$E(t)f = \begin{cases} \sum\limits_{\{j:\lambda_j \leqslant t\}} P_j f & \text{for} \quad t<0, \\ \sum\limits_{\{j:\lambda_j \leqslant t\}} P_j f + P_0 f & \text{for} \quad t \geqslant 0 \end{cases}$$

defines the spectral family of T, since

$$\hat{E}(\mathrm{id})f = \sum_j \lambda_j P_j f + 0P_0 f = Tf \quad \text{for all} \quad f \in H$$

by Theorem 7.1.

Theorem 7.18 (Spectral representation theorem). *Let T be a self-adjoint operator on H. Then there exist a family $\{\rho_\alpha : \alpha \in A\}$ of right continuous non-decreasing functions and a unitary operator U from H onto $\oplus_{\alpha \in A} L_2(\mathbb{R}, \rho_\alpha)$ for which*

$$T = U^{-1} T_{\mathrm{id}} U,$$

where T_{id} denotes the maximal operator of multiplication by the function id *on* $\oplus_{\alpha\in A}L_2(\mathbb{R},\rho_\alpha)$. *For the spectral family* E *of* T *we have* $E(t)=U^{-1}\chi_{(-\infty,t]}U$.

REMARK. In place of the assertion of Theorem 7.18 we can briefly say the following: Every self-adjoint operator is unitary equivalent to a multiplication operator by the function $u=\mathrm{id}$; these are operators that we already know quite well. The cardinality of the set A is at most equal to the dimension of H. The spaces $L_2(\mathbb{R},\rho_\alpha)$ have to be chosen to be real when H is real. One disadvantage of this theorem compared to Theorem 7.17 is that this representation is not unique.

PROOF. If E is the spectral family of T, and $\oplus_{\alpha\in A}L_2(\mathbb{R},\rho_\alpha)$, U, and F are constructed as in Theorem 7.16, then by Theorem 7.16 and Theorem 7.17

$$T=\hat{E}(\mathrm{id})=U^{-1}\hat{F}(\mathrm{id})U=U^{-1}T_{\mathrm{id}}U,$$
$$E(t)=U^{-1}F(t)U=U^{-1}\chi_{(-\infty,t]}U. \qquad \square$$

If E is the spectral family of the self-adjoint operator T, and u is an E-measurable function, then we write $u(T)$ for $\hat{E}(u)$. We already know that $u_z(T)=(z-T)^{-1}$ for $u_z(t)=(z-t)^{-1}$. The following theorem gives further justification for this notation.

Theorem 7.19. *If* $u(t)=\sum_{j=0}^n c_jt^j$, *then* $u(T)=\sum_{j=0}^n c_jT^j$, *where we set* $T^0=I$.

PROOF. The assertion obviously holds for $n=0$ (cf. Theorem 7.14(d)). Let us assume that it holds for polynomials of degree $\leq n-1$. Then $v(T)=\sum_{j=1}^n c_jT^{j-1}$ for $v(t)=\sum_{j=1}^n c_jt^{j-1}$. Because of the equality $u=v\cdot\mathrm{id}+c_0$, it follows from Theorem 7.14(h) that

$$u(T)\supset v(T)T+c_0I.$$

Since for $n\geq 1$ we moreover have $D(T)=D(\hat{E}(\mathrm{id}))\supset D(u(T))$, it follows from Theorem 7.14(h) that

$$D(u(T))=D(T)\cap D(u(T))=D(v(T)T),$$

and thus

$$u(T)=v(T)T+c_0I=\left(\sum_{j=1}^n c_jT^{j-1}\right)T+c_0I=\sum_{j=0}^n c_jT^j. \qquad \square$$

Let E be a spectral family on H. A subset M of $\mathbb{R}$ is said to be *E-measurable* if its characteristic function χ_M is E-measurable. We write $E(M)$ for $\hat{E}(\chi_M)$. We have $E(\mathbb{R}\setminus M)=\hat{E}(1-\chi_M)=I-\hat{E}(\chi_M)=I-E(M)$. If E is the spectral family of T, then $E(M)=\chi_M(T)$. The operators $E(M)$ are orthogonal projections.

Proposition. *Let* T *be a self-adjoint operator on* H, *and let* E *be its spectral family.*

(1) *For any E-measurable subset M of* $\mathbb{R}$ *the subspace* $R(E(M))$ *is a reducing subspace of T, i.e.,* $E(M)T \subset TE(M)$.
(2) *Let* $D(M, T) = E(M)D(T)$. *We have* $\langle f, Tf\rangle < \gamma\|f\|^2$ *for all* $f \in D((-\infty, \gamma), T)$, $f \neq 0$, *and* $\langle f, Tf\rangle \leq \gamma\|f\|^2$ *for all* $f \in D((-\infty, \gamma], T)$. *Similar statements hold for* (γ, ∞) *and* $[\gamma, \infty)$.
(3) *We have* $\langle f, Tf\rangle \leq \gamma\|f\|^2$ *for all* $f \in D(T)$ *if and only if* $E(t) = I$ *for all* $t \geq \gamma$. *We have* $\langle f, Tf\rangle \geq \gamma\|f\|^2$ *for all* $f \in D(T)$ *if and only if* $E(t) = 0$ *for all* $t < \gamma$.
(4) *For every bounded interval J the subspace* $R(E(J))$ *is contained in* $D(T)$ *and* $TE(J) = \hat{E}(\mathrm{id}\,\chi_J) \in B(H)$.
(5) *Assume that* $t \in \mathbb{R}$ *and* $s > 0$. *We have* $f \in D(T)$ *and* $\|(T-t)f\| \leq s\|f\|$ *for every* $f \in R(E(t+s) - E(t-s))$.
(6) *If u is a real-valued E-measurable function, then we have for the spectral family F of* $u(T)$ *that* $F(t) = E(\{s \in \mathbb{R} : u(s) \leq t\})$ *for every* $t \in \mathbb{R}$ *and* $F(M) = E(\{s \in \mathbb{R} : u(s) \in M\})$ *for every Borel set M.*

The *proofs* are obvious when T is the operator T_{id} on $\oplus_{\alpha\in A} L_2(\mathbb{R}, \rho_\alpha)$. In the general case we use the spectral representation theorem 7.18. As to Part 6, observe that $F(M) = \chi_M(u(T)) = (\chi_M \circ u)(T)$, $\chi_M \circ u = \chi_{\{s \in \mathbb{R} : u(s) \in M\}}$ and use Section 7.2, Example 2.

Now we are in a position to define the nth root of an arbitrary non-negative self-adjoint operator and give the *polar decomposition* of unbounded operators.

Theorem 7.20.
(a) *Every non-negative self-adjoint operator T possesses exactly one non-negative self-adjoint* n*th root* $T^{1/n}$. *If E is the spectral family of T, then* $T^{1/n} = \int t^{1/n}\,\mathrm{d}E(t)$ *(here* $t^{1/n} \geq 0$ *for* $t \geq 0$; *for* $t < 0$ *the value of* $t^{1/n}$ *is immaterial, as the* ρ_f*-measure of* $(-\infty, 0)$ *vanishes for every* $f \in H$). *If T is compact, then* $T^{1/n}$ *is also compact.*
(b) *Let T be a densely defined closed operator from* H_1 *into* H_2. *The operator T can be uniquely represented in the form* $T = US$, *where S is a non-negative self-adjoint operator on* H_1 *and U is a partial isometry with initial domain* $\overline{R(S)}$ *and final domain* $\overline{R(T)}$. *We have* $S = (T^*T)^{1/2}$; *we again write* $|T|$ *for* $(T^*T)^{1/2}$.

PROOF.
(a) By Theorem 7.14(f), (h) and (i) the given operator $T^{1/n}$ is a non-negative nth root of T. By Part 6 of the last proposition the spectral family of $T^{1/n}$ is

$$E_{1/n}(t) = \begin{cases} 0 & \text{for } t < 0, \\ E(t^n) & \text{for } t \geq 0. \end{cases}$$

If S is an arbitrary non-negative nth root of T with spectral family F, then the spectral family of $T = S^n$ is

$$F_n(t) = \begin{cases} 0 & \text{for} \quad t<0, \\ F(t^{1/n}) & \text{for} \quad t \geq 0. \end{cases}$$

The equality $E = F_n$ follows because of the uniqueness of the spectral family of T. Consequently, $F = E_{1/n}$, and thus $S = T^{1/n}$.

If T is compact, then there exists a compact non-negative nth root of T by Theorem 7.4. The operator $T^{1/n}$ constructed here is then compact, because of the uniqueness of the nth root.

(b) If $T = US$ is such a representation, then $T^*T = S^*U^*US = SU^*US = S^2$, since U^*U is the orthogonal projection onto $\overline{R(S)}$. The equalities $S = (T^*T)^{1/2} = |T|$ follow from part (a). This proves the uniqueness, since U is uniquely determined by the equality $U|T|f = Tf$. It remains to prove the existence of such a representation. By Theorem 5.40 we have $D(|T|) = D(T)$ and $\| \, |T|f \| = \|Tf\|$ for all $f \in D(T)$. The mapping $V : R(|T|) \to R(T)$, $|T|f \mapsto Tf$ is therefore isometric, and $T = V|T|$. The operator V can be uniquely extended to an isometric mapping $\overline{V}$ acting from $\overline{R(|T|)}$ onto $\overline{R(T)}$. The equality $U(f+g) = \overline{V}f$ for $f \in \overline{R(|T|)}$ and $g \in R(|T|)^{\perp}$ proves the assertion. □

The boundedness and the norm of self-adjoint operators can be seen from their spectral family.

Theorem 7.21. *A self-adjoint T on H is bounded if and only if there exist real numbers γ_1 and γ_2 for which*

$$E(t) = \begin{cases} 0 & \text{for} \quad t<\gamma_1, \\ I & \text{for} \quad t \geq \gamma_2. \end{cases}$$

We can then choose

$$\gamma_1 = m = \inf\{\langle f, Tf\rangle : f \in D(T), \|f\| = 1\},$$
$$\gamma_2 = M = \sup\{\langle f, Tf\rangle : f \in D(T), \|f\| = 1\}.$$

For $m < t < M$ we have $E(t) \neq 0$ and $E(t) \neq I$.

Proof. By Theorem 4.4 the operator T is bounded if and only if m and M are finite. By Part 3 of the last proposition this is equivalent to the assertions that $E(t) = 0$ for $t < m$ and $E(t) = I$ for $t \geq M$. If we had $E(t_0) = 0$ for some $t_0 > m$, then we would have $\langle f, Tf\rangle \geq t_0\|f\|^2$ for all $f \in D(T)$, on account of Part 3 of the last proposition. This contradicts the definition of m. The relation $E(t) \neq I$ for $t < M$ follows similarly. □

Exercises

7.19. Assume that T is a self-adjoint operator, E is its spectral family, and M is a closed subspace of H. We have $M = R(E(t))$ if and only if M is a reducing subspace of T and $\langle f, Tf\rangle \leq t\|f\|^2$ for $f \in D(T) \cap M$ and $\langle f, Tf\rangle > t\|f\|^2$ for $f \in D(T) \cap M^{\perp}$.

7.20. Suppose that T is self-adjoint with spectral family E, $a, b \in \rho(T) \cap \mathbb{R}$, and Γ is a positively oriented Jordan curve for which $(a, b) \cap \sigma(T)$ lies inside Γ and all other points of the spectrum of T lie outside Γ. Then $E(b) - E(a) = (2\pi i)^{-1} \int_\Gamma R(z, T)\, dz$ in the sense of the Riemann integral.

7.21. Assume that T is self-adjoint, $0 \leq p \leq 1$, and $f \in D(T)$, $T^p = \int t^p\, dE(t)$ with an arbitrary choice of the branch of the function $\mathbb{R} \to \mathbb{C}$, $t \mapsto t^p$. Then $\|T^p f\| \leq \|Tf\|^p \|f\|^{1-p}$.
Hint: Hölder's inequality for $\|T^p f\|^2 = \int |t|^{2p}\, d\|E(t)f\|^2$.

7.22. Let T be a non-negative self-adjoint operator, and let $\lambda > 0$. Then $(\lambda + T)^{-1} = (1/\lambda) \int_0^\infty e^{-s} \cos(\lambda^{-1/2} s T^{1/2})\, ds$, in the sense of the improper Riemann-Stieltjes integral.

7.23. (a) Let T_1 and T_2 be densely defined and closed. Assume that $D(T_1) = D(T_2)$ and $\|T_1 f\| = \|T_2 f\|$ for all $f \in D(T_1)$. Then $|T_1| = |T_2|$. If T_1 and T_2 are self-adjoint and non-negative, then $T_1 = T_2$.
(b) A densely defined closed operator is normal if and only if $D(T^*T) = D(TT^*)$ and $\|T^*Tf\| = \|TT^*f\|$ for $f \in D(T^*T)$.

7.24. If T is self-adjoint and $u, v : \mathbb{R} \to \mathbb{R}$ are Borel functions, then $u(v(T)) = (u \circ v)(T)$.
Hint: Use the spectral representation theorem 7.18.

7.25. Assume that H is a real Hilbert space, T is a self-adjoint operator on H; $H_\mathbb{C}$ and $T_\mathbb{C}$ are the complexifications of H and T, respectively (cf. Exercise 5.32).
(a) The mapping $K : H_\mathbb{C} \to H_\mathbb{C}$, $(f, g) \mapsto (f, -g)$ for $f, g \in H$ has the properties $KK = I$ and $K(ah_1 + bh_2) = a^*Kh_1 + b^*Kh_2$ for all $h_1, h_2 \in H_\mathbb{C}$ (K is called a *conjugation*; cf. also Section 8.1). We have $H = \{h \in H_\mathbb{C} : Kh = h\}$.
(b) We have $KT_\mathbb{C} = T_\mathbb{C}K$ and $K(z - T_\mathbb{C})^{-1} = (z^* - T_\mathbb{C})^{-1}K$ for all $z \in \rho(T_\mathbb{C})$.
(c) If $E_\mathbb{C}$ is the spectral family of $T_\mathbb{C}$, then $KE_\mathbb{C}(t) = E_\mathbb{C}(t)K$ for all $t \in \mathbb{R}$.
Hint: Use (7.22).
(d) The formula $E(t) = E_\mathbb{C}(t)|_H$, $t \in \mathbb{R}$ defines a spectral family on H such that $T = \hat{E}(\mathrm{id})$.
(e) E is the only spectral family for which $T = \hat{E}(\mathrm{id})$. (This proves Theorem 7.17 for real Hilbert spaces.)

7.26. Let T be a densely defined closed operator on H and let $T = U|T|$ be its polar decomposition.
(a) We have $N(|T|) = N(T)$ and $\overline{R(|T|)} = \overline{R(T^*)}$.
(b) We have $T^* = |T|U^*$ and $|T^*| = U|T|U^*$.
Hint: $TT^* = (U|T|U^*)(U|T|U^*)$, and $U|T|U^*$ is non-negative and self-adjoint.
(c) Prove, furthermore, that $T = U|T| = |T^*|U = UT^*U$, $T^* = U^*|T^*| = |T|U^* = U^*TU^*$, $|T| = U^*T = T^*U = U^*|T^*|U$, $|T^*| = UT^* = TU^* = U|T|U^*$.
(d) If T is normal, then $|T| = |T^*|$, $U|T| = |T|U$, $U^*|T| = |T|U^*$, and $\overline{R(T)} = \overline{R(T^*)} = \overline{R(|T|)}$.
(e) If T is normal, then the operators $T^j(T^*)^k$ are normal for all $j, k \in \mathbb{N}_0$; furthermore, $D(T^j(T^*)^k) = D(T^{j+k})$, $(T^j(T^*)^k)^* = T^k(T^*)^j$ and $\|T^j(T^*)^k f\| = \|T^{j+k} f\|$ for $f \in D(T^{j+k})$.

7.27. Let S be closed and symmetric but not self-adjoint, and let $T = |S|$. Then

$T+\lambda S$ is self-adjoint for $\lambda\in(-1, 1)$, closed and not self-adjoint for $|\lambda|>1$, and not closed for $|\lambda|=1$. (S is T-bounded with T-bounded 1.)

7.28. Let T be a self-adjoint operator on $L_2(M)$ and let $(z_0-T)^{-n}$ be a Carleman operator for some $z_0\in\rho(T)$.
(a) $(z-T)^{-n}$ is a Carleman operator for every $z\in\rho(T)$.
(b) $E(b)-E(a)$ is a Carleman operator for all $a, b\in\mathbb{R}$.
Hint: $(z-T)^n(z_0-T)^{-n}$ and $(z-T)^n(E(b)-E(a))$ are bounded.

7.29. (a) If T is a non-negative self-adjoint operator, then $D(T)$ is a core of $T^{1/2}$.
(b) If A is symmetric and non-negative, T is the Friedrichs extension of A, and S is an arbitrary non-negative self-adjoint extension of A, then $D(T^{1/2})\subset D(S^{1/2})$.
Hint: $D(T^{1/2})$ is the completion of $D(A)$ with respect to the norm $\{\|f\|^2+\langle f, Af\rangle\}^{1/2}$.

7.30. Let S and T be non-negative self-adjoint operators. We write $T\leqslant S$ if $D(S^{1/2})\subset D(T^{1/2})$ and $\|T^{1/2}f\|\leqslant\|S^{1/2}f\|$ for all $f\in D(S^{1/2})$.
(a) If $0\in\rho(T)$, then $T\leqslant S$ if and only if $S^{-1}\leqslant T^{-1}$ (i.e., $\langle f, S^{-1}f\rangle\leqslant\langle f, T^{-1}f\rangle$ for all $f\in H$).
Hint: Show that $T\leqslant S\Leftrightarrow\|T^{1/2}S^{-1/2}\|\leqslant 1\Leftrightarrow\{\|S^{-1/2}T^{1/2}f\|\leqslant\|f\|$ for all $f\in D(T^{1/2})\}\Leftrightarrow S^{-1}\leqslant T^{-1}$.
(b) If A is symmetric and non-negative, T is the Friedrichs extension of A, and S is an arbitrary non-negative self-adjoint extension of A, then $S\leqslant T$.

7.30′ Let T be self-adjoint on a complex Hilbert space, and let A be T-bounded.
(a) The relative bound of A equals $\lim_{t\to-\infty}\|A(it-T)^{-1}\|$.
(b) If T is bounded from below, then the relative bound of A equals $\lim_{t\to\infty}\|A(t+T)^{-1}\|$.

7.4 Spectra of self-adjoint operators

We know from Section 5.3 that the spectrum of a self-adjoint operator is a closed subset of the real axis (of course, this is a non-trivial statement only in the complex case). In this section we want to study how the spectral points of a self-adjoint operator may be characterized by means of its spectral family.

Let T be a self-adjoint operator on H. The *spectrum* $\sigma(T)$ and the *point spectrum* $\sigma_p(T)$ are defined as in Section 5.2.

Theorem 7.22. *Let T be a self-adjoint operator on H, let E be the spectral family of T, and let T_0 be a restriction of T for which $\overline{T_0}=T$. Then the following statements are equivalent*:
(i) $s\in\sigma(T)$;
(ii) *there exists a sequence* (f_n) *from* $D(T)$ *for which* $\liminf\|f_n\|>0$ *and* $(s-T)f_n\to 0$;
(iii) *there exists a sequence* (g_n) *from* $D(T_0)$ *for which* $\liminf\|g_n\|>0$ *and* $(s-T_0)g_n\to 0$;
(iv) $E(s+\epsilon)-E(s-\epsilon)\neq 0$ *for every* $\epsilon>0$.

PROOF. The equivalence of (i) and (ii) immediately follows from Theorem 5.24.

(ii) *implies* (iii): Since $D(T_0)$ is a core of T, for every $n \in \mathbb{N}$ there exists a $g_n \in D(T_0)$ for which $\|g_n - f_n\| \leqslant n^{-1}$ and $\|T_0 g_n - Tf_n\| \leqslant n^{-1}$. Hence, $\liminf \|g_n\| > 0$ and $(s - T_0)g_n \to 0$.

(iii) *implies* (ii): This is clear because of the inclusion $D(T_0) \subset D(T)$.

(ii) *implies* (iv): Assume that (iv) does not hold, i.e., that there exists an $\epsilon > 0$ such that $E(s+\epsilon) - E(s-\epsilon) = 0$. If (f_n) is the sequence from (ii), then

$$\|(s-T)f_n\|^2 = \int |s-t|^2 \,\mathrm{d}\|E(t)f_n\|^2 \geqslant \epsilon^2 \int \mathrm{d}\|E(t)f_n\|^2$$
$$= \epsilon^2 \|f_n\|^2$$

(here we have used the fact that $|s-t| \geqslant \epsilon$ almost everywhere relative to the measure induced by $\rho_{f_n} = \|E(\,\cdot\,)f_n\|^2$). This is a contradiction because $(s-T)f_n \to 0$ and $\liminf \|f_n\| > 0$.

(iv) *implies* (ii): We have $E(s+n^{-1}) - E(s-n^{-1}) \neq 0$ for every $n \in \mathbb{N}$, i.e., there exists an $f_n \in R(E(s+n^{-1}) - E(s-n^{-1}))$ such that $\|f_n\| = 1$. For this sequence we have $\liminf \|f_n\| = 1$ and

$$\|(s-T)f_n\|^2 = \int |s-t|^2 \,\mathrm{d}\|E(t)f_n\|^2 \leqslant n^{-2} \int \mathrm{d}\|E(t)f_n\|^2$$
$$= n^{-2}\|f_n\|^2 \to 0. \qquad \square$$

Corollary 1. *Let $a < b$. If $E(b) - E(a) \neq 0$ then $(a, b] \cap \sigma(T) \neq \varnothing$. We have $(a, b) \cap \sigma(T) \neq \varnothing$ if and only if $E(b-) - E(a) \neq 0$.*

PROOF.

(a) Assume that $(a, b] \subset \rho(T)$. Then by Theorem 7.22 the spectral family E is constant in some neighborhood of s for every $s \in (a, b]$. Consequently, E is constant in $(a, b]$, and thus $E(b) - E(a) = s-\lim_{\epsilon \to 0+}(E(b) - E(a+\epsilon)) = 0$.

(b) If $(a, b) \cap \sigma(T) = \varnothing$, then we can prove, as in part (a), that E is constant in (a, b), i.e., that $E(b-) - E(a) = s-\lim_{\epsilon \to 0+}(E(b-\epsilon) - E(a+\epsilon)) = 0$. If $\lambda \in (a, b) \cap \sigma(T)$, and $\epsilon > 0$ is so small that $(\lambda - \epsilon, \lambda + \epsilon] \subset (a, b)$, then $E(\lambda+\epsilon) - E(\lambda-\epsilon) \neq 0$, so that $E(b-) - E(a) \neq 0$. $\square$

Corollary 2. *A self-adjoint operator T is bounded from below if and only if its spectrum is bounded from below. The greatest lower bound of T is equal to $\min \sigma(T)$.*

PROOF. By Part 3 of the proposition preceding Theorem 7.20, we have $\langle f, Tf \rangle \geqslant \gamma \|f\|^2$ for all $f \in D(T)$ if and only if $E(t) = 0$ for $t < \gamma$. If $E(t) = 0$ for $t < \gamma$, then by Theorem 7.22 no spectral point of T can lie in $(-\infty, \gamma)$. Therefore, $\min \sigma(T) \geqslant \gamma$. If $\sigma(T)$ is bounded from below, then E is constant in $(-\infty, \min \sigma(T))$. Consequently, $E(t) = 0$ for $t < \min \sigma(T)$, and thus $\gamma \geqslant \min \sigma(T)$. $\square$

Theorem 7.23. *Let T, T_0 and E be as in Theorem 7.22. Then the following assertions are equivalent:*

(i) $s \in \sigma_p(T)$;

(ii) *there exists a Cauchy sequence* (f_n) *from* $D(T)$ *for which* $\lim\|f_n\| > 0$ *and* $(s-T)f_n \to 0$;

(iii) *there exists a Cauchy sequence* (g_n) *from* $D(T_0)$ *for which* $\lim\|g_n\| > 0$ *and* $(s-T_0)g_n \to 0$;

(iv) $E(s) - E(s-) \neq 0$.

We have $N(s-T) = R(E(s) - E(s-))$.

PROOF. (i) *implies* (ii): If f is an eigenelement of T belonging to the eigenvalue s, then we can choose the constant sequence $f_n = f$.

(ii) *implies* (iii): Since $D(T_0)$ is a core of T, for every $n \in \mathbb{N}$ there exists a $g_n \in D(T_0)$ for which $\|g_n - f_n\| \leqslant n^{-1}$ and $\|T_0 g_n - Tf_n\| \leqslant n^{-1}$. Everything follows from this.

(iii) *implies* (i): Let $f = \lim g_n$. Then $(s-T)f = \lim(s-T_0)g_n = 0$.

(i) *implies* (iv): Let f be an eigenelement of T belonging to the eigenvalue s. Then

$$0 = \|(s-T)f\|^2 = \int |s-t|^2 \, d\|E(t)f\|^2,$$

i.e., $|s-t| = 0$ almost everywhere with respect to the measure induced by $\rho_f = \|E(.)f\|^2$. Therefore, $E(t)f$ is constant in $(-\infty, s)$ and in (s, ∞), and thus $E(s-)f = 0$, $E(s)f = E(s+)f = f$. Hence, $E(s) - E(s-) \neq 0$.

(iv) *implies* (i): For every $f \in R(E(s) - E(s-))$ we obviously have $\|(s-T)f\|^2 = \int |s-t|^2 \, d\|E(t)f\|^2 = 0$.

It follows from the last two steps that $N(s-T) = R(E(s) - E(s-))$. □

Proposition. *Any isolated point λ of the spectrum of a self-adjoint operator T is an eigenvalue of T.*

PROOF. There is an $\epsilon > 0$ such that $[\lambda-\epsilon, \lambda+\epsilon] \cap \sigma(T) = \{\lambda\}$. Hence, by Corollary 1 to Theorem 7.22, E is constant in $[\lambda-\epsilon, \lambda)$ and in $(\lambda, \lambda+\epsilon]$. Since $\lambda \in \sigma(T)$, by Theorem 7.22 we have

$$E(\lambda) - E(\lambda-) = E(\lambda+\epsilon) - E(\lambda-\epsilon) \neq 0,$$

i.e., λ is an eigenvalue of T. □

The *essential spectrum* $\sigma_e(T)$ of a self-adjoint operator T is the set of those points (of $\sigma(T)$) that are either accumulation points of $\sigma(T)$ or isolated eigenvalues of infinite multiplicity. The set $\sigma_d(T) = \sigma(T) \setminus \sigma_e(T)$ is called the *discrete spectrum* of T. By the last proposition $\sigma_d(T)$ is the set of those eigenvalues of finite multiplicity that are isolated points of $\sigma(T)$. We say that T has a *pure discrete spectrum* if $\sigma_e(T)$ is empty.

Theorem 7.24. *Let T, T_0 and E be as in Theorem 7.22. Then the following statements are equivalent*:

(i) $s \in \sigma_e(T)$;

(ii) *there exists a sequence* (f_n) *from* $D(T)$ *for which* $f_n \xrightarrow{w} 0$, $\liminf \|f_n\| > 0$ *and* $(s-T)f_n \to 0$;

(iii) *there exists a sequence* (g_n) *from* $D(T_0)$ *for which* $g_n \xrightarrow{w} 0$, $\liminf \|g_n\| > 0$ *and* $(s-T_0)g_n \to 0$;

(iv) *for every* $\epsilon > 0$ *we have* $\dim R(E(s+\epsilon)-E(s-\epsilon)) = \infty$.

PROOF. (i) *implies* (ii): If s is an eigenvalue of infinite multiplicity, then there exists an orthonormal sequence (f_n) in $N(s-T)$; this sequence has the properties required in (ii). If s is an accumulation point of $\sigma(T)$, then there exists a sequence (s_n) from $\sigma(T)$ such that $s_n \neq s$, $s_n \neq s_m$ for $n \neq m$, and $s_n \to s$ as $n \to \infty$. Let us now choose $\epsilon_n > 0$ so small that the intervals $(s_n-\epsilon_n, s_n+\epsilon_n)$ be mutually disjoint. Since $s_n \in \sigma(T)$, we have $E(s_n+\epsilon_n) - E(s_n-\epsilon_n) \neq 0$. Let us choose a normed element f_n from $R(E(s_n+\epsilon_n) - E(s_n-\epsilon_n))$. Then we obviously have $\langle f_n, f_m\rangle = \delta_{nm}$ and $(s-T)f_n \to 0$.

(ii) *implies* (iii): For every $n \in \mathbb{N}$ there exists a $g_n \in D(T_0)$ for which $\|g_n - f_n\| < n^{-1}$ and $\|T_0 g_n - Tf_n\| < n^{-1}$. All the properties required in (iii) follow from this.

(iii) *implies* (iv): Assume that we have $\dim R(E(s+\epsilon)-E(s-\epsilon)) < \infty$ for some $\epsilon > 0$, i.e., that the projection $E(s+\epsilon)-E(s-\epsilon)$ is compact. For the sequence (g_n) from (iii) we then have $(E(s+\epsilon)-E(s-\epsilon))g_n \to 0$. Consequently,

$$\begin{aligned}
\|(s-T)g_n\|^2 &= \int |s-t|^2 \, \mathrm{d}\|E(t)g_n\|^2 \\
&\geq \epsilon^2\left[\int \mathrm{d}\|E(t)g_n\|^2 - \int \chi_{(s-\epsilon,s+\epsilon]}(t)\, \mathrm{d}\|E(t)g_n\|^2\right] \\
&= \epsilon^2\left[\|g_n\|^2 - \|(E(s+\epsilon)-E(s-\epsilon))g_n\|^2\right],
\end{aligned}$$

and thus

$$\liminf \|(s-T)g_n\|^2 \geq \epsilon^2 \liminf \|g_n\|^2 > 0.$$

This is in contradiction with (iii).

(iv) *implies* (i): If $\dim R(E(s)-E(s-)) = \infty$, then s is an eigenvalue of infinite multiplicity (Theorem 7.23). Therefore, $s \in \sigma_e(T)$. Let $R(E(s)-E(s-))$ be finite-dimensional, but let $\dim R(E(s+\epsilon)-E(s-\epsilon)) = \infty$ for all $\epsilon > 0$. Then the set $(s-\epsilon, s) \cup (s, s+\epsilon]$ contains at least one spectral point for every $\epsilon > 0$ by Corollary 1 to Theorem 7.22; hence s is an accumulation point of $\sigma(T)$. □

Theorems 7.22 and 7.24 provide natural characterizations of $\sigma_d(T)$. We will not give these explicitly here. They are partly contained in the following propositions.

Proposition. *If* $a<b$ *and* $\dim R(E(b-)-E(a))=m$ $(m\in\mathbb{N})$, *then* $\sigma(T)\cap(a,b)$ *consists of only isolated eigenvalues of finite multiplicity. The sum of the multiplicities of these eigenvalues equals* m.

PROOF. By Theorem 7.24 we have $(a,b)\cap\sigma_e(T)=\varnothing$, i.e., $(a,b)\cap\sigma(T)\subset\sigma_d(T)$. Let $\lambda_1,\lambda_2,\ldots$ be the eigenvalues of T in (a,b) (there are at most countably many of these, since they cannot accumulate in the interior of (a,b)). Then

$$E(b-)-E(a)\geqslant\sum_{j=1}^{n}\left(E(\lambda_j)-E(\lambda_j-)\right)\quad\text{for all}\quad n\in\mathbb{N}.$$

Therefore, only finitely many eigenvalues $\lambda_1,\lambda_2,\ldots,\lambda_k$ can lie in (a,b), and

$$\dim R(E(b-)-E(a))=\sum_{j=1}^{k}\dim\left(E(\lambda_j)-E(\lambda_j-)\right).$$

The right-hand side equals the sum of the multiplicities of the eigenvalues $\lambda_1,\ldots,\lambda_k$. □

Proposition. *If* $\dim R(E(b))=m<\infty$ *for some* $b\in\mathbb{R}$, *then* T *is bounded from below.*

PROOF. By the previous proposition the interval (a, b) contains at most m spectral points for any $a<b$; hence $(-\infty,b)$ contains at most m spectral points. The smallest of these finitely many eigenvalues is a lower bound for T by Corollary 2 to Theorem 7.22.

Proposition. *If* $\dim R(E(b)-E(a))=\infty$, *then* $\sigma_e(T)\cap[a,b]\neq\varnothing$.

PROOF. If we had $\sigma_e(T)\cap[a,b]=\varnothing$, then for every $s\in[a,b]$ there would be an $\epsilon>0$ such that $\dim R(E(s+\epsilon)-E(s-\epsilon))<\infty$. The interval $[a,b]$ could be covered by finitely many intervals of this kind. This implies that $\dim R(E(b)-E(a))<\infty$, which is a contradiction. □

Theorem 7.25. *Let* T *be a self-adjoint operator on* H, *and let* $H=H_1\oplus H_2\oplus H_3$ *with* $\dim H_3=m<\infty$. *Assume that the orthogonal projection* P_j *onto* H_j *maps* $D(T)$ *into itself for* $j=1,2$.[5] *If*

$$\langle f,Tf\rangle\begin{cases}\leqslant a\|f\|^2 & \text{for}\quad f\in P_1D(T),\\ \geqslant b\|f\|^2 & \text{for}\quad f\in P_2D(T),\end{cases}$$

then $(a,b)\cap\sigma(T)$ *consists of only isolated eigenvalues; the sum of the multiplicities of these eigenvalues is at most* m.

[5]Then P_3 also maps $D(T)$ into itself.

Proof. Let us assume that $\dim R(E(b-)-E(a)) \geq m+1$. Then there exists an $f \in R(E(b-)-E(a)) \cap H_3^{\perp}$ such that $f \neq 0$. Hence, $f = P_1 f + P_2 f$, and, putting $c=(a+b)/2$, we have

$$\|(T-c)f\|^2 = \int \chi_{(a,b)}(t)(t-c)^2 \,\mathrm{d}\|E(t)f\|^2 < \left(\frac{b-a}{2}\right)^2 \|f\|^2.$$

With properly chosen $\alpha_j \in \mathbb{K}$, $|\alpha_j| = 1$ it follows from this that

$$\begin{aligned}
&|\langle (T-c)f, P_1 f\rangle| + |\langle P_2 f, (T-c)f\rangle| \\
&= \alpha_1^* \langle P_1 f, (T-c)f\rangle + \alpha_2^* \langle P_2 f, (T-c)f\rangle \\
&= \langle \alpha_1 P_1 f + \alpha_2 P_2 f, (T-c)f\rangle \leq \|\alpha_1 P_1 f + \alpha_2 P_2 f\| \, \|(T-c)f\| \\
&= \left\{\|P_1 f\|^2 + \|P_2 f\|^2\right\}^{1/2} \|(T-c)f\| = \|f\| \, \|(T-c)f\| < \frac{b-a}{2}\|f\|^2.
\end{aligned}$$

Consequently,

$$\begin{aligned}
b\|P_2 f\|^2 &\leq \langle P_2 f, TP_2 f\rangle = \langle P_2 f, Tf\rangle - \langle Tf, P_1 f\rangle + \langle P_1 f, TP_1 f\rangle \\
&\leq c(\|P_2 f\|^2 - \|P_1 f\|^2) + \langle P_2 f, (T-c)f\rangle \\
&\quad - \langle (T-c)f, P_1 f\rangle + a\|P_1 f\|^2 \\
&\leq b\|P_2 f\|^2 - \frac{b-a}{2}\|f\|^2 + |\langle P_2 f, (T-c)f\rangle| \\
&\quad + |\langle (T-c)f, P_1 f\rangle| \\
&< b\|P_2 f\|^2.
\end{aligned}$$

This is a contradiction. Therefore, $\dim R(E(b-)-E(a)) \leq m$, and the assertion follows from the first proposition after Theorem 7.24. □

A corresponding theorem holds in the case when (a, b) is a half-line.

Theorem 7.26.

(a) *Let T be a self-adjoint operator on H, and let H_1 be a closed subspace of H such that $\dim H_1^{\perp} = m < \infty$. Assume that $P_1 D(T) \subset D(T)$ for the orthogonal projection P_1 onto H_1, and that*

$$\langle f, Tf\rangle \geq b\|f\|^2 \quad \text{for} \quad f \in P_1 D(T).$$

Then $(-\infty, b) \cap \sigma(T)$ consists of only isolated eigenvalues; the sum of the multiplicities of these eigenvalues is at most m. The operator T is bounded from below.

(b) *Let T be self-adjoint on H, and let H_1 be an m-dimensional subspace of $D(T)$. Assume that $\langle f, Tf\rangle \leq a\|f\|^2$ for all $f \in H_1$. Then $\dim R(E(a)) \geq m$.*

Proof.

(a) With $H_2 = \{0\}$ and $H_3 = H_1^{\perp}$ the assumptions of Theorem 7.25 are fulfilled for every $a < b$. Consequently, $(-\infty, b) \cap \sigma(T)$ contains only isolated eigenvalues of finite multiplicity, with total multiplicity less than or equal to m. In particular, $\sigma(T)\cap(-\infty, b)$ is a finite set; therefore, $s_0=\min\sigma(T)$ exists. Hence, T is bounded from below by Corollary 2 to Theorem 7.22.

(b) If we had $\dim R(E(a))<m$, then there would be an $f\in H_1$ for which $f\neq 0$ and $f\perp R(E(a))$. For this f

$$\langle f, Tf\rangle \leqslant a\|f\|^2 < \langle f, Tf\rangle,$$

by assumption and by part (2) of the proposition preceding Theorem 7.20. This is a contradiction. □

In some investigations another partition of the spectrum is useful. For that we first need some definitions. Let T again be a self-adjoint operator on H with spectral family E. Let H_p denote the closed linear hull of all eigenelements of T. We call $H_p=H_p(T)$ the *discontinuous subspace* of H with respect to T. The orthogonal complement of H_p is called the *continuous subspace* of H with respect to T. This is denoted by $H_c=H_c(T)$. The *singular continuous subspace* $H_{sc}=H_{sc}(T)$ of H with respect to T is the set of those $f\in H_c$ for which there exists a Borel set $N\subset\mathbb{R}$ of Lebesgue measure zero (briefly: a Borel null set) such that $E(N)f=f$. The subspace H_{sc} is closed. This can be seen in the following way. If (f_n) is a sequence from H_{sc}, $f_n\to f$, and the N_n are Borel null sets such that $E(N_n)f_n=f_n$, then $N=\cup_n N_n$ is also a null set and $E(N)f=\lim E(N)f_n=\lim E(N_n)f_n=\lim f_n=f$. Since H_c is closed, f lies in H_c, hence in H_{sc}. The orthogonal complement of H_{sc} relative to H_c (i.e., $H_c\ominus H_{sc}$) is called the *absolutely continuous subspace* of H relative to T. This is denoted by $H_{ac}=H_{ac}(T)$. The *singular subspace* H_s of H with respect to T is defined by the equality $H_s=H_s(T)=H_p\oplus H_{sc}$. Let P_p, P_c, P_{sc}, P_{ac}, and P_s denote the orthogonal projections onto these subspaces.

Theorem 7.27. *Let T be a self-adjoint operator on H with spectral family E. Denote, for every $f\in H$, by ρ_f the measure induced on $\mathbb{R}$ by means of $\|E(\,.\,)f\|^2$.*

(a) *H_p equals the set of those $f\in H$ for which there exists an at most countable set $A\subset\mathbb{R}$ such that $\rho_f(\mathbb{R}\setminus A)=0$, i.e., for which the measure ρ_f is concentrated on (at most) countably many points.*

(b) *H_c is the set of those $f\in H$ for which $\rho_f(\{t\})=0$ for every $t\in\mathbb{R}$, i.e., for which the function $t\mapsto\|E(t)f\|^2$ is continuous. (For $f\in H_c$ we obviously have $\rho_f(A)=0$ for every at most countable set $A\subset\mathbb{R}$.)*

(c) *H_s equals the set of those $f\in H$ for which there exists a Borel null set N such that $\rho_f(\mathbb{R}\setminus N)=0$, i.e., for which ρ_f is singular with respect to Lebesgue measure.*

(d) *H_{ac} equals the set of those $f\in H$ for which $\rho_f(N)=0$ for every Borel null set N, i.e., for which ρ_f is absolutely continuous with respect to Lebesgue measure.*

PROOF.

(a) If f_j is an eigenelement for the eigenvalue λ_j, and $f=\sum_{j=1}^{\infty}c_jf_j$, then we obviously have $E(\{\lambda_j : j\in\mathbb{N}\})f=f$. If $A=\{\lambda_j : j\in\mathbb{N}\}$ and $E(A)f=f$, then

$$f=\sum_{j=1}^{\infty}E(\{\lambda_j\})f=\sum_{j=1}^{\infty}(E(\lambda_j)-E(\lambda_j-))f.$$

Since $(E(\lambda_j)-E(\lambda_j-))f$ (when it is different from zero) is an eigenelement belonging to the eigenvalue λ_j, the element f lies in the closed hull of the eigenvectors.

(b) We have $E(\{t\})f\in H_p$ and $\langle f, E(\{t\})f\rangle=\|E(\{t\})f\|^2$ for every $f\in H$ and every $t\in\mathbb{R}$. If $f\in H_c=H_p^{\perp}$, then we have $\rho_f(\{t\})=\|E(\{t\})f\|^2=\langle f, E(\{t\})f\rangle=0$ for every $t\in\mathbb{R}$. Let $\rho_f(\{t\})=0$ for every $t\in\mathbb{R}$. Then $\rho_f(A)=0$ for every at most countable set $A\subset\mathbb{R}$; hence $\|E(A)f\|^2=\rho_f(A)=0$. If $g\in H_p$ and A is an at most countable set such that $E(A)g=g$, then $\langle f,g\rangle=\langle f, E(A)g\rangle=\langle E(A)f, g\rangle=\langle 0, g\rangle=0$, i.e., $f\in H_p^{\perp}=H_c$.

(c) We have $H_s=H_p+H_{sc}$. If $f=f_p+f_{sc}$ with $f_p\in H_p$ and $f_{sc}\in H_{sc}$, then there exist an at most countable set $A\subset\mathbb{R}$ and a Borel null set N for which $E(A)f_p=f_p$, $E(N)f_{sc}=f_{sc}$, and thus $E(A\cup N)f=f$; the set $A\cup N$ is a Borel null set. Conversely, let $E(N)f=f$ for some Borel null set N. The set A of jump points of the non-decreasing function $t\mapsto\|E(t)f\|^2$ is at most countable, and $E(A)f\in H_p$. Let $g\in H_p$ be arbitrary. Then there exists an at most countable set A' such that $E(A')=g$. Since $E(\{t\})(f-E(A)f)=E(\{t\})f-E(\{t\}\cap A)f=0$ for every $t\in\mathbb{R}$ (as $\{t\}=\{t\}\cap A$ for $t\in A$ and $E(\{t\})f=E(\{t\}\cap A)f=0$ for $t\neq A$), it follows that $E(A')(f-E(A)f)=0$, and thus

$$\begin{aligned}\langle g, f-E(A)f\rangle &= \langle E(A')g, f-E(A)f\rangle\\ &= \langle g, E(A')(f-E(A)f)\rangle = 0,\end{aligned}$$

i.e., $f-E(A)f\in H_p^{\perp}=H_c$. Because of the equality $E(N)(f-E(A)f)=f-E(A)f$ we have $f-E(A)f\in H_{sc}$; therefore, $f=E(A)f+(f-E(A)f)\in H_p+H_{sc}=H_s$.

(d) We have $H_{ac}=H_c\ominus H_{sc}=H_s^{\perp}$. Assume that $f\in H_{ac}=H_s^{\perp}$. Then $\|E(N)f\|^2=\langle f, E(N)f\rangle=0$ for every Borel null set N, since $E(N)f\in H_s$. Let us now assume that $E(N)f=0$ for every Borel null set N. If $g\in H_s$, then by part c) there exists a Borel null set N such that $g=E(N)g$. It follows from this that

$$\langle f,g\rangle=\langle f,E(N)g\rangle=\langle E(N)f,g\rangle=0,$$

and thus $f\in H_s^{\perp}=H_{ac}$. □

Let M be a closed subspace of H, and let P be the orthogonal projection onto M. We say that M *reduces* the operator T if $PT\subset TP$ (this obviously implies $PD(T)\subset D(T)$, as well as $(I-P)D(T)\subset D(T)$ and $(I-P)T\subset$

$T(I-P)$; cf. also Exercise 5.39 and Section 7.3, Proposition 1). The formulae $D(T_M)=M\cap D(T)$ and $T_Mf=Tf$ for $f\in D(T_M)$ define an operator on M. The subspace M is a reducing subspace of T if and only if $M^\perp$ is a reducing subspace of T. Then $D(T)=D(T_M)+D(T_{M^\perp})$.

Theorem 7.28. *Let T be a self-adjoint operator on H, with the spectral family E, and let M be a reducing subspace of T. Then T_M and $T_{M^\perp}$ are self-adjoint on M and $M^\perp$, respectively. We have $\sigma(T)=\sigma(T_M)\cup\sigma(T_{M^\perp})$. The subspace M reduces T if and only if $PE(t)=E(t)P$ for every $t\in\mathbb{R}$, where P denotes the orthogonal projection onto M.*

PROOF. $D(T_M)$ is dense in M, since $M\ominus\overline{D(T_M)}=D(T)^\perp\cap M=\{0\}$ because of the equality $D(T)=D(T_M)+D(T_{M^\perp})$. As a restriction of a self-adjoint operator, T_M is surely Hermitian. Therefore, T_M is symmetric on M. It remains to prove that $D((T_M)^*)\subset D(T_M)$. Let $g\in D((T_M)^*)$. Then

$$\langle (T_M)^*g,f\rangle=\langle (T_M)^*g,f_1\rangle=\langle g,T_Mf_1\rangle=\langle g,T_Mf_1+T_{M^\perp}f_2\rangle=\langle g,Tf\rangle$$

for all $f=f_1+f_2\in D(T_M)+D(T_{M^\perp})=D(T)$; i.e., $g\in M\cap D(T^*)=M\cap D(T)=D(T_M)$. We can show the self-adjointness of $T_{M^\perp}$ analogously. We have

$$\|(z-T)f\|^2=\|(z-T_M)f_1\|^2+\|(z-T_{M^\perp})f_2\|^2$$

for every $z\in\mathbb{K}$ and for $f=f_1+f_2\in D(T_M)+D(T_{M^\perp})=D(T)$. It follows from this by Theorem 5.24 that $z\in\rho(T)$ if and only if $z\in\rho(T_M)\cap\rho(T_{M^\perp})$, i.e., $\sigma(T)=\sigma(T_M)\cup\sigma(T_{M^\perp})$.

If M reduces the operator T, then $R(z,T)P=R(z,T)P(z-T)R(z,T)\subset R(z,T)(z-T)PR(z,T)=PR(z,T)$. Therefore, $R(z,T)P=PR(z,T)$, because $D(R(z,T)P)=H$. It follows, by formula (7.22) for the spectral family E, that $E(t)P=PE(t)$ for all $t\in\mathbb{R}$. Now let the equalities $E(t)P=PE(t)$ $(t\in\mathbb{R})$ hold true. Then

$$\int|t|^2\,\mathrm{d}\|E(t)Pf\|^2=\int|t|^2\,\mathrm{d}\|PE(t)f\|^2\leq\int|t|^2\,\mathrm{d}\|E(t)f\|^2<\infty$$

for every $f\in D(T)$. Consequently, $Pf\in D(T)$. This implies that $D(TP)\supset D(T)=D(PT)$. If $f\in D(T)$ and (u_n) is a sequence of step functions such that u_n tends to id in $L_2(\mathbb{R},\rho_f)$, then u_n tends to id in $L_2(\mathbb{R},\rho_{Pf})$ also, and

$$\begin{aligned}PTf&=P\hat{E}(\mathrm{id})f=P\lim\hat{E}(u_n)f=\lim P\hat{E}(u_n)f=\lim\hat{E}(u_n)Pf\\&=\hat{E}(\mathrm{id})Pf=TPf.\end{aligned}$$

□

Theorem 7.29. *Let T be a self-adjoint operator on H. The subspaces H_p, H_c, H_{sc}, H_{ac}, and H_s reduce the operator T.*

PROOF. It is obviously sufficient to show that H_p and H_s reduce the operator T. The remaining assertions follow from this, because $H_c=H_p^\perp$, $H_{sc}=H_s\ominus H_p$, and $H_{ac}=H_s^\perp$.

For any $f\in H_p$ there is an at most countable subset A of $\mathbb{R}$ such that $E(A)f=f$. Consequently, $E(t)f=E(t)E(A)f=E(A)E(t)f\in H_p$ for all $t\in\mathbb{R}$, i.e., $E(t)P_p = P_pE(t)P_p$. It follows from this that

$$P_pE(t) = (E(t)P_p)^* = (P_pE(t)P_p)^* = P_pE(t)P_p = E(t)P_p.$$

We can show in an entirely analogous way that $P_sE(t) = E(t)P_s$ if we replace A by the null set N for which $E(N)f = f$. □

We denote by T_p, T_c, T_{sc}, T_{ac}, and T_s the restrictions of T to H_p, H_c, H_{sc}, H_{ac}, and H_s. These operators are called the *(spectral) discontinuous, continuous, singular continuous, absolutely continuous*, and *singular parts* of T.

The *continuous spectrum* $\sigma_c(T)$, *singular continuous spectrum* $\sigma_{sc}(T)$, *absolutely continuous spectrum* $\sigma_{ac}(T)$, and the *singular spectrum* $\sigma_s(T)$ of T are defined as the spectrum of T_c, T_{sc}, T_{ac}, and T_s, respectively. In contrast with this, the point spectrum $\sigma_p(T)$ is defined as the set of eigenvalues of T (these are also the eigenvalues of T_p; however, in general we only have $\sigma(T_p) = \overline{\sigma_p(T)}$; cf. Exercise 7.33). The sets $\sigma_c(T)$, $\sigma_{sc}(T)$, $\sigma_{ac}(T)$, and $\sigma_s(T)$ are closed (as they are spectra). We obviously have $\sigma(T) = \overline{\sigma_p(T)} \cup \sigma_{sc}(T) \cup \sigma_{ac}(T) = \sigma_s(T) \cup \sigma_{ac}(T) = \overline{\sigma_p(T)} \cup \sigma_c(T)$.

We say that T has a *pure point spectrum, pure continuous spectrum, pure singular continuous spectrum, pure absolutely continuous spectrum* and *pure singular spectrum* if $H_p = H$, $H_c = H$, $H_{sc} = H$, $H_{ac} = H$, and $H_s = H$, respectively. We then have $\sigma(T) = \overline{\sigma_p(T)}$, $\sigma(T) = \sigma_c(T)$, $\sigma(T) = \sigma_{sc}(T)$, $\sigma(T) = \sigma_{ac}(T)$, and $\sigma(T) = \sigma_s(T)$, respectively.

EXAMPLE 1. Let ρ_p, ρ_{sc}, and ρ_{ac} be measures on $\mathbb{R}$. Let ρ_p be concentrated on a countable set (i.e., there exists a countable set A such that $\rho_p(\mathbb{R}\setminus A) = 0$), let ρ_{sc} be singular continuous (i.e., there exists a Borel null set N such that $\rho_{sc}(\mathbb{R}\setminus N) = 0$ and $\rho_{sc}(\{t\}) = 0$ for every $t \in \mathbb{R}$), let ρ_{ac} be absolutely continuous (i.e., $\rho_{ac}(N) = 0$ for every Borel null set N). Let T be the operator of multiplication by the function id on $L_2(\mathbb{R}, \rho_p) \oplus L_2(\mathbb{R}, \rho_{sc}) \oplus L_2(\mathbb{R}, \rho_{ac})$. Then $H_p = L_2(\mathbb{R}, \rho_p)$, $H_c = L_2(\mathbb{R}, \rho_{sc}) \oplus L_2(\mathbb{R}, \rho_{ac})$, $H_{sc} = L_2(\mathbb{R}, \rho_{sc})$, $H_{ac} = L_2(\mathbb{R}, \rho_{ac})$, and $H_s = L_2(\mathbb{R}, \rho_p) \oplus L_2(\mathbb{R}, \rho_{sc})$. The proof will be left to the reader. We shall show in Exercise 7.34 that every self-adjoint operator is the orthogonal sum of operators of this type.

EXERCISES

7.31. If T is a self-adjoint operator on H with pure point spectrum, and $\{e_\alpha : \alpha \in A\}$ is an orthonormal basis of eigenelements with corresponding eigenvalues $\{\lambda_\alpha : \alpha \in A\}$, then $D(T)$ is equal to the set of those $f \in H$ for which $\sum_\alpha |\lambda_\alpha \langle e_\alpha, f\rangle|^2 < \infty$; we have $Tf = \sum_\alpha \lambda_\alpha \langle e_\alpha, f\rangle e_\alpha$ for $f \in D(T)$.

7.32. Let T be a self-adjoint operator on the infinite-dimensional Hilbert space H.
(a) If T is bounded, then $\sigma_e(T) \neq \emptyset$.
(b) T is compact if and only if T is bounded and $\sigma_e(T) = \{0\}$.
(c) If H is separable, then $B_\infty(H)$ is the only closed ideal in $B(H)$.

7.33. For any self-adjoint operator T we have the following:
(a) $\overline{\sigma_p(T)} = \sigma(T_p)$; however, $\sigma_p(T)$ is in general not closed, and thus $\sigma_p(T) \neq \sigma(T_p)$ in general.
(b) $\sigma(T) = \overline{\sigma_p(T)} \cup \sigma_c(T) = \overline{\sigma_p(T)} \cup \sigma_{sc}(T) \cup \sigma_{ac}(T) = \sigma_s(T) \cup \sigma_{ac}(T)$.
(c) $\sigma_e(T) = \sigma_c(T) \cup (\overline{\sigma_p(T)} \setminus \sigma_d(T))$.

7.34. Every self-adjoint operator is unitarily equivalent to a maximal operator of multiplication by the function id on $(\oplus_{\alpha \in A} L_2(\mathbb{R}, \rho_\alpha)) \oplus (\oplus_{\beta \in B} L_2(\mathbb{R}, \sigma_\beta)) \oplus (\oplus_{\gamma \in \Gamma} L_2(\mathbb{R}, \tau_\gamma))$; where the measures $\rho_\alpha, \sigma_\beta, \tau_\gamma$ have the following properties: for every ρ_α there exists a countable set A_α such that $\rho_\alpha(\mathbb{R} \setminus A_\alpha) = 0$; $\sigma_\beta(\{t\}) = 0$ for every $t \in \mathbb{R}$, and there exists a Borel null set N_β such that $\sigma_\beta\ (\mathbb{R} \setminus N_\beta) = 0\ (\beta \in B)$; all τ_γ are absolutely continuous with respect to Lebesgue measure.
Hint: Apply Theorem 7.18 to T_p, T_{sc}, and T_{ac}.

7.35. Let T be a self-adjoint with spectral family E, and let $u : \mathbb{R} \to \mathbb{R}$ be E-measurable. Then we have the following:
(a) $\sigma(u(T)) \subset \overline{u(\sigma(T))}$; if u is continuous on $\sigma(T)$, then $\sigma(u(T)) = \overline{u(\sigma(T))}$; if T is bounded and u is continuous, then $\sigma(u(T)) = u(\sigma(T))$;
(b) $H_p(T) \subset H_p(u(T))$, $u(\sigma_p(T)) \subset \sigma_p(u(T))$;
(c) $H_{ac}(u(T)) \subset H_{ac}(T)$, $\sigma_{ac}(u(T)) \subset \overline{u(\sigma_{ac}(T))}$.

7.36. Let T be self-adjoint with spectral family E, and let $u, v : \mathbb{R} \to \mathbb{C}$ be E-measurable.
(a) If $u(t) = v(t)$ for all $t \in \sigma(T)$, then $u(T) = v(T)$.
(b) If T has a pure point spectrum, then it is sufficient to assume that $u(t) = v(t)$ for all $t \in \sigma_p(T)$ in (a).

7.37. Let T be a self-adjoint operator with spectral family E, and let M be a subspace of $D(T)$ such that $\|(\lambda - T)f\| \leq c\|f\|$ for all $f \in M$.
(a) $\dim R(E(\lambda + c) - E(\lambda - c -)) \geq \dim M$.
(b) If $\dim M = \infty$, then $\sigma_e(T) \cap [\lambda - c, \lambda + c] \neq \emptyset$.

7.38. Let T be a self-adjoint operator on H. The operator T has a pure discrete spectrum if and only if $(\lambda - T)^{-1}$ is compact for every $\lambda \in \rho(T)$. If H is not separable, then $\sigma_e(T) \neq \emptyset$.

7.39. Let A be a symmetric semi-bounded operator, let T be the Friedrichs extension of A, and let S be an arbitrary semi-bounded self-adjoint extension of A.
(a) $\dim E_T(t) \leq \dim E_S(t)$ for every $t \in \mathbb{R}$.
Hint: Exercise 7.30c and Theorem 7.26b.
(b) If S has a pure discrete spectrum, then T also has a pure discrete spectrum.

7.5 The spectral theorem for normal operators

We have shown in Section 7.3 that every self-adjoint operator can be represented in the form $\int_{\mathbb{R}} t \, dE(t)$, where E is a (real) spectral family defined on $\mathbb{R}$. Here we show that every normal operator can be represented in a corresponding way as an "integral" $\int_{\mathbb{C}} z \, dG(z)$, where G is a spectral family defined on $\mathbb{C}$.

Let H be a Hilbert space. A function $G : \mathbb{C} \to B(H)$ is called a *complex spectral family* if there are real spectral families E and F such that

$$G(t+\mathrm{i}s) = E(t)F(s) = F(s)E(t) \quad \text{for all} \quad s, t \in \mathbb{R}.$$

Theorem 7.30. *Let G be a complex spectral family on H with $G(t+\mathrm{i}\,s) = E(t)F(s)$.*

(a) *$G(t+\mathrm{i}\,s)G(t'+\mathrm{i}\,s') = G(\min\{t, t'\} + \mathrm{i}\min\{s, s'\})$ for all $s, s', t, t' \in \mathbb{R}$; in particular, $G(z) \leqslant G(z')$ for all $z, z' \in \mathbb{C}$ such that $\operatorname{Re} z \leqslant \operatorname{Re} z'$ and $\operatorname{Im} z \leqslant \operatorname{Im} z'$.*

(b) *$z_n \to z$, $\operatorname{Re} z_n \geqslant z$, and $\operatorname{Im} z_n \geqslant z$ imply $G(z_n) \xrightarrow{s} G(z)$.*

(c) *If (z_n) is a sequence from $\mathbb{C}$ for which $\operatorname{Re} z_n \to -\infty$ or $\operatorname{Im} z_n \to -\infty$, then $G(z_n) \xrightarrow{s} 0$; if (z_n) is a sequence from $\mathbb{C}$ for which $\operatorname{Re} z_n \to \infty$ and $\operatorname{Im} z_n \to \infty$, then $G(z_n) \xrightarrow{s} I$.*

(d) *The spectral families E and F are uniquely determined by G; we have $E(t) = s-\lim_{s\to\infty} G(t+\mathrm{i}s)$ and $F(s) = s-\lim_{t\to\infty} G(t+\mathrm{i}s)$.*

(e) *If we set $G(J) = E(J_1)F(J_2)$ for an arbitrary interval $J = J_1 \times J_2 = \{z \in \mathbb{C} : \operatorname{Re} z \in J_1, \operatorname{Im} z \in J_2\}$, then $G(J)G(J') = 0$ for $J \cap J' = \emptyset$.*

(f) *The equality $\gamma_f(J) = \|G(J)f\|^2$ for all intervals J in $\mathbb{C}$ defines a regular interval function on $\mathbb{C}$ for every $f \in H$ (cf. Appendix A 1).*

PROOF.

(a) For all $s, s', t, t' \in \mathbb{R}$

$$\begin{aligned} G(t+\mathrm{i}s)G(t'+\mathrm{i}s') &= E(t)F(s)E(t')F(s') \\ &= E(t)E(t')F(s)F(s') = E(\min\{t, t'\})F(\min\{s, s'\}) \\ &= G(\min\{t, t'\} + \mathrm{i}\min\{s, s'\}). \end{aligned}$$

(b) Let (ϵ_n) and (η_n) be null-sequences from $[0, \infty)$. Then

$$\begin{aligned} &\lim_{n\to\infty} \|(G(z+\epsilon_n+\mathrm{i}\eta_n) - G(z))f\|^2 \\ &\qquad = \lim_{n\to\infty} \langle E(t+\epsilon_n)F(s+\eta_n)f - E(t)F(s)f, f\rangle \\ &\qquad = \lim_{n\to\infty} \{\langle F(s+\eta_n)f, E(t+\epsilon_n)f\rangle - \langle F(s)f, E(t)f\rangle\} = 0 \end{aligned}$$

for all $z = t + \mathrm{i}s$ and $f \in H$.

(c) Assume that $z_n = t_n + \mathrm{i}s_n$, and $t_n \to -\infty$ or $s_n \to -\infty$. Then for all $f \in H$

$$\begin{aligned} \lim_{n\to\infty} \|G(z_n)f\|^2 &= \lim_{n\to\infty} \langle E(t_n)F(s_n)f, f\rangle \\ &= \lim_{n\to\infty} \langle F(s_n)f, E(t_n)f\rangle = 0, \end{aligned}$$

since in $\langle F(s_n)f, E(t_n)f\rangle$ at least one factor tends to zero, while the other remains bounded. If $t_n \to \infty$ and $s_n \to \infty$, then it follows for all $f \in H$ that

$$\lim_{n\to\infty} \|f - G(z_n)f\|^2 = \lim_{n\to\infty} \{\|f\|^2 - \langle F(s_n)f, E(t_n)f\rangle\} = 0.$$

(d) It follows from the formula $F(s)\overset{s}{\to} I$ as $s\to\infty$ that

$$E(t)f = \lim_{s\to\infty} F(s)E(t)f = \lim_{s\to\infty} G(t+is)f$$

for all $f\in H$. The assertion for F follows in a similar way.

(e) If $J=J_1\times J_2$, $J'=J_1'\times J_2'$ and $J\cap J'=\varnothing$, then $J_1\cap J_1'=\varnothing$ or $J_2\cap J_2'=\varnothing$, and thus

$$G(J)G(J') = E(J_1)E(J_1')F(J_2)F(J_2') = 0.$$

(f) If J_1 is an arbitrary interval in $\mathbb{R}$, then we can show easily that there exists a sequence $(J_{1,n})$ of open intervals for which $J_1\subset J_{1,n}$ and $E(J_{1,n})\overset{s}{\to} E(J_1)$ (the sequence $(J_{1,n})$ has to be chosen so that we have $\chi_{J_{1,n}}(t)\to\chi_{J_1}(t)$ for all $t\in\mathbb{R}$). Let $J_{2,n}$ be a similar sequence with the property $F(J_{2,n})\overset{s}{\to} F(J_2)$. Then

$$\begin{aligned}\gamma_f(J_1\times J_2) &= \langle F(J_2)f, E(J_1)f\rangle = \lim_{n\to\infty}\langle F(J_{2,n})f, E(J_{1,n})f\rangle\\ &= \lim_{n\to\infty}\|G(J_{1,n}\times J_{2,n})f\|^2 = \lim_{n\to\infty}\gamma_f(J_{1,n}\times J_{2,n})\end{aligned}$$

for all $f\in H$. Since the intervals $J_{1,n}\times J_{2,n}$ are open and $J_1\times J_2\subset J_{1,n}\times J_{2,n}$, this proves the regularity of γ_f. □

If for the step function $u:\mathbb{C}\to\mathbb{K}$,

$$u(z) = \sum_{j=1}^{n} c_j\chi_{J_j}(z)$$

we define the integral with respect to the complex spectral family G by the equality

$$\int u(z)\,\mathrm{d}G(z) = \sum_{j=1}^{n} c_j G(J_j),$$

we can use the same arguments as we did in Section 7.2 when we discussed integration with respect to a real spectral family. Let us also denote by γ_f the measure that is induced by the interval function γ_f. We say that a function $u:\mathbb{C}\to\mathbb{K}$ is G-measurable if it is γ_f-measurable for all $f\in H$. If $u\in L_2(\mathbb{C},\gamma_f)$, then we can define the integral

$$\int_{\mathbb{C}} u(z)\,\mathrm{d}G(z)f$$

just as in Section 7.2. For every G-measurable function $u:\mathbb{C}\to\mathbb{K}$ the formulae

$$D(\hat{G}(u)) = \{f\in H: u\in L_2(\mathbb{C},\gamma_f)\},$$

$$\hat{G}(u)f = \int_{\mathbb{C}} u(z)\,\mathrm{d}G(z)f \quad \text{for} \quad f\in D(\hat{G}(u))$$

define a normal operator on H. We also write

$$\hat{G}(u) = \int_{\mathbb{C}} u(z)\, \mathrm{d}G(z).$$

The assertions of Theorem 7.14 (and their proofs) remain valid and will be used in the following without further explanation.

Theorem 7.31 (The spectral theorem for bounded normal operators). *Let H be a complex Hilbert space, and let $T \in B(H)$ be a normal operator. Then there exists exactly one complex spectral family G for which*

$$T = \int_{\mathbb{C}} z\, \mathrm{d}G(z).$$

We have $G(t+\mathrm{i}s) = E(t)F(s) = F(s)E(t)$ for $s, t \in \mathbb{R}$; where E and F are the spectral families of the self-adjoint operators $A = (T+T^*)/2$ and $B = (T-T^*)/2\mathrm{i}$, respectively. $G(z) = I$ for $z \in \mathbb{C}$ such that $\operatorname{Re} z \geq \|T\|$ and $\operatorname{Im} z \geq \|T\|$, and $G(z) = 0$ for $z \in \mathbb{C}$ such that $\operatorname{Re} z < -\|T\|$ or $\operatorname{Im} z < -\|T\|$. Moreover, $T = A + \mathrm{i}B$ and $AB = BA$. The operators A and B are called the real part *and the* imaginary part *of T.*

PROOF. If A and B are defined as in the theorem, then it is obvious that $A^* = A$, $B^* = B$, $T = A + \mathrm{i}B$, and

$$AB = \frac{1}{4\mathrm{i}}(T^2 - TT^* + T^*T - T^{*2}) = \frac{1}{4\mathrm{i}}(T^2 - T^{*2}) = BA.$$

Then we also have

$$\begin{aligned} R(z, A)R(z', B) &= \left[(z' - B)(z - A)\right]^{-1} = \left[(z - A)(z' - B)\right]^{-1} \\ &= R(z', B)R(z, A) \end{aligned}$$

for all $z, z' \in \mathbb{C}\setminus\mathbb{R}$. For the spectral families E and F of A and B, respectively, it now follows with the aid of formula (7.22) that

$$E(t)F(s) = F(s)E(t) \quad \text{for all} \quad s, t \in \mathbb{R}.$$

Now we define the complex spectral family G by the formula

$$G(t+\mathrm{i}s) = E(t)F(s) \quad \text{for all} \quad s, t \in \mathbb{R}.$$

Then

$$G(t+\mathrm{i}s) = E(t)F(s) = I \quad \text{for} \quad t \geq \|T\| \text{ and } s \geq \|T\|,$$

since $\|A\| \leq \|T\|$ and $\|B\| \leq \|T\|$. Similarly,

$$G(t+\mathrm{i}s) = E(t)F(s) = 0 \quad \text{for} \quad t < -\|T\| \text{ or } s < -\|T\|.$$

If for (u_n) we choose a bounded sequence of step functions defined on $\mathbb{R}$

which converges to id uniformly for $|t| \leq \|T\|$, then

$$A = \int_{\mathbf{R}} t \, \mathrm{d}E(t) = \lim_{n\to\infty} \int_{\mathbf{R}} u_n(t) \, \mathrm{d}E(t) = \lim_{n\to\infty} \int_{\mathbf{C}} u_n(\mathrm{Re}\, z) \, \mathrm{d}G(z)$$
$$= \int_{\mathbf{C}} \mathrm{Re}\, z \, \mathrm{d}G(z),$$

where we have used the fact that $G(\{z \in \mathbb{C} : \mathrm{Re}\, z \in J\}) = E(J)$. We can obtain similarly that

$$B = \int_{\mathbf{C}} \mathrm{Im}\, z \, \mathrm{d}G(z).$$

Consequently, it follows that

$$T = A + \mathrm{i}B = \int_{\mathbf{C}} \mathrm{Re}\, z \, \mathrm{d}G(z) + \mathrm{i} \int_{\mathbf{C}} \mathrm{Im}\, z \, \mathrm{d}G(z)$$
$$= \int_{\mathbf{C}} (\mathrm{Re}\, z + \mathrm{i}\, \mathrm{Im}\, z) \, \mathrm{d}G(z) = \int_{\mathbf{C}} z \, \mathrm{d}G(z).$$

It only remains to prove the uniqueness of G. Let G' be a complex spectral family for which

$$T = \int_{\mathbf{C}} z \, \mathrm{d}G'(z) \quad \text{and} \quad G'(t + \mathrm{i}s) = E'(t)F'(s) = F'(s)E'(t).$$

Then

$$A = \tfrac{1}{2}(T + T^*) = \tfrac{1}{2} \int_{\mathbf{C}} (z + z^*) \, \mathrm{d}G'(z) = \int_{\mathbf{C}} \mathrm{Re}\, z \, \mathrm{d}G'(z) = \int_{\mathbf{R}} t \, \mathrm{d}E'(t).$$

The unicity of the spectral family of self-adjoint operators gives that $E = E'$. We can show similarly that $F = F'$. Therefore, $G = G'$. □

If G is the spectral family of the operator T in the sense of Theorem 7.31, then we also write $u(T)$ for $\hat{G}(u)$. Theorem 7.19 also holds for normal operators.

EXAMPLE 1. Assume T is a compact normal operator on the complex Hilbert space H, $\{\lambda_1, \lambda_2, \ldots\}$ are its non-zero eigenvalues, P_j is the orthogonal projection onto $N(\lambda_j - T)$, $\lambda_0 = 0$, and P_0 is the orthogonal projection onto $N(T)$. Then the formula

$$G(z) = \Sigma\{P_j : j \in \mathbb{N}_0, \mathrm{Re}\, \lambda_j \leq \mathrm{Re} z, \mathrm{Im}\, \lambda_j \leq \mathrm{Im} z\} \quad \text{for} \quad z \in \mathbb{C}$$

defines a complex spectral family, and we have

$$T = \int_{\mathbf{C}} z \, \mathrm{d}G(z).$$

The proof goes as in the self-adjoint case.

Theorem 7.32 (The spectral theorem for normal operators). *Let T be a normal operator on the complex Hilbert space H. Then there exists a unique complex spectral family G for which*

$$T = \int_{\mathbb{C}} z \, \mathrm{d}G(z).$$

The operators $A = \overline{(T+T^*)/2}$ and $B = \overline{(T-T^*)/2\mathrm{i}}$ are self-adjoint. For the spectral families E and F of A and B, respectively, we have

$$G(t+\mathrm{i}s) = E(t)F(s) = F(s)E(t).$$

We have $T = A + \mathrm{i}B$ and $T^* = A - \mathrm{i}B$. The operators A and B are called the real part *and* the imaginary *part of T (cf. Exercise 7.49).*

PROOF. Since this theorem has little significance in the applications, we shall not work out its proof in detail. The operator $S = T^*T$ is self-adjoint by Theorem 5.39. Let E_0 denote its spectral family. $TS = ST$ holds, because T is normal. Hence $R(z,S)T \subset TR(z,S)$ for all $z \in \mathbb{C} \setminus \mathbb{R}$. It follows from this by (7.22) that $E_0(t)T \subset TE_0(t)$. This implies that the subspaces

$$H_n = R\big(E_0((-n, -n+1] \cup (n-1, n])\big), \quad n \in \mathbb{N}$$

reduce the operator T (cf. Section 7.4). The same holds for T^*. Let $T_n = T|_{H_n}$. We have $H_n \subset D(T^*T) \subset D(T)$ by Section 7.3, Proposition 4. Therefore, $D(T_n) = H_n$. For all $f \in H_n$ we have

$$\|T_n f\|^2 = \langle T^*Tf, f\rangle \leqslant n\|f\|^2,$$

i.e., $T_n \in B(H_n)$. Similarly, $(T_n)^* \in B(H_n)$, $(T_n)^* = T^*|_{H_n}$ and $\|(T_n)^*f\| = \|T^*f\| = \|Tf\| = \|T_n f\|$ for all $f \in H_n$, i.e., T_n is a bounded normal operator on H_n. Consequently, by Theorem 7.31 there is a complex spectral family G_n such that

$$T_n = \int_{\mathbb{C}} z \, \mathrm{d}G_n(z),$$

where $G_n(t+\mathrm{i}s) = E_n(t)F_n(s) = F_n(s)E_n(t)$ for $s, t \in \mathbb{R}$ with the spectral families E_n and F_n of

$$A_n = \tfrac{1}{2}(T_n + T_n^*) = \tfrac{1}{2}(T+T^*)\Big|_{H_n} \quad \text{and} \quad B_n = \frac{1}{2\mathrm{i}}(T_n - T_n^*) = \frac{1}{2\mathrm{i}}(T-T^*)\Big|_{H_n},$$

respectively. In what follows we consider the operators $G_n(z)$, $E_n(t)$, and $F_n(s)$ to be defined on H (more precisely, we should write $I_n G_n(z) P_n$, etc., where I_n is the embedding of H_n into H and P_n is the orthogonal projection onto H_n). Since $H = \oplus_{n\in\mathbb{N}} H_n$, the sums

$$G(z) = \sum_{n\in\mathbb{N}} G_n(z),$$

$$E(t) = \sum_{n\in\mathbb{N}} E_n(t) \quad \text{and} \quad F(s) = \sum_{n\in\mathbb{N}} F_n(s)$$

exist for all $z \in \mathbb{C}$ and $s, t \in \mathbb{R}$ in the strong sense. It is easy to see that E and F are (real) spectral families and $G(t+\mathrm{i}s) = E(t)F(s) = F(s)E(t)$.

Now let $D_0 = L\{H_n : n \in \mathbb{N}\}$. Then D_0 is a core of T. For every $f \in D_0$ there exists an $N \in \mathbb{N}$ such that $f \in \oplus_{n=1}^N H_n$. Then

$$Tf = \sum_{n=1}^{N} T_n P_n f = \sum_{n=1}^{N} \int_{\mathbb{C}} z \,\mathrm{d}G_n(z) f$$

$$= \int_{\mathbb{C}} z \,\mathrm{d}\left(\sum_{n=1}^{N} G_n(z)\right) f = \int_{\mathbb{C}} z \,\mathrm{d}G(z) f = \hat{G}(\mathrm{id}) f.$$

The restriction T_0 of T to D_0 is therefore contained in the normal operator $\hat{G}(\mathrm{id})$. Then we also have $T = \overline{T}_0 \subset \hat{G}(\mathrm{id})$, and it follows by Section 5.6, Proposition 1 that $T = \hat{G}(\mathrm{id})$.

Similarly,

$$Af = \tfrac{1}{2}(T+T^*)f = \int_{\mathbb{R}} t \,\mathrm{d}E(t) f, \quad Bf = \frac{1}{2\mathrm{i}}(T-T^*)f = \int_{\mathbb{R}} s \,\mathrm{d}F(s) f$$

for $f \in D_0$. As the restrictions of $(T+T^*)/2$ and $(T-T^*)/2\mathrm{i}$ to D_0 are essentially self-adjoint by Exercise 5.43, it follows from this that

$$\tfrac{1}{2}\overline{(T+T^*)} = A = \int t \,\mathrm{d}E(t), \quad \frac{1}{2\mathrm{i}}\overline{(T-T^*)} = B = \int s \,\mathrm{d}F(s).$$

We have $T \subset A + \mathrm{i}B$, by construction. In order to prove that $T = A + \mathrm{i}B$, we therefore have to prove that $D(A+\mathrm{i}B) = D(A) \cap D(B) \subset D(T)$. If $f \in D(A) \cap D(B)$, then (observe that $G(\{z \in \mathbb{C} : \mathrm{Re}\, z \in J\}) = E(J)$, $G(\{z \in \mathbb{C} : \mathrm{Im}\, z \in J\}) = F(J)$)

$$\int_{\mathbb{C}} |z|^2 \,\mathrm{d}\|G(z)f\|^2 = \int_{\mathbb{C}} (|\mathrm{Re}\, z|^2 + |\mathrm{Im}\, z|^2) \,\mathrm{d}\|G(z)f\|^2$$

$$= \int_{\mathbb{R}} t^2 \,\mathrm{d}\|E(t)f\|^2 + \int_{\mathbb{R}} s^2 \,\mathrm{d}\|F(s)f\|^2 < \infty;$$

consequently, $f \in D(T)$. □

EXAMPLE 2. Let γ be a measure on $\mathbb{C}(=\mathbb{R}^2)$ and let T be the maximal operator of multiplication by the function id, i.e., let

$$D(T) = \{f \in L_2(\mathbb{C}, \gamma) : \mathrm{id}\, f \in L_2(\mathbb{C}, \gamma)\},$$
$$Tf = \mathrm{id}\, f \quad \text{for} \quad f \in D(T).$$

Then T is normal, and the spectral family G of T is given by

$$G(z)f = \chi_z f \quad \text{for} \quad f \in L_2(\mathbb{C}, \gamma),$$

where χ_z is the characteristic function of the set $\{w \in \mathbb{C} : \mathrm{Re}\, w \leq \mathrm{Re}\, z$ and $\mathrm{Im}\, w \leq \mathrm{Im}\, z\}$.

We can show in an entirely analogous way as for self-adjoint operators that every normal operator is unitarily equivalent to an orthogonal sum of operators of the kind considered in Example 2. We give the theorem without proof.

Theorem 7.33 (The spectral representation theorem for normal operators). *If T is a normal operator on a complex Hilbert space, then there exist a family $\{\gamma_\alpha : \alpha \in A\}$ of measures on $\mathbb{C}$ and a unitary operator $U: H \to \oplus_{\alpha \in A} L_2(\mathbb{C}, \gamma_\alpha)$ for which*

$$T = U^{-1} T_{\mathrm{id}} U;$$

where T_{id} is the maximal operator of multiplication by the function id *on $\oplus_{\alpha \in A} L^2(\mathbb{C}, \gamma_\alpha)$, i.e.,*

$$D(T_{\mathrm{id}}) = \Big\{ (f_\alpha)_{\alpha \in A} \in \bigoplus_{\alpha \in A} L_2(\mathbb{C}, \gamma_\alpha) : (\mathrm{id}\, f_\alpha)_{\alpha \in A} \in \bigoplus_{\alpha \in A} L_2(\mathbb{C}, \gamma_\alpha) \Big\},$$

$$T_{\mathrm{id}}(f_\alpha)_{\alpha \in A} = (\mathrm{id}\, f_\alpha)_{\alpha \in A} \quad \text{for} \quad (f_\alpha)_{\alpha \in A} \in D(T_{\mathrm{id}}).$$

The spectral points of a normal operator can be classified similarly to those of a self-adjoint operator, and they can be characterized by means of the spectral family. We do not go into details and mention only the following result.

Theorem 7.34. *Let T be a normal operator on a complex Hilbert space, and let G be the spectral family of T.*

(a) *$z \in \sigma(T)$ if and only if*

$$G(z+\epsilon+i\epsilon) + G(z-\epsilon-i\epsilon) - G(z+\epsilon-i\epsilon) - G(z-\epsilon+i\epsilon) \neq 0$$

for every $\epsilon > 0$.

(b) *$\|R(z, T)\| = d(z, \sigma(T))^{-1}$ for $z \in \rho(T)$, where $d(z, \sigma(T))$ denotes the distance of the point z from $\sigma(T)$.*

For the *proof*: (a) The proof is analogous to that of Theorem 7.22. Observe that $\hat{G}(\chi_M) = G(b_1 + ib_2) + G(a_1 + ia_2) - G(a_1 + ib_2) - G(b_1 + ia_2)$ for $M = \{z \in \mathbb{C} : a_1 < \operatorname{Re} z \leq b_1,\ a_2 < \operatorname{Im} z \leq b_2\}$.

(b) $\|R(z, T)\| \geq d(z, \sigma(T))^{-1}$ by Theorem 5.15. Since we have $R(z, T) = u_z(T)$ with $u_z(w) = (z-w)^{-1}$, and as $|u_z(w)| \leq d(z, \sigma(T))^{-1}$ G-almost everywhere, it follows that $\|R(z, T)\| \leq d(z, \sigma(T))^{-1}$.

Theorem 7.35. *Assume that T_n ($n \in \mathbb{N}$) and T are bounded normal operators on the complex Hilbert space H and $\|T - T_n\| \to 0$ as $n \to \infty$. Then*

$$\begin{aligned} \sigma(T) &= \lim_{n\to\infty} \sigma(T_n) \\ &= \{ z \in \mathbb{C} : \text{there exists a sequence } (z_n) \text{ from } \mathbb{C} \text{ for which } \\ &\qquad z_n \in \sigma(T_n) \text{ and } z_n \to z \}. \end{aligned}$$

(This is an *assertion about* the *continuity* of the set-valued function $T \mapsto \sigma(T)$ defined on the set of bounded normal operators. We cannot allow all bounded operators here, and $\|T-T_n\| \to 0$ cannot be replaced by $T_n \xrightarrow{s} T$; cf. Exercises 7.41 and 7.42.)

Proof. If $z \notin \sigma(T)$, i.e., $z \in \rho(T)$, then (second Corollary to Theorem 5.11) $z \in \rho(T_n)$ for sufficiently large n and $\|(z-T_n)^{-1}-(z-T)^{-1}\| \to 0$. Hence, $\|(z-T_n)^{-1}\| \to \|(z-T)^{-1}\|$ as $n \to \infty$. Since $d(z, \sigma(T_n)) = \|R(z, T_n)\|^{-1}$, we have $\lim_{n\to\infty} d(z, \sigma(T_n)) = \|R(z, T)\|^{-1} > 0$. Consequently, the point z is not contained in $\lim \sigma(T_n)$.

Assume $z \in \sigma(T)$. Then by Theorem 5.43 there is a sequence (f_n) from H for which $\|f_n\| = 1$ and $(z-T)f_n \to 0$. Then

$$(z-T_n)f_n = (z-T)f_n - (T_n - T)f_n \to 0;$$

hence $d(z, \sigma(T_n)) = \|R(z, T_n)\|^{-1} \leq \|(z-T_n)f_n\| \to 0$. It follows from this that $z \in \lim \sigma(T_n)$. □

Theorem 7.36. *If U is a unitary operator on a complex Hilbert space, then there exists a real spectral family E for which $E(t) = 0$ for $t < 0$, $E(t) = I$ for $t \geq 2\pi$ and $U = \int e^{it}\, dE(t)$ (cf. also Exercise 7.46).*

Proof. By Section 5.2, Example 2 the spectrum of U is contained in $\{z \in \mathbb{C} : |z| = 1\} = \{e^{it} : 0 \leq t < 2\pi\}$, i.e., $G(\{e^{it} : 0 \leq t < 2\pi\} = I$, where G denotes the complex spectral family of U. Then the formulae

$$E(t) = \begin{cases} 0 & \text{for} \quad t < 0, \\ G(\{e^{is} : 0 \leq s \leq t\}) & \text{for} \quad 0 \leq t < 2\pi, \\ I & \text{for} \quad t \geq 2\pi \end{cases}$$

define a real spectral family. We can verify easily that $U = \int e^{it}\, dE(t)$. □

Exercises

7.40. A function $G : \mathbb{C} \to B(H)$ is a complex spectral family if and only if $G(z)$ is an orthogonal projection for every $z \in \mathbb{C}$ and properties (a), (b) and (c) of Theorem 7.30 are satisfied.

7.41. We cannot replace norm convergence by strong convergence in Theorem 7.35.
Hint: Let the operators $T_m \in B(l_2)$ be defined by $T_m(f_n)_{n \in \mathbb{N}} = (f_1, f_2, \ldots, f_m, 0, 0, 0, \ldots)$. Then $T_m \xrightarrow{s} I$, $\sigma(T_m) = \{0, 1\}$ and $\sigma(I) = \{1\}$.

7.42. Let the operators T_m, $m \in \mathbb{N}$, and T from $B(l_2(\mathbb{Z}))$ be defined by the formulae

$$T_m(f_n)_{n \in \mathbb{Z}} = (g_n)_{n \in \mathbb{Z}} \text{ with } g_n = \begin{cases} f_{n+1} & \text{for} \quad n \neq -1, \\ \dfrac{1}{m} f_0 & \text{for} \quad n = -1, \end{cases}$$

$$T(f_n)_{n \in \mathbb{Z}} = (g_n)_{n \in \mathbb{Z}} \text{ with } g_n = \begin{cases} f_{n+1} & \text{for} \quad n \neq -1, \\ 0 & \text{for} \quad n = -1. \end{cases}$$

(a) We have $\|T\| = \|T_m\| = 1$. Therefore, $\sigma(T) \subset \{z \in \mathbb{C} : |z| \leq 1\}$ and $\sigma(T_m) \subset \{z \in \mathbb{C} : |z| \leq 1\}$.
(b) For every $z \in \mathbb{C}$ such that $|z| < 1$, the vector $(f_n)_{n \in \mathbb{Z}}$ defined by $f_n = 0$ for $n < 0$ and $f_n = z^n$ for $n \geq 0$ is an eigenelement of T belonging to the eigenvalue z; hence $\sigma(T) = \{z \in \mathbb{C} : |z| \leq 1\}$.
(c) $T_m^{-1} \in B(l_2(\mathbb{Z}))$ for all $m \in \mathbb{N}$, and $r(T_m^{-1}) = 1$. Consequently, $\{z \in \mathbb{C} : |z| > 1\} \subset \rho(T_m^{-1})$, and thus $\{z \in \mathbb{C} : |z| < 1\} \subset \rho(T_m)$ (cf. Exercise 5.27).
(d) Theorem 7.35 does not hold in general if the operators T_m and T are not normal.

7.43. Assume that T is a bounded normal operator on the complex Hilbert space H, $r > \|T\|$, and u is a function, holomorphic in $\{z \in \mathbb{C} : |z| < r\}$ with $u(z) = \sum_{j=0}^{\infty} u_j z^j$ for $|z| < r$. Then $u(T) = \sum_{j=0}^{\infty} u_j T^j$, where the series converges in $B(H)$. For self-adjoint operators this statement holds in real Hilbert spaces, as well.

7.44. Let T_1 and T_2 be *similar* bounded normal operators on the complex Hilbert space H, i.e., let there exist a bijective operator $S \in B(H)$ such that $ST_1 = T_2 S$. Prove that T_1 and T_2 are unitarily equivalent.
(a) $S = e^{zT_2} S\, e^{-zT_1}$ for all $z \in \mathbb{C}$.
Hint: Exercise 7.43.
(b) $e^{izT_2^*} S e^{-izT_1^*} = e^{i(zT_2^* + z^* T_2)} S\, e^{-i(zT_1^* + z^* T_1)}$. Consequently, $\|e^{izT_2^*} S\, e^{-izT_1^*}\| \leq \|S\|$ for all $z \in \mathbb{C}$.
(c) $S = e^{izT_2^*} S\, e^{-izT_1^*}$ for all $z \in \mathbb{C}$.
Hint: The Liouville theorem and part (b).
(d) $ST_1^* = T_2^* S$.
Hint: Differentiate in (c) with respect to z and substitute $z = 0$.
(e) $T_1 S^* S = S^* S T_1$, $T_1 |S| = |S| T_1$ and $T_1 |S|^{-1} = |S|^{-1} T_1$.
(f) If $S = U|S|$ is the polar decomposition of S, then U is unitary, $|S|$ is bijective, and $U = S|S|^{-1}$.
(g) We have $T_1 = U^{-1} T_2 U$ with the operator U from part (f). Hence T_1 and T_2 are unitarily equivalent.

7.45. Assume that H is a complex Hilbert space, $S, T \in B(H)$, $ST = TS$, and T is normal.
(a) Then $ST^* = T^* S$.
Hint: Use the reasoning of Exercise 7.44(b), (c) and (d) for $T_1 = T_2 = T$.
(b) If G is the spectral family of T, then $SG(z) = G(z)S$ for all $z \in \mathbb{C}$.

7.46. Prove the uniqueness of the spectral family E in Theorem 7.36.

7.47. If T is a normal operator on a complex Hilbert space and $n \in \mathbb{N}$, then there exists exactly one normal operator A for which $A^n = T$ and $G(\{z = r\, e^{i\varphi} : r > 0, 0 \leq \varphi < 2\pi/n\}) = I$ for the spectral family G of A.

7.48. If A and B are self-adjoint (not necessarily bounded) operators on a complex Hilbert space with spectral families E and F, and $E(t)F(s) = F(s)E(t)$ for all $s, t \in \mathbb{R}$, then $T = A + iB$ is normal.

7.49. Decomposition in *real-* and *imaginary parts* for arbitrary operators (cf. Exercise 5.38 and Theorem 7.32):
(a) If T is a densely defined operator on H such that $D(T) \subset D(T^*)$, then the

operators $A=(T+T^*)/2$ and $B=(T-T^*)/2\mathrm{i}$ are symmetric, $D(A)=D(B)$, and $T=A+\mathrm{i}B$.

(b) If A and B are symmetric, $D(A)=D(B)$, and $T=A+\mathrm{i}B$, then $A=(T+T^*)/2$ and $B=(T-T^*)/2\mathrm{i}$.

(c) Even if T is closed and $D(T)=D(T^*)$, we cannot expect that A and B in (a) are essentially self-adjoint.
Hint: Choose $T=|C|+(\mathrm{i}C/2)$ or $T=\mathrm{i}|C|+(C/2)$, where C is a closed symmetric but not self-adjoint operator.

(d) Even if A and B are self-adjoint, we cannot expect that T in (b) is closed.
Hint: Take an unbounded self-adjoint operator C. Define A and B by the formulae $D(A)=D(B)=D(C)\oplus D(C)$, $A(f,g)=(Cg,Cf)$ and $B(f,g)=\mathrm{i}(Cg,Cf)$ for $f,g\in D(C)$.

7.6 One-parameter unitary groups

One example of the significance of the theory of self-adjoint operators is shown by Stone's theorem (cf. Theorem 7.37 and Theorem 7.38). We shall learn more about this later in this section.

Let H be a Hilbert space. A family $\{B(t) : t\in\mathbb{R}\}$ of operators out of $B(H)$ is called a *one-parameter group* if

$$B(0)=I \quad\text{and}\quad B(s)B(t)=B(s+t) \quad\text{for all}\quad s,t\in\mathbb{R}.$$

(This is then a "representation" of the additive group $\mathbb{R}$ by operators on H.) The one-parameter group $\{B(t) : t\in R\}$ is said to be *strongly continuous* if the function

$$B(\cdot)f : \mathbb{R}\to H,\quad t\mapsto B(t)f$$

is continuous for every $f\in H$.

Let $\{B(t) : t\in\mathbb{R}\}$ be a one-parameter group of operators on H. The operator A defined by the formulae

$$D(A)=\left\{f\in H : \lim_{t\to0}\frac{1}{t}(B(t)-I)f \quad\text{exists}\right\},$$

$$Af=\lim_{t\to0}\frac{1}{t}(B(t)-I)f \quad\text{for}\quad f\in D(A)$$

is called the *infinitesimal generator* of $\{B(t) : t\in\mathbb{R}\}$.

It can be proved that every strongly continuous one-parameter group possesses a densely defined infinitesimal generator. In the following we only consider (one-parameter) *unitary groups* (i.e., one-parameter groups of unitary operators). The situation is somewhat simpler in that case.

Theorem 7.37. *Let T be a self-adjoint operator on the complex Hilbert space H, let E be the spectral family of T, and let*

$$U(t)=\mathrm{e}^{\mathrm{i}tT}=\int\mathrm{e}^{\mathrm{i}ts}\,\mathrm{d}E(s)\quad\textit{for}\quad t\in\mathbb{R}.$$

Then $\{U(t): t \in \mathbb{R}\}$ *is a strongly continuous (one-parameter) unitary group. The infinitesimal generator is* $\mathrm{i}T$. *We have* $U(t)f \in D(T)$ *for all* $f \in D(T)$ *and* $t \in \mathbb{R}$.

Proof. By Theorem 7.14 we have

$$U(t) \in B(H) \quad \text{and} \quad U(t)^* = U(-t) = U(t)^{-1} \quad \text{for all } t \in \mathbb{R},$$

i.e., $U(t)$ is unitary for all $t \in \mathbb{R}$ (cf. Theorem 4.34). For all $x, y \in \mathbb{R}$

$$|e^{ix} - e^{iy}| = |e^{i(x-y)/2} - e^{-i(x-y)/2}| = 2\left|\sin\frac{x-y}{2}\right|.$$

It follows from this that

$$\begin{aligned}\|U(t)f - U(t')f\|^2 &= \left\|\int (e^{its} - e^{it's})\,dE(s)f\right\|^2 \\ &= \int |e^{its} - e^{it's}|^2\,d\|E(s)f\|^2 \\ &= 4\int \left|\sin\frac{(t-t')s}{2}\right|^2 d\|E(s)f\|^2\end{aligned}$$

for all $f \in H$ and $t, t' \in \mathbb{R}$. Because of the relations

$$\left|\sin\frac{(t-t')s}{2}\right| \leq 1 \quad \text{and} \quad \sin\frac{(t-t')s}{2} \to 0 \text{ as } t' \to t,$$

it follows by Lebesgue's theorem that

$$\|U(t)f - U(t')f\| \to 0 \quad \text{as} \quad t' \to t.$$

This proves the strong continuity of $U(t)$. For all $f \in H$ and $t \in \mathbb{R}$ we have

$$\frac{1}{t}(U(t) - I)f = \frac{1}{t}\int (e^{its} - 1)\,dE(s)f.$$

Because of the relations

$$\frac{1}{t}(e^{its} - 1) \to is \quad \text{as} \quad t \to 0,\ s \in \mathbb{R},$$

and

$$\left|\frac{1}{t}(e^{its} - 1)\right| \leq |s| \quad \text{for} \quad s, t \in R,\ t \neq 0,$$

the right-hand side converges, as $t \to 0$, if and only if the function $u(s) = |s|$ belongs to $L_2(\mathbb{R}, \rho_f)$ (with $\rho_f(s) = \|E(s)f\|^2$), i.e., if and only if $f \in D(T)$. The limit equals $\mathrm{i}Tf$. Consequently, $\mathrm{i}T$ is the infinitesimal generator of $\{U(t): t \in \mathbb{R}\}$.

If $f \in D(T)$, then for every $t \in \mathbb{R}$

$$\int |s|^2\,d_s\|E(s)U(t)f\|^2 = \int |s|^2\,d_s\|U(t)E(s)f\|^2 = \int |s|^2\,d\|E(s)f\|^2 < \infty,$$

i.e., $U(t)f \in D(T)$. □

Actually, every strongly continuous (one-parameter) unitary group can be represented in this form.

Theorem 7.38 (Stone). *Let $\{U(t) : t \in \mathbb{R}\}$ be a strongly continuous (one-parameter) unitary group on the complex Hilbert space H. Then there exists a uniquely determined self-adjoint operator T on H for which*

$$U(t) = \mathrm{e}^{\mathrm{i}tT} \quad \textit{for all} \quad t \in \mathbb{R}.$$

If H is separable, then strong continuity can be replaced by weak measurability, i.e., it is sufficient to require that the function

$$\langle f, U(\cdot)g \rangle : \mathbb{R} \to \mathbb{C},\ t \mapsto \langle f, U(t)g \rangle$$

is measurable (with respect to Lebesgue measure on $\mathbb{R}$) for all $f, g \in H$.

PROOF. The equality $U(t) = \mathrm{e}^{\mathrm{i}tT}$ implies that $\mathrm{i}T$ is the infinitesimal generator of $\{U(t) : t \in \mathbb{R}\}$. This proves the unicity of T and, at the same time, provides an opportunity to construct T. In what follows let A be the infinitesimal generator of $\{U(t) : t \in \mathbb{R}\}$, and let $T = -\mathrm{i}A$. We show that T is essentially self-adjoint and $U(t) = \mathrm{e}^{\mathrm{i}t\overline{T}}$. Since $\mathrm{i}\overline{T}$ is then the infinitesimal generator of $\{U(t) : t \in \mathbb{R}\}$, it follows that $T = \overline{T}$. First, let us assume that the group is strongly continuous.

$D(A)$ *is dense*: For every $\varphi \in C_0^\infty(\mathbb{R})$ and every $f \in H$ the equality

$$f_\varphi = \int \varphi(s) U(s) f \,\mathrm{d}s$$

defines an $f_\varphi \in H$ (the integral is extended only to the support of φ and can be understood as a Riemann integral). We have

$$\begin{aligned}
\frac{1}{t}(U(t) - I) f_\varphi &= \frac{1}{t} \int \varphi(s)(U(s+t) - U(s)) f \,\mathrm{d}s \\
&= \frac{1}{t} \int [\varphi(s) U(s+t) - \varphi(s) U(s)] f \,\mathrm{d}s \\
&= \frac{1}{t} \int [\varphi(s-t) U(s) - \varphi(s) U(s)] f \,\mathrm{d}s \\
&= \int \frac{1}{t}(\varphi(s-t) - \varphi(s)) U(s) f \,\mathrm{d}s \to -\int \varphi'(s) U(s) f \,\mathrm{d}s
\end{aligned}$$

as $t \to 0$. The set $D_0 = \{f_\varphi : f \in H,\ \varphi \in C_0^\infty(\mathbb{R})\}$ is therefore contained in $D(A)$. If (φ_n) is a sequence from $C_0^\infty(\mathbb{R})$ such that

$$\begin{aligned}
\varphi_n(s) &= 0 \quad \text{for} \quad |s| > \frac{1}{n}, \\
\varphi_n(s) &\geq 0 \quad \text{for all} \quad s \in \mathbb{R}, \\
\int \varphi_n(s) \,\mathrm{d}s &= 1 \quad \text{for all} \quad n \in \mathbb{N},
\end{aligned}$$

then $f_{\varphi_n} \to f$ as $n \to \infty$, since

$$\|f_{\varphi_n} - f\| = \left\| \int \varphi_n(s)(U(s) - I)f \, ds \right\|$$
$$\leqslant \sup\left\{ \|(U(s) - I)f\| : -\frac{1}{n} < s < \frac{1}{n} \right\}.$$

Hence D_0, and thus $D(A)$, too, is dense in H.

$T = -iA$ is symmetric: If $f, g \in D(T) = D(A)$, then

$$\begin{aligned}\langle g, Tf\rangle &= -i\langle g, Af\rangle = -\lim_{t\to 0} i\left\langle g, \frac{1}{t}(U(t) - I)f\right\rangle \\ &= -\lim_{t\to 0} i\left\langle \frac{1}{t}(U(-t) - I)g, f\right\rangle \\ &= \lim_{t\to 0} i\left\langle \frac{1}{-t}(U(-t) - I)g, f\right\rangle = i\langle Ag, f\rangle \\ &= \langle Tg, f\rangle.\end{aligned}$$

$R(\pm i - T)$ is dense in H: Assume $g \in R(i - T)^\perp = N(i + T^*)$. Since for every $f_\varphi \in D_0$ and all $t \in \mathbb{R}$

$$U(t)f_\varphi = U(t)\int \varphi(s)U(s)f \, ds = \int \varphi(s - t)U(s)f \, ds \in D_0,$$

it follows that

$$\begin{aligned}\frac{d}{dt}\langle g, U(t)f_\varphi\rangle &= \langle g, AU(t)f_\varphi\rangle = \langle A^*g, U(t)f_\varphi\rangle \\ &= \langle -iT^*g, U(t)f_\varphi\rangle = -\langle g, U(t)f_\varphi\rangle.\end{aligned}$$

The function $h(t) = \langle g, U(t)f_\varphi\rangle$ is therefore a solution of the differential equation $h' = -h$, i.e., we have $h(t) = e^{-t}h(0)$. Since $U(t)$ is unitary, h is bounded; however this is possible only if $\langle g, f_\varphi\rangle = \langle g, U(0)f_\varphi\rangle = h(0) = 0$. Since this holds for all $f_\varphi \in D_0$, we have $g = 0$. Consequently, $\overline{R(i - T)} = H$. We can show similarly that $\overline{R(-i - T)} = H$. Hence, T is essentially self-adjoint.

We have $U(t) = e^{it\bar{T}}$: Let $V(t) = e^{it\bar{T}}$ and $f \in D_0$. Because of the relation $f \in D(\bar{T})$ we also have $V(t)f \in D(\bar{T})$ by Theorem 7.37 and $d/dt\, V(t)f = i\bar{T}V(t)f$. Since, moreover, $U(t)f \in D_0 \subset D(T)$ for all $t \in \mathbb{R}$, it follows that

$$\frac{d}{dt}(U(t)f - V(t)f) = iTU(t)f - i\bar{T}V(t)f = i\bar{T}(U(t)f - V(t)f).$$

Consequently, because $\bar{T}$ is self-adjoint,

$$\frac{d}{dt}\|U(t)f - V(t)f\|^2 = 2\,\mathrm{Re}\langle U(t)f - V(t)f, i\bar{T}(U(t)f - V(t)f)\rangle = 0.$$

It follows from this that $U(t)f = V(t)f$ for all $t \in \mathbb{R}$ and all $f \in D_0$, because $U(0)f = V(0)f$. Since D_0 is dense, this implies that $U(t) = V(t) = e^{it\bar{T}}$.

It remains to prove that weak measurability implies strong continuity in the separable case. Let $f \in H$. As $t \mapsto \langle U(t)f, g\rangle$ is bounded (with bound $\|f\|\,\|g\|$) and measurable for all $g \in H$, the function

$$g \mapsto \int_0^a \langle U(t)f, g\rangle \,dt$$

is a continuous linear functional with norm $\leqslant a\|f\|$ for every $a > 0$. By the Riesz representation theorem (Theorem 4.8) there exists an $f_a \in H$ for which

$$\langle f_a, g\rangle = \int_0^a \langle U(t)f, g\rangle \,dt.$$

Therefore,

$$\begin{aligned}\langle U(s)f_a, g\rangle &= \langle f_a, U(-s)g\rangle = \int_0^a \langle U(t)f, U(-s)g\rangle \,dt\\ &= \int_0^a \langle U(t+s)f, g\rangle \,dt = \int_s^{a+s} \langle U(t)f, g\rangle \,dt,\end{aligned}$$

and thus

$$\begin{aligned}|\langle U(s)f_a, g\rangle - \langle f_a, g\rangle| &\leqslant \left|\int_0^s \langle U(t)f, g\rangle \,dt\right| + \left|\int_a^{a+s} \langle U(t)f, g\rangle \,dt\right|\\ &\leqslant 2|s|\,\|f\|\,\|g\|.\end{aligned}$$

Hence,

$$\langle U(s)f_a, g\rangle \to \langle f_a, g\rangle \quad \text{as} \quad s \to 0,$$

i.e., $U(\cdot)f_a$ is weakly continuous at the origin. Because of the equality $\|U(s)f_a\| = \|f_a\|$, the continuity of $U(\cdot)f_a$ at the origin follows from this, since

$$\|U(s)f_a - f_a\|^2 = \|U(s)f_a\|^2 - 2\,\mathrm{Re}\langle U(s)f_a, f_a\rangle + \|f_a\|^2 \to 0 \quad \text{as} \quad s \to 0.$$

We show, in addition, that the set of elements f_a is dense in H. It follows from this that $U(s)$ is strongly continuous at the origin, and thus everywhere. Let h be orthogonal to all f_a, and let $\{e_n : n \in \mathbb{N}\}$ be an orthonormal basis of H. Then

$$\int_0^a \langle U(t)e_n, h\rangle \,dt = \langle e_{n,a}, h\rangle = 0$$

for all $n \in \mathbb{N}$ and all $a > 0$. It follows from this that for all $n \in \mathbb{N}$ we have $\langle U(t)e_n, h\rangle = 0$ almost everywhere in $(0, \infty)$ (cf. Theorem A16(c)). Consequently, there is a $t_0 > 0$ such that

$$\langle U(t_0)e_n, h\rangle = 0 \quad \text{for all} \quad n \in \mathbb{N}.$$

As $U(t_0)$ is unitary, $\{U(t_0)e_n : n \in \mathbb{N}\}$ is an orthonormal basis. Hence, we must have $h = 0$. □

Corollary. *Let $\{U(t) : t \in \mathbb{R}\}$ be a strongly continuous unitary group, and let $\mathrm{i}T$ be its infinitesimal generator. Then the initial value problem*

$$\frac{1}{\mathrm{i}}\frac{\mathrm{d}}{\mathrm{d}t}u(t) = Tu(t), \quad u(0) = f$$

is uniquely solvable for every $f \in D(T)$ and the solution is $u(t) = U(t)f$. (A solution is a continuously differentiable function defined on $\mathbb{R}$ with values in $D(T)$ that satisfies the differential equation.)

Proof. As $U(t)f \in D(T)$ for all $f \in D(T)$ and all $t \in \mathbb{R}$, the function $u(t) = U(t)f$ is a solution of the initial value problem. If u and v are solutions, then $u(0) - v(0) = 0$, and $\mathrm{d}/\mathrm{d}t\ \|u(t) - v(t)\|^2 = 2\ \mathrm{Re}\langle u(t) - v(t), \mathrm{i}T(u(t) - v(t))\rangle = 0$, i.e., $u(t) = v(t)$ for all $t \in \mathbb{R}$. □

A few more words about the significance of Stone's theorem: In quantum mechanics the states of a system are described by the normalized elements of a Hilbert space. If $u(t)$ is the state of the system at time t, then we write $u(t) = U(t)u(0)$; for reasons derived from physics, $U(t)$ has to be a linear operator. Of course, every state has to be possible at any time; since, moreover, $U(t)$ has to preserve the norm of the states, it follows that $U(t)$ is unitary. If, in addition, we require strong continuity (weak measurability in the separable case), which is quite plausible on a physical basis, then it follows that there exists a self-adjoint operator T such that $U(t) = \mathrm{e}^{\mathrm{i}tT}$; in particular,

$$\frac{1}{\mathrm{i}}\frac{\mathrm{d}}{\mathrm{d}t}U(t)f = TU(t)f$$

for all $f \in D(T)$. Since the time dependent Schrödinger equation is of this form, this implies that the Schrödinger operators must be self-adjoint.

Example 1. Let $H = L_2(\mathbb{R})$. The formula

$$U(t)f(x) = f(x+t), \quad f \in L_2(\mathbb{R})$$

defines a strongly continuous unitary group. The infinitesimal generator is $A = \mathrm{i}T$, where

$$D(T) = W_{2,1}(\mathbb{R}) \quad \text{and} \quad Tf = -\mathrm{i}f'$$

(for $W_{2,1}(\mathbb{R})$ see Section 6.4). It is obvious that T contains the operator $T_{1,0}$ defined by the equalities

$$D(T_{1,0}) = C_0^\infty(\mathbb{R}) \quad \text{and} \quad T_{1,0}f = -\mathrm{i}f'.$$

As by Theorem 6.30 $T_{1,0}$ is essentially self-adjoint, the assertion now follows.

EXAMPLE 2. Assume that $H = L_2(M)$ and $g : M \to \mathbb{R}$ is a measurable function. The equality

$$U(t)f(x) = \exp(\mathrm{i}tg(x))f(x), \quad f \in L_2(M)$$

defines a strongly continuous unitary group. It is easy to see that the infinitesimal generator is $A = \mathrm{i}T$, where T is the maximal operator of multiplication by the function g.

Theorem 7.39. *Let T be a self-adjoint operator on the complex Hilbert space H, and let M be a closed subspace of H. If $\mathrm{e}^{\mathrm{i}sT}f \in M$ for all $f \in M$ and $s \in \mathbb{R}$, then M reduces T and $\mathrm{e}^{\mathrm{i}sT}M = M$, $\mathrm{e}^{\mathrm{i}sT}M^{\perp} = M^{\perp}$ for all $s \in \mathbb{R}$.*

PROOF. We have $\mathrm{e}^{\mathrm{i}sT}M = M$ for all $s \in \mathbb{R}$, since

$$\mathrm{e}^{\mathrm{i}sT}M \subset M \quad \text{for all} \quad s \in \mathbb{R}$$

by assumption, and because every $f \in M$ can be written in the form

$$f = \mathrm{e}^{\mathrm{i}sT}(\mathrm{e}^{-\mathrm{i}sT}f) \quad \text{with} \quad \mathrm{e}^{-\mathrm{i}sT}f \in M$$

for every $s \in \mathbb{R}$.

For all $f \in M$, $g \in M^{\perp}$ and $s \in \mathbb{R}$ we have

$$\langle \mathrm{e}^{\mathrm{i}sT}g, f \rangle = \langle g, \mathrm{e}^{-\mathrm{i}sT}f \rangle = 0.$$

Consequently, $\mathrm{e}^{\mathrm{i}sT}M^{\perp} \subset M^{\perp}$, and thus $\mathrm{e}^{\mathrm{i}sT}M^{\perp} = M^{\perp}$. If P is the orthogonal projection onto M, then

$$P\,\mathrm{e}^{\mathrm{i}sT} = \mathrm{e}^{\mathrm{i}sT}P \quad \text{for all} \quad s \in \mathbb{R},$$

because of what we have just shown. We have to show that $PT \subset TP$. Let $f \in D(T)$. Then

$$PTf = P \lim_{t \to 0} \frac{-\mathrm{i}}{t}(\mathrm{e}^{\mathrm{i}tT}f - f) = \lim_{t \to 0} \frac{-\mathrm{i}}{t}(\mathrm{e}^{\mathrm{i}tT}Pf - Pf).$$

Hence, $Pf \in D(T)$ and $TPf = PTf$. □

We next prove the following special case of a theorem of *H. Trotter* for unitary groups that are generated by the sum of two self-adjoint operators.

Theorem 7.40. *Let T, S, and $T + S$ be self-adjoint operators on the complex Hilbert space H. Then*

$$\mathrm{e}^{\mathrm{i}t(T+S)} = s - \lim_{n \to \infty} \left[\mathrm{e}^{\mathrm{i}(t/n)T}\mathrm{e}^{\mathrm{i}(t/n)S}\right]^n$$

for all $t \in \mathbb{R}$.

Proof. For every $f \in D(T+S) = D(T) \cap D(S)$

$$t^{-1}(e^{itT}\,e^{itS} - e^{it(T+S)})f$$
$$= t^{-1}(e^{itT} - I)f + t^{-1}e^{itT}(e^{itS} - I)f - t^{-1}(e^{it(T+S)} - I)f$$
$$\xrightarrow{s} iTf + iSf - i(T+S)f = 0 \quad \text{as} \quad t \to 0.$$

In particular, for every $f \in D(T+S)$ there exists a $C(f) \geq 0$ for which

$$\|t^{-1}(e^{itT}\,e^{itS} - e^{it(T+S)})f\| \leq C(f) \quad \text{for all} \quad t \in \mathbb{R} \setminus \{0\}.$$

Since the space $D(T+S)$ is a Hilbert space with the $(T+S)$-scalar product $\langle .\,, .\rangle_{T+S}$, by the uniform boundedness principle (Theorem 4.22) there exists a $C \geq 0$ such that

$$\|t^{-1}(e^{itT}\,e^{itS} - e^{it(T+S)})f\| \leq C\|f\|_{T+S} \quad \text{for} \quad f \in D(T+S), t \in \mathbb{R} \setminus \{0\}.$$

If E is the spectral family of $T+S$, then for $f \in D(T+S)$

$$\|e^{is(T+S)}f - e^{is'(T+S)}f\|^2_{T+S} = \int (t^2+1)|e^{ist} - e^{is't}|^2 \, d\|E(t)f\|^2$$
$$\leq 4\int (t^2+1)\left|\sin\left(\frac{s-s'}{2}t\right)\right|^2 d\|E(t)f\|^2 \to 0 \quad \text{as} \quad s' \to s,$$

since the integrand is bounded by (t^2+1) (because of $f \in D(T+S)$ this function is integrable) and tends to 0 pointwise. Hence, for fixed $f \in D(T+S)$ and $r > 0$ the family of functions

$$\varphi_t : [-r, r] \to H, t \in \mathbb{R} \setminus \{0\},$$
$$\varphi_t(s) = t^{-1}(e^{itT}\,e^{itS} - e^{it(T+S)})e^{is(T+S)}f$$

is equicontinuous. Since $e^{is(T+S)}f \in D(T+S)$, we moreover have $\varphi_t(s) \to 0$ as $t \to 0$ for an arbitrary $s \in [-r, r]$. Consequently, φ_t uniformly converges to 0 on $[-r, r]$ as $t \to 0$. Let us now make the following estimates

$$\left\|\left[(e^{i(t/n)T}e^{i(t/n)S})^n - e^{it(T+S)}\right]f\right\|$$
$$= \left\|\sum_{k=0}^{n-1}(e^{i(t/n)T}\,e^{i(t/n)S})^k\left[e^{i(t/n)T}\,e^{i(t/n)S} - e^{i(t/n)(T+S)}\right](e^{i(t/n)(T+S)})^{n-1-k}f\right\|$$
$$\leq n \max_{|s| \leq |t|}\left\|\left[e^{i(t/n)T}\,e^{i(t/n)S} - e^{i(t/n)(T+S)}\right]e^{is(T+S)}f\right\|$$
$$= |t| \max_{|s| \leq |t|}\left\|\left(\frac{t}{n}\right)^{-1}\left[e^{i(t/n)T}\,e^{i(t/n)S} - e^{i(t/n)(T+S)}\right]e^{is(T+S)}f\right\|$$
$$\leq |t| \max\{\varphi_{t/n}(s) : s \in [-|t|, |t|]\}.$$

For fixed t the last expression tends to zero as $n \to \infty$, as we have already proved (choose $[-r, r] = [-|t|, |t|]$). This proves the convergence for $f \in D(T+S)$. Since $D(T+S)$ is dense and all operators have norm 1, the required strong convergence follows (cf. Theorem 4.23). □

If the operators S and T are bounded from below, then an analogous result holds for e^{-tT}, e^{-tS} and $e^{-t(T+S)}$ with $t \geqslant 0$.

Theorem 7.41. *Let T, S, and $T+S$ be self-adjoint operators on the Hilbert space H. Assume that these operators are bounded from below. Then*

$$e^{-t(T+S)} = s - \lim_{n\to\infty} \left[e^{-(t/n)T}\, e^{-(t/n)S}\right]^n$$

for all $t \geqslant 0$.

The proof follows that of Theorem 7.40. We consider only nonnegative t and s; the details can be left to the reader.

EXERCISES

7.50. Let U be a unitary operator.

(a) There exists a strongly continuous unitary group $\{U(t) : t \in \mathbb{R}\}$ for which $U(1) = U$ and whose infinitesimal generator has norm $\leqslant 2\pi$.
Hint: If G is the (complex) spectral family of U and we define $z^t = r^t\, e^{it\varphi}$ for $z = r\, e^{i\varphi}$ $(0 < r,\ 0 \leqslant \varphi < 2\pi$, and $t \in \mathbb{R})$, then we can choose $U(t) = \int z^t\, dG(z)$.

(b) Prove Theorem 7.36 with the aid of part (a) and Stone's theorem.

7.51. If T is self-adjoint and V is a T-bounded operator, then $t \mapsto V\, e^{itT} f$ is continuous for every $f \in D(T)$.

7.52. Let T_1 and T_2 be self-adjoint operators on H, and assume that the strong limit $W = s - \lim_{t\to\infty} e^{itT_2}\, e^{-itT_1}$ exists (W is called a *wave operator*; cf. Section 11.1). We have:

(a) W is isometric.

(b) $R(W)$ reduces e^{isT_2}.

(c) $R(W)$ reduces T_2.

(d) T_1 and $T_2|_{R(W)}$ are unitarily equivalent; $T_1 = W^{-1} T_2|_{R(W)} W$.

Self-adjoint extensions of symmetric operators 8

In Sections 5.4 and 5.5 we have already learned that certain symmetric operators (the semi-bounded and continuously invertible ones) possess self-adjoint extensions. The question of whether all (or which) symmetric operators have self-adjoint extensions could not be answered there. The key to our studies was the fact that $\lambda - T$ was continuously invertible for some $\lambda \in \mathbb{R}$; however, this is not always the case. In this chapter we develop the *von Neumann extension theory*, which completely answers this question. Moreover, we shall prove certain theorems about the spectra of all self-adjoint extensions of a symmetric operator.

8.1 Defect indices and Cayley transforms

First let T be an arbitrary linear operator on a Hilbert space H. The set

$$\Gamma(T) = \{z \in \mathbb{K} : \text{there exists a } k(z) > 0 \text{ such that} \\ \|(z-T)f\| \geq k(z)\|f\| \quad \text{for all} \quad f \in D(T)\}$$

is called the *regularity domain* of T.

Proposition.
1. *We have $z \in \Gamma(T)$ if and only if $(z - T)$ is continuously invertible. Then $\|(z-T)^{-1}\| \leq k(z)^{-1}$ (observe that $(z-T)^{-1}$ need not belong to $B(H)$).*
2. *If H is complex and T is Hermitian, then $\mathbb{C} \setminus \mathbb{R} \subset \Gamma(T)$.*
3. *If T is isometric, then $\mathbb{K} \setminus \{z \in \mathbb{K} : |z| = 1\} \subset \Gamma(T)$.*
4. *$\Gamma(T)$ is open.*

PROOF.
1. It follows from the inequality $\|(z-T)f\| \geq k(z)\|f\|$ for all $f \in D(T)$ that $z-T$ is injective and $\|(z-T)^{-1}\| \leq k(z)^{-1}$. If $(z-T)$ is injective and $(z-T)^{-1}$ is bounded, then
$$\|(z-T)f\| \geq \|(z-T)^{-1}\|^{-1}\|(z-T)^{-1}(z-T)f\| = \|(z-T)^{-1}\|^{-1}\|f\|.$$
Consequently, $z \in \Gamma(T)$ and we can choose $k(z) = \|(z-T)^{-1}\|^{-1}$.
2. For $z = a+ib$ and $f \in D(T)$
$$\|(z-T)f\|^2 = \|(a-T)f\|^2 + b^2\|f\|^2 \geq b^2\|f\|^2.$$
Hence, $z \in \Gamma(T)$ and we can choose $k(z) = |b|$.
3. If $|z| \neq 1$, then
$$\|(z-T)f\| \geq |\, \|Tf\| - |z|\, \|f\| \,| = |1-|z||\, \|f\|.$$
Consequently, $z \in \Gamma(T)$ and we can choose $k(z) = |1-|z||$.
4. Let $z_0 \in \Gamma(T)$. If $z \in \mathbb{K}$ such that $|z-z_0| < k(z_0)$, then
$$\|(z-T)f\| \geq \|(z_0-T)f\| - |z-z_0|\, \|f\| \geq (k(z_0) - |z-z_0|)\|f\|$$
for all $f \in D(T)$. Therefore, $z \in \Gamma(T)$, i.e., $\Gamma(T)$ is open. □

The subspace $R(z-T)^{\perp}$ is called the *defect space* of T and z. The cardinal number $\beta(T, z) = \dim R(z-T)^{\perp}$ is called the *defect index* of T and z.

Theorem 8.1. *The defect index $\beta(T, z)$ is constant on each connected subset of $\Gamma(T)$. If T is Hermitian, then the defect index is constant in the upper and lower half-planes.*

PROOF. It is sufficient to show that $\beta(T, z)$ is *locally constant* in $\Gamma(T)$, i.e., that for every $z_0 \in \Gamma(T)$ there exists an $\epsilon > 0$ such that $\beta(T, z) = \beta(T, z_0)$ for all $z \in \Gamma(T)$ with the property $|z-z_0| < \epsilon$. Replace A by $z_0 - T$ and B by I in Theorem 5.25, and write Q_z for the orthogonal projection onto $R(z-T)$. Then $\|Q_z - Q_{z_0}\| \to 0$ as $z \to z_0$. If P_z is the orthogonal projection onto $R(z-T)^{\perp}$, then we have
$$\|P_z - P_{z_0}\| = \|Q_z - Q_{z_0}\| \to 0 \quad \text{as} \quad z \to z_0.$$
If we choose $\epsilon > 0$ such that $\|P_z - P_{z_0}\| < 1$ for $|z-z_0| < \epsilon$, then it follows from Theorem 4.35 that
$$\beta(T, z) = \beta(T, z_0) \quad \text{for} \quad |z-z_0| < \epsilon.$$
If T is Hermitian, then the upper and lower half-planes are connected subsets of $\Gamma(T)$ (cf. Proposition 2); therefore, the defect index is constant there. □

If T is a Hermitian operator on a complex Hilbert space, then we define

$$\gamma_+(T) = \beta(T, \mathrm{i}), \quad \gamma_-(T) = \beta(T, -\mathrm{i})$$

(of course, here i can be replaced by an arbitrary $z \in \mathbb{C}$ for which $\operatorname{Im} z > 0$ and $-\mathrm{i}$ by an arbitrary $z \in \mathbb{C}$ for which $\operatorname{Im} z < 0$). The pair $(\gamma_+(T), \gamma_-(T)) = (\gamma_+, \gamma_-)$ is called the *defect indices* of T.

Remark. We can reformulate the result of Theorem 5.21 in terms of this definition: *A symmetric operator is essentially self-adjoint if and only if its defect indices are equal to* (0, 0). *A closed symmetric operator is self-adjoint if and only if its defect indices are equal to* (0, 0).

Let T be a symmetric operator on a complex Hilbert space. The *Cayley transform* of T is defined by the equality

$$V = (\mathrm{i} - T)(-\mathrm{i} - T)^{-1} = -(\mathrm{i} - T)(\mathrm{i} + T)^{-1}.$$

V is therefore a linear operator from $R(-\mathrm{i} - T)$ onto $R(\mathrm{i} - T)$.

Theorem 8.2. *Let T be a symmetric operator on the complex Hilbert space H. The Cayley transform of T is an isometric mapping of $R(-\mathrm{i} - T)$ onto $R(\mathrm{i} - T)$. The range $R(I - V)$ is dense in H, and $T = \mathrm{i}(I + V)(I - V)^{-1}$. In particular, T is uniquely determined by V.*

Proof. For every $g = (-\mathrm{i} - T)f \in R(-\mathrm{i} - T) = D(V)$ we have

$$\|Vg\|^2 = \|(\mathrm{i} - T)(-\mathrm{i} - T)^{-1}g\|^2 = \|(\mathrm{i} - T)f\|^2$$
$$= \|f\|^2 + \|Tf\|^2 = \|(-\mathrm{i} - T)f\|^2 = \|g\|^2.$$

Consequently, V is isometric. It is clear that $R(V) = R(\mathrm{i} - T)$, since $(-\mathrm{i} - T)^{-1}$ maps $D(V) = R(-\mathrm{i} - T)$ *onto* $D(T)$ and $(\mathrm{i} - T)$ maps $D(T)$ *onto* $R(\mathrm{i} - T)$. We have

$$I - V = I + (\mathrm{i} - T)(\mathrm{i} + T)^{-1} = [(\mathrm{i} + T) + (\mathrm{i} - T)](\mathrm{i} + T)^{-1} = 2\mathrm{i}(\mathrm{i} + T)^{-1},$$

$$I + V = I - (\mathrm{i} - T)(\mathrm{i} + T)^{-1} = 2T(\mathrm{i} + T)^{-1}.$$

In particular, $R(I - V) = D(T)$ is dense, $I - V$ is injective, and

$$T = \mathrm{i}(I + V)(I - V)^{-1}. \qquad \square$$

Remark. We could define a (generalized) Cayley transform

$$V_z = (z - T)(z^* - T)^{-1}$$

for every $z \in \mathbb{C}$ such that $\operatorname{Im} z > 0$. Then V_z is an isometric mapping of $R(z^* - T)$ onto $R(z - T)$, and

$$T = (z - z^* V_z)(I - V_z)^{-1}.$$

In what follows V can be replaced by V_z. We use $z = \mathrm{i}$, as this saves us from using unnecessary indices.

Theorem 8.3. *An operator V on the complex Hilbert space H is the Cayley transform of a symmetric operator T if and only if V has the following properties*:
(i) *V is an isometric mapping of $D(V)$ onto $R(V)$,*
(ii) *$R(I - V)$ is dense in H.*
The symmetric operator T is given by the equality $T = \mathrm{i}(I + V)(I - V)^{-1}$.

PROOF. If V is the Cayley transform of T, then V has properties (i) and (ii) by Theorem 8.2. We also have then that $T = \mathrm{i}(I + V)(I - V)^{-1}$. Let V now be an operator with properties (i) and (ii). Then $I - V$ is injective, since the equality $Vg = g$ implies that

$$\begin{aligned}\langle g, f - Vf\rangle &= \langle g, f\rangle - \langle g, Vf\rangle = \langle g, f\rangle - \langle Vg, Vf\rangle \\ &= \langle g, f\rangle - \langle g, f\rangle = 0 \quad \text{for all} \quad f \in D(V),\end{aligned}$$

i.e., that $g \in R(I - V)^{\perp}$, and thus $g = 0$. Therefore, we can define an operator T by the equality

$$T = \mathrm{i}(I + V)(I - V)^{-1}.$$

By assumption, $D(T) = R(I - V)$ is dense. For all $f = (I - V)f_1$ and $g = (I - V)g_1$ from $D(T) = R(I - V)$ we have

$$\begin{aligned}\langle Tf, g\rangle &= -\mathrm{i}\langle (I + V)(I - V)^{-1}f, g\rangle = -\mathrm{i}\langle (I + V)f_1, (I - V)g_1\rangle \\ &= -\mathrm{i}\{\langle f_1, g_1\rangle + \langle Vf_1, g_1\rangle - \langle f_1, Vg_1\rangle - \langle Vf_1, Vg_1\rangle\} \\ &= -\mathrm{i}\{\langle Vf_1, Vg_1\rangle + \langle Vf_1, g_1\rangle - \langle f_1, Vg_1\rangle - \langle f_1, g_1\rangle\} \\ &= \mathrm{i}\langle (I - V)f_1, (I + V)g_1\rangle = \mathrm{i}\langle f, (I + V)(I - V)^{-1}g\rangle \\ &= \langle f, Tg\rangle.\end{aligned}$$

Consequently, T is symmetric.

It remains to prove that V is the Cayley transform of T. This immediately follows from

$$\begin{aligned}(\mathrm{i} - T) &= \mathrm{i} - \mathrm{i}(I + V)(I - V)^{-1} = \mathrm{i}[(I - V) - (I + V)](I - V)^{-1} \\ &= -2\mathrm{i}V(I - V)^{-1}, \\ (-\mathrm{i} - T) &= -\mathrm{i}[(I - V) + (I + V)](I - V)^{-1} = -2\mathrm{i}(I - V)^{-1}. \qquad \square\end{aligned}$$

Theorem 8.4. *Let T be a symmetric operator on a complex Hilbert space, and let V denote its Cayley transform.*
(a) *The following statements are equivalent*:
  (i) *T is closed,*
  (ii) *V is closed,*

(iii) $D(V)=R(\mathrm{i}+T)$ *is closed,*
(iv) $R(V)=R(\mathrm{i}-T)$ *is closed.*

(b) *T is self-adjoint if and only if V is unitary.*

PROOF.
(a) (i) *is equivalent to* (iii) *and to* (iv): T is closed if and only if $(\pm\mathrm{i}-T)^{-1}$ is closed. The bounded operator $(\pm\mathrm{i}-T)^{-1}$ is closed if and only if its domain $D((\mathrm{i}-T)^{-1})=R(\mathrm{i}-T)=R(V)$ or $D((-\mathrm{i}-T)^{-1})=R(\mathrm{i}+T)=D(V)$ is closed (Theorem 5.6).
(ii) *is equivalent to* (iii): The bounded operator V is closed if and only if its domain is closed.

(b) T is self-adjoint if and only if $R(\mathrm{i}-T)=R(-\mathrm{i}-T)=H$, i.e., if and only if $D(V)=R(V)=H$. This is equivalent to the statement that V is unitary. □

For the construction of symmetric or self-adjoint extensions of a symmetric operator the following theorem is essential. The proof of this theorem is obvious.

Theorem 8.5. *Let T_1 and T_2 be symmetric operators on a complex Hilbert space, and let V_1 and V_2 denote their Cayley transforms. Then $T_1\subset T_2$ if and only if $V_1\subset V_2$.*

Consequently, we can obtain all symmetric extensions of a symmetric operator T if we determine all those extensions V' of the Cayley transform V of T which possess property (i) of Theorem 8.3 (since V has property (ii), V' automatically does, too) and we calculate the corresponding symmetric operators $T'=\mathrm{i}(I+V')(I-V')^{-1}$ (Theorems 8.5 and 8.3). We can obtain all self-adjoint extensions (provided that such exist) if we determine all unitary extensions V' (Theorems 8.5 and 8.4b). In particular, T has self-adjoint extensions if and only if V has unitary extensions. The following theorem makes it possible to explicitly construct the extensions V' of V. Here we assume, without loss of generality, that T is closed.

Theorem 8.6. *Let T be a closed symmetric operator on a complex Hilbert space, and let V denote its Cayley transform.*

(a) *V' is the Cayley transform of a closed symmetric extension T' of T if and only if the following holds: There exist closed subspaces F_- of $R(\mathrm{i}-T)^\perp$ and F_+ of $R(-\mathrm{i}-T)^\perp$ and an isometric mapping $\tilde{V}$ of F_+ onto F_- for which*

$$D(V')=R(-\mathrm{i}-T')=R(-\mathrm{i}-T)\oplus F_+.$$
$$V'(f+g)=Vf+\tilde{V}g\quad \textit{for}\quad f\in R(-\mathrm{i}-T),\, g\in F_+,$$
$$R(V')=R(\mathrm{i}-T')=R(\mathrm{i}-T)\oplus F_-.$$

The spaces F_- and F_+ have the same dimension.

(b) *The operator V' in part* (a) *is unitary* (*i.e., T' is self-adjoint*) *if and only if $F_- = R(i - T)^\perp$ and $F_+ = R(-i - T)^\perp$.*
(c) *T possesses self-adjoint extensions if and only if its defect indices are equal.*

PROOF.
(a) If V' has the given form, then V' is obviously an isometric mapping of $R(-i - T) \oplus F_+$ onto $R(i - T) \oplus F_-$. Consequently, V' satisfies assumption (i) of Theorem 8.3. Since $R(I - V)$ is dense, $R(I - V')$ is also dense, so that V' also satisfies (ii) of Theorem 8.3. Therefore, V' is the Cayley transform of a symmetric extension T' of T. Since $\tilde{V}$ is an isomorphism of F_+ onto F_-, we have $\dim F_+ = \dim F_-$. If V' is the Cayley transform of a symmetric extension T' of T, then put $F_- = R(i - T') \ominus R(i - T)$, $F_+ = R(-i - T') \ominus R(-i - T)$, and $\tilde{V} = V'|_{F_+}$.
(b) V' is unitary if and only if $D(V') = H = R(V')$, i.e., if and only if $F_+ = R(-i - T)^\perp$ and $F_- = R(i - T)^\perp$.
(c) By (a) and (b) the operator V possesses a unitary extension if and only if there exists an isometric mapping $\tilde{V}$ of $R(-i - T)^\perp$ onto $R(i - T)^\perp$. This happens if and only if $\dim R(-i - T)^\perp = \dim R(i - T)^\perp$. □

Theorem 8.7. *Let T be a symmetric operator on a complex Hilbert space. The operator T is essentially self-adjoint if and only if T has exactly one self-adjoint extension.*
(For the real case, compare Exercise 8.4.)

PROOF. If T is essentially self-adjoint, then $\bar{T}$ is the only self-adjoint extension of T by Theorem 5.31(c). We show: If T is not essentially self-adjoint, i.e., if $\bar{T}$ is not self-adjoint, then T has either no or infinitely many self-adjoint extensions. If the defect indices of $\bar{T}$ are different, then $\bar{T}$ (and thus T) has no self-adjoint extension. If the defect indices are equal (>0, as $\bar{T}$ is not self-adjoint), then there are infinitely many unitary mappings $\tilde{V} : R(-i - T)^\perp \to R(i - T)^\perp$ (proof!), and therefore infinitely many self-adjoint extensions. □

Now we are in a position to define certain classes of symmetric operators that have self-adjoint extensions.

Theorem 8.8. *Let T be a symmetric operator on a complex Hilbert space.*
(a) *If $\Gamma(T) \cap \mathbb{R} \neq \emptyset$, then T has self-adjoint extensions.*
(b) *If T is semibounded, then T has self-adjoint extensions.*
(The statements of this theorem have already been proved in another way in Sections 5.4 and 5.5.)

PROOF.
(a) $\Gamma(T)$ is connected, since $\Gamma(T) \cap \mathbb{R} \neq \emptyset$. Then $\gamma_+(T) = \gamma_-(T)$ by Theorem 8.1.
(b) Let T be bounded, for example, from below, and let c be a lower

bound of T. Then

$$\|(\lambda - T)f\| \geq \langle f, (T-\lambda)f\rangle \|f\|^{-1} \geq (c-\lambda)\|f\|$$

for $\lambda < c$ and all $f \in D(T)$, $f \neq 0$. Consequently, in this case we also have $\Gamma(T) \cap \mathbb{R} \neq \emptyset$. □

Let H be a complex Hilbert space. A mapping K of H onto itself is called a *conjugation* if

(a) $K(af + bg) = a^* Kf + b^* Kg$ for $f, g \in H$, $a, b \in \mathbb{C}$,
(b) $K^2 = I$. (8.1)
(c) $\langle Kf, Kg\rangle = \langle g, f\rangle$ for $f, g \in H$.

An operator T on H is said to be *K-real* if

(a) $KD(T) \subset D(T)$,
(b) $TKf = KTf$ for $f \in D(T)$. (8.2)

Theorem 8.9. *Let H be a complex Hilbert space, and let K be a conjugation on H. If T is a K-real symmetric operator on H, then T possesses self-adjoint extensions.*

Proof. It follows from (8.1(b)) and (8.2(a)) that $D(T) \supset KD(T) \supset K^2 D(T) = D(T)$. Consequently, $KD(T) = D(T)$. *If* $f \in R(\mathrm{i} - T)^\perp$, then

$$\langle Kf, (-\mathrm{i} - T)Kg\rangle = \langle Kf, K(\mathrm{i} - T)g\rangle = \langle (\mathrm{i} - T)g, f\rangle = 0$$

for all $g \in D(T)$. Therefore, $Kf \in R(-\mathrm{i} - T)^\perp$. We can show similarly that if $f \in R(-\mathrm{i} - T)^\perp$, then $Kf \in R(\mathrm{i} - T)^\perp$. As $K^2 = I$, we have $R(-\mathrm{i} - T)^\perp = KR(\mathrm{i} - T)^\perp$. Since $\{e_\alpha : \alpha \in A\}$ is an orthonormal basis of $R(\mathrm{i} - T)^\perp$ if and only if $\{Ke_\alpha : \alpha \in A\}$ is an orthonormal basis of $R(-\mathrm{i} - T)^\perp$, it follows that $\dim R(-\mathrm{i} - T)^\perp = \dim R(\mathrm{i} - T)^\perp$. □

Example 1. The formula

$$Kf(x) = f(x)^*, \quad f \in L_2(M)$$

defines a conjugation (the *natural conjugation*) on $L_2(M)$. The conditions given by (8.1) are obviously satisfied.

(8.3) *Let $G \subset \mathbb{R}^m$ be open and let T be defined on $L_2(G)$ by the equalities*

$$D(T) = C_0^\infty(G), \quad \textit{and} \quad Tf = -\Delta f + qf \quad \textit{for} \quad f \in D(T)$$

(here $\Delta = \sum_{j=1}^m (\partial^2/\partial x_j^2)$ denotes the Laplace differential form). Assume that the function q is real-valued and measurable on G and belongs to $L_{2,\mathrm{loc}}(G)$ (i.e., it is square integrable over every compact subset of G). Then T is obviously symmetric and K-real. Consequently, T has self-adjoint extensions by Theorem 8.9.

(8.4) *Let $K(.,.) : M \times M \to \mathbb{R}$, be a Hermitian Carleman kernel. Then the operator $T_{K,0}$ from Section 6.2 is symmetric and K-real. $T_{K,0}$ therefore possesses self-adjoint extensions.*

EXAMPLE 2. The formula

$$Kf(x) = f(-x)^*$$

defines a conjugation on $L_2(\mathbb{R}^m)$ and on $L_2\{x \in \mathbb{R}^m : -a_i < x_i < a_i\}$.

(8.5) *Let T be defined on $L_2(\mathbb{R})$ or on $L_2(-a, a)$ by the formulae*

$$D(T) = C_0^\infty(\mathbb{R}) \quad \textit{or} \quad D(T) = C_0^\infty(-a, a)$$

$$Tf = \frac{1}{\mathrm{i}} f' \quad \textit{for} \quad f \in D(T).$$

Then we obviously have $Kf \in D(T)$ and

$$TKf(x) = \frac{1}{\mathrm{i}} \frac{\mathrm{d}}{\mathrm{d}x} f(-x)^* = \frac{1}{-\mathrm{i}} f'(-x)^* = \frac{1}{-\mathrm{i}} (Kf')(x) = KTf(x)$$

for all $f \in D(T)$. Hence T is K-real, and thus possesses self-adjoint extensions.
(In Section 6.4 we could prove this only for the case of $L_2(\mathbb{R})$.)

EXAMPLE 3. The formula

$$K(f_1, f_2) = (f_1^*, f_2^*) \quad \text{for} \quad (f_1, f_2) \in L_2(M) \oplus L_2(M)$$

defines a conjugation on $L_2(M) \oplus L_2(M)$. The operator T defined by

$$D(T) = C_0^\infty(M) \oplus C_0^\infty(M),$$

$$T(f_1, f_2) = (f_2', -f_1') \quad \text{for} \quad (f_1, f_2) \in D(T)$$

is symmetric, since

$$\langle T(f_1, f_2), (g_1, g_2)\rangle = \int f_2'^* g_1 \, \mathrm{d}x - \int f_1'^* g_2 \, \mathrm{d}x$$

$$= -\int f_2^* g_1' \, \mathrm{d}x + \int f_1^* g_2' \, \mathrm{d}x = \langle (f_1, f_2), T(g_1, g_2)\rangle$$

for all $(f_1, f_2), (g_1, g_2) \in D(T)$. Moreover, T is obviously K-real. Therefore, T possesses self-adjoint extensions.

Theorem 8.10. *Let T be a closed symmetric operator on the complex Hilbert space H with equal finite defect indices (m, m). If T_1 and T_2 are self-adjoint extensions of T, then $(z - T_1)^{-1} - (z - T_2)^{-1}$ is of rank at most m for every $z \in \rho(T_1) \cap \rho(T_2)$. (Therefore, it is in $B_p(H)$ for all $p > 0$.)*

PROOF. Every $z \in \rho(T_1) \cap \rho(T_2)$ obviously belongs to $\Gamma(T)$; consequently, $R(z - T)^\perp$ is m-dimensional. Since $(z - T_1)^{-1} f = (z - T_2)^{-1} f = (z - T)^{-1} f$ for $f \in R(z - T)$, we have

$$(z - T_1)^{-1} - (z - T_2)^{-1} = ((z - T_1)^{-1} - (z - T_2)^{-1}) P,$$

where P denotes the projection onto $R(z-T)^{\perp}$. Hence, dim $R((z-T_1)^{-1} - (z-T_2)^{-1}) < m$. □

Exercises

8.1. For every self-adjoint operator T on a complex Hilbert space there exists a conjugation K for which T is K-real.
Hint: Use a spectral representation of T (Theorem 7.18) and the natural conjugation on $\oplus_\alpha L_2(\mathbb{R}, \rho_\alpha)$.

8.2. Let K denote the natural conjugation on $L_2(M)$. If T is a K-real self-adjoint operator on $L_2(M)$ and $(z-T)^{-1}$ is a Carleman operator for all $z \in \rho(T)$ (cf. Exercise 6.12) with kernel $k_z(x,y)$, then $k_z(x,y) = k_z(y,x)$ almost everywhere in $M \times M$.

8.2 Construction of self-adjoint extensions

In this section we wish to give the explicit construction of the self-adjoint extensions of a symmetric operator with equal defect indices. For a closed symmetric operator T on the complex Hilbert space H we set

$$N_+ = N(\mathrm{i} - T^*) = R(-\mathrm{i} - T)^{\perp},$$
$$N_- = N(-\mathrm{i} - T^*) = R(\mathrm{i} - T)^{\perp}.$$

Theorem 8.11 (The first formula of von Neumann). *Let T be a closed symmetric operator on a complex Hilbert space. Then*

$$D(T^*) = D(T) \dotplus N_+ \dotplus N_- \quad (\textit{direct sum}),$$
$$T^*(f_0 + g_+ + g_-) = Tf_0 + \mathrm{i}g_+ - \mathrm{i}g_- \quad \textit{for} \quad f_0 \in D(T), g_+ \in N_+, g_- \in N_-.$$

Proof. Since $N_+ \subset D(T^*)$ and $N_- \subset D(T^*)$, we obviously have $D(T) + N_+ + N_- \subset D(T^*)$. We show that we have equality here, i.e., every $f \in D(T^*)$ can be written in the form $f = f_0 + g_+ + g_-$ with $f_0 \in D(T)$, $g_+ \in N_+$, and $g_- \in N_-$. To this end, let $f \in D(T^*)$. Then by the projection theorem we can decompose $(-\mathrm{i} - T^*)f$ into its components in N_+ and in $N_+^{\perp} = R(-\mathrm{i} - T)$,

$$(-\mathrm{i} - T^*)f = (-\mathrm{i} - T)f_0 + g, \quad (-\mathrm{i} - T)f_0 \in R(-\mathrm{i} - T), g \in N_+.$$

Since $T^*f_0 = Tf_0$ and $T^*g = \mathrm{i}g$, we then have (with $g_+ = \mathrm{i}g/2$)

$$T^*(f - f_0 - g_+) = T^*f - Tf_0 + \frac{1}{2}g = -\mathrm{i}f + \mathrm{i}f_0 - \frac{1}{2}g$$
$$= -\mathrm{i}(f - f_0) + \mathrm{i}g_+ = -\mathrm{i}(f - f_0 - g_+).$$

If we set $g_- = f - f_0 - g_+$, then $g_- \in N_-$ and $f = f_0 + g_+ + g_-$.

It remains to prove that the sum is direct, i.e., that the relations $0=f_0+g_++g_-$, $f_0\in D(T)$, $g_+\in N_+$, and $g_-\in N_-$ imply $f_0=g_+=g_-=0$. It follows from the equality $0=f_0+g_++g_-$ that

$$0 = T^*(f_0+g_++g_-) = Tf_0 + \mathrm{i}g_+ - \mathrm{i}g_-.$$

We obtain from this that

$$(\mathrm{i}-T)f_0 = -\mathrm{i}(g_++g_-)+\mathrm{i}(g_+-g_-) = -2\mathrm{i}g_-$$

and

$$(-\mathrm{i}-T)f_0 = 2\mathrm{i}g_+;$$

consequently, $g_-\in N_-\cap R(\mathrm{i}-T)=\{0\}$, $g_+\in N_+\cap R(-\mathrm{i}-T)=\{0\}$. Therefore, $g_-=g_+=0$, and thus $f_0=0$, also. □

Theorem 8.12 (The second formula of von Neumann). *Let T be a closed symmetric operator on a complex Hilbert space.*

(a) *T' is a closed symmetric extension of T if and only if the following holds: There are closed subspaces F_+ of N_+ and F_- of N_- and an isometric mapping $\hat{V}$ of F_+ onto F_- such that*

$$D(T') = D(T) + \{g+\hat{V}g : g\in F_+\}$$

and

$$\begin{aligned} T'(f_0+g+\hat{V}g) &= Tf_0 + \mathrm{i}g - \mathrm{i}\hat{V}g \\ &= T^*(f_0+g+\hat{V}g) \quad \text{for} \quad f_0\in D(T),\, g\in F_+. \end{aligned}$$

(b) *T' is self-adjoint if and only if $F_+=N_+$ and $F_-=N_-$.*

Proof. This theorem immediately follows from Theorem 8.6 if we show that the operator T' of Theorem 8.6 can be represented in the above form. We have (with $\tilde{V}$ as in Theorem 8.6)

$$\begin{aligned} D(T') &= R(I-V') = (I-V')D(V') = (I-V')(D(V)+F_+) \\ &= (I-V)D(V) + (I-\tilde{V})F_+ \\ &= D(T) + \{g-\tilde{V}g : g\in F_+\}. \end{aligned}$$

The sum is direct, as $\{g-\tilde{V}g : g\in F_+\}\subset F_++F_-\subset N_++N_-$. Since $T'\subset T^*$, we have moreover that

$$T'(f_0+g-\tilde{V}_g) = T^*(f_0+g-\tilde{V}g) = Tf_0 + \mathrm{i}g + \mathrm{i}\tilde{V}g$$

for all $f_0\in D(\mathrm{T})$ and $g\in F_+$. The assertion follows by taking $\hat{V}=-\tilde{V}$. □

As long as the subspaces N_+ and N_- are known, this theorem enables us to determine all closed symmetric extensions (in particular, all self-adjoint extensions) of a symmetric operator.

Let T and T' be linear operators such that $T \subset T'$; then we say that T' is a *finite-dimensional* (*m-dimensional*) *extension* of T if the quotient space $D(T')/D(T)$ is finite-dimensional (m-dimensional). We also say that T is a *finite-dimensional* (*m-dimensional*) *restriction* of T'.

Theorem 8.13. *Let T be a closed symmetric operator on a complex Hilbert space, and let T' be a symmetric extension of T.*
(a) *T' is an m-dimensional extension if and only if F_+ (defined in Theorem 8.12) is m-dimensional.*
(b) *If T has defect indices (m, m), then a symmetric extension T' of T is self-adjoint if and only if T' is an m-dimensional extension of T.*

PROOF.
(a) As $D(T') = D(T) \dotplus (I + \hat{V})F_+$ is a direct sum, we have $\dim D(T')/D(T) = \dim(I + \hat{V})F_+$. Since $\hat{V}F_+ = F_- \subset N_-$, $F_+ \subset N_+$ and $N_- \cap N_+ = \{0\}$, we obviously have $\dim(I + V)F_+ = \dim F_+$.
(b) T' is an m-dimensional extension if and only if $\dim F_+ = \dim F_- = m$. Since $F_+ \subset N_+$ and $F_- \subset N_-$, this holds if and only if $F_+ = N_+$ and $F_- = N_-$. □

An operator T on H is said to be *maximal symmetric* if we have $T = A$ for every symmetric operator A such that $T \subset A$. As the closure of a symmetric operator is symmetric, every maximal symmetric operator is closed.

Theorem 8.14.
(a) *A closed symmetric operator T is maximal symmetric if and only if at least one of its defect indices is equal to 0.*
(b) *Every self-adjoint operator is maximal symmetric.*
(c) *Let T be a closed symmetric operator with equal finite defect indices. Then every maximal symmetric extension of T is self-adjoint.*

PROOF.
(a) By Theorem 8.12(a) we can construct proper symmetric extensions if and only if both defect indices are different from zero.
(b) This follows from (a), since for every self-adjoint operator both defect indices are equal to 0.
(c) An extension T' is maximal symmetric if and only if it has the form given in Theorem 8.12(a) with $F_+ = N_+$ or $F_- = N_-$. Dimensionality arguments then show that $F_+ = N_+$ and $F_- = N_-$, so that T' is self-adjoint. □

EXAMPLE 1. Consider the operator $\overline{T_{1,0}}$ of Section 6.4 (cf. Theorems 6.29 and 6.31) defined on $L_2(0, \infty)$ by

$$D(\overline{T_{1,0}}) = \{f \in W_{2,1}(0, \infty) : f(0) = 0\} \text{ and } \overline{T_{1,0}}f = \frac{1}{i}f' \text{ for } f \in D(\overline{T_{1,0}}).$$

We have $\overline{T_{1,0}^*} = T_{1,0}^* = T_1$, where

$$D(T_1) = W_{2,1}(0, \infty) \quad \text{and} \quad T_1 f = \frac{1}{i} f' \quad \text{for } f \in D(T_1).$$

Then $N_+ = N(i - T_1)$ is the set of those solutions of the differential equation $if - (f'/i) = 0$, hence of the differential equation $f + f' = 0$, that lie in $L_2(0, \infty)$. As the solutions of this differential equation are given by $f(x) = ce^{-x}$, we have $N_+ \neq \{0\}$. The subspace $N_- = N(-i - T_1)$ is the set of those solutions of the differential equation $-f + f' = 0$ that lie in $L_2(0, \infty)$. Consequently, $N_- = \{0\}$. Therefore, the defect indices of $\overline{T_{1,0}}$ are different, and thus $\overline{T_{1,0}}$ possesses no self-adjoint extension.

Example 2. Consider the operator $\overline{T_{1,0}}$ from Section 6.4 (cf. Theorem 6.31) defined on $L_2(a, b)$, $-\infty < a < b < \infty$ by the formulae

$$D(\overline{T_{1,0}}) = \{f \in W_{2,1}(a, b) : f(a) = f(b) = 0\}$$

and

$$\overline{T_{1,0}} f = \frac{1}{i} f' \text{ for } f \in D(\overline{T_{1,0}}).$$

We have $\overline{T_{1,0}^*} = T_{1,0}^* = T_1$, where

$$D(T_1) = W_{2,1}(a, b) \quad \text{and} \quad T_1 f = \frac{1}{i} f' \quad \text{for } f \in D(T_1).$$

We also have

$$N_+ = N(i - T_1) = L(e_+) \quad \text{with} \quad e_+(x) = \exp(b - x),$$

and

$$N_- = N(-i - T_1) = L(e_-) \quad \text{with} \quad e_-(x) = \exp(x - a).$$

The defect indices are therefore equal, and thus $\overline{T_{1,0}}$ has self-adjoint extensions. We want to construct these extensions. It is obvious that $\|e_+\| = \|e_-\|$, so that all unitary mappings of N_+ onto N_- are given by formula

$$V_\vartheta(ce_+) = ce^{i\vartheta} e_- \quad \text{for all} \quad c \in \mathbb{C} \quad (0 \leq \vartheta < 2\pi).$$

Consequently, all self-adjoint extensions S_ϑ of $\overline{T_{1,0}}$ are given by

$$D(S_\vartheta) = D(\overline{T_{1,0}}) + L\{e_+ + e^{i\vartheta} e_-\},$$

$$S_\vartheta f = T_1 f = \frac{1}{i} f, \quad \text{for} \quad f \in D(S_\vartheta), \quad (0 \leq \vartheta < 2\pi).$$

It is usual and convenient to describe the domains of differential operators as the restrictions of the maximal operators (here T_1) with the aid of boundary conditions.

(8.6) *We have*

$$D(S_\vartheta) = \{f \in D(T_1) : f(a) = \Theta(\vartheta) f(b)\},$$

where $\Theta(\vartheta) = (1 + e^{i\vartheta} e^{b-a})^{-1}(e^{b-a} + e^{i\vartheta})$. *The mapping* $\Theta : [0, 2\pi) \to \mathbb{C}$ *is bijective as a map from* $[0, 2\pi)$ *onto the unit circle.*

Proof. For every $f = f_0 + ce_+ + ce^{i\vartheta}e_- \in D(S_\vartheta)$ with $f_0 \in D(\overline{T_{1,0}})$ we have

$$\frac{f(a)}{f(b)} = \frac{ce_+(a) + ce^{i\vartheta}e_-(a)}{ce_+(b) + ce^{i\vartheta}e_-(b)} = \frac{e^{b-a} + e^{i\vartheta}}{1 + e^{i\vartheta}e^{b-a}}.$$

Consequently, $f(a) = \Theta(\vartheta) f(b)$. Now let $f(a) = \Theta(\vartheta) f(b)$; then we have

$$\begin{aligned}(f - ce_+ - ce^{i\vartheta}e_-)(a) &= f(a) - c(e^{b-a} + e^{i\vartheta}) = 0,\\ (f - ce_+ - ce^{i\vartheta}e_-)(b) &= \Theta(\vartheta)^{-1} f(a) - c(1 + e^{i\vartheta}e^{b-a})\\ &= \Theta(\vartheta)^{-1} f(a) - f(a)\Theta(\vartheta)^{-1} = 0\end{aligned}$$

with $c = f(a)(e^{b-a} + e^{i\vartheta})^{-1}$. Hence $f - ce_+ - ce^{i\vartheta}e_- \in D(\overline{T_{1,0}})$, and thus $f \in D(S_\vartheta)$. It is clear that Θ is a bijective map of $[0, 2\pi)$ onto the unit circle. □

(8.7) *The eigenvalues* λ_n *and the normalized eigenelements* f_n *of* S_ϑ *are given by the formulae*

$$\begin{aligned}\lambda_n &= (a-b)^{-1}(\alpha + 2\pi n),\\ f_n(x) &= c_n \exp(i\lambda_n x), \quad (n \in \mathbb{Z})\end{aligned}$$

where α *is chosen so that* $e^{i\alpha} = \Theta(\vartheta)$ *and the* c_n *are normalizing factors.*

Proof. λ and f are an eigenvalue and a corresponding eigenelement of S_ϑ if and only if

$$\lambda f = \frac{1}{i} f', \quad \text{and} \quad f(a) = \Theta(\vartheta) f(b).$$

As all solutions of the equation $\lambda f = f'/i$ have the form $f(x) = ce^{i\lambda x}$, we obtain from the boundary condition that $e^{i\lambda a} = \Theta(\vartheta) e^{i\lambda b}$. Therefore, $e^{i\lambda(a-b)} = \Theta(\vartheta) = e^{i\alpha}$. It follows from this that

$$\lambda(a-b) = \alpha \quad \text{modulo } 2\pi;$$

consequently,

$$\lambda_n = (a-b)^{-1}(\alpha + 2n\pi)$$

and

$$f_n(x) = c_n \exp(i\lambda_n x) \quad \text{for} \quad n \in \mathbb{Z}.$$ □

In the reasoning of Section 8.1 *K*-real symmetric operators played an important role; they possess self-adjoint extensions. Now we can show that they also possess *K*-real self-adjoint extensions.

Theorem 8.15. *Let T be a K-real symmetric operator on the complex Hilbert space H. For every K-real self-adjoint extension T' of T there exists an orthonormal basis $\{e_\alpha : \alpha \in A\}$ of N_+ such that*

$$V'\left(\sum c_\alpha e_\alpha\right) = \sum c_\alpha K e_\alpha \quad \textit{for} \quad \sum |c_\alpha|^2 < \infty$$

holds for the Cayley transform V' of T'. If $\{e_\alpha : \alpha \in A\}$ is an arbitrary orthonormal basis of N_+, then the unitary operator

$$\hat{V} : N_+ \to N_-, \hat{V}\left(\sum c_\alpha e_\alpha\right) = \sum c_\alpha K e_\alpha \quad \text{for} \quad \sum |c_\alpha|^2 < \infty$$

induces a K-real self-adjoint extension of T in the sense of Theorem 8.12.

PROOF. If $\{e_\alpha : \alpha \in A\}$ is an orthonormal basis of N_+, $\hat{V}$ is defined as in the theorem, and T' denotes the self-adjoint extension of T defined by $\hat{V}$, then

$$K\left(f_0 + \sum c_\alpha(e_\alpha + K e_\alpha)\right) = K f_0 + \sum c_\alpha^*(e_\alpha + K e_\alpha) \in D(T')$$

and

$$\begin{aligned} T'K\left(f_0 + \sum c_\alpha(e_\alpha + K e_\alpha)\right) &= TKf_0 + \sum c_\alpha^*(\mathrm{i}e_\alpha - \mathrm{i}Ke_\alpha) \\ &= K\left(Tf_0 + \sum c_\alpha(-\mathrm{i}Ke_\alpha + \mathrm{i}e_\alpha)\right) \\ &= KT'\left(f_0 + \sum c_\alpha(e_\alpha + Ke_\alpha)\right) \end{aligned}$$

for all $f_0 + \sum c_\alpha(e_\alpha + Ke_\alpha) \in D(T')$. Consequently, T' is K-real.

Let T' now be a K-real self-adjoint extension of T, and let V' be the Cayley transform of T'.[1] With the aid of Zorn's lemma we can show the existence of a maximal orthonormal system $\{e_\alpha : \alpha \in A\}$ in N_+ with the property

$$V'\left(\sum c_\alpha e_\alpha\right) = \sum c_\alpha K e_\alpha \quad \text{for} \quad \sum |c_\alpha|^2 < \infty.$$

Then the formulae

$$\begin{aligned} D(S) &= D(T) + \left\{\sum c_\alpha(e_\alpha + Ke_\alpha) : \sum |c_\alpha|^2 < \infty\right\}, \\ Sf &= T'f \quad \text{for} \quad f \in D(S) \end{aligned}$$

define a K-real symmetric extension of T (this can be proved as above). If we assume that $\{e_\alpha : \alpha \in A\}$ is not an orthonormal basis of N_+, then there is a non-zero element f of $R(-\mathrm{i} - S)^\perp$. Then $V'f \in R(\mathrm{i} - S)^\perp$, $Kf \in R(\mathrm{i} - S)^\perp$, and $KV'f \in R(-\mathrm{i} - S)^\perp$; consequently, $f + KV'f \in R(-\mathrm{i} - S)^\perp$. If $f + KV'f = 0$, then the orthonormal system $\{e_\alpha : \alpha \in A\}$ can be enlarged by taking the element $e = \mathrm{i}\|f\|^{-1}f$, since then $Kf = -V'f$ and thus $Ke = V'e$. If $f + KV'f \neq 0$, then we can choose $e = \|f + KV'f\|^{-1}(f + KV'f)$, since

$$V'(f + KV'f) = V'f + V'KV'f = V'f + Kf = K(f + KV'f).$$

[1] I am indebted to Dr. Jürgen Voigt for the following proof.

Here we have used the fact that $V' = (\mathrm{i} - T')(-\mathrm{i} - T')^{-1}$, $K(\mathrm{i} - T') = (-\mathrm{i} - T')K$, and $K(-\mathrm{i} - T')^{-1} = (\mathrm{i} - T')^{-1}(\mathrm{i} - T')K(-\mathrm{i} - T')^{-1} = (\mathrm{i} - T')^{-1}K(-\mathrm{i} - T')(-\mathrm{i} - T')^{-1} = (\mathrm{i} - T')^{-1}K$, so that $V'KV' = K$. Therefore, in both cases we obtain a contradiction to the maximality of the system $\{e_\alpha : \alpha \in A\}$. □

EXERCISES

8.3. Assume that T is a symmetric operator on a real Hilbert space H, the space $H_{\mathbb{C}}$ is the complexification of H, and K is the conjugation defined in $H_{\mathbb{C}}$ as in Exercise 7.25.
   (a) The complexification $T_{\mathbb{C}}$ of T is symmetric and K-real; $T_{\mathbb{C}}$ therefore possesses K-real self-adjoint extensions.
   (b) The K-real self-adjoint extensions S of $T_{\mathbb{C}}$ have the form $S = (T')_{\mathbb{C}}$, where the T' are self-adjoint extensions of T.
   (c) Every symmetric operator on a real Hilbert space has self-adjoint extensions.

8.4. A symmetric operator (on a real or complex Hilbert space) is essentially self-adjoint if and only if it has a unique self-adjoint extension. (The complex case was considered in Theorem 8.7.)

8.5. Let T be a symmetric operator. If T^n is maximal symmetric for some $n \in \mathbb{N}$, $n > 1$, then T^m is essentially self-adjoint for $m \in \mathbb{N}$, $m < n$, and $\overline{T^m} = \overline{T}^m$.
   Hint: First consider the complex case. $\overline{T}$ is self-adjoint by Theorem 5.22; the assumption and the inclusion $\overline{T^n} \subset \overline{T}^n$ imply that $\overline{T^n} = \overline{T}^n$; $D(T^n)$ is a core of $\overline{T}^m$ for $m < n$; therefore, $D(T^m)$ is a core of $\overline{T}^m$, too.

8.6 (a) If T is a K-real operator, then T^* is also K-real.
   (b) If T' is a symmetric extension of a K-real operator T, then T' is K-real if and only if $KD(T') \subset D(T')$.
   (c) Let T be a K-real symmetric operator, and let $\{e_\alpha : \alpha \in A\}$, $\{f_\alpha : \alpha \in A\}$ be orthonormal bases of N_+. The operators $\hat{V}_1, \hat{V}_2 : N_+ \to N_-$ defined by the formulae

$$\hat{V}_1\left(\sum c_\alpha e_\alpha\right) = \sum c_\alpha K e_\alpha, \quad \hat{V}_2\left(\sum c_\alpha f_\alpha\right) = \sum c_\alpha K f_\alpha \quad \text{for} \quad \sum |c_\alpha|^2 < \infty$$

   are equal (i.e., the K-real self-adjoint extensions induced by V_1 and V_2 are equal) if and only if $\langle e_\alpha, f_\beta \rangle$ is real for all $\alpha, \beta \in A$.

8.7. Let T be a symmetric K-real operator with defect (1, 1). Then every self-adjoint extension of T is also K-real.
   Hint: cf. Theorem 8.15.

8.3 Spectra of self-adjoint extensions of a symmetric operator

In this section we study what can be said about the spectra of the self-adjoint extensions of a given symmetric operator (with equal defect indices).

In what follows T will always be a closed symmetric operator. Let us set $n(T,\lambda)=\dim N(\lambda-T)$. If λ is an eigenvalue of T, then $n(T,\lambda)$ is the multiplicity of λ. If λ is not an eigenvalue, then $n(T,\lambda)=0$.

Theorem 8.16. *If T' is an m-dimensional extension of T, then*

$$\dim(N(\lambda-T')\ominus N(\lambda-T))\leqslant m.$$

If, in addition, $n(T,\lambda)<\infty$, then $n(T',\lambda)-n(T,\lambda)\leqslant m$.

PROOF. It is obvious that $N(\lambda-T)\subset N(\lambda-T')$. The formula

$$(N(\lambda-T')\ominus N(\lambda-T))\cap D(T)=\{0\}$$

implies

$$(N(\lambda-T')\ominus N(\lambda-T))\dotplus D(T)\subset D(T').$$

Therefore,

$$\dim(N(\lambda-T')\ominus N(\lambda-T))\leqslant\dim D(T')/D(T)=m.$$

If $n(T,\lambda)<\infty$, then the second assertion follows from this. □

Let $H_z=N(z-T)^{\perp}$ for every $z\in\mathbb{K}$. The operator T obviously maps $N(z-T)$ into itself. We also have $T(H_z\cap D(T))\subset H_z$, since for all $f\in H_z\cap D(T)$ and $g\in N(z-T)$

$$\langle g, Tf\rangle=\langle Tg,f\rangle=z^*\langle g,f\rangle=0,$$

i.e., $Tf\in N(z-T)^{\perp}=H_z$. Consequently, H_z is a reducing subspace of T. We denote by T_z the restriction of T to H_z, i.e.,

$$D(T_z)=H_z\cap D(T)\quad\text{and}\quad T_zf=Tf\quad\text{for}\quad f\in D(T_z).$$

It is obvious that $z-T_z$ is injective (as we have excluded exactly the null space). Being a restriction of a symmetric operator, T_z is Hermitian. $D(T_z)$ is dense in H_z, since any $f\in H_z$ that is orthogonal to $D(T_z)$ is also orthogonal to $D(T)=D(T_z)+N(z-T)$. Therefore, T_z is a symmetric operator on H_z. The operator T_z is closed.

In the following we call $S(T)=\mathbb{K}\setminus\Gamma(T)$ the *spectral kernel* of T. The set

$$S_e(T)=\{z\in\mathbb{K}:(z-T_z)^{-1}\text{ is unbounded or }\ n(T,z)=\infty\}$$

is called the *essential spectral kernel* of T.

Theorem 8.17. *Let T be a closed symmetric operator.*

(a) *We have $S_e(T)\subset S(T)\subset\mathbb{R}$ and $S(T)\subset\sigma(T)$.*
(b) *If T' is a closed symmetric extension of T, then $S(T)\subset S(T')$ and $S_e(T)\subset S_e(T')$.*
(c) *If T' is a finite-dimensional symmetric extension of T, then $S_e(T')=S_e(T)$.*
(d) *If T is self-adjoint, then $S(T)=\sigma(T)$ and $S_e(T)=\sigma_e(T)$.*

PROOF.

(a) If $\lambda \in S_e(T)$, then $\dim N(\lambda - T) = \infty$ or $(\lambda - T_\lambda)^{-1}$ is unbounded. It is clear that in both cases λ does not lie in $\Gamma(T)$, i.e., $\lambda \in S(T)$. It is also evident that $S(T) \subset \mathbb{R}$, since $\Gamma(T)$ contains the upper and lower half-planes. $(\lambda - T)$ is not continuously invertible for $\lambda \in S(T)$; therefore, $S(T) \subset \sigma(T)$.

(b) The inclusion $\Gamma(T') \subset \Gamma(T)$ is evident because of the definition of $\Gamma(T)$. Hence, $S(T) \subset S(T')$. We show that $S_e(T) \subset S_e(T')$. To this end, let $\lambda \in S_e(T)$. If $n(T', \lambda) = \infty$, then $\lambda \in S_e(T')$. Consequently, we can assume without loss of generality that $n(T, \lambda) \leqslant n(T', \lambda) < \infty$, i.e., that $(\lambda - T_\lambda)^{-1}$ is unbounded. Then there exists a sequence (f_n) in $D(T_\lambda)$ for which $\|f_n\| = 1$ and $(\lambda - T_\lambda)f_n \to 0$. The sequence (f_n) contains no convergent subsequence, since $f_{n_k} \to f$ would imply $\|f\| = 1$ and $T_\lambda f_{n_k} \to \lambda f$, i.e., $f \in D(T_\lambda)$ (as T_λ is closed) and $(\lambda - T_\lambda)f = 0$; this would contradict the injectivity of $\lambda - T_\lambda$. Let P now denote the orthogonal projection onto the finite-dimensional subspace $N(\lambda - T')$, and let $g_n = (I - P)f_n$. Since P is compact, there exist a subsequence (f_{n_k}) of (f_n) and an $h \in H$ for which $Pf_{n_k} \to h$. The sequence (g_{n_k}) with $g_{n_k} = (I - P)f_{n_k}$ is therefore not convergent, and

$$(\lambda - T'_\lambda)g_{n_k} = (\lambda - T')f_{n_k} - (\lambda - T')Pf_{n_k} = (\lambda - T)f_{n_k} \to 0.$$

Hence, $(\lambda - T'_\lambda)^{-1}$ is unbounded, and thus $\lambda \in S_e(T')$.

(c) It is sufficient to prove that $S_e(T') \subset S_e(T)$, since we already have $S_e(T) \subset S_e(T')$. Assume that $\lambda \notin S_e(T)$. Then we have to prove that $\lambda \notin S_e(T')$. It follows from $n(T, \lambda) < \infty$ by Theorem 8.16 that $n(T', \lambda) \leqslant n(T, \lambda) + m < \infty$. The operator $(\lambda - T_\lambda)^{-1}$ is continuous and closed; therefore, $R(\lambda - T) = R(\lambda - T_\lambda) = D((\lambda - T_\lambda)^{-1})$ is closed. Since T' is a finite-dimensional extension of T, there are finitely many elements $f_1, \ldots, f_l$ such that

$$R(\lambda - T') = R(\lambda - T) + L\{f_1, \ldots, f_l\}.$$

Therefore, $D(\lambda - T'_\lambda)^{-1} = R(\lambda - T'_\lambda) = R(\lambda - T')$ is closed by Theorem 3.4, and thus the closed operator $(\lambda - T'_\lambda)^{-1}$ is continuous. Hence, $\lambda \notin S_e(T')$.

(d) The equality $S(T) = \sigma(T)$ follows from the characterization of the spectral points of a self-adjoint operator given in Theorem 5.24. We show that $S_e(T) = \sigma_e(T)$. Let $\lambda \in S_e(T)$. If $n(T, \lambda) = \infty$, then $\lambda \in \sigma_e(T)$. If $(\lambda - T_\lambda)^{-1}$ is not continuous, then we show that λ is an accumulation point of the spectrum of T. If this were not the case, then there would be an $\epsilon > 0$ for which $(\lambda - \epsilon, \lambda + \epsilon) \cap \sigma(T) \subset \{\lambda\}$, and with the spectral family E of T we would then have

$$\|(\lambda - T_\lambda)f\|^2 = \|(\lambda - T)f\|^2 = \int_{|t-\lambda| \geqslant \epsilon} |\lambda - t|^2 \, d\|E(t)f\|^2 \geqslant \epsilon^2 \|f\|^2$$

for all $f \in N(\lambda - T)^{\perp} \cap D(T) = R(E(\{\lambda\}))^{\perp} \cap D(T)$, which contradicts the discontinuity of $(\lambda - T_\lambda)^{-1}$. Hence we have $\lambda \in \sigma_e(T)$ in this case, also, and thus $S_e(T) \subset \sigma_e(T)$.

Assume that $\lambda \in \sigma_e(T)$. If λ is an eigenvalue of infinite multiplicity, then $\lambda \in S_e(T)$. If λ is an accumulation point of $\sigma(T)$, then there is a sequence (f_n) for which

$$f_n \in R\Big(E\Big(\lambda + \frac{1}{n}\Big) - E(\lambda) + E(\lambda -) - E\Big(\lambda - \frac{1}{n}\Big)\Big), \quad \|f_n\| = 1.$$

We have $f_n \in R(E(\{\lambda\}))^{\perp} \cap D(T) = H_\lambda \cap D(T)$ and $(\lambda - T_\lambda)f_n = (\lambda - T)f_n \to 0$; consequently, $(\lambda - T_\lambda)^{-1}$ is not continuous, so that $\lambda \in S_e(T)$. □

Theorem 8.18. *Let T be a closed symmetric operator on a complex Hilbert space with equal finite defect indices. Then all self-adjoint extensions of T have the same essential spectrum. If some self-adjoint extension of T has a pure discrete spectrum, then all self-adjoint extensions of T do, too.*

Proof. The first assertion immediately follows from Theorem 8.17(c) and (d). The second assertion follows from the fact that the spectrum is discrete if and only if the essential spectrum is empty. □

Theorem 8.19. *Let T be a closed symmetric operator on a complex Hilbert space with equal finite defect indices (m, m) and assume that*

$$\|(\lambda - T)f\| \geq c\|f\| \quad \text{for all} \quad f \in D(T)$$

with some $\lambda \in \mathbb{R}$ and $c > 0$. Then every self-adjoint extension T' of T has the following property: $\sigma(T') \cap (\lambda - c, \lambda + c)$ contains only isolated eigenvalues with total multiplicity $\leq m$.

Proof. By the first proposition after Theorem 7.24 we only have to prove that $\dim R(E(\lambda + c -) - E(\lambda - c)) \leq m$ for the spectral family E of T'. Assume that $\dim R(E(\lambda + c -) - E(\lambda - c)) > m$. Since

$$\dim D(T')/D(T) = m \quad \text{and} \quad R(E(\lambda + c -) - E(\lambda - c)) \subset D(T'),$$

there exists an $f \in R(E(\lambda + c -) - E(\lambda - c)) \cap D(T)$, $f \neq 0$. For this f we have

$$c\|f\| \leq \|(\lambda - T)f\| = \left\{\int_{|\lambda - t| < c} |\lambda - t|^2 \, d\|E(t)f\|^2\right\}^{1/2} < c\|f\|,$$

which is a contradiction. □

Corollary 1. *Let T be a closed symmetric operator on a complex Hilbert space with finite defect indices (m, m), and let T_1 and T_2 be self-adjoint extensions of T. If $\sigma(T_1) \cap (a, b) = \varnothing$, then $\sigma(T_2) \cap (a, b)$ consists of only isolated eigenvalues of total multiplicity $\leq m$.*

PROOF. If $-\infty < a < b < \infty$, then T satisfies the assumptions of Theorem 8.19 with $\lambda = (a+b)/2$ and $c = (b-a)/2$, since for all $f \in D(T)$

$$\|(\lambda - T)f\|^2 = \|(\lambda - T_1)f\|^2 = \int_{|\lambda - t| \geq c} |\lambda - t|^2 \, \mathrm{d}\|E_1(t)f\|^2 \geq c^2 \|f\|^2,$$

where E_1 is the spectral family of T_1. The assertion therefore follows by taking $T' = T_2$. If (a, b) is unbounded, then Theorem 8.19 can be applied to every bounded subinterval $(a', b') \subset (a, b)$. □

Corollary 2. *If T is a closed symmetric operator on a complex Hilbert space, bounded from below with lower bound γ and finite defect indices (m, m), and T' is a self-adjoint extension of T, then $\sigma(T') \cap (-\infty, \gamma)$ consists of only isolated eigenvalues with total multiplicity $\leq m$.*

PROOF. Theorem 8.19 can be applied with any $\lambda < \gamma$ and $c = \gamma - \lambda$, since

$$\|(\lambda - T)f\| \geq \langle f, (T-\lambda)f \rangle \|f\|^{-1} \geq (\gamma - \lambda)\|f\|$$

for all $f \in D(T), f \neq 0$. □

EXERCISE

8.8. Let T be a closed symmetric operator with equal finite defect indices (m, m), and let T_1 and T_2 be self-adjoint extensions of T with spectral families E_1 and E_2. Then

$$\dim R(E_2(b-) - E_2(a)) \leq m + \dim R(E_1(b-) - E_1(a)).$$

Hint: Use Exercise 7.37.

8.4 Second order ordinary differential operators

In this section we would like to apply the results of Sections 8.1 to 8.3 to second order ordinary differential operators. This way we obtain part of the theory of Sturm-Lioville operators developed by Weyl, Titchmarsh, and Kodaira. For further results and examples we refer the reader to *Hellwig* [15] and *Jörgens-Rellich* [20].

Let (a, b) be a bounded or unbounded interval in $\mathbb{R}$, and let $r : (a, b) \to \mathbb{R}$ be a measurable and almost everywhere positive locally integrable function (i.e., let it be integrable over every compact subinterval of (a, b)). In the following we consider the Hilbert space $L_2(a, b, r)$. This is the space of (equivalence classes of) measurable functions f defined on (a, b) for which $\int_a^b |f(x)|^2 r(x) \, \mathrm{d}x < \infty$. The scalar product on $L_2(a, b, r)$ is

$$\langle f, g \rangle = \int_a^b f(x)^* g(x) r(x) \, \mathrm{d}x.$$

We denote the corresponding norm by $\| \cdot \|$. The formula $U_r : L_2(a, b, r)$

$\to L_2(a, b)$, $U_r f = r^{1/2} f$ defines an isomorphism of $L_2(a, b, r)$ onto $L_2(a, b)$; this shows, in particular, that $L_2(a, b, r)$ is a Hilbert space.

First we consider differential forms L of the type

$$Lf = \frac{1}{r}\{-(pf')' + \mathrm{i}sf' + \mathrm{i}(sf)' + qf\}, \tag{8.8}$$

where the coefficients p, q, r, s satisfy the following assumptions:

(8.9) (a) *p, q, r, and s are real-valued continuous functions defined on (a, b); moreover, p and s are continuously differentiable.*
(b) *$p(x) > 0$ and $r(x) > 0$ for all $x \in (a, b)$.*

L is said to be *regular at a* if $a > -\infty$ and the coefficients p, q, r and s can be continuously extended to $[a, b)$ with $p(a) > 0$ and $r(a) > 0$. *Regularity at b* ($b < \infty$) can be defined in a corresponding way. L is said to be *regular* if L is regular at a and b. L is said to be *singular at a* (*singular at b, singular*) if L is not regular at a (at b, at a or b).

We now define operators on $L_2(a, b, r)$ with the aid of a differential form L such as the one given by (8.8). The *maximal* operator T induced by L is defined by the formulae

$$D(T) = \{f \in L_2(a, b, r)\colon f \text{ is continuously differentiable, } f' \text{ is absolutely continuous on } (a, b), \text{ and } Lf \in L_2(a, b, r)\}^2$$

and

$$Tf = Lf \quad \text{for} \quad f \in D(T).$$

The *minimal* operator T_0 induced by L is defined by the formulae

$$D(T_0) = \{f \in D(T)\colon \textit{the support of } f \textit{ is compact and contained in } (a, b)\},$$

and

$$T_0 f = Tf \quad \text{for} \quad f \in D(T_0).$$

Theorem 8.20. *Let L be as in (8.8). The operator T_0 is symmetric. If $s = 0$, then T_0 has equal defect indices, i.e., T_0 has self-adjoint extensions.*

PROOF. The Hermitian character of T_0 follows by integration by parts. $D(T_0)$ is dense, because $C_0^\infty(a, b) \subset D(T_0)$. Therefore, T_0 is symmetric. If $s = 0$, then T_0 is K-real for the natural conjugation on $L_2(a, b, r)$ ($Kf = f^*$). The assertion follows from this by Theorem 8.9. □

If $z \in \mathbb{C}$ and $g : (a, b) \to \mathbb{C}$ is a locally integrable function, then we say that $f : (a, b) \to \mathbb{C}$ is a solution of the equation $(L - z)f = g$ if f is continuously differentiable, f' is absolutely continuous, and $(L - z)f(x) = g(x)$

[2] Since f' is absolutely continuous, pf' is also absolutely continuous. Let $(pf')'$ be the derivative of pf' in the sense of Appendix A5.

almost everywhere in (a, b). Every solution f of the homogeneous equation $(L-z)f=0$ is obviously twice continuously differentiable, since $(pf')'$ is continuous in this case.

The solutions of the homogeneous equation $(L-z)f=0$ constitute a two-dimensional (complex) vector space.[3] Two solutions u_1, u_2 constitute a fundamental system (i.e., they are linearly independent) if the *modified Wronskian determinant*

$$W(u_1, u_2, x) = p(x)\det\begin{pmatrix} u_1(x) & u_2(x) \\ u_1'(x) & u_2'(x) \end{pmatrix} = p(x)(u_1(x)u_2'(x) - u_1'(x)u_2(x))$$

does not vanish for some (and then for all) $x \in (a, b)$.

If $g:(a, b) \to \mathbb{C}$ is locally integrable and u_1, u_2 is a fundamental system for the equation $(L-z)u=0$, then the solutions h of the equation $(L-z)h = g$ are given by the formula

$$h(x) = c_1u_1(x) + c_2u_2(x) + u_1(x)\int_c^x W(u_1, u_2, y)^{-1}u_2(y)g(y)r(y)\,dy$$

$$-u_2(x)\int_c^x W(u_1, u_2, y)^{-1}u_1(y)g(y)r(y)\,dy, \tag{8.10}$$

where $c \in (a, b)$ and $c_1, c_2 \in \mathbb{C}$.

For continuously differentiable functions $f, g:(a, b) \to \mathbb{C}$ we define

$$[f, g]_x = p(x)(f'(x)^*g(x) - f(x)^*g'(x)) + 2is(x)f(x)^*g(x)$$

for $x \in (a, b)$. If, in addition, f' and g' are absolutely continuous, then

$$\int_\alpha^\beta \{f(x)^*Lg(x) - (Lf(x))^*g(x)\}r(x)\,dx = [f, g]_\beta - [f, g]_\alpha \tag{8.11}$$

for $[\alpha, \beta] \subset (a, b)$. It follows from this that for $f, g \in D(T)$ the limits $[f, g]_a = \lim_{x\to a+}[f, g]_x$ and $[f, g]_b = \lim_{x\to b-}[f, g]_x$ exist. We have

$$\langle f, Tg\rangle - \langle Tf, g\rangle = [f, g]_b - [f, g]_a \quad \text{for all } f, g \in D(T). \tag{8.12}$$

Theorem 8.21. *Let $L_{2,0}(a, b, r)$ be the subspace of those functions in $L_2(a, b, r)$ that vanish almost everywhere near a and b. Then*

$$R(T_0) = \Big\{k \in L_{2,0}(a, b, r) : \int_a^b u(x)^*k(x)r(x)\,dx = 0$$

$$\textit{for every solution } u \textit{ of the equation } Lu = 0\Big\}.$$

Proof. We denote the subspace on the right hand side by R. For $f \in D(T_0)$ and for every solution u of the equation $Lu=0$ we obtain via integration

[3] Concerning the results mentioned here about ordinary differential equations we refer to the textbooks on this subject, for example, *Knobloch-Kappel* [23], Chapter I.

by parts that

$$\int_a^b u(x)^*(T_0 f)(x)r(x)\,dx = \int_a^b (Lu)(x)^* f(x)r(x)\,dx = 0.$$

Therefore, $R(T_0) \subset R$. Now let $k \in R$, and let $[\alpha, \beta]$ be a compact subinterval of (a, b) with the property that k vanishes outside $[\alpha, \beta]$. For $c \in (a, \alpha)$ and $c_1 = c_2 = 0$ let h be the solution of the equation $Lh = k$ given by (8.10) for $z = 0$. Then h' is absolutely continuous, and $h(x) = 0$ for $x \in (a, \alpha)$. For every solution u of the equation $Lu = 0$ and for every $x_0 \in (a, \alpha)$, $x \in (\beta, b)$

$$\begin{aligned}[u, h]_x &= [u, h]_x - [u, h]_{x_0} \\ &= \int_{x_0}^x \{u(y)^* k(y) - (Lu(y))^* h(y)\} r(y)\,dy = 0.\end{aligned}$$

As this holds for every solution u of the equation $Lu = 0$, it follows that $h(x) = 0$ for all $x \in (\beta, b)$ (if we choose the solution u for which $u(x) = 0$ and $u'(x) = 1$). Therefore, $h \in D(T_0)$ and $T_0 h = k$. □

Theorem 8.22. *We have $T_0^* = T$. The operator T_0 is essentially self-adjoint if and only if T is symmetric. Then $\overline{T}_0 = T$.*

Proof. Integration by parts shows that T_0 and T are formal adjoints of each other. To prove that $T_0^* = T$, it remains to prove that $D(T_0^*) \subset D(T)$. Let $f \in D(T_0^*)$. Then $g = T_0^* f$ is locally integrable. Let h be a solution of the equation $Lh = g$. Then

$$\begin{aligned}&\int_a^b (f(x) - h(x))^*(T_0 k)(x)r(x)\,dy \\ &\qquad = \int_a^b ((T_0^* f)(x) - (Lh)(x))^* k(x)r(x)\,dx = 0\end{aligned}$$

for all $k \in D(T_0)$. Hence, $N(F) \supset R(T_0)$ for the functional

$$F : L_{2,0}(a, b, r) \to \mathbb{C}, \quad l \mapsto \int_a^b (f(x) - h(x))^* l(x)r(x)\,dx.$$

Consequently, by Theorem 8.21 and Theorem 4.1 we have $F = c_1 F_1 + c_2 F_2$ with appropriate $c_1, c_2 \in \mathbb{C}$ and

$$F_j : L_{2,0}(a, b, r) \to \mathbb{C}, \quad l \mapsto \int_a^b u_j(x)^* l(x)r(x)\,dx \quad \text{for} \quad j = 1, 2,$$

where u_1, u_2 is a fundamental system of the differential equation $Lu = 0$. This implies (compare with the proof of Theorem 6.29) that

$$f(x) - h(x) = c_1 u_1(x) + c_2 u_2(x) \quad \text{almost everywhere in} \quad (a, b).$$

Hence, f is a solution of the equation $Lf = g$. Since $f \in L_2(a, b, r)$, it follows that $f \in D(T)$. The rest follows from Theorem 5.20. □

Theorem 8.23. *The defect index* $\gamma_+ = \gamma_+(T_0)$ $(\gamma_- = \gamma_-(T_0))$ *is equal to the number of linearly independent solutions of the equation* $(L+\mathrm{i})u = 0$ $((L-\mathrm{i})u=0)$ *that lie in* $L_2(a, b, r)$. *If L is regular, then the defect indices of* T_0 *are equal to* (2, 2).

PROOF. We have $R(\mathrm{i}-T_0)^\perp = N(\mathrm{i}+T)$ and $R(-\mathrm{i}-T_0)^\perp = N(-\mathrm{i}+T)$. Furthermore, $N(\pm\mathrm{i}+T)$ is equal to the set of those solutions of the equation $(L\pm\mathrm{i})u=0$ that lie in $L_2(a, b, r)$. If L is regular, then every solution of the equation $(L\pm\mathrm{i})u=0$ is in $L_2(a, b, r)$. Consequently, $\dim N(\pm\mathrm{i}+T)=2$. □

Hence for the defect indices γ_+ and γ_- of T_0 there are only three possible values: 0, 1, and 2. If the defect indices are (0, 0), then $\overline{T}_0 = T$ is self-adjoint. If the defect indices are equal and different from zero, then $(\gamma_+, \gamma_-) = (\gamma, \gamma) = (1, 1)$ or (2, 2). Consequently, by Theorem 8.13(b) every γ-dimensional symmetric extension of $\overline{T}_0$ is self-adjoint.

Theorem 8.18 and Corollary 2 to Theorem 8.19 immediately imply

Theorem 8.24. *Let L be a differential form such as in* (8.8). *All self-adjoint extensions of* T_0 *have the same essential spectrum. If* T_0 *is semibounded, then all self-adjoint extensions of* T_0 *are semibounded.*

Theorem 8.25. *Let L be a regular differential form of the kind* (8.8). *Then we have the following:*

(a) *For every* $f \in D(T)$ *the functions f and f′ are continuously extendible to* $[a, b]$. *For* $f, g \in D(T)$ *we have*

$$[f, g]_x = p(x)(f'(x)^* g(x) - f(x)^* g'(x)) + 2isf(x)^* g(x) \quad \text{for all} \quad x \in [a, b]$$

(b) *We have* $D(\overline{T}_0) = \{f \in D(T) : f(a) = f'(a) = f(b) = f'(b) = 0\}$.

PROOF.

(a) If $f \in D(T)$ and $g = Tf$, then f can be represented in the form (8.10) with a fundamental system u_1, u_2 of the equation $Lu=0$. As the functions u_j and u_j' are continuously extendible to $[a, b]$, the same follows also for f from this representation. The rest can be obtained from the definitions of $[.\,,.]_a$ and $[.\,,.]_b$.

(b) There exists a $\varphi \in D(T)$ such that $\varphi(a)=0$, $\varphi'(a)=1$ and $\varphi(x)=0$ for x near b (it is enough to choose φ twice continuously differentiable). Then for every $f \in D(\overline{T}_0)$ it follows from part (a) that

$$0 = \langle \varphi, \overline{T}_0 f\rangle - \langle T\varphi, f\rangle = [\varphi, f]_b - [\varphi, f]_a = -p(a)f(a),$$

and thus $f(a)=0$. If we now choose φ such that $\varphi(a)=1$ and $\varphi'(a)=0$, then we find that $f'(a)=0$, too. We can show similarly that $f(b)=f'(b)=0$. Since $\overline{T}_0\subset T$, we therefore obtain that

$$D(\overline{T}_0)\subset D_0 = \{f\in D(T): f(a)=f'(a)=f(b)=f'(b)=0\}.$$

Let $\tilde{T}_0$ be defined by the formulae $D(\tilde{T}_0)=D_0$ and $\tilde{T}_0 f=Tf$ for $f\in D(\tilde{T}_0)$. Then by part (a)

$$\langle f, Tg\rangle - \langle \tilde{T}_0 f, g\rangle = [f,g]_b - [f,g]_a = 0$$

for all $f\in D(\tilde{T}_0)$ and $g\in D(T)$, i.e., $\tilde{T}_0$ and T are formal adjoints of each other. Hence, $\tilde{T}_0\subset T^* = \overline{T}_0$, and thus $D_0\subset D(\overline{T}_0)$. □

We perform the construction of self-adjoint extensions only for the case $s=0$. Hence, in the following we only consider *Sturm-Liouville differential forms*

$$Lf = \frac{1}{r}\{-(pf')'+qf\}, \tag{8.13}$$

where p, q, and r satisfy the assumptions (8.9(a)) and (8.9(b)). In this case T_0 always possesses self-adjoint extensions by Theorem 8.20.

We have $(d/dx)W(u_1, u_2, x)=0$ for any two solutions u_1, u_2 of the equation $(L-z)u=0$, as can be easily verified. Therefore, $W(u_1, u_2, x)$ is constant in (a, b). We briefly write $W(u_1, u_2)$ for this value.

For any two continuously differentiable functions $f, g:(a,b)\rightarrow\mathbb{C}$ and $x\in(a,b)$ we have now

$$[f,g]_x = p(x)(f'(x)^*g(x)-f(x)^*g'(x)) = -W(f^*, g, x). \tag{8.14}$$

Corresponding assertions hold in the regular case for $x=a$ and $x=b$, respectively.

Theorem 8.26. *Let L be a regular Sturm-Liouville differential form as in* (8.13).

(a) *The formulae*

$$D(T_{\alpha,\beta}) = \left\{ f\in D(T): \begin{matrix} f(a)\cos\alpha - f'(a)\sin\alpha = 0 \\ f(b)\cos\beta - f'(b)\sin\beta = 0 \end{matrix}\right\},$$

$$T_{\alpha,\beta}f = Tf \quad \text{for} \quad f\in D(T_{\alpha,\beta})$$

define a self-adjoint extension of T_0 for arbitrary $\alpha, \beta\in[0,\pi)$.[4]

[4] The boundary conditions occurring here are called "separated boundary conditions", since every boundary condition affects only one boundary point. There are also "mixed boundary conditions" that define self-adjoint extensions of T_0 (cf. Exercises 8.10 and 8.11).

(b) *For every* $z \in \rho(T_{\alpha,\beta})$ *the resolvent* $R_z = (z - T_{\alpha,\beta})^{-1}$ *has the form*

$$R_z g(x) = W(u_a, u_b)^{-1}\Big\{u_b(x)\int_a^x u_a(y)g(y)r(y)\,dy + u_a(x)\int_x^b u_b(y)g(y)r(y)\,dy\Big\},$$

where u_a *and* u_b *are non-trivial solutions of the equation* $(L-z)u=0$ *that satisfy the boundary condition at* a *and* b*, respectively (hence, for example,* $u_a(a)=\sin\alpha$, $u_a'(a)=\cos\alpha$, $u_b(b)=\sin\beta$, *and* $u_b'(b)=\cos\beta$*).*

(c) *The operators* $T_{\alpha,\beta}$ *have pure discrete spectrum. Every eigenvalue is simple.*

PROOF.

(a) We can verify easily that $T_{\alpha,\beta}$ is symmetric. Hence, it is sufficient to show that $T_{\alpha,\beta}$ is an (at least) two-dimensional extension of $\bar{T}_0$. To this end, let us choose φ_a and φ_b from $D(T)$ in such a way that we have

$$\varphi_a(a)=\sin\alpha,\ \varphi_a'(a)=\cos\alpha,\quad \varphi_a(x)=0 \text{ near } b,$$
$$\varphi_b(b)=\sin\beta,\ \varphi_b'(b)=\cos\beta,\quad \varphi_b(x)=0 \text{ near } a.$$

These elements obviously lie in $D(T_{\alpha,\beta})$ and are linearly independent modulo $D(\bar{T}_0)$, i.e., no non-trivial linear combination lies in $D(\bar{T}_0)$.

(b) The functions u_a and u_b are linearly independent, since otherwise u_a would fulfill the boundary conditions at *a and b*, and z would be an eigenvalue of $T_{\alpha,\beta}$, which would contradict the relation $z\in\rho(T_{\alpha,\beta})$. Therefore, $W(u_a,u_b)\neq 0$. Let K be the integral operator given in the theorem. If we define U_r as at the beginning of this section, then $U_rKU_r^{-1}$ is an integral operator on $L_2(a,b)$ with kernel

$$k(x,y)=\begin{cases} W(u_a,u_b)^{-1}r^{1/2}(x)u_b(x)u_a(y)r^{1/2}(y) & \text{for } x>y,\\ W(u_a,u_b)^{-1}r^{1/2}(x)u_a(x)u_b(y)r^{1/2}(y) & \text{for } x<y.\end{cases}$$

The function k obviously belongs to $L_2[(a,b)\times(a,b)]$, i.e., K is a Hilbert-Schmidt operator. Hence, K belongs to $B(L_2(a,b,r))$. For all $g\in L_2(a,b,r)$ we have

$$Kg(x) = cu_a(x) + W(u_a,u_b)^{-1}\Big\{u_b(x)\int_a^x u_a(y)g(y)r(y)\,dy - u_a(x)\int_a^x u_b(y)g(y)r(y)\,dy\Big\} \tag{8.15}$$

with $c=W(u_a,u_b)^{-1}\int_a^b u_b(y)g(y)r(y)\,dy$. Therefore, by (8.10) Kg is a solution of the equation $(z-L)u=g$. We can infer from (8.15) that for $g\in L_{2,0}(a,b,r)$ the function Kg is a multiple of u_a in a neighborhood of a and a multiple of u_b in a neighborhood of b. Consequently, $Kg\in D(T_{\alpha,\beta})$. The operator K therefore coincides with R_z on $L_{2,0}(a,b,r)$. Since $L_{2,0}(a,b,r)$ is dense and since the operators K and R_z are continuous, it follows that $K=R_z$.

(c) R_z is a normal and injective Hilbert-Schmidt operator (cf. the proof of part (b)). Consequently, there exist an orthonormal basis $\{f_n : n \in \mathbb{N}\}$ in $L_2(a, b, r)$ and a null-sequence (z_n) for which $z_n \neq 0$ $(n \in \mathbb{N})$ and

$$R_z f = \sum_{n \in \mathbb{N}} z_n \langle f_n, f \rangle f_n \quad \text{for all} \quad f \in L_2(a, b, r).$$

It follows from this that

$$T_{\alpha, \beta} f = \sum_{n \in \mathbb{N}} (z - z_n^{-1}) \langle f_n, f \rangle f_n \quad \text{for all} \quad f \in D(T_{\alpha, \beta}).$$

Consequently, $\sigma(T_{\alpha, \beta}) = \{z - z_n^{-1} : n \in \mathbb{N}\}$. As every solution of the equation $(\lambda - T_{\alpha, \beta})u = 0$ is determined by the boundary condition at one boundary point up to a constant factor, every eigenvalue is simple. Hence, $\sigma(T_{\alpha, \beta}) = \sigma_d(T_{\alpha, \beta})$. □

Let us now turn to the singular case (more precisely, the not necessarily regular case).

Theorem 8.27 (The Weyl alternative). *Let L be a Sturm-Liouville differential form defined on (a, b), and let $c \in (a, b)$. Either every solution u of the equation $(L - z)u = 0$ lies in $L_2(c, b, r)$ for every $z \in \mathbb{C}$ or for every $z \in \mathbb{C}$ there exists at least one solution u of the equation $(L - z)u = 0$ for which $u \notin L_2(c, b, r)$. In the second case, for every $z \in \mathbb{C} \setminus \mathbb{R}$ there exists (up to a factor) exactly one solution u of the equation $(L - z)u = 0$ for which $u \in L_2(c, b, r)$.*

According to H. Weyl we say in the first case that we have the *limit circle case* (LCC) at b; in the second case we say that we have the *limit point case* (LPC) at b. The terminology can be explained from the original construction of Weyl (cf. *H. Weyl* [56]; cf. also *Hellwig* [15] and *Jörgens-Rellich* [20]). A corresponding theorem holds for the boundary point a. The limit circle case at a and limit point case at a are defined similarly.

PROOF. In order to prove the alternative, it is sufficient to show the following: If there exists a $z_0 \in \mathbb{C}$ such that $u \in L_2(c, b, r)$ for every solution u of the equation $(L - z_0)u = 0$, then this holds for all $z \in \mathbb{C}$. Let v_1, v_2 be a fundamental system of the equation $(L - z_0)v = 0$. (We can assume, without loss of generality, that $W(v_1, v_2) = 1$.) We have $(L - z_0)u = (z - z_0)u$ for every solution u of the equation $(L - z)u = 0$. It follows from this by (8.10) that

$$\begin{aligned} u(x) = {} & c_1 v_1(x) + c_2 v_2(x) \\ & + (z - z_0) \int_c^x \{v_1(x) v_2(y) - v_2(x) v_1(y)\} u(y) r(y)\, dy. \end{aligned}$$

With $v=|v_1|+|v_2|$, $c=\max\{|c_1|, |c_2|\}$ and $M=2|z-z_0|^2\int_c^b v(y)^2 r(y)\,dy$ it follows that

$$|u(x)|^2 \leq 2c^2 v(x)^2 + 2|z-z_0|^2 v(x)^2 \int_c^x v(y)^2 r(y)\,dy \int_c^x |u(y)|^2 r(y)\,dy$$
$$\leq 2c^2 v(x)^2 + M v(x)^2 \int_c^x |u(y)|^2 r(y)\,dy.$$

As $v \in L_2(c, b, r)$, there exists a $d \in (c, b)$ such that

$$\int_d^b v(x)^2 r(x)\,dx \leq (2M)^{-1}.$$

Consequently, for all $x_1 \in (d, b)$

$$\int_d^{x_1} |u(x)|^2 r(x)\,dx \leq 2c^2 \int_d^{x_1} v(x)^2 r(x)\,dx$$
$$+ M \int_d^{x_1} v(x)^2 r(x) \left\{ \int_c^{x_1} |u(y)|^2 r(y)\,dy \right\} dx$$
$$\leq 2c^2 \int_d^b v(x)^2 r(x)\,dx + \tfrac{1}{2} \int_c^{x_1} |u(y)|^2 r(y)\,dy,$$

and thus

$$\int_d^{x_1} |u(x)|^2 r(x)\,dx \leq 4c^2 \int_d^b v(x)^2 r(x)\,dx + \int_c^d |u(y)|^2 r(y)\,dy.$$

This implies that $u \in L_2(d, b, r)$, and thus $u \in L_2(c, b, r)$, as well, since u is continuous on (a, b).

Let us now assume that we have the limit point case at b, i.e., that for every $z \in \mathbb{C}$ there exists (up to a constant factor) at most one solution u of the equation $(L-z)u=0$ such that $u \in L_2(c, b, r)$. It remains to prove that for every $z \in \mathbb{C} \setminus \mathbb{R}$ there exists at least one solution with this property. For this we consider the differential form L on the interval (c, b). The form L is obviously regular at c. Let T_0 and T be the minimal and the maximal operators on $L_2(c, b, r)$ induced by L. It is sufficient to show that T is not symmetric, since this implies that T_0 has positive defect indices, i.e., that $N(z-T) \neq \{0\}$ for every $z \in \mathbb{C} \setminus \mathbb{R}$. In order to prove this we use two twice continuously differentiable functions $\varphi_1, \varphi_2 : [c, b) \to \mathbb{C}$ for which

$$\varphi_1(c) = 1,\ \varphi_1'(c) = 0,\ \varphi_1(x) = 0 \quad \text{near} \quad b,$$
$$\varphi_2(c) = 0,\ \varphi_2'(c) = 1,\ \varphi_2(x) = 0 \quad \text{near} \quad b,$$

Then $\varphi_1, \varphi_2 \in D(T)$ and

$$\langle \varphi_1, T\varphi_2 \rangle - \langle T\varphi_1, \varphi_2 \rangle = -[\varphi_1, \varphi_2]_c = p(c) \neq 0,$$

i.e., T is not symmetric. □

Auxiliary theorem 8.28. *Let L be a Sturm-Liouville differential form on (a, b).*

(a) $[f, g]_a = 0$ *for* $f \in D(\bar{T}_0)$ *and* $g \in D(T)$.

(b) *Let us have the limit circle case at a. If u is a solution of the equation* $(L-z)u=0$ *for some* $z\in\mathbb{C}$, *the function* u_0 *is twice continuously differentiable on* (a, b), *and we have* $u_0(x)=u(x)$ *near a and* $u_0(x)=0$ *near b, then* $u_0\in D(T)\setminus D(\overline{T}_0)$.
(c) *If we have the limit point case at a, then* $[f, g]_a=0$ *for all* $f, g\in D(T)$.
Corresponding assertions hold for b.

PROOF.
(a) Assume $f\in D(\overline{T}_0)$ and $g\in D(T)$. Then there is a $g_0\in D(T)$ such that $g_0(x)=g(x)$ near a and $g_0(x)=0$ near b (proof!). Therefore,

$$[f, g]_a=[f, g_0]_a=-\{\langle f, Tg_0\rangle-\langle\overline{T}_0 f, g_0\rangle\}=0.$$

(b) u_0 obviously lies in $D(T)$. If v is a further solution of the equation $(L-z)u=0$ such that $W(u, v)\neq 0$, and v_0 is defined analogously to u_0, then we also have $v_0^*\in D(T)$ and

$$[u_0, v_0^*]_a=[u, v^*]_a=-W(u^*, v^*)=-W(u, v)^*\neq 0.$$

Thus, $u_0\notin D(\overline{T}_0)$ by part (a).
(c) We may assume without loss of generality that L is regular at b (otherwise we consider L on (a, c) with some $c\in(a, b)$). Then the defect indices of T_0 are $(1, 1)$ by Theorems 8.27 and 8.23. Let u_1, u_2 be linearly independent solutions of the equation $Lu=0$, and let v_1, v_2 be twice continuously differentiable functions for which $v_j(x)=u_j(x)$ near b and $v_j(x)=0$ near a. By part (b) the elements v_1, v_2 belong to $D(T)$ and are linearly independent modulo $D(\overline{T}_0)$. Consequently, $D(T)=D(\overline{T}_0)+L\{v_1, v_2\}$. This implies that for arbitrary $f, g\in D(T)$ there are elements $f_0, g_0\in D(\overline{T}_0)$ that coincide with f and g, respectively, in a neighborhood of a. It follows by part (a) that $[f, g]_a=[f_0, g_0]_a=0$. □

We are now in a position to give self-adjoint extensions of T_0 in the singular case, as well.

Theorem 8.29. *Let L be a Sturm-Liouville differential form* (8.13). *Moreover, let* $\lambda\in\mathbb{R}$, *and let v and w be real solutions of the equation* $(L-\lambda)u=0$.
(a) *The operator* $T_{v,w}$ *defined by the formulae*

$$D(T_{v,w})=\left\{f\in D(T):\begin{matrix}[v, f]_a=0 \text{ if we have the LCC at } a\\ [w, f]_b=0 \text{ if we have the LCC at } b\end{matrix}\right\},$$

$$T_{v,w}f=Tf \quad \text{for} \quad f\in D(T_{v,w})^5$$

define a self-adjoint extension of T_0.

[5] If we have the limit point case at a and/or b, then the index v and/or w has no significance. Cf. also footnote 4.

(b) *For $z \in \mathbb{C}\setminus\mathbb{R}$ the resolvent $R_z = (z - T_{v,w})^{-1}$ is of the form*

$$R_z g(x) = W(u_a, u_b)^{-1}\left\{u_b(x)\int_a^x u_a(y)g(y)r(y)\,\mathrm{d}y + u_a(x)\int_x^b u_b(y)g(y)r(y)\,\mathrm{d}y\right\},$$

where u_a and u_b are the solutions of the equation $(L - z)u = 0$, uniquely determined up to a factor by the conditions

$[v, u_a]_a = 0$ *if we have the* LCC *at a, respectively*
$u_a \in L_2(a, c, r)$ *if we have the* LPC *at a,*
$[w, u_b]_b = 0$ *if we have the* LCC *at b, respectively*
$u_b \in L_2(c, b, r)$ *if we have the* LPC *at b.*

(c) *If we have the limit circle case at both a and b, then $T_{v,w}$ has a pure discrete spectrum.*
(d) *All eigenvalues of $T_{v,w}$ are simple.*

PROOF.
(a) If we have the limit point case at both boundary points, then

$$\langle f, Tg\rangle - \langle Tf, g\rangle = [f, g]_b - [f, g]_a = 0 \quad \text{for all} \quad f, g \in D(T)$$

by Auxiliary theorem 8.28. Consequently, T is symmetric and thus self-adjoint by Theorem 8.22. This is the assertion in this case.

If we have the limit circle case at a and the limit point case at b, then the defect indices are (1, 1), as immediately follows from Theorem 8.27 (the Weyl alternative). If v_0 is a twice continuously differentiable function for which $v_0(x) = v(x)$ near a and $v_0(x) = 0$ near b, then $v_0 \in D(T)\setminus D(\overline{T_0})$ by Auxiliary theorem 8.28(b). Moreover, we obviously have $D(\overline{T_0}) + L\{v_0\} \subset D(T_{v,w})$. If u is a solution of the equation $(L - \lambda)u = 0$ that is linearly independent of v and if u_0 is defined analogously to v_0, then $u_0 \notin D(T_{v,w})$. Therefore $T_{v,w}$ is a proper restriction of T, and thus $D(T_{v,w}) = D(\overline{T_0}) + L\{v_0\}$ (since T is a two-dimensional extension of $\overline{T_0}$). With the aid of this representation of $D(T_{v,w})$ we can immediately see that $T_{v,w}$ is symmetric. Hence, $T_{v,w}$ is a one-dimensional symmetric extension of $\overline{T_0}$, and thus it is self-adjoint.

If we have the limit circle case at both boundary points, then the defect indices are (2, 2). We can show in an entirely analogous way as in the case just treated that $T_{v,w}$ is a two-dimensional symmetric extension of $\overline{T_0}$. Consequently, $T_{v,w}$ is self-adjoint.

(b) If we have the limit point case at a (respectively at b), then u_a (respectively u_b) is determined up to a factor. If we have the limit circle case at a and u_1, u_2 is a fundamental system of the equation $(L-z)u=0$, then because of the equality $[v, cu_1+du_2]_a=c[v, u_1]_a+d[v, u_2]_a$, there is at least one non-trivial linear combination $u_a=cu_1+du_2$ for which $[v, u_a]_a=0$. On the other hand, we do not have $[v, u_1]_a=[v, u_2]_a=0$, since otherwise there would exist at least one non-trivial solution of the equation $(T_{v,w}-z)f=0$, which contradicts the relation $z\in\mathbb{C}\setminus\mathbb{R}\subset\rho(T_{v,w})$. We can handle the boundary point b in a similar way.

Let K_0 be the restriction, to $L_{2,0}(a,b,r)$, of the integral operator given in the formulation of our theorem. As in Theorem 8.26(b), we can show that K_0 coincides with the restriction of R_z to $L_{2,0}(a,b,r)$. If K is the maximal integral operator defined by the formula in our theorem, then K is a closed operator (as $U_rKU_r^{-1}$ is a Carleman operator on $L_2(a,b)$). Since K_0 is continuous and densely defined, K is also continuous, and $K=R_z$.

(c) This can be proved in exactly the same way as in Theorem 8.26(c), because $u_a, u_b\in L_2(a,b,r)$.

(d) If we have the limit point case at at least one boundary point, then the assertion is clear, since the space of solutions of the equation $(\lambda-T)u=0$ is at most one-dimensional (Theorem 8.27). Let us now assume that we have the limit circle case at both boundary points. Assume that λ is an eigenvalue of multiplicity 2 (the multiplicity cannot be greater), and let u_1, u_2 be linearly independent eigenelements. Let $u_{j,a}$ and $u_{j,b}$ be twice continuously differentiable functions for which

$$u_{j,a}(x)=u_j(x)\quad\text{near } a,\ u_{j,a}(x)=0\quad\text{near } b,$$
$$u_{j,b}(x)=0\quad\text{near } a,\ u_{j,b}(x)=u_j(x)\quad\text{near } b$$

for $j=1,2$. These are four elements from $D(T_{v,w})$ that are linearly independent modulo $D(\overline{T_0})$. This is a contradiction to

$$\dim D(T_{v,w})/D(\overline{T_0})=2. \qquad \square$$

EXERCISES

8.9. Let L be a regular differential form of the form (8.8). The formulae $D(T_1)=\{f\in D(T): f(a)=f(b)=0\}$ and $T_1f=Tf$ for $f\in D(T_1)$ define a self-adjoint extension of T_0.

8.10. Let $Lf(x)=f''(x)+q(x)$ for $x\in(0,1)$. Assume that q is continuously extendible to $[0,1]$.

(a) The operator T_ϑ defined by the formulae

$$D(T_\vartheta)=\{f\in D(T): f(0)=\vartheta f(1), f'(0)=\vartheta f'(1)\},$$
$$T_\vartheta f=Tf\quad\text{for}\quad f\in D(T_\vartheta)$$

is a self-adjoint extension of T_0 for every $\vartheta\in\mathbb{C}$ such that $|\vartheta|=1$ (the boundary conditions are mixed).

(b) Prove, with the aid of (a) (for the case $q=0$) that the eigenvalues are in general not simple in the case of mixed boundary conditions.

8.11. Let L be a Sturm-Liouville differential form with limit circle case at a and b.
(a) If λ is real, and v, w are real linearly independent solutions of the equation $(L-\gamma)u=0$, then the formulae $D(T_{v,w})=D(\overline{T_0})+L\{v, w\}$, $T_{(v,w)}f=Tf$ for $f\in D(T_{(v,w)})$ define a self-adjoint extension of T_0.
(b) We have $D(T_{(v,w)})=\{f\in D(T):[v,f]_b-[v,f]_a=[w,f]_b-[w,f]_a=0\}$ (these are mixed boundary conditions).

8.12. The representation of R_z given in Theorem 8.29(b) also holds for $z\in\mathbf{R}\cap\rho(T_{v,w})$.

8.13. (a) Let L be a Sturm-Liouville differential form on (a, b) such that $(q(x)/r(x))\geqslant\gamma$ for $x\geqslant x_0$. If $g\notin L_2(x_0, b, r)$ for $g(x)=\int_{x_0}^{x}p(s)^{-1}\,ds$, then we have the limit point case at b.
Hint: Consider that solution u of the equation $(L-\gamma)u=0$ for which $u(x_0)=u'(x_0)=1$.
(b) Let L be a Sturm-Liouville differential form on (0, 1) such that $p(x)=r(x)=1$ and $q(x)\geqslant cx^{-2}$ with $c\geqslant 3/4$. Then we have the limit point case at 0.

8.14. Let L be a differential form of the form (8.8) with $p=1$.
(a) Consider the unitary operator U defined on $L_2(a, b, r)$ by the formula $(Uf)(x)=\exp(-i\int_c^x s(t)\,dt)f(x)$. Then

$$ULU^{-1}g(x)=\frac{1}{r(x)}\{-g''(x)+(q(x)-s^2(x))g(x)\}.$$

(b) T_0 has equal defect indices.

8.15. (a) The self-adjoint extensions of T_0 given in Theorem 8.26 and Theorem 8.29 are K-real ($Kf=f^*$).
(b) The operators T_ϑ from Exercise 8.10 are not K-real for $\vartheta\notin\mathbf{R}$.
(c) If we have the limit point case at at least one boundary point, then Theorem 8.29 provides all K-real self-adjoint extensions of T_0. (These are all self-adjoint extensions of T_0 by Exercise 8.7.)

8.5 Analytic vectors and tensor products of self-adjoint operators

With the aid of the results of Section 8.1 we can also prove the criterion of Nelson for the essential self-adjointness of symmetric operators. For this we need the notion of analytic vectors.

Let S be a symmetric operator on the Hilbert space H. Introduce the notation $C^\infty(S)=\bigcap_{n=0}^{\infty}D(S^n)$. An element $f\in C^\infty(S)$ is called an *analytic vector* of S if there exists a $t(f)>0$ such that

$$\sum_{n=0}^{\infty}\frac{|t|^n}{n!}\|S^nf\|<\infty \quad \text{for} \quad |t|<t(f)$$

Theorem 8.30. *If S is a self-adjoint operator and f is an analytic vector of S, then*

$$f \in D(e^{zS}) \quad \text{and} \quad e^{zS}f = \sum_{n=0}^{\infty} \frac{z^n}{n!} S^n f$$

for every $z \in \mathbb{K}$ such that $|z| < t(f)$.

PROOF. Let E denote the spectral family of S. Then

$$\left\{ \int_{-M}^{M} |e^{zs}|^2 \, d\|E(s)f\|^2 \right\}^{1/2} = \left\| \int_{-M}^{M} e^{zs} \, dE(s)f \right\|$$

$$= \left\| \int_{-M}^{M} \sum_{n=0}^{\infty} \frac{(zs)^n}{n!} \, dE(s)f \right\|$$

$$\leqslant \sum_{n=0}^{\infty} \frac{|z|^n}{n!} \left\| \int_{-M}^{M} s^n \, dE(s)f \right\|$$

$$\leqslant \sum_{n=0}^{\infty} \frac{|z|^n}{n!} \|S^n f\|$$

for every $M > 0$ and $z \in \mathbb{K}$ such that $|z| < t(f)$. Letting M tend to ∞, we obtain that $f \in D(e^{zS})$. Furthermore,

$$e^{zS}f = \int_{-\infty}^{\infty} e^{zs} \, dE(s)f = \int_{-\infty}^{\infty} \sum_{n=0}^{\infty} \frac{(zs)^n}{n!} \, dE(s)f$$

$$= \sum_{n=0}^{m} \frac{z^n}{n!} \int_{-\infty}^{\infty} s^n \, dE(s)f + \int_{-\infty}^{\infty} \sum_{n=m+1}^{\infty} \frac{(zs)^n}{n!} \, dE(s)f$$

$$= \lim_{m \to \infty} \sum_{n=0}^{m} \frac{z^n}{n!} S^n f = \sum_{n=0}^{\infty} \frac{z^n}{n!} S^n f,$$

since

$$\left\| \int_{-\infty}^{\infty} \sum_{n=m+1}^{\infty} \frac{(zs)^n}{n!} \, dE(s)f \right\| = \lim_{M \to \infty} \left\| \int_{-M}^{M} \sum_{n=m+1}^{\infty} \frac{(zs)^n}{n!} \, dE(s)f \right\|$$

$$= \lim_{M \to \infty} \left\| \sum_{n=m+1}^{\infty} \frac{z^n}{n!} \int_{-M}^{M} s^n \, dE(s)f \right\|$$

$$\leqslant \sum_{n=m+1}^{\infty} \frac{|z|^n}{n!} \left\| \int_{-\infty}^{\infty} s^n \, dE(s)f \right\|$$

$$= \sum_{n=m+1}^{\infty} \frac{|z|^n}{n!} \|S^n f\| \to 0 \quad \text{as} \quad m \to \infty. \qquad \square$$

Theorem 8.31 (Nelson). *Let T be a symmetric operator on the Hilbert space H. Assume that the set of analytic vectors of T is dense. Then T is essentially self-adjoint.*

PROOF.
(a) First we consider the complex case. Introduce the notations $\tilde{H} = H \oplus H$ and $\tilde{T} = T \oplus (-T)$ (i.e., $D(\tilde{T}) = D(T) \oplus D(T)$ and $\tilde{T}(f, g) = (Tf, -Tg)$ for $(f, g) \in D(\tilde{T})$). If (γ_+, γ_-) denote the defect indices of T, then the defect indices of $\tilde{T}$ are $(\gamma_+ + \gamma_-, \gamma_+ + \gamma_-)$. Therefore, $\tilde{T}$ possesses self-adjoint extensions. We can also see from the defect indices of $\tilde{T}$ that T is essentially self-adjoint if and only if $\tilde{T}$ is essentially self-adjoint. Hence, it is enough to prove that $\tilde{T}$ is essentially self-adjoint.

The set of analytic vectors of $\tilde{T}$ is dense: If f and g are analytic vectors of T, then (f, g) is an analytic vector of $\tilde{T}$. Consequently, it is sufficient to show the following (where we write T instead of $\tilde{T}$): If T is a symmetric operator with equal defect indices and a dense set of analytic vectors, then T is essentially self-adjoint.

Let S be a self-adjoint extension of T. If f is an analytic vector of T, then f is also an analytic vector of S. Because the formula $\|S^n \mathrm{e}^{\mathrm{i}sS} f\| = \|\mathrm{e}^{\mathrm{i}sS} S^n f\| = \|S^n f\| = \|T^n f\|$, the element $\mathrm{e}^{\mathrm{i}sS} f$ is also an analytic vector of S for every $s \in \mathbb{R}$, and $t(\mathrm{e}^{\mathrm{i}sS} f) = t(f)$. Then for every $s \in \mathbb{R}$, $z \in \mathbb{C}$ such that $|z - s| < t(f)$ and for every $g \in H$ we have by Theorem 8.30 that

$$f \in D(\mathrm{e}^{\mathrm{i}(z-s)S}) = D(\mathrm{e}^{\mathrm{i}zS})$$

and

$$F(z) = \langle g, \mathrm{e}^{\mathrm{i}zS} f\rangle = \langle g, \mathrm{e}^{\mathrm{i}(z-s)S} \mathrm{e}^{\mathrm{i}sS} f\rangle$$
$$= \sum_{n=0}^{\infty} \frac{[\mathrm{i}(z-s)]^n}{n!} \langle g, S^n \mathrm{e}^{\mathrm{i}sS} f\rangle.$$

Consequently, F is holomorphic in $\{z \in \mathbb{C} : |\mathrm{Im}\, z| < t(f)\}$.

If $g \in R(\mathrm{i} - T)^{\perp} = N(-\mathrm{i} - T^*)$, then $(T^*)^n g = (-\mathrm{i})^n g$ for all $n \in \mathbb{N}_0$. Therefore, $\langle T^n f, g\rangle = (-\mathrm{i})^n \langle f, g\rangle$. This implies for all $n \in \mathbb{N}_0$ that

$$F^{(n)}(0) = \langle g, (\mathrm{i}S)^n f\rangle = \langle g, (\mathrm{i}T)^n f\rangle = (-1)^n \langle g, f\rangle = (-1)^n F(0).$$

Hence, for all $s \in \mathbb{R}$

$$\langle g, \mathrm{e}^{\mathrm{i}sS} f\rangle = F(s) = \mathrm{e}^{-s} F(0) = \mathrm{e}^{-s} \langle g, f\rangle.$$

Since $\mathrm{e}^{-\mathrm{i}sS}$ is unitary, the function F is bounded. Consequently, $\langle g, f\rangle = 0$ for every analytic vector f of T. Hence, $R(\mathrm{i} - T)^{\perp} = \{0\}$. The equality $R(-\mathrm{i} - T)^{\perp} = \{0\}$ follows similarly.

(b) Let H now be a real Hilbert space. Then the set of analytic vectors of the complexification $T_{\mathbb{C}}$ is dense, namely it is equal to the complex linear hull of the analytic vectors of T (observe that the set of analytic vectors is a vector space). $T_{\mathbb{C}}$ is therefore essentially self-adjoint. Then T is also essentially self-adjoint by Exercise 5.32(b). □

Corollary. *A closed symmetric operator T is self-adjoint if and only if the set of analytic vectors of T is dense.*

PROOF. If the set of analytic vectors of T is dense, then T is essentially self-adjoint by Theorem 8.31. As T is closed, it is self-adjoint.

Let T now be self-adjoint, and let E denote the spectral family of T. Then all elements of $R(E(t)-E(-t))$ are analytic vectors of T for every $t>0$, since $\|T^n f\| \leq t^n \|f\|$ for $f \in R(E(t)-E(-t))$. As $\cup_{t>0} R(E(t)-E(-t))$ is dense in H, the assertion follows. □

The above results enable us to prove the essential self-adjointness of tensor products of operators. Let H_1 and H_2 be Hilbert spaces, and let us consider the space $H = H_1 \hat{\otimes} H_2$ (cf. Section 3.4). If T_1 and T_2 are operators on H_1 and H_2, respectively, then we define the operator $T_1 \otimes T_2$ on $H_1 \hat{\otimes} H_2$ by the formulae

$$D(T_1 \otimes T_2) = D(T_1) \otimes D(T_2),$$

and

$$(T_1 \otimes T_2)\left(\sum_{j=1}^{n} c_j f_j \otimes g_j\right) = \sum_{j=1}^{n} c_j T_1 f_j \otimes T_2 g_j.$$

In order to prove that this is a linear operator, it is sufficient to prove that this definition is independent of the representation of the elements from $D(T_1) \otimes D(T_2)$ as linear combinations of simple tensors; the linearity then follows directly from the definition. In order to prove this, we have to show that $\sum_{j=1}^{n} c_j f_j \otimes g_j = 0$ implies $\sum_{j=1}^{n} c_j T_1 f_j \otimes T_2 g_j = 0$. By (3.3) we have $\sum_{j=1}^{n} c_j f_j \otimes g_j = 0$ if and only if this sum can be written as a finite linear combination of elements of the form

$$\sum_{j=1}^{n} \sum_{k=1}^{m} a_j b_k \psi_j \otimes \gamma_k - \left(\sum_{j=1}^{n} a_j \psi_j\right) \otimes \left(\sum_{k=1}^{n} b_k \gamma_k\right).$$

Then $\sum_{j=1}^{n} c_j T_1 f_j \otimes T_2 g_j$ is also a linear combination of elements of the same form, and thus it is equal to zero.

In what follows we study, for two given operators T_1 and T_2, the operators

$$A = T_1 \otimes T_2 \quad \text{and} \quad B = T_1 \otimes I_2 + I_1 \otimes T_2.$$

We have $D(A) = D(B) = D(T_1) \otimes D(T_2)$.

Theorem 8.32. *Let H_1, H_2, T_1, T_2, A and B be as above.*

(a) *A is different from zero (i.e., there exists an $f \in D(A)$ such that $Af \neq 0$) if and only if T_1 and T_2 are different from zero. If A is different from zero, then A is bounded if and only if T_1 and T_2 are bounded. Then $\|A\| = \|T_1\|\,\|T_2\|$.*

(b) *If $D(B) \neq \{0\}$, then B is bounded if and only if T_1 and T_2 are bounded. Then $\|B\| \leq \|T_1\| + \|T_2\|$.*

(c) *If T_1 and T_2 are densely defined, then A and B are also densely defined, and $A^* \supset T_1^* \otimes T_2^*$ and $B^* \supset T_1^* \otimes I_2 + I_1 \otimes T_2^*$. If T_1 and T_2 are symmetric, then A and B are symmetric.*

PROOF.
(a) If T_1 and T_2 are different from zero, then there are elements $f_1 \in D(T_1)$ and $f_2 \in D(T_2)$ for which $T_1 f_1 \neq 0$ and $T_2 f_2 \neq 0$. Hence $A(f_1 \otimes f_2) \neq 0$, i.e., $A \neq 0$. If one of the operators T_1 and T_2 is zero, then $A(f_1 \otimes f_2) = T_1 f_1 \otimes T_2 f_2 = 0$ for all $f_1 \in D(T_1)$ and $f_2 \in D(T_2)$. Because of the equality $D(A) = L\{f_1 \otimes f_2 : f_1 \in D(T_1), f_2 \in D(T_2)\}$, it follows from this that $Af = 0$ for all $f \in D(A)$, i.e., that $A = 0$.

Assume now that $A \neq 0$ is bounded. Then for all $f_1 \in D(T_1)$, $f_2 \in D(T_2)$ such that $\|f_1\| = \|f_2\| = 1$ we have

$$\|T_1 f_1\| \, \|T_2 f_2\| = \|T_1 f_1 \otimes T_2 f_2\| = \|A(f_1 \otimes f_2)\| \leqslant \|A\|. \quad (8.16)$$

As $T_2 \neq 0$, there is an $f_2 \in D(T_2)$ such that $\|f_2\| = 1$ and $T_2 f_2 \neq 0$. Therefore, it follows that

$$\|T_1 f_1\| \leqslant \|T_2 f_2\|^{-1} \|A\| \quad \text{for all} \quad f_1 \in D(T_1),\ \|f_1\| = 1.$$

Consequently, T_1 is bounded. We can prove in just the same way that T_2 is bounded. The left side of (8.16) assumes values arbitrarily close to $\|T_1\| \, \|T_2\|$; thus $\|T_1\| \, \|T_2\| \leqslant \|A\|$.

Now let T_1 and T_2 be bounded. We show that A is bounded and $\|A\| \leqslant \|T_1\| \, \|T_2\|$. Because of the formula $A = (T_1 \otimes I_2)(I_1 \otimes T_2)$ it is sufficient to prove that the operators $T_1 \otimes I_2$ and $I_1 \otimes T_2$ are bounded and $\|T_1 \otimes I_2\| \leqslant \|T_1\|$ and $\|I_1 \otimes T_2\| \leqslant \|T_2\|$. We show the first inequality. To do this, we use the fact that according to (3.5) every element $f \in D(T_1) \otimes D(I_2) = D(T_1) \otimes H_2$ can be written in the form $f = \sum_{j=1}^n f_j \otimes e_j$ with orthonormal elements e_j, $j = 1, \ldots, n$. For such an f we have

$$\|(T_1 \otimes I_2) f\|^2 = \left\| \sum_{j=1}^n T_1 f_j \otimes e_j \right\|^2 = \sum_{j=1}^n \|T_1 f_j\|^2$$

$$\leqslant \|T_1\|^2 \sum_{j=1}^n \|f_j\|^2 = \|T_1\|^2 \|f\|^2.$$

(b) If T_1 and T_2 are bounded, then the boundedness of $T_1 \otimes I_2$ and $I_1 \otimes T_2$ follows as in (a). Therefore, $\|B\| \leqslant \|T_1\| + \|T_2\|$. If T_1 is unbounded and $D(T_2)$ is different from zero, then there exist a sequence (f_n) from $D(T_1)$ for which $\|f_n\| = 1$ and $\|T_1 f_n\| \to \infty$ and a $g \in D(T_2)$ for which $\|g\| = 1$. Then

$$\|B(f_n \otimes g)\| \geqslant \|T_1 f_n \otimes g\| - \|f_n \otimes T_2 g\| = \|T_1 f_n\| - \|T_2 g\| \to \infty$$

as $n \to \infty$. Hence, B is unbounded. The same follows if T_2 is unbounded and $D(T_1)$ is different from zero.

(c) We can easily verify that A and $T_1^*\otimes T_2^*$ (respectively B and $T_1^*\otimes I_2+I_1\otimes T_2^*$) are formal adjoints of each other. This implies that $A^*\supset T_1^*\otimes T_2^*$ and $B^*\supset T_1^*\otimes I_2+I_1\otimes T_2^*$. The last assertion immediately follows from this. □

Theorem 8.33. *Let T_1 and T_2 be essentially self-adjoint operators on H_1 and H_2, respectively. Then the operators $A=T_1\otimes T_2$ and $B=T_1\otimes I_2+I_1\otimes T_2$ are essentially self-adjoint on $H=H_1\hat{\otimes} H_2$.*

PROOF.

(a) First we assume that T_1 and T_2 are self-adjoint. Then A and B are symmetric according to Theorem 8.32(c). We construct total sets of analytic vectors for A and B. As the linear combinations of analytic vectors are again analytic vectors, the assertion follows from this by Theorem 8.31.

First we consider the operator A. Let M_1 and M_2 be the sets of analytic vectors of T_1^2 and T_2^2, respectively. We show that all $f_1\otimes f_2$ such that $f_j\in M_j$ are analytic vectors of A. If

$$\sum_{n=0}^{\infty}\frac{t^n}{n!}\|T_j^{2n}f_j\|<\infty \quad \text{for} \quad j=1,2 \quad \text{and} \quad 0\leqslant t<t_0,$$

then

$$\begin{aligned}\sum_{n=0}^{\infty}\frac{t^n}{n!}\|A^n(f_1\otimes f_2)\| &= \sum_{n=0}^{\infty}\frac{t^n}{n!}\|T_1^nf_1\otimes T_2^nf_2\| \\ &= \sum_{n=0}^{\infty}\frac{t^n}{n!}\|T_1^nf_1\|\,\|T_2^nf_2\| \\ &\leqslant \left\{\sum_{n=0}^{\infty}\frac{t^n}{n!}\|T_1^nf_1\|^2\sum_{n=0}^{\infty}\frac{t^n}{n!}\|T_2^nf_2\|^2\right\}^{1/2} \\ &= \left\{\sum_{n=0}^{\infty}\frac{t^n}{n!}\langle T_1^{2n}f_1,f_1\rangle\sum_{n=0}^{\infty}\frac{t^n}{n!}\langle T_2^{2n}f_2,f_2\rangle\right\}^{1/2} \\ &\leqslant \|f_1\|^{1/2}\|f_2\|^{1/2}\left\{\sum_{n=0}^{\infty}\frac{t^n}{n!}\|T_1^{2n}f_1\|\sum_{n=0}^{\infty}\frac{t^n}{n!}\|T_2^{2n}f_2\|\right\}^{1/2}<\infty\end{aligned}$$

for $0\leqslant t<t_0$. Consequently, $f_1\otimes f_2$ is an analytic vector of A. As the M_j are dense subsets of H_j $(j=1,2)$, the set of these analytic vectors is total in $H_1\hat{\otimes} H_2$.

Now we consider B. Let M_1 and M_2 now be the sets of analytic vectors of T_1 and T_2, respectively. Assume that for $f_j\in M_j$ we have

$$\sum_{n=0}^{\infty}\frac{t^n}{n!}\|T_j^nf_j\|<\infty \quad \text{for} \quad j=1,2 \quad \text{and} \quad 0\leqslant t<t_0.$$

Then

$$\sum_{n=0}^{\infty} \frac{t^n}{n!} \|(T_1 \otimes I_2 + I_1 \otimes T_2)^n f_1 \otimes f_2\|$$

$$\leqslant \sum_{n=0}^{\infty} \frac{t^n}{n!} \left\| \sum_{k=0}^{n} \binom{n}{k} T_1^k f_1 \otimes T_2^{n-k} f_2 \right\|$$

$$\leqslant \sum_{n=0}^{\infty} \frac{t^n}{n!} \sum_{k=0}^{n} \frac{n!}{k!(n-k)!} \|T_1^k f_1\| \, \|T_2^{n-k} f_2\|$$

$$= \sum_{n=0}^{\infty} \sum_{k=0}^{n} \frac{t^k}{k!} \|T_1^k f_1\| \frac{t^{n-k}}{(n-k)!} \|T_2^{n-k} f_2\|$$

$$= \sum_{k=0}^{\infty} \left\{ \frac{t^k}{k!} \|T_1^k f_1\| \sum_{n=k}^{\infty} \frac{t^{n-k}}{(n-k)!} \|T_2^{n-k} f_2\| \right\}$$

$$= \left\{ \sum_{k=0}^{\infty} \frac{t^k}{k!} \|T_1^k f_1\| \right\} \left\{ \sum_{m=0}^{\infty} \frac{t^m}{m!} \|T_2^m f_2\| \right\}$$

$$< \infty$$

for $0 \leqslant t < t_0$. Consequently, $f_1 \otimes f_2$ is an analytic vector of B. The set of these analytic vectors is total in $H_1 \hat{\otimes} H_2$.

(b) Let T_1 and T_2 now be essentially self-adjoint, i.e., let $\overline{T}_1$ and $\overline{T}_2$ be self-adjoint. The operators A and B are symmetric, and thus closable. We can verify easily that

$$\overline{A} = \overline{T_1 \otimes T_2} \supset \overline{T}_1 \otimes \overline{T}_2,$$
$$\overline{B} = \overline{T_1 \otimes I_2 + I_1 \otimes T_2} \supset \overline{T}_1 \otimes I_2 + I_1 \otimes \overline{T}_2.$$

The essential self-adjointness of A and B now follows from part (a) of the proof. □

Theorem 8.34. *Let T_1 and T_2 be self-adjoint on H_1 and H_2 with spectral families E_1 and E_2, respectively. Then*

$$\langle f_1 \otimes f_2, E(t)(g_1 \otimes g_2)\rangle = \int_{-\infty}^{\infty} \langle f_1, E_1(t-s) g_1\rangle \, \mathrm{d}_s \langle f_2, E_2(s) g_2\rangle$$

for the spectral family E of $\overline{B} = \overline{T_1 \otimes I_2 + I_1 \otimes T_2}$ and for all $f_1, g_1 \in H_1$, $f_2, g_2 \in H_2$.

Proof. The formulae $F_1(t) = \overline{E_1(t) \otimes I_2}$ and $F_2(t) = \overline{I_1 \otimes E_2(t)}$, $t \in \mathbb{R}$ define spectral families on $H = H_1 \hat{\otimes} H_2$. We show this for F_1:

(a) Because of the formulae $F_1^* = (E_1 \otimes I_2)^* \supset \overline{E_1^* \otimes I_2} = \overline{E_1 \otimes I_2}$ we have $F_1^* = F_1$. It is obvious that $F_1(t)F_1(s)f = F_1(s)F_1(t)f = F_1(t)f$ for $f \in H_1 \otimes H_2$ and $t \leqslant s$. Due to continuity arguments (cf. Theorem 8.32(a)) this holds for all $f \in H_1 \hat{\otimes} H_2$. In particular, $F_1(t)^2 = F_1(t)$, i.e., all $F_1(t)$ are orthogonal projections.

(b) We have $F_1(t)F_1(s) = F_1(s)F_1(t) = F_1(t)$ for $t < s$ by part (a); this proves that $\overline{F}_1$ is increasing.
(c) For $f = f_1 \otimes f_2$ we have

$$\|F_1(t+\epsilon)f - F_1(t)f\| = \|(E_1(t+\epsilon) - E_1(t))f_1\| \, \|f_2\| \to 0 \quad \text{as } \epsilon \to 0+.$$

This proves the right continuity of $F_1(t)f$ for all $f \in H_1 \hat{\otimes} H_2$ (cf. Theorem 4.23(b)).
(d) For $f = f_1 \otimes f_2$ we have $F_1(t)f = E_1(t)f_1 \otimes f_2 \to 0$ as $t \to -\infty$ and $F_1(t)f \to f$ as $t \to \infty$. This implies the corresponding assertion for all $f \in H_1 \hat{\otimes} H_2$ (cf. Theorem 4.23(b)).

The spectral families F_1 and F_2 commute, since

$$F_1(t)F_2(s)f = E_1(t)f_1 \otimes E_2(s)f_2 = F_2(s)F_1(t)f$$

for all $f = f_1 \otimes f_2$. This then holds for all $f \in H_1 \hat{\otimes} H_2$. Consequently, the equality

$$G(t + \mathrm{i}s) = F_1(t)F_2(s), \quad s, t \in \mathbb{R}$$

defines a complex spectral family on $H_1 \hat{\otimes} H_2$. We obviously have $G(t + \mathrm{i}s) = \overline{E_1(t) \otimes E_2(s)}$. We show that

$$\overline{B} = \hat{G}(u) = \int (\operatorname{Re} z + \operatorname{Im} z) \, \mathrm{d}G(z),$$

where u is the function defined by the formulae $u : \mathbb{C} \to \mathbb{R}$, $u(z) = \operatorname{Re} z + \operatorname{Im} z$. As $\overline{B}$ and $\hat{G}(u)$ are self-adjoint (u is real-valued), it is enough to prove that $B \subset \hat{G}(u)$. It will follow from this that $\overline{B} \subset \hat{G}(u)$, and thus $\overline{B} = \hat{G}(u)$. For $f = f_1 \otimes f_2 \in D(T_1) \otimes D(T_2)$ we have

$$\begin{aligned}
\int |u(z)|^2 \, \mathrm{d}\|G(z)f\|^2 &< 2 \int \{(\operatorname{Re} z)^2 + (\operatorname{Im} z)^2\} \, \mathrm{d}\|G(z)f\|^2 \\
&= 2\left\{ \int (\operatorname{Re} z)^2 \, \mathrm{d}\|G(z)f\|^2 + \int (\operatorname{Im} z)^2 \, \mathrm{d}\|G(z)f\|^2 \right\} \\
&= 2\left\{ \int t^2 \, \mathrm{d}\|E_1(t)f_1\|^2 \|f_2\|^2 + \int s^2 \, \mathrm{d}\|E_2(s)f_2\|^2 \|f_1\|^2 \right\} \\
&= 2\{\|T_1 f_1\|^2 \|f_2\|^2 + \|f_1\|^2 \|T_2 f_2\|^2\} < \infty.
\end{aligned}$$

Consequently, $f \in D(\hat{G}(u))$ and

$$\begin{aligned}
\hat{G}(u)f &= \int (\operatorname{Re} z + \operatorname{Im} z) \, \mathrm{d}G(z)f = \int \operatorname{Re} z \, \mathrm{d}G(z)f + \int \operatorname{Im} z \, \mathrm{d}G(z)f \\
&= \left(\int t \, \mathrm{d}E_1(t)f_1 \right) \otimes f_2 + f_1 \otimes \left(\int s \, \mathrm{d}E_2(s)f_2 \right) = Bf.
\end{aligned}$$

If we consider linear combinations of such elements, then it follows that $B \subset \hat{G}(u)$.

Now we can give the spectral family E of $\overline{B} = \hat{G}(u)$. We have

$$E(t) = G(\{z \in \mathbb{C} : \operatorname{Re} z + \operatorname{Im} z < t\}) = \hat{G}(\chi_{\{z \in \mathbb{C} : \operatorname{Re} z + \operatorname{Im} z < t\}}).$$

$E(.)$ is obviously a spectral family. For every $f \in H_1 \hat{\otimes} H_2$ we have

$$\int t^2 \,\mathrm{d}\|E(t)f\|^2 = \int (\operatorname{Re} z + \operatorname{Im} z)^2 \,\mathrm{d}\|G(z)f\|^2.$$

Consequently, $D(\hat{E}(\mathrm{id})) = D(\hat{G}(u)) = D(\bar{B})$. For all $f \in D(\bar{B})$

$$\int t \,\mathrm{d}E(t)f = \int u(z)\,\mathrm{d}G(z)f = \bar{B}f.$$

Hence, $\hat{E}(\mathrm{id}) = \bar{B}$, i.e., E is the spectral family of $\bar{B}$.

By Fubini's theorem we have for all $f_1 \otimes f_2 \in H_1 \otimes H_2$ that

$$\begin{aligned}\|E(t)(f_1 \otimes f_2)\|^2 &= \int_{\operatorname{Re} z + \operatorname{Im} z < t} \mathrm{d}\|G(z)(f_1 \otimes f_2)\|^2 \\ &= \int_{-\infty}^{\infty}\int_{-\infty}^{t-s} \mathrm{d}_u\|E_1(u)f_1\|^2 \,\mathrm{d}_s\|E_2(s)f_2\|^2 \\ &= \int_{-\infty}^{\infty} \|E_1(t-s)f_1\|^2 \,\mathrm{d}_s\|E_2(s)f_2\|^2.\end{aligned}$$

The assertion follows from this by means of the polarization identity. □

Theorem 8.35. *Let T_1 and T_2 be self-adjoint operators on complex Hilbert spaces, and let $B = T_1 \otimes I_2 + I_1 \otimes T_2$. Then*

$$\exp(\mathrm{i}t\bar{B}) = \overline{\exp(\mathrm{i}tT_1) \otimes \exp(\mathrm{i}tT_2)}.$$

PROOF. For all simple tensors $f \otimes g$ such that $f \in D(T_1)$ and $g \in D(T_2)$ we have $f \otimes g \in D(\bar{B})$, and thus

$$\frac{\mathrm{d}}{\mathrm{d}t}\left[\exp(\mathrm{i}t\bar{B})(f \otimes g)\right] = \mathrm{i}\bar{B}\exp(\mathrm{i}t\bar{B})(f \otimes g).$$

Hence, $u(t) = \exp(\mathrm{i}t\bar{B})(f \otimes g)$ is a solution of the initial value problem

$$\frac{\mathrm{d}}{\mathrm{d}t}u(t) = \mathrm{i}\bar{B}u(t), \quad u(0) = f \otimes g.$$

On the other hand, it is easy to prove that

$$u(t) = [\exp(\mathrm{i}tT_1)f] \otimes g + f \otimes [\exp(\mathrm{i}tT_2)g]$$

is also a solution of this initial value problem. The solution is uniquely determined by the Corollary to Theorem 7.38. This proves the assertion. □

EXERCISES

8.16. Let T be a self-adjoint multiplication operator on $L_2(\mathbf{R}, \rho)$ (respectively on $\oplus_{\alpha \in A} L_2(\mathbf{R}, \rho_\alpha)$). Give a dense set of analytic vectors of T.

8.17. There are essentially self-adjoint operators whose sets of analytic vectors are not dense.
Hint: Let T be defined by the equalities $D(T) = C_0^\infty(\mathbf{R})$, $Tf = (1/\mathrm{i})f' + qf$,

where $q(x)=0$ for $x\leqslant 0$, and $q(x)=1$ for $x>0$. If f is an analytic vector of T, then $f(0)=0$. If the set of analytic vectors of T were dense, then the same would hold for the operator $T_0\subset T$ with $D(T_0)=\{f\in C_0^\infty(\mathbb{R}):f(0)=0\}$; the defect indices of T_0 are equal to $(1, 1)$.

8.18. Let T_1 and T_2 be the operators of multiplication by the variables on $\oplus_{\alpha\in A}L_2(\mathbb{R}, \rho_\alpha)$ and on $\oplus_{\beta\in B}L_2(\mathbb{R}, \sigma_\beta)$, respectively.
(a) $\overline{T_1\otimes T_2}$ is the operator of multiplication by x_1x_2 on $\oplus_{\alpha\in A,\beta\in B}L_2(\mathbb{R}^2, \rho_\alpha\times\sigma_\beta)$.
(b) $\overline{T_1\otimes I_2+I_1\otimes T_2}$ is the operator of multiplication by x_1+x_2 on $\oplus_{\alpha\in A,\beta\in B}L_2(\mathbb{R}^2, \rho_\alpha\times\sigma_\beta)$.
(c) Prove Theorem 8.34 with the aid of (b) and the spectral representation theorem.

8.19. (a) If $T_1\otimes T_2$ is different from zero, then $T_1\otimes T_2$ is symmetric if and only if there exists a $c\in\mathbb{K}$, $c\neq 0$ for which cT_1 and $c^{-1}T_2$ are symmetric.
(b) $T_1\otimes I_2+I_1\otimes T_2$ is symmetric if and only if there exists a $c\in\mathbb{R}$ for which T_1-icI_1 and T_2+icI_2 are symmetric; we have $T_1\otimes I_2+I_1\otimes T_2=(T_1-icI_1)\otimes I_2+I_1\otimes(T_2+icI_2)$. (If H_1 and H_2 are real Hilbert spaces, then $T_1\otimes I_2+I_1\otimes T_2$ is symmetric if and only if T_1 and T_2 are symmetric.)

8.20. Let T_1 and T_2 be self-adjoint. Then for $B=T_1\otimes I_2+I_1\otimes T_2$ we have

$$\sigma(\bar{B}) = \overline{\{\lambda\in\mathbb{R}:\lambda=\lambda_1+\lambda_2,\ \lambda_j\in\sigma(T_j)\}},$$
$$\sigma_p(\bar{B}) = \{\lambda\in\mathbb{R}:\lambda=\lambda_1+\lambda_2,\ \lambda_j\in\sigma_p(T_j)\}.$$

The multiplicity $n(\lambda)$ of an eigenvalue λ of B is equal to $\Sigma_{\lambda_1+\lambda_2=\lambda}n_1(\lambda_1)n_2(\lambda_2)$, where $n_j(\lambda_j)$ is the multiplicity of the eigenvalue λ_j of T_j.

8.21. Assume that T_1, T_2 and B are as in Exercise 8.20, T_2 has a pure point spectrum, and P_s denotes the projection onto $N(T_2-s)$. Then

$$E(t) = \sum_{s\in\sigma_p(T_2)} \overline{E_1(t-s)\otimes P_s},$$

where E and E_1 denote the spectral families of B and T_1, respectively.

8.22. Let T_1 and T_2 be self-adjoint operators with spectral families E_1 and E_2. If $E_1(t)E_2(s)=E_2(s)E_1(t)$ for all $t, s\in\mathbb{R}$, then the operators T_1+T_2 and T_1T_2 are essentially self-adjoint.
Hint: For all bounded intervals J_1 and J_2 the set $R(E_1(J_1)E_2(J_2))$ consists of analytic vectors of T_1+T_2 and T_1T_2.

Perturbation theory for self-adjoint operators 9

Here we will deal almost exclusively with the perturbation theory for self-adjoint and essentially self-adjoint operators. Essentially two questions arise:

(9.1) *Let T be a self-adjoint or essentially self-adjoint operator on the Hilbert space H. Let V be a symmetric operator, a perturbation. Is $T+V$ also self-adjoint or essentially self-adjoint?*

(9.2) *Let T be a self-adjoint operator and assume that we know certain properties of its spectrum. Can we say anything about the spectral properties of $T+V$ (or $\overline{T+V}$)?*

We already answered question (9.1) in Section 5.3. Now we turn our attention to question (9.2). We shall study the question of whether T and $T+V$ have the same essential spectrum and whether the semi-boundedness of T implies that of $T+V$. Moreover, we obtain results concerning the perturbation of the discrete spectrum and concerning the continuous dependence, on the given operator, of the spectrum and the spectral family. Further results of this kind are also included in Section 7.5. For the absolutely continuous spectrum, see also Chapter 11.

9.1 Relatively bounded perturbations

First we consider the case where the unperturbed operator T is bounded from below.

Theorem 9.1. *Let T be self-adjoint and bounded from below with lower bound γ_T. Let V be symmetric and T-bounded with T-bound < 1. Then $T+V$ is* *self-adjoint and bounded from below. If*

$$\|Vf\| \leq a\|f\| + b\|Tf\| \quad \textit{for all} \quad f \in D(T)$$

with some $b < 1$, *then*

$$\gamma = \gamma_T - \max\left\{\frac{a}{1-b}, a + b|\gamma_T|\right\}$$

is a lower bound of $T + V$.

Proof. By Corollary 2 to Theorem 7.22 it is sufficient to show that $(-\infty, \gamma)$ is contained in $\rho(T+V)$, i.e., that the operator $T+V-\lambda = (T-\lambda)+V$ is bijective for every $\lambda < \gamma$. By Theorem 5.11, this is surely the case if $\|VR(\lambda, T)\| < 1$. We obtain from the spectral theorem that

$$\begin{aligned}
\|VR(\lambda, T)\| &\leq a\|R(\lambda, T)\| + b\|TR(\lambda, T)\| \\
&\leq a(\gamma_T - \lambda)^{-1} + b \sup\left\{|t|(t-\lambda)^{-1} : t \geq \gamma_T\right\} \\
&= a(\gamma_T - \lambda)^{-1} + b \max\left\{1, |\gamma_T|(\gamma_T - \lambda)^{-1}\right\} \\
&= \max\left\{a(\gamma_T - \lambda)^{-1} + b, a(\gamma_T - \lambda)^{-1} \right. \\
&\qquad \left. + b|\gamma_T|(\gamma_T - \lambda)^{-1}\right\}.
\end{aligned}$$

The last expression is obviously less than 1 for $\lambda < \gamma$. □

Theorem 9.2. *Let T be self-adjoint and bounded from below, and let V be symmetric and T-bounded. If $T + \mu V$ is closed for all $\mu \in [0, 1]$, then $T + V$ is self-adjoint and bounded from below.*

Proof. The operator $T + V$ is self-adjoint by Theorem 5.27. For every $\mu \in [0, 1]$ the operator V is relatively bounded with respect to $T + \mu V$, i.e., there exist $a_\mu \geq 0$ and $b_\mu \geq 0$ for which

$$\|Vf\| \leq a_\mu\|f\| + b_\mu\|(T + \mu V)f\|.$$

Consequently, for $|\mu - \mu'| < (2b_\mu)^{-1}$

$$\begin{aligned}
\|Vf\| &\leq a_\mu\|f\| + b_\mu(\|(T + \mu' V)f\| + |\mu' - \mu| \, \|Vf\|) \\
&< a_\mu\|f\| + b_\mu\|(T + \mu' V)f\| + \tfrac{1}{2}\|Vf\|,
\end{aligned}$$

and thus

$$\|Vf\| < 2a_\mu\|f\| + 2b_\mu\|(T + \mu' V)f\|.$$

The segment $[0, 1]$ is covered by the open intervals $(\mu - (2b_\mu)^{-1}, \mu + (2b_\mu)^{-1})$, $\mu \in [0, 1]$. Consequently, there are finitely many $\mu_1, \ldots, \mu_n$ for which the corresponding intervals cover the whole interval $[0, 1]$. The operator V is therefore $(T + \mu V)$-bounded for all $\mu \in [0, 1]$ with relative bound $b = \max\{2b_{\mu_j} : j = 1, \ldots, n\}$. If we choose $m \in \mathbb{N}$ such that $b/m < 1$ holds, then by successive applications of Theorem 9.1 we obtain the semi-boundedness of $T + (1/m)V, T + 2(1/m)V, \ldots, T + V$. □

Theorem 9.1 also enables us to prove the following useful inequalities.

Theorem 9.3 (Heinz). *Assume that T is self-adjoint, non-negative, S is symmetric, $D(T) \subset D(S)$, and $\|Sf\| \leq \|Tf\|$ for all $f \in D(T)$. Then*

$$|\langle f, Sf\rangle| \leq \langle f, Tf\rangle \quad \text{for all} \quad f \in D(T).$$

PROOF. Theorem 9.1 applies with $V = \kappa S$ for every $\kappa \in (-1, 1)$ if we take $a = 0$, $b = |\kappa|$, and $\gamma_T = 0$. Then $\gamma = 0$. Consequently, $T + \kappa S$ is self-adjoint and non-negative for every $\kappa \in (-1, 1)$. For $\kappa \to \pm 1$ we get

$$\left.\begin{array}{l}\langle f, (T+S)f\rangle \geq 0 \\ \langle f, (T-S)f\rangle \geq 0\end{array}\right\} \quad \text{for all} \quad f \in D(T),$$

and thus

$$\langle f, Tf\rangle \geq |\langle f, Sf\rangle| \quad \text{for all} \quad f \in D(T). \qquad \square$$

Theorem 9.4. *Let S and T be self-adjoint and non-negative.*
(a) *$D(T) \subset D(S)$ and $\|Sf\| \leq \|Tf\|$ for all $f \in D(T)$ imply $D(T^{1/2}) \subset D(S^{1/2})$ and $\|S^{1/2}f\| \leq \|T^{1/2}f\|$ for all $f \in D(T^{1/2})$.*
(b) *$D(T) \subset D(S)$ implies $D(T^{1/2}) \subset D(S^{1/2})$. The equality $D(T) = D(S)$ implies $D(T^{1/2}) = D(S^{1/2})$.*

PROOF.
(a) It follows from Theorem 9.3 that

$$\|S^{1/2}f\|^2 = \langle f, Sf\rangle \leq \langle f, Tf\rangle = \|T^{1/2}f\|^2 \quad \text{for all} \quad f \in D(T).$$

Let $f \in D(T^{1/2})$. Since $D(T)$ is a core of $T^{1/2}$, there is a sequence (f_n) from $D(T)$ for which $f_n \to f$ and $T^{1/2}f_n \to T^{1/2}f$. Then $(S^{1/2}f_n)$ is also a Cauchy sequence. Therefore, $f \in D(S^{1/2})$ and $S^{1/2}f = \lim_{n\to\infty} S^{1/2}f_n$. Consequently,

$$\|S^{1/2}f\| = \lim_{n\to\infty} \|S^{1/2}f_n\| \leq \lim_{n\to\infty} \|T^{1/2}f_n\| = \|T^{1/2}f\|.$$

(b) Because of the inclusion $D(T) \subset D(S)$, the operator S is T-bounded by Theorem 5.9, i.e., there exists a $c \geq 0$ such that

$$\begin{aligned}\|Sf\| &\leq c(\|f\| + \|Tf\|) \leq \sqrt{2}\, c(\|f\|^2 + \|Tf\|^2)^{1/2} \\ &\leq \sqrt{2}\, c(\|f\|^2 + 2\langle f, Tf\rangle + \|Tf\|^2)^{1/2} = \sqrt{2}\, c\|(I+T)f\|.\end{aligned}$$

By part (a), $D((I+T)^{1/2})$ is therefore contained in $D(S^{1/2})$. We can immediately infer from the spectral theorem that $D(T^{1/2}) = D((I+T)^{1/2})$. Consequently, $D(T^{1/2}) \subset D(S^{1/2})$. If $D(T) = D(S)$, then it also follows that $D(S^{1/2}) \subset D(T^{1/2})$, and thus that $D(S^{1/2}) = D(T^{1/2})$. $\square$

Now we prove a result concerning the continuous dependence, on the given operator, of the spectrum and the essential spectrum.

Theorem 9.5. *Let T and T_n $(n\in\mathbb{N})$ be self-adjoint, and assume that $D(T)=D(T_n)$. Assume, furthermore, that there are null-sequences (a_n) and (b_n) from $\mathbb{R}$ for which*

$$\|(T-T_n)f\| \leqslant a_n\|f\| + b_n\|Tf\| \quad \textit{for all} \quad f\in D(T).$$

Then $\sigma(T)=\lim_{n\to\infty}\sigma(T_n)$ and $\sigma_e(T)=\lim_{n\to\infty}\sigma_e(T_n)$.

PROOF. We have to prove that $\lambda\in\sigma_e(T)$ (respectively $\lambda\in\sigma(T)$) if and only if there is a sequence (λ_n) for which $\lambda_n\in\sigma_e(T_n)$ (respectively $\lambda_n\in\sigma(T_n)$) and $\lambda_n\to\lambda$. There is no loss of generality in considering only the point $\lambda=0$. Let E and E_n denote the spectral families of T and T_n.

(a) If $0\in\sigma(T)$, then there is a sequence (f_n) from $D(T)$ for which $\|f_n\|=1$ and $Tf_n\to 0$. Then

$$\begin{aligned} d(0,\sigma(T_n)) &\leqslant \|T_nf_n\| \leqslant \|Tf_n\| + \|(T-T_n)f_n\| \\ &\leqslant (1+b_n)\|Tf_n\| + a_n\|f_n\| \to 0 \quad \text{as} \quad n\to\infty. \end{aligned}$$

Therefore, $d(0,\sigma(T_n))\to 0$, i.e., $0\in\lim_{n\to\infty}\sigma(T_n)$.

If $0\notin\sigma(T)$, then T is bijective. Since

$$\|(T-T_n+\lambda)T^{-1}\| \leqslant a_n\|T^{-1}\| + b_n + |\lambda|\,\|T^{-1}\|,$$

by Theorem 5.11 $T_n-\lambda$ is also bijective for $|\lambda|<(2\|T^{-1}\|)^{-1}$ and sufficiently large $n\in\mathbb{N}$. Hence, $0\notin\lim_{n\to\infty}\sigma(T_n)$.

(b) Assume that $0\in\sigma_e(T)$. Then $\dim R(E(\epsilon)-E(-\epsilon))=\infty$ for every $\epsilon>0$, by Theorem 7.24. If $n_0\in\mathbb{N}$ is such that $a_n+b_n\epsilon<\epsilon$ for all $n\geqslant n_0$, then it follows for all $n\geqslant n_0$ and $f\in R(E(\epsilon)-E(-\epsilon))$, $f\neq 0$ that

$$\begin{aligned} \|T_nf\| &\leqslant \|Tf\| + \|(T-T_n)f\| \leqslant \epsilon\|f\| + a_n\|f\| + b_n\|Tf\| \\ &< 2\epsilon\|f\|. \end{aligned}$$

It follows from this that $\dim R(E_n(2\epsilon)-E_n(-2\epsilon))=\infty$, since otherwise there would exist an

$$f\in R(E_n(2\epsilon)-E_n(-2\epsilon))^{\perp}\cap R(E(\epsilon)-E(-\epsilon)), \quad f\neq 0;$$

for this f we would have

$$2\epsilon\|f\| \leqslant \|T_nf\| < 2\epsilon\|f\|.$$

By the proposition preceding Theorem 7.25 we have $\sigma_e(T_n)\cap[-2\epsilon,2\epsilon]\neq\emptyset$ for every $n\geqslant n_0$. As $\epsilon>0$ was arbitrary, it follows that $0\in\lim_{n\to\infty}\sigma_e(T_n)$.

Let us assume that $0\notin\sigma_e(T)$, i.e., that there exists an $\epsilon>0$ such that $\dim R(E(\epsilon)-E(-\epsilon))<\infty$. If $n_0\in\mathbb{N}$ is such that $a_n<\epsilon/3$ and $b_n<1/3$ for all $n\geqslant n_0$, then

$$\begin{aligned} \|T_nf\| &\geqslant \|Tf\| - \|(T-T_n)f\| > \|Tf\| - \tfrac{1}{3}\epsilon\|f\| - \tfrac{1}{3}\|Tf\| \\ &= \tfrac{2}{3}\|Tf\| - \tfrac{1}{3}\epsilon\|f\| \geqslant \tfrac{1}{3}\epsilon\|f\| \end{aligned}$$

for all $n \geq n_0$ and $f \in R(E(\epsilon)-E(-\epsilon))^{\perp}, f \neq 0$. It follows from this that $\dim R(E_n(\epsilon/3)-E_n(-\epsilon/3)) < \infty$, since otherwise there would exist an

$$f \in R\big(E_n(\tfrac{1}{3}\epsilon)-E_n(-\tfrac{1}{3}\epsilon)\big) \cap R(E(\epsilon)-E(-\epsilon))^{\perp}, \quad f \neq 0;$$

for this f we would have

$$\tfrac{1}{3}\epsilon\|f\| \geq \|T_n f\| > \tfrac{1}{3}\epsilon\|f\|.$$

Consequently, $(-\epsilon/3, \epsilon/3) \cap \sigma_e(T_n) = \emptyset$ for all $n \geq n_0$, and thus $0 \notin \lim_{n\to\infty} \sigma_e(T_n)$. □

Corollary. *Assume that T is self-adjoint, V is symmetric and T-bounded, and Ω denotes the set of those real μ for which $T+\mu V$ is self-adjoint. Then the set-valued functions $\mu \mapsto \sigma(T+\mu V)$ and $\mu \mapsto \sigma_e(T+\mu V)$ are continuous on Ω (i.e., if $\mu_0, \mu_n \in \Omega$, and $\mu_n \to \mu_0$, then $\sigma(T+\mu_0 V) = \lim_{n\to\infty}\sigma(T+\mu_n V)$ and $\sigma_e(T+\mu_0 V) = \lim_{n\to\infty}\sigma_e(T+\mu_n V)$).*

PROOF. V is $(T+\mu_0 V)$-bounded for every $\mu_0 \in \Omega$. Therefore,

$$\begin{aligned}\|[(T+\mu_0 V)-(T+\mu_n V)]f\| &= |\mu_0-\mu_n|\,\|Vf\| \\ &\leq |\mu_0-\mu_n| a\|f\| + |\mu_0-\mu_n| b\|(T+\mu_0 V)f\|.\end{aligned}$$

Consequently, the operators $T+\mu_0 V$ and $T+\mu_n V$ satisfy the assumptions of Theorem 9.5. □

EXERCISES

9.1. The converse of Theorem 9.4(a) does not hold. If we consider the operators induced by the matrices

$$S = 3\begin{pmatrix}1 & 1\\ 1 & 1\end{pmatrix} \quad \text{and} \quad T = 4\begin{pmatrix}1 & 0\\ 0 & 3\end{pmatrix}$$

on $\mathbb{C}^2$ (with the usual scalar product), then $S \leq T$, but not $S^2 \leq T^2$.

9.2. Let the assumptions of Theorem 9.5 be satisfied. Assume that $\alpha, \beta \in \mathbb{R} \cap \rho(T)$ and $\alpha < \beta$. Then $\alpha, \beta \in \rho(T_n)$ for large n and $\|(E_n(\beta)-E_n(\alpha))-(E(\beta)-E(\alpha))\| \to 0$ as $n \to \infty$.
Hint: The second resolvent identity, Exercise 7.20, and Theorem 9.5.

9.2 Relatively compact perturbations and the essential spectrum

By Theorem 7.24 the number $\lambda \in \mathbb{R}$ belongs to the essential spectrum of a self-adjoint operator T if and only if there exists a sequence (f_n) from $D(T)$ for which

$$f_n \xrightarrow{w} 0, \ \liminf_{n\to\infty} \|f_n\| \neq 0, \ (\lambda - T)f_n \to 0. \tag{9.3}$$

Such a sequence is called a *singular sequence* for T and λ. With the aid of this characterization, we can prove the following theorem.

Theorem 9.6. *Let T_1 and T_2 be self-adjoint operators with the spectral families E_1 and E_2.*

(a) *If $R(E_1(J)) \subset D(T_2)$ and $(T_1 - T_2)E_1(J)$ is compact for every bounded interval J, then $\sigma_e(T_1) \subset \sigma_e(T_2)$.*

(b) *If the assumptions of* (a) *are satisfied and $D(T_1) \subset D(T_2)$, then every sequence that is singular for T_1 and λ is also singular for T_2 and λ.*

PROOF.

(a) Assume that $\lambda \in \sigma_e(T_1)$. As in the proof of Theorem 7.24 (part (i) *implies* (ii)) we can show that there exists a singular sequence (f_n) for T_1 and λ that is contained in $R(E_1(\lambda+1) - E_1(\lambda-1))$ (cf. also part (b) of this proof). As (f_n) tends to 0 weakly and as $(T_1 - T_2)(E_1(\lambda+1) - E_1(\lambda-1))$ is compact, we have

$$(T_1 - T_2)f_n = (T_1 - T_2)(E_1(\lambda+1) - E_1(\lambda-1))f_n \to 0$$

as $n \to \infty$ (cf. Theorem 6.3). We obtain from this that

$$\|(T_2 - \lambda)f_n\| \leq \|(T_2 - T_1)f_n\| + \|(T_1 - \lambda)f_n\| \to 0$$

as $n \to \infty$. Therefore, (f_n) is a singular sequence for T_2 and λ, and thus $\lambda \in \sigma_e(T_2)$.

(b) Due to the inclusion $D(T_1) \subset D(T_2)$, the operator T_2 is relatively bounded with respect to T_1 (cf. Theorem 5.9). Let (f_n) be a singular sequence for T_1 and λ. Then

$$\begin{aligned}\|(I - E_1(\lambda+1) + E_1(\lambda-1))f_n\|^2 &= \int_{(-\infty,\lambda-1]} + \int_{(\lambda+1,\infty)} 1 \, \mathrm{d}\|E_1(t)f_n\|^2 \\ &\leq \int_{-\infty}^{\infty} |\lambda - t|^2 \, \mathrm{d}\|E_1(t)f_n\|^2 \\ &= \|(\lambda - T_1)f_n\|^2 \to 0 \qquad (9.4)\end{aligned}$$

and

$$\begin{aligned}&\|T_1(I - E_1(\lambda+1) + E_1(\lambda-1))f_n\|^2 \\ &\quad = \int_{(-\infty,\lambda-1]} + \int_{(\lambda+1,\infty)} |t|^2 \, \mathrm{d}\|E_1(t)f_n\|^2 \\ &\quad \leq \int_{(-\infty,\lambda-1]} + \int_{(\lambda+1,\infty)} (|\lambda| + |\lambda - t|)^2 \, \mathrm{d}\|E_1(t)f_n\|^2 \\ &\quad \leq 2|\lambda|^2\|(I - E_1(\lambda+1) + E_1(\lambda-1))f_n\|^2 \\ &\qquad + 2\|(\lambda - T_1)f_n\|^2 \to 0 \quad \text{as} \quad n \to \infty. \qquad (9.5)\end{aligned}$$

Hence, $((E_1(\lambda+1)-E_1(\lambda-1))f_n)$ is a singular sequence for T_1 and λ that is contained in $R(E_1(\lambda+1)-E_1(\lambda-1))$. As in part (a), it follows that

$$(\lambda-T_2)(E_1(\lambda+1)-E_1(\lambda-1))f_n \to 0 \quad \text{as} \quad n\to\infty.$$

Since T_2 is bounded with respect to T_1, we can derive from (9.4) and (9.5) that

$$T_2(I-E_1(\lambda+1)+E_1(\lambda-1))f_n \to 0 \quad \text{as} \quad n\to\infty,$$

and thus

$$(\lambda-T_2)f_n\to 0 \quad \text{as} \quad n\to\infty.$$

Consequently, (f_n) is a singular sequence for T_2 and λ. □

If in Theorem 9.6 we make the assumptions symmetrical with respect to T_1 and T_2, then we obtain a criterion for the coincidence of the essential spectra and for the coincidence of the singular sequences (in the following we say that T_1 and T_2 have the *same singular sequences* if (f_n) is a singular sequence for T_1 and λ if and only if (f_n) is a singular sequence for T_2 and λ). However, the result is not very useful in this form since too many properties of T_1 *and* T_2 are explicitly assumed; usually only one of these operators (the *unperturbed* one) is known accurately. In what follows we give conditions that imply the assumptions of Theorem 9.6. First we need some preparation.

Let H_1, H_2, and H_3 be Hilbert spaces. Let A be an operator from H_1 into H_2. An operator B from H_1 into H_3 is said to be *A-compact* if $D(A)\subset D(B)$ and B, as a mapping from $(D(A), \|\cdot\|_A)$ into H_3, is compact. If A is bounded and $D(A)=H_1$, then it is obvious that B is A-compact if and only if B is compact (since the norms $\|\cdot\|$ and $\|\cdot\|_A$ are then equivalent).

Proposition. *If A is an A_1-bounded operator and B is A-compact, then B is also A_1-compact. If A is densely defined and closed, then B is A-compact if and only if B is $|A|$-compact.*

Proof. Any $\|\cdot\|_{A_1}$-bounded set is also $\|\cdot\|_A$-bounded. Consequently, every $\|\cdot\|_{A_1}$-bounded set is mapped by B onto a relatively compact set. The second assertion follows from the equalities $D(A)=D(|A|)$ and $\|\cdot\|_A=\|\cdot\|_{|A|}$. □

If B is an A-compact operator, then B, as an operator from $(D(A), \|\cdot\|_A)$ into H_3, is bounded, i.e., B is A-bounded. In fact, much more is true.

Theorem 9.7. *Let A be an operator from H_1 into H_2, and let B be an A-compact operator from H_1 into H_3. If A or B is closable, then B is A-bounded with A-bound zero.*

PROOF. Let us assume that the A-bound of B is positive. Then there exists an $\epsilon > 0$ with the property that for every $n \in \mathbb{N}$ there is an $f_n \in D(A)$ such that $\|Bf_n\| > n\|f_n\| + \epsilon\|Af_n\|$ (we can choose any positive number ϵ that is less than the A-bound of B). If we put $g_n = \|Bf_n\|^{-1}f_n$, then $\|Bg_n\| = 1$ and

$$\epsilon(\|g_n\| + \|Ag_n\|) \leqslant n\|g_n\| + \epsilon\|Ag_n\| < \|Bg_n\| = 1$$

for all $n \in \mathbb{N}$ such that $n \geqslant \epsilon$. It follows from this that $g_n \to 0$ and $\|g_n\|_A \leqslant \|g_n\| + \|Ag_n\| < 1/\epsilon$.

First let B be closable. As B is A-compact, there exists a subsequence (g_{n_k}) of (g_n) for which $Bg_{n_k} \to h \in H_3$. The formula $\|Bg_n\| = 1$ implies $\|h\| = 1$. This is a contradiction because of $g_n \to 0$ and the closability of B.

Now let A be closable. Without loss of generality we can assume that $D(B) = D(A)$. By Theorem 6.2 the operator B can be extended to an $\bar{A}$-compact operator $\tilde{B}$ on $D(\bar{A})$. Consequently, we can assume without loss of generality that A is closed. Since $\|Ag_n\| < 1/\epsilon$, there exists a subsequence (g_{n_k}) of (g_n) such that $Ag_{n_k} \xrightarrow{w} h \in H_2$ (cf. Theorem 4.25). Hence, $(0, h) = w-\lim_{k\to\infty}(g_{n_k}, Ag_{n_k}) \in \overline{G(A)} = G(A)$, and thus $h = 0$. It follows by Theorem 6.3 that $Bg_{n_k} \to 0$, which contradicts the equality $\|Bg_{n_k}\| = 1$. □

Corollary. *Let B be closable. B is A-compact if and only if it is $(A+B)$-compact.*

PROOF. Let B be A-compact. Since $D(A) \subset D(B)$, we have $D(A+B) = D(A) \subset D(B)$. By Theorem 9.7 there is an $a \geqslant 0$ such that $\|Bf\| \leqslant a\|f\| + \|Af\|/2$ for all $f \in D(A)$. Consequently, for all $f \in D(A)$

$$\|Af\| \leqslant 2(\|Af\| - \|Bf\| + a\|f\|) \leqslant 2(\|(A+B)f\| + a\|f\|),$$

i.e., A is $(A+B)$-bounded. The $(A+B)$-compactness of B follows by the above proposition. We can prove similarly the other direction. □

Theorem 9.8. *Let A be a closed operator from H_1 into H_2, and let B be an operator from H_1 into H_3. Then the following assertions are equivalent.*

(i) *B is A-compact.*

(ii) *$f_n \in D(A)$, $f_n \xrightarrow{w} 0$ and $Af_n \xrightarrow{w} 0$ imply that $Bf_n \to 0$.*

If A is self-adjoint on H_1 and E denotes its spectral family, then these assertions are equivalent to

(iii) *$BE(J)$ is compact for every bounded interval J, and B is A-bounded with A-bound zero.*

PROOF. (i) *implies* (ii): If $f_n \xrightarrow{w} 0$ and $Af_n \xrightarrow{w} 0$, then

$$\langle f_n, g\rangle_A = \langle f_n, g\rangle + \langle Af_n, Ag\rangle \to 0$$

for all $g \in D(A)$, i.e., (f_n) weakly tends to zero in the Hilbert space $(D(A), \langle . , .\rangle_A)$. It follows from this by Theorem 6.3 that $Bf_n \to 0$.

(ii) *implies* (i): Let (f_n) be a weak null-sequence in $(D(A), \langle . , .\rangle_A)$, i.e., assume that $\langle f_n, g\rangle + \langle Af_n, h\rangle \to 0$ for all $(g, h) \in G(A)$. By Theorem 6.3 it is sufficient to show that (Bf_n) tends to zero as $n \to \infty$. Since $\langle f_n, g \rangle + \langle Af_n, h \rangle = 0$ for all $(g, h) \in G(A)^{\perp}$, we have

$$\langle f_n, g \rangle + \langle Af_n, h \rangle \to 0 \quad \text{for all} \quad (g, h) \in H_1 \oplus H_2.$$

It follows from this that $f_n \xrightarrow{w} 0$ and $Af_n \xrightarrow{w} 0$, and thus $Bf_n \to 0$ (because of (ii)).

(i) *implies* (iii): By Theorem 9.7 the operator B is A-bounded with A-bound zero. Let (f_n) be a bounded sequence in H_1. Then the sequence $(AE(J)f_n)$ is also bounded, because

$$\|AE(J)f_n\|^2 = \int_J t^2 \, \mathrm{d}\|E(t)f_n\|^2 \leqslant \|f_n\|^2 \sup\{t^2 : t \in J\}.$$

Consequently, $(E(J)f_n)$ is a bounded sequence in $(D(A), \langle . , . \rangle_A)$, and by (i) there exists a subsequence (f_{n_k}) for which $(BE(J)f_{n_k})$ is convergent. Hence, $BE(J)$ is compact.

(iii) *implies* (i): Let (f_n) be a weak null-sequence from $(D(A), \langle . , . \rangle_A)$, i.e., assume that $\langle f_n, g \rangle + \langle Af_n, h \rangle \to 0$ for all $(g, h) \in H_1 \oplus H_1$ (cf. the "(ii) *implies* (i)" part of the proof). Then $f_n \xrightarrow{w} 0$, and thus $B(E(N) - E(-N))f_n \to 0$ for all $N > 0$. Since the A-bound of B is equal to 0, for every $\epsilon > 0$ there is a $C > 0$ such that

$$\|Bf\| \leqslant \frac{\epsilon}{2}\|Af\| + C\|f\| \quad \text{for all} \quad f \in D(A).$$

Therefore, for all $n \in \mathbb{N}$ and sufficiently large N

$$\begin{aligned}\|B(I - E(N) + E(-N))f_n\| &\leqslant \frac{\epsilon}{2}\|Af_n\| + C\|(I - E(N) + E(-N))f_n\| \\ &\leqslant \frac{\epsilon}{2}\|Af_n\| + \frac{C}{N}\|Af_n\| \leqslant \epsilon\|Af_n\|,\end{aligned}$$

and thus

$$\limsup_{n\to\infty} \|Bf_n\| \leqslant \epsilon \limsup_{n\to\infty} \|Af_n\|.$$

Since the sequence (Af_n) is bounded and since $\epsilon > 0$ was arbitrary, it follows that $Bf_n \to 0$. Consequently, *B is A-compact.* □

Now we can prove an old result that is essentially due to H. Weyl.

Theorem 9.9. *Let T be a self-adjoint operator on the Hilbert space H, and let V be a symmetric T-compact operator. Then $T + V$ is self-adjoint, T and $T + V$ have the same singular sequences, and $\sigma_e(T) = \sigma_e(T + V)$.*

PROOF. By Theorem 9.7 the operator V is T-bounded with T-bound 0. Therefore, $T + V$ is self-adjoint by Theorem 5.28. V is also $(T + V)$-compact by the corollary to Theorem 9.7. Now it follows from Theorem 9.8 that $T_1 = T$ and $T_2 = T + V$ satisfy the assumptions of Theorem 9.6(b), just as $T_1 = T + V$ and $T_2 = T$ do. Consequently, T and $T + V$ have the same singular sequences, and $\sigma_e(T+V) = \sigma_e(T)$. □

In applications (particularly to differential operators) it is important that the assumptions of Theorem 9.9 can be somewhat weakened. First we prove some preparatory theorems.

Theorem 9.10. *Let T be a self-adjoint operator on H such that $\rho(T) \neq \emptyset$, and let $p > 0$. An operator V is T^p-compact (respectively T^p-bounded) if and only if $V(z-T)^{-p}$ is compact (respectively bounded) for some (and then for all) $z \in \rho(T)$.*

(The operators T^p and $(z-T)^{-p}$ are defined with the aid of the spectral theorem by the formulae $T^p = \int t^p \, \mathrm{d}E(t)$ and $(z-T)^{-p} = \int (z-t)^{-p} \, \mathrm{d}E(t)$, where $t \mapsto t^p$ and $t \mapsto (z-t)^{-p}$ are chosen to be continuous on $\sigma(T)$.)

Proof. We obviously have $D(T^p) = D((z-T)^p)$; the T^p-norm and the $(z-T)^p$-norm are equivalent. Consequently, V is T^p-compact (T^p-bounded) if and only if it is $(z-T)^p$-compact ($(z-T)^p$-bounded). Since $(z-T)^p : D(T^p) \to H$ is continuous and continuously invertible, $V : D(T^p) \to H$ is compact (bounded) if and only if $V(z-T)^{-p}$ is compact (bounded). □

Theorem 9.11. *Let T be a self-adjoint operator with spectral family E, and let V be a T-bounded operator. Then*

(a) *V is T^p-bounded with T^p-bound zero for all $p > 1$.*
(b) *If V is T^p-compact for some $p \geq 0$, then $VE(J)$ is compact for every bounded interval J.*
(c) *If $VE(J)$ is compact for every bounded interval J, then V is T^p-compact for every $p > 1$.*
(d) *V is T-compact if and only if it is T^2-compact and T-bounded with T-bound zero.*

Proof.

(a) There are numbers $a, b \geq 0$ such that $\|Vf\| \leq a\|f\| + b\|Tf\|$ for all $f \in D(T)$. We have $D(T^p) \subset D(T)$ for $p > 1$, and thus

$$\begin{aligned}\|Vf\| &\leq \|V(E(N)-E(-N))f\| + \|V(I-E(N)+E(-N))f\| \\ &\leq a\|f\| + bN\|f\| + a\|f\| + b\|T(I-E(N)+E(-N))f\| \\ &\leq (2a+bN)\|f\| + bN^{1-p}\|T^p f\|\end{aligned}$$

for all $f \in D(T^p)$ and $N > 0$. Since N can be chosen arbitrarily large, the assertion follows from this.

(b) Let (f_n) be a bounded sequence. Then $(E(J)f_n)$ is a bounded sequence in $(D(T^p), \langle . , . \rangle_{T^p})$. Since V is T^p-compact, there exists a subsequence $(E(J)f_{n_k})$ for which $(VE(J)f_{n_k})$ is convergent. Hence, $VE(J)$ is compact.

(c) The operator

$$V(1+|T|)^{-p}(E(N)-E(-N)) = V(E(N)-E(-N))(1+|T|)^{-p}$$

is compact for every $N \geq 0$, and

$$\|V(1+|T|)^{-p}(I-E(N)+E(-N))\| \leq \|V(1+|T|)^{-1}\|(1+N)^{1-p}.$$

Consequently, $V(1+|T|)^{-p} = \lim_{N\to\infty} V(1+|T|)^{-p}(E(N)-E(-N))$, and thus $V(1+|T|)^{-p}$ is compact. V is therefore T^p-compact by Theorem 9.10 and the proposition preceding Theorem 9.7.

(d) This follows from Theorem 9.8 and parts (b) and (c) of this theorem. □

Theorem 9.12. *Let T_1 and T_2 be self-adjoint operators and assume that $D(T_1) = D(T_2)$. Put $V = T_2 - T_1$.*

(a) V is T_1^2-compact if and only if it is T_2^2-compact.

(b) If V is T_1^2-compact (or T_2^2-compact), then $(z-T_1)^{-2}-(z-T_2)^{-2}$ is compact for every $z \in \rho(T_1)\cap\rho(T_2)$.

(c) If V is T_1^2-compact, then for every T_1-bounded operator W we have the following: W is T_1^2-compact if and only if it is T_2^2-compact.

PROOF. Write $R_j = (z-T_j)^{-1}$ for $z \in \rho(T_1)\cap\rho(T_2)$. Then the operators VR_j are bounded for $j = 1, 2$. (If H is real and $\sigma(T_1)\cup\sigma(T_2) = \mathbb{R}$, then H must be complexified in order that we may have $\rho(T_1)\cap\rho(T_2) \neq \emptyset$.)

(a) If V is T_1^2-compact, then VR_1^2 is compact by Theorem 9.10. It follows from the resolvent identity $R_2 - R_1 = R_1VR_2 = R_2VR_1$ that

$$R_2^2 = (R_1 + R_2VR_1)(R_1 + R_1VR_2) = (I+R_2V)R_1^2(I+VR_2), \tag{9.6}$$

and thus

$$VR_2^2 = (I+VR_2)VR_1^2(I+VR_2). \tag{9.7}$$

It follows from this that VR_2^2 is compact. Therefore, V is T_2^2-compact. We can prove similarly the reverse direction.

(b) Let V be T_1^2-compact. (9.6) implies that

$$\begin{aligned}R_2^2 - R_1^2 &= R_1^2VR_2 + R_2VR_1^2(I+VR_2)\\ &= \left[(z^*-T_2)^{-1}V(z^*-T_1)^{-2}\right]^* + R_2VR_1^2(I+VR_2);\end{aligned}$$

here we have used the equality

$$\langle f, (z^*-T_2)^{-1}V(z^*-T_1)^{-2}g\rangle = \langle R_1^2VR_2f, g\rangle$$

for all $f, g \in H$. As $V(z^*-T_1)^{-2}$ and VR_1^2 are compact, the compactness of $R_2^2 - R_1^2$ follows.

(c) Let W be T_1^2-compact. Then WR_1^2 is compact. It follows from (9.6) that

$$WR_2^2 = (WR_1^2 + WR_2VR_1^2)(I+VR_2).$$

As WR_1^2 and VR_1^2 are compact, WR_2^2 is also compact, i.e., W is T_2^2-compact. We can prove similarly the other direction. □

Theorem 9.13. *Let T be a self-adjoint operator on H, and denote its spectral family by E. Assume that V is symmetric, $D(T) \subset D(V)$, and $T+V$ is self-adjoint. Assume, furthermore, that $VE(J)$ is compact for every bounded interval J (this condition can be replaced by the following: V is T^p-compact for some $p \geqslant 0$). Then T and $T+V$ have the same singular sequences. In particular, $\sigma_e(T) = \sigma_e(T+V)$.*

Proof. It follows from the assumptions that V is T^2-compact, and thus also $(T+V)^2$-compact by Theorem 9.12. Therefore $VE'(J)$ is also compact for every bounded interval J, where E' denotes the spectral family of $T+V$. The assumptions of Theorem 9.6(b) are therefore satisfied for $T_1 = T$, $T_2 = T+V$ and for $T_1 = T+V$, $T_2 = T$. If V is T^p-compact for some $p > 0$, then the compactness of $VE(J)$ follows from Theorem 9.11(b) for every bounded interval J. □

The assumptions of Theorem 9.13 do not guarantee that $T+V$ is semi-bounded in case T is semi-bounded.

Example 1. Let T be a semi-bounded self-adjoint operator with discrete spectrum, i.e., assume that there exist an orthonormal basis $\{e_n : n \in \mathbb{N}\}$ and a sequence (λ_n) for which $\lambda_n \to \infty$ and

$$D(T) = \left\{ f \in H : \sum_{n \in \mathbb{N}} |\lambda_n|^2 |\langle e_n, f \rangle|^2 < \infty \right\},$$

$$Tf = \sum_{n \in \mathbb{N}} \lambda_n \langle e_n, f \rangle e_n \quad \text{for} \quad f \in D(T).$$

Furthermore, write $V = -2T$. Then V is T-bounded and $-T = T+V$ is self-adjoint. Since the space $R(E(J))$ is finite-dimensional for every bounded interval J, the operator $VE(J)$ is compact. Consequently, all assumptions of Theorem 9.13 are satisfied, T is bounded from below, and $T+V$ is not bounded from below.

The following theorem studies the behavior of gaps in the essential spectrum of T in the case of a non-negative T^2-compact perturbation.

Theorem 9.14. *Let T be self-adjoint operator on H such that $\sigma_e(T) \cap (a, b) = \varnothing$. Assume that the point b is not an accumulation point of those eigenvalues of T that belong to (a, b). Assume, furthermore, that V is symmetric, non-negative, T^2-compact and T-bounded with T-bound < 1. Then $\sigma_e(T+V) \cap (a, b) = \varnothing$, and b is not an accumulation point of those eigenvalues of $T+V$ that belong to (a, b). (If $V \leqslant 0$, then a similar result holds for the point a.)*

Proof. We only have to prove the second assertion. We can assume, without loss of generality, that $(a, b)=(-1, 1)$. We use Theorem 7.25.

Since, by assumption, V is T-bounded with T-bound <1, there exists a $c\geqslant 0$ for which

$$\|Vf\| \leqslant c(\|f\|+\|(T+sV)f\|) \quad \text{for all} \quad f\in D(T) \quad \text{and} \quad s\in[0,1].$$

Let $s_0\in[0,1]$ be chosen such that 1 is not an accumulation point of eigenvalues of $T+s_0V$ belonging to $(-1, 1)$. By assumption, this holds for $s_0=0$ in any event. Let E_0 denote the spectral family of $T+s_0V$. Since $\sigma_e(T+s_0V)\cap(-1,1)=\varnothing$ (cf. Theorem 9.13), $E_0(1-)-E_0(0)$ is of finite rank.

For all $f\in D(T)$ we have

$$\|Vf\| \leqslant c(\|f\|+\|(T+s_0V)f\|) \leqslant 2c\|(1+|T+s_0V|)f\|.$$

It follows from this by Theorem 9.3 that

$$\langle f, Vf\rangle \leqslant 2c\langle f, (1+|T+s_0V|)f\rangle \quad \text{for} \quad f\in D(T),$$

and thus

$$\langle f, Vf\rangle \leqslant 2c\langle f, (1-T-s_0V)f\rangle \quad \text{for} \quad f\in R(E_0(0))\cap D(T).$$

If $s>s_0$ and $s-s_0\leqslant(1/4c)$, then $T+sV=(T+s_0V)+(s-s_0)V$ is self-adjoint and

$$\begin{aligned}\langle f, (T+sV)f\rangle &= \langle f, (T+s_0V)f\rangle+(s-s_0)\langle f, Vf\rangle\\ &\leqslant \langle f, (T+s_0V)f\rangle+(1/4c)\langle f, Vf\rangle\\ &\leqslant \langle f, (T+s_0V)f\rangle+\tfrac{1}{2}\langle f, (1-T-s_0V)f\rangle\\ &= \tfrac{1}{2}\langle f, (T+s_0V)f\rangle+\tfrac{1}{2}\|f\|^2 \leqslant \tfrac{1}{2}\|f\|^2\end{aligned}$$

for $f\in R(E_0(0))\cap D(T)$. It is obvious that

$$\langle f, (T+sV)f\rangle \geqslant \langle f, (T+s_0V)f\rangle \geqslant \|f\|^2$$

for $f\in R(I-E_0(1-))\cap D(T)$.

Since $R(E_0(1-)-E_0(0))$ is finite-dimensional, Theorem 7.25 implies the following: The interval $(\frac{1}{2}, 1)$ contains at most finitely many points of the spectrum of $T+sV$, and thus 1 is not an accumulation point of the eigenvalues of $T+sV$ from $(-1, 1)$.

If we choose $m\in\mathbb{N}$ such that $m\geqslant 4c$, and $\mu=1/m$, then in this way we can prove the assertion step by step for $T+\mu V, T+2\mu V, \ldots, T+m\mu V=T+V$, starting with $s_0=0$. □

Remark. In Theorem 9.14 the T-boundedness of V with T-bound <1 is not necessary. Instead, it is enough to assume that V is T-bounded and $T+sV$ is self-adjoint for all $s\in[0,1]$. For the proof see the technique used in Theorem 9.2.

EXERCISES

9.3. Let T_1 and T_2 be self-adjoint such that $D(T_1) \supset D(T_2)$.
(a) If T_1 has a pure discrete spectrum, then so does T_2.
(b) If $\{\lambda_j^{(1)}\}$ and $\{\lambda_j^{(2)}\}$ are the eigenvalues of T_1 and T_2, respectively (each eigenvalue counted according to its multiplicity), then $|\lambda_j^{(2)}| \geq a|\lambda_j^{(1)}| + b$ for all j, with appropriate numbers $a > 0$ and $b \geq 0$.
Hint: Use the equality $(\mathrm{i} - T_2)^{-1} = (\mathrm{i} - T_2)^{-1}(\mathrm{i} - T_1)(\mathrm{i} - T_1)^{-1}$, the boundedness of $(\mathrm{i} - T_2)^{-1}(\mathrm{i} - T_1)$, and (7.3).

9.4. Let T be self-adjoint, and denote by E its spectral family. If V is T-bounded, and $VE(J)$ is compact for every bounded interval J, then V is $f(T)$-compact for every E-measurable function f such that $\lim_{|t|\to\infty}|t^{-1}f(t)| = \infty$.

9.5. For a normal operator A let us define the *essential spectrum* $\sigma_e(A)$ to be the set of accumulation points of $\sigma(A)$ plus the set of eigenvalues of infinite multiplicity. (This definition extends the definition for self-adjoint operators.)
(a) $\lambda \in \sigma_e(A)$ if and only if there exists a sequence (f_n) from $D(A)$ such that $\liminf_{n\to\infty}\|f_n\| > 0$, $f_n \xrightarrow{w} 0$ and $(\lambda - A)f_n \to 0$ ((f_n) is a *singular sequence* for A and λ).
(b) If A_1 and A_2 are normal, $D(A_1) = D(A_2)$, and $\rho(A_1) \cap \rho(A_2) \neq \emptyset$, and if $A_1 - A_2$ is A_1^p-compact for some $p > 0$, then $\sigma_e(A_1) \subset \sigma_e(A_2)$.
(c) Let T_1 and T_2 be normal. If $\sigma_e((z - T_1)^{-1}) = \sigma_e((z - T_2)^{-1})$ for some $z \in \rho(T_1) \cap \rho(T_2)$, then $\sigma_e(T_1) = \sigma_e(T_2)$. If $(z - T_1)^{-1} - (z - T_2)^{-1}$ is compact for some $z \in \rho(T_1) \cap \rho(T_2)$, then $\sigma_e(T_1) = \sigma_e(T_2)$.
(d) With the aid of (c) and Theorem 8.10 prove that all self-adjoint extensions of a symmetric operator with finite defect indices have the same essential spectrum.

9.6. Let T be a self-adjoint operator such that $(a, b) \cap \sigma(T) = \emptyset$ or $(a, b) \cap \sigma_e(T) = \emptyset$. If $V \geq 0$ is symmetric, $D(T) \subset D(V)$, $T + V$ is self-adjoint, and $\langle f, Vf\rangle \leq -\langle f, Tf\rangle + (a + \eta)\|f\|^2$ for all $f \in E(a)D(T)$ with some $\eta < b - a$, then $(a + \eta, b) \cap \sigma(T + V) = \emptyset$ or $(a + \eta, b) \cap \sigma_e(T + V) = \emptyset$, respectively. This holds in particular if V is bounded, symmetric, and $0 \leq V \leq \eta$.
Hint: Use Theorem 7.25.

9.7. Let H be a Hilbert space, and let A be an unbounded linear functional on H. (A is a non-closable operator from H into $\mathbb{K}$.) A is A-compact; however, the A-bound of A equals 1. (In Theorem 9.7 the assumption that A or B are closable cannot be dropped.)

9.3 Strong resolvent convergence

If $T_n (n \in \mathbb{N})$ and T are self-adjoint operators on the complex Hilbert space H, then we say that the sequence (T_n) *converges to T in the sense of the strong resolvent convergence* if $(z - T_n)^{-1} \xrightarrow{s} (z - T)^{-1}$ for some $z \in \mathbb{C}\setminus\mathbb{R}$. Then this holds for all $z \in \mathbb{C}\setminus\mathbb{R}$ on the basis of the following theorem.

Theorem 9.15. *Let T_n ($n \in \mathbb{N}$) and T be self-adjoint operators on the complex Hilbert space H. If $(z_0 - T_n)^{-1} \xrightarrow{s} (z_0 - T)^{-1}$ for some $z_0 \in \mathbb{C} \setminus \mathbb{R}$, then $(z - T_n)^{-1} \xrightarrow{s} (z - T)^{-1}$ for all $z \in \mathbb{C} \setminus \mathbb{R}$.*

PROOF. If $z \in \mathbb{C}$ and $|z - z_0| < |\mathrm{Im}\, z_0|$, then by Theorem 5.14 we have

$$(z - T_n)^{-1} f - (z - T)^{-1} f = \sum_{k=0}^{\infty} (z_0 - z)^k \left[(z_0 - T_n)^{-k-1} - (z_0 - T)^{-k-1} \right] f$$

for all $f \in H$. Therefore,

$$\|(z - T_n)^{-1} f - (z - T)^{-1} f\| \leq \sum_{k=0}^{N} |z_0 - z|^k \|(z_0 - T_n)^{-k-1} f$$
$$- (z_0 - T)^{-k-1} f\| + 2 \sum_{k>N} |z_0 - z|^k |\mathrm{Im}\, z_0|^{-k-1} \|f\|$$

for every $N \in \mathbb{N}$. The second sum will be arbitrarily small if N is chosen large enough. The first sum tends to 0 for fixed N as $n \to \infty$, since $(z_0 - T_n)^{-k-1} \xrightarrow{s} (z_0 - T)^{-k-1}$. Consequently, the assertion follows for all $z \in \mathbb{C}$ such that $|z - z_0| < |\mathrm{Im}\, z_0|$. An iterative application of this step provides the assertion for all z in the half-plane where z_0 lies. The limit relation $(z - T_n)^{-1} \xrightarrow{s} (z - T)^{-1}$ implies that $(z^* - T_n)^{-1} \xrightarrow{w} (z^* - T)^{-1}$ and (as $(z - T_n)^{-1}$ and $(z - T)^{-1}$ are normal) that

$$\|(z^* - T_n)^{-1} f\| = \|(z - T_n)^{-1} f\| \to \|(z - T)^{-1} f\| = \|(z^* - T)^{-1} f\|.$$

Therefore, for all $f \in H$

$$\begin{aligned}
&\|(z^* - T_n)^{-1} f - (z^* - T)^{-1} f\|^2 \\
&= \|(z^* - T_n)^{-1} f\|^2 - 2\,\mathrm{Re}\langle (z^* - T_n)^{-1} f, (z^* - T)^{-1} f \rangle + \|(z^* - T)^{-1} f\|^2 \\
&\to \|(z^* - T)^{-1} f\|^2 - 2\,\mathrm{Re}\langle (z^* - T)^{-1} f, (z^* - T)^{-1} f \rangle + \|(z^* - T)^{-1} f\|^2 \\
&= 0.
\end{aligned}$$

Hence, the assertion holds for all $z \in \mathbb{C} \setminus \mathbb{R}$. □

Now we prove a few sufficient conditions for strong resolvent convergence.

Theorem 9.16. *Let T_n ($n \in \mathbb{N}$) and T be self-adjoint operators on the complex Hilbert space H. The sequence (T_n) converges to T in the sense of the strong resolvent convergence if one of the following assumptions is satisfied:*

(i) *There is a core D_0 of T such that for every $f \in D_0$ there exists an $n_0 \in \mathbb{N}$ with the properties that $f \in D(T_n)$ for $n \geq n_0$ and $T_n f \to Tf$.*
(ii) *The operators T_n and T are bounded and $T_n \xrightarrow{s} T$.*
(iii) *$D(T_n) = D(T)$ for all $n \in \mathbb{N}$ and there are null sequences (a_n) and (b_n) such that*

$$\|(T - T_n) f\| \leq a_n \|f\| + b_n \|Tf\| \quad \text{for all } f \in D(T).$$

(iv) $G(T) = \lim_{n\to\infty} G(T_n)$, *i.e.,* $G(T)$ *is the set of those elements* (f, g) *from* $H \oplus H$ *for which there exists a sequence* (f_n) *such that* $f_n \in D(T_n)$ *and* $(f_n, T_n f_n) \to (f, g)$ *(graph convergence).*

PROOF.

(i) We have

$$(\mathrm{i} - T_n)^{-1} f - (\mathrm{i} - T)^{-1} f$$
$$= (\mathrm{i} - T_n)^{-1}(T_n - T)(\mathrm{i} - T)^{-1} f \to 0 \quad \text{as} \quad n \to \infty$$

for all $f \in H$ such that $(\mathrm{i} - T)^{-1} f \in D_0$. As D_0 is a core of T, the set of these f is dense in H. Therefore, $(\mathrm{i} - T_n)^{-1} \xrightarrow{s} (\mathrm{i} - T)^{-1}$ by Theorem 4.23.

(ii) *or* (iii) *implies* (i); these cases are therefore also proved.

(iv) It obviously follows from the formula $G(T) = \lim_{n\to\infty} G(T_n)$ that $G(\mathrm{i} - T) = \lim_{n\to\infty} G(\mathrm{i} - T_n)$. Let $g \in H$ be arbitrary. Then there is an $f \in D(T)$ such that $g = (\mathrm{i} - T)f$. Furthermore, there is a sequence (f_n) for which $f_n \in D(T_n)$, $f_n \to f$, and $(\mathrm{i} - T_n) f_n \to (\mathrm{i} - T) f = g$. Due to the inequality $\|(\mathrm{i} - T_n)^{-1}\| \leq 1$ it follows from this that

$$\|(\mathrm{i} - T_n)^{-1} g - (\mathrm{i} - T)^{-1} g\| \leq \|(\mathrm{i} - T_n)^{-1} g - f_n\| + \|f_n - (\mathrm{i} - T)^{-1} g\|$$
$$= \|(\mathrm{i} - T_n)^{-1}(g - (\mathrm{i} - T_n) f_n)\| + \|f_n - f\|$$
$$\to 0.$$

Therefore, $(\mathrm{i} - T_n)^{-1} \xrightarrow{s} (\mathrm{i} - T)^{-1}$. □

Theorem 9.17. *Let* T_n $(n \in \mathbb{N})$ *and* T *be self-adjoint operators on the complex Hilbert space* H, *and assume that* $(\mathrm{i} - T_n)^{-1} \xrightarrow{s} (\mathrm{i} - T)^{-1}$. *Then* $u(T_n) \xrightarrow{s} u(T)$ *for every continuous bounded function defined on* $\mathbb{R}$.

PROOF. First we assume that the limits $\lim_{t\to\pm\infty} u(t)$ exist, and $\lim_{t\to-\infty} u(t) = \lim_{t\to\infty} u(t)$. These are the functions that can be considered as continuous functions defined on the Alexandroff-compactification $\hat{\mathbb{R}}$ of $\mathbb{R}$. We consider the space $C(\hat{\mathbb{R}})$ with the maximum norm. The polynomials in $(\mathrm{i} - t)^{-1}$ and $(-\mathrm{i} - t)^{-1}$ can be considered as elements of $C(\hat{\mathbb{R}})$. The set P of these polynomials has the following properties: (i) the constant functions lie in P, (ii) the elements of P separate the points of $\hat{\mathbb{R}}$ (i.e., for $x, y \in \hat{\mathbb{R}}$ such that $x \neq y$ there exists a $u \in P$ for which $u(x) \neq u(y)$), and (iii) if $u \in P$, then $u^* \in P$. By the complex form of the Stone-Weierstrass theorem (cf. *Hewitt-Stromberg* [18], Theorem (7.34)) P is therefore dense in $C(\hat{\mathbb{R}})$. Consequently, for every $u \in C(\hat{\mathbb{R}})$ there exists a sequence (u_m) from P such that $\max_{t\in\mathbb{R}} |u(t) - u_m(t)| \to 0$ as $m \to \infty$.

Now let $f \in H$ and $\epsilon > 0$ be given. Then there is an $m_0 \in \mathbb{N}$ such that $|u(t) - u_m(t)| \leq (\epsilon/3\|f\|)$ for all $t \in \hat{\mathbb{R}}$ and $m \geq m_0$. As $u_m(t)$ is a polynomial in $(\mathrm{i} - t)^{-1}$ and $(-\mathrm{i} - t)^{-1}$, we have

$$u_m(T_n) \xrightarrow{s} u_m(T) \quad \text{as} \quad n \to \infty$$

for all $m \in \mathbb{N}$. Consequently, there exists an $n_0 \in \mathbb{N}$ such that

$$\|u_{m_0}(T_n)f - u_{m_0}(T)f\| \leq \frac{\epsilon}{3} \quad \text{for} \quad n \geq n_0.$$

Hence, it follows for $n \geq n_0$ that

$$\|u(T_n)f - u(T)f\| \leq \|u(T_n)f - u_{m_0}(T_n)f\| + \|u_{m_0}(T_n)f - u_{m_0}(T)f\| + \|u_{m_0}(T)f - u(T)f\| \leq \epsilon,$$

since $\|u(T_n) - u_{m_0}(T_n)\| \leq (\epsilon/3\|f\|)$ and $\|u(T) - u_{m_0}(T)\| \leq (\epsilon/3\|f\|)$. Therefore, $u(T_n) \to u(T)$.

Now let u be an arbitrary continuous and bounded function defined on $\mathbb{R}$. Let (φ_m) be a sequence of continuous functions with compact supports that tends to 1 non-decreasing. Then

$$\|\varphi_m(T)f - f\|^2 = \int |\varphi_m(t) - 1|^2 \, \mathrm{d}\|E(t)f\|^2 \to 0 \quad \text{as} \quad m \to \infty$$

for every $f \in H$. Because of $u\varphi_m \in C(\hat{\mathbb{R}})$ we have

$$u(T_n)\varphi_m(T_n) \xrightarrow{s} u(T)\varphi_m(T) \quad \text{as} \quad n \to \infty$$

for every $m \in \mathbb{N}$, by the first part of the proof. For all $n, m \in \mathbb{N}$

$$\begin{aligned}
&\|u(T_n)f - u(T)f\| \\
&\leq \|u(T_n)f - u(T_n)\varphi_m(T)f\| + \|u(T_n)\varphi_m(T)f - u(T_n)\varphi_m(T_n)f\| \\
&\quad + \|u(T_n)\varphi_m(T_n)f - u(T)\varphi_m(T)f\| + \|u(T)\varphi_m(T)f - u(T)f\| \\
&\leq \|u(T_n)\| \, \|f - \varphi_m(T)f\| + \|u(T)\| \, \|f - \varphi_m(T)f\| \\
&\quad + \|u(T_n)\| \, \|\varphi_m(T)f - \varphi_m(T_n)f\| + \|u(T_n)\varphi_m(T_n)f - u(T)\varphi_m(T)f\|.
\end{aligned}$$

The first two terms on the right side will be small for sufficiently large m (observe that $\|u(T_n)\| \leq \sup\{|u(t)| : t \in \mathbb{R}\}$). The last two terms will be small for fixed m if n is chosen sufficiently large. Consequently, the assertion is proved. □

Now we can prove, in particular, that the unitary group induced by a self-adjoint operator depends on this operator continuously in the strong sense.

Theorem 9.18. *Let T_n ($n \in \mathbb{N}$) and T be self-adjoint operators on the complex Hilbert space H. Assume that $(\mathrm{i} - T_n)^{-1} \xrightarrow{s} (\mathrm{i} - T)^{-1}$.*

(a) $\mathrm{e}^{\mathrm{i}tT_n} \xrightarrow{s} \mathrm{e}^{\mathrm{i}tT}$ *for all* $t \in \mathbb{R}$.

(b) *If $T_n \geq \gamma$ and $T \geq \gamma$ for some $\gamma \in \mathbb{R}$, then we also have $\mathrm{e}^{-tT_n} \xrightarrow{s} \mathrm{e}^{-tT}$ for all $t \geq 0$.*

PROOF.

(a) The function $s \mapsto \mathrm{e}^{\mathrm{i}ts}$ is continuous and bounded on $\mathbb{R}$. The assertion therefore follows from Theorem 9.17.

(b) With $u(s)=e^{-ts}$ for $s \geq \gamma$ and $u(s)=e^{-t\gamma}$ for $s<\gamma$ we have $u(T)=e^{-tT}$ and $u(T_n)=e^{-tT_n}$. As, on the other hand, u is continuous and bounded, Theorem 9.17 can be applied again. □

Now we shall investigate the influence of the strong resolvent convergence on the spectral family.

Theorem 9.19. *Let $T_n (n \in \mathbb{N})$ and T be self-adjoint operators on the complex Hilbert space H, and assume that $(\mathrm{i}-T_n)^{-1} \xrightarrow{s} (\mathrm{i}-T)^{-1}$. If E_n and E are the spectral families of T_n and T, respectively, then, as $n \to \infty$,*

$$\left.\begin{array}{l} E_n(t) \xrightarrow{s} E(t) \\ E_n(t-) \xrightarrow{s} E(t) \end{array}\right\} \quad \text{for all} \quad t \in \mathbb{R} \quad \text{such that} \quad E(t)=E(t-).$$

Proof. Assume that $E(t)=E(t-)$. Let (φ_m) respectively (ψ_m) be non-decreasing respectively non-increasing sequences of continuous functions such that $\varphi_m(s) \to \chi_{(-\infty,t)}(s)$, $\psi_m(s) \to \chi_{(-\infty,t]}(s)$, $|\varphi_m(s)| \leq 1$, and $|\psi_m(s)| \leq 1$ for all $s \in \mathbb{R}$. Then for all $f \in H$

$$\|(\varphi_m(T_n)-E_n(t-))f\|^2$$
$$= \int |\varphi_m(s)-\chi_{(-\infty,t)}(s)|^2 \, \mathrm{d}\|E_n(s)f\|^2 \to 0 \quad \text{as} \quad m \to \infty$$

(Lebesgue's theorem). Therefore,

$$\varphi_m(T_n) \xrightarrow{s} E_n(t-).$$

It follows similarly that

$$\psi_m(T_n) \xrightarrow{s} E_n(t),$$

and (because $E(t)=E(t-)$)

$$\varphi_m(T) \xrightarrow{s} E(t) \quad \text{and} \quad \psi_m(T) \xrightarrow{s} E(t).$$

Hence, for every $f \in H$ and every $\epsilon>0$ there are continuous functions $\varphi \leq \chi_{(-\infty,t)}$ and $\psi \geq \chi_{(-\infty,t]}$ for which

$$\|\psi(T)f-\varphi(T)f\| \leq \frac{\epsilon}{5};$$

we can choose $\varphi=\varphi_{m_0}$ and $\psi=\psi_{m_0}$ with a sufficiently large m_0. By Theorem 9.17 there is an $n_0 \in \mathbb{N}$ such that

$$\left.\begin{array}{l} \|\varphi(T)f-\varphi(T_n)f\| \leq \frac{\epsilon}{5} \\ \|\psi(T)f-\psi(T_n)f\| \leq \frac{\epsilon}{5} \end{array}\right. \quad \text{for all} \quad n \geq n_0.$$

We therefore have

$$\|\psi(T_n)f-\varphi(T_n)f\| \leq 3\frac{\epsilon}{5} \quad \text{for all} \quad n \geq n_0.$$

Since

$$\|E(t)f-\varphi(T)f\| = \left\{\int |\chi_{(-\infty,t]}(s)-\varphi(s)|^2 \,\mathrm{d}\|E(s)f\|^2\right\}^{1/2}$$
$$\leqslant \left\{\int |\psi(s)-\varphi(s)|^2 \,\mathrm{d}\|E(s)f\|^2\right\}^{1/2}$$
$$= \|\psi(T)f-\varphi(T)f\| < \frac{\epsilon}{5}$$

and

$$\|E_n(t-)f-\varphi(T_n)f\| \leqslant \|E_n(t)f-\varphi(T_n)f\| \leqslant \|\psi(T_n)f-\varphi(T_n)f\| < 3\frac{\epsilon}{5},$$

we obtain that

$$\|E(t)f-E_n(t)f\|$$
$$\leqslant \|E(t)f-\varphi(T)f\|+\|\varphi(T)f-\varphi(T_n)f\|+\|\varphi(T_n)f-E_n(t)f\|$$
$$\leqslant \|\psi(T)f-\varphi(T)f\|+\|\varphi(T)f-\varphi(T_n)f\|+\|\varphi(T_n)f-\psi(T_n)f\|$$
$$<\epsilon \quad \text{for all} \quad n \geqslant n_0.$$

It follows similarly that

$$\|E(t)f-E_n(t-)f\| \leqslant \epsilon \quad \text{for all} \quad n \geqslant n_0.$$

□

It is worth noting that the results of Theorems 9.5 and 9.19 are not comparable. It is clear that $\sigma(T)=\lim_{n\to\infty}\sigma(T_n)$ does not imply $E(t)= s-\lim_{n\to\infty}E_n(t)$.

Conversely, from $E_n(t)\xrightarrow{s}E(t)$ (for all $t\in\mathbb{R}$) we cannot infer $\sigma(T_n)\to\sigma(T)$, as the following example shows.

EXAMPLE 1. Let H be a separable infinite dimensional Hilbert space, and let $\{e_m : m\in\mathbb{N}\}$ be an orthonormal basis of H. For every $n\in\mathbb{N}$ let T_n be the orthogonal projection onto $L\{e_m : 1\leqslant m\leqslant n\}$. Then $T_n\xrightarrow{s}I$ as $n\to\infty$. The spectral family E_n of T_n is given by the equality

$$E_n(t) = \begin{cases} 0 & \text{for} \quad t<0, \\ I-T_n & \text{for} \quad 0\leqslant t<1, \\ I & \text{for} \quad t\geqslant 1. \end{cases}$$

Therefore, $E_n(t)\xrightarrow{s}E(t)$ for all $t\in\mathbb{R}$, where E denotes the spectral family of I, i.e.,

$$E(t) = \begin{cases} 0 & \text{for} \quad t<1, \\ I & \text{for} \quad t\geqslant 1. \end{cases}$$

On the other hand, $\sigma(T_n)=\{0, 1\}$ for all $n\in\mathbb{N}$, while $\sigma(I)=\{1\}$ (cf. Exercise 7.41).

Exercise

9.8. Let H be a *real* Hilbert space, and let T and $T_n (n \in \mathbb{N})$ be self-adjoint operators. Assume that one of the assumptions (i)-(iv) of Theorem 9.16 is fulfilled.

(a) For the spectral families E_n and E of T_n and T, respectively, we have $E_n(t) \xrightarrow{s} E(t)$ for every $t \in \mathbb{R}$ such that E is continuous at t.

(b) If there exists a $\gamma \in \mathbb{R}$ such that $T \geq \gamma$ and $T_n \geq \gamma$ for all $n \in \mathbb{N}$, then $e^{-tT_n} \xrightarrow{s} e^{-tT}$ for all $t > 0$.

Hint: Complexify.

Differential operators on $L_2(\mathbb{R}^m)$ 10

10.1 The Fourier transformation on $L_2(\mathbb{R}^m)$

In what follows we shall use so-called multiindices. A *multiindex* (of m components) is an m-tuple $\alpha = (\alpha_1, \ldots, \alpha_m)$ of non-negative integers $\alpha_j \in \mathbb{N}_0$, $j = 1, 2, \ldots, m$. The *absolute value* of α is defined by the formula

$$|\alpha| = \sum_{j=1}^{m} \alpha_j.$$

We set, for every $x \in \mathbb{R}^m$,

$$x^\alpha = \prod_{j=1}^{m} x_j^{\alpha_j}.$$

Correspondingly, we write

$$D^\alpha = \prod_{j=1}^{m} \left(\frac{1}{i} \frac{\partial}{\partial x_j} \right)^{\alpha_j} = (-i)^{|\alpha|} \prod_{j=1}^{m} \frac{\partial^{\alpha_j}}{\partial x_j^{\alpha_j}}.$$

The space of *rapidly decreasing functions* (the *Schwartz space*) $S(\mathbb{R}^m)$ is the vector space of arbitrarily many times continuously differentiable functions $f: \mathbb{R}^m \to \mathbb{C}$ for which we have the following: For every multiindex α and for every $p \in \mathbb{N}_0$ there exists a $c_{\alpha p} > 0$ such that

$$|x|^p |D^\alpha f(x)| \leq c_{\alpha p} \quad \text{for all} \quad x \in \mathbb{R}^m.$$

It is obvious that this assumption can also be formulated in the following way: For every multiindex α and for arbitrary $p, q \in \mathbb{N}_0$ there exists a $c_{\alpha pq} > 0$ such that

$$(1+|x|)^p |D^\alpha f(x)| \leq c_{\alpha pq}(1+|x|)^{-q} \quad \text{for all} \quad x \in \mathbb{R}^m.$$

This formulation shows that $S(\mathbb{R}^m) \subset L_p(\mathbb{R}^m)$ for all $p \in [1, \infty]$. In particular, for all $f \in S(\mathbb{R}^m)$ we can define the *Fourier transformation* F_0 by the integral

$$(F_0 f)(x) = (2\pi)^{-m/2} \int e^{-ixy} f(y)\, dy,^1$$

where

$$xy = \sum_{j=1}^{m} x_j y_j \quad \text{for} \quad x, y \in \mathbb{R}^m.$$

Theorem 10.1. *We have $F_0 S(\mathbb{R}^m) \subset S(\mathbb{R}^m)$. For every $f \in S(\mathbb{R}^m)$ and every multiindex α*

$$D^\alpha F_0 f = (-1)^{|\alpha|} F_0 M_\alpha f, \quad M_\alpha F_0 f = F_0 D^\alpha f,$$

where $(M_\alpha f)(x) = x^\alpha f(x)$.

PROOF. It is easy to see that the function

$$(F_0 f)(x) = (2\pi)^{-m/2} \int e^{-ixy} f(y)\, dy$$

is arbitrarily many times continuously differentiable. The differentiation can be done under the integral sign, i.e.,

$$D^\alpha (F_0 f)(x) = (2\pi)^{-m/2} (-1)^{|\alpha|} \int e^{-ixy} [y^\alpha f(y)]\, dy.$$

It follows from this for any multiindex β that

$$\begin{aligned} x^\beta (D^\alpha F_0 f)(x) &= (2\pi)^{-m/2} (-1)^{|\alpha| + |\beta|} \int D_y^\beta (e^{-ixy}) [y^\alpha f(y)]\, dy \\ &= (2\pi)^{-m/2} (-1)^{|\alpha|} \int e^{-ixy}\, D^\beta [y^\alpha f(y)]\, dy. \end{aligned}$$

Since $D^\beta M_\alpha f$ is in $S(\mathbb{R}^m)$, too, $x^\beta (D^\alpha F_0 f)(x)$ is bounded for all α and β. Therefore, $F_0 f \in S(\mathbb{R}^m)$, and

$$M_\beta D^\alpha F_0 f = (-1)^{|\alpha|} F_0 D^\beta M_\alpha f.$$

Both formulae follow from this. □

Theorem 10.2. *The function $\vartheta : \mathbb{R}^m \to \mathbb{R}$ defined by the equality*

$$\vartheta(x) = \exp\left(-\tfrac{1}{2}|x|^2\right) \quad \text{for all} \quad x \in \mathbb{R}^m$$

is in $S(\mathbb{R}^m)$. We have $F_0 \vartheta = \vartheta$.

[1] If no domain of integration is indicated, then the integral always is to be taken over $\mathbb{R}^m$.

Proof. The reader can easily verify that $\vartheta \in S(\mathbb{R}^m)$. To prove that $F_0\vartheta = \vartheta$, we first consider the case $m = 1$. In this case we obviously have the first order differential equation

$$\vartheta'(x) + x\vartheta(x) = 0.$$

It follows from this by Theorem 10.1 that

$$(F_0\vartheta)' + M_1F_0\vartheta = -\mathrm{i}F_0M_1\vartheta - \mathrm{i}F_0\vartheta' = -\mathrm{i}F_0(M_1\vartheta + \vartheta') = 0,$$

i.e., $F_0\vartheta$ satisfies the same differential equation as ϑ. Due to the equalities

$$(F_0\vartheta)(0) = (2\pi)^{-1/2}\int \vartheta(x)\,\mathrm{d}x = 1 = \vartheta(0)$$

we obtain that $F_0\vartheta = \vartheta$. The result can be derived for an arbitrary m by taking products. □

Theorem 10.3. *The Fourier transformation F_0 is a bijective linear mapping of $S(\mathbb{R}^m)$ onto itself. We have*

$$(F_0^{-1}g)(x) = (2\pi)^{-m/2}\int \mathrm{e}^{\mathrm{i}xy}g(y)\,\mathrm{d}y, \quad g \in S(\mathbb{R}^m).$$

Moreover, $(F_0f)(x) = (F_0^{-1}f)(-x)$ for every $f \in S(\mathbb{R}^m)$, and we have $F_0^4 = I$.

Proof. For all $f, g \in S(\mathbb{R}^m)$

$$\begin{aligned}\int g(y)(F_0f)(y)\mathrm{e}^{\mathrm{i}xy}\,\mathrm{d}y &= (2\pi)^{-m/2}\int g(y)\mathrm{e}^{\mathrm{i}xy}\int \mathrm{e}^{-\mathrm{i}yz}f(z)\,\mathrm{d}z\,\mathrm{d}y\\ &= (2\pi)^{-m/2}\int f(z)\int \mathrm{e}^{-\mathrm{i}(z-x)y}g(y)\,\mathrm{d}y\,\mathrm{d}z\\ &= \int f(z)(F_0g)(z-x)\,\mathrm{d}z\\ &= \int (F_0g)(z)f(z+x)\,\mathrm{d}z.\end{aligned}$$

With $g_\epsilon(x) = g(\epsilon x)$ we have for $g \in S(\mathbb{R}^m)$ and $\epsilon > 0$ that

$$\begin{aligned}(F_0g_\epsilon)(x) &= (2\pi)^{-m/2}\int \mathrm{e}^{-\mathrm{i}xy}g(\epsilon y)\,\mathrm{d}y\\ &= (2\pi)^{-m/2}\epsilon^{-m}\int \mathrm{e}^{-\mathrm{i}xy/\epsilon}g(y)\,\mathrm{d}y\\ &= \epsilon^{-m}(F_0g)\left(\frac{x}{\epsilon}\right);\end{aligned}$$

therefore,

$$\begin{aligned}\int g(\epsilon y)(F_0f)(y)\mathrm{e}^{\mathrm{i}xy}\,\mathrm{d}y &= \int g_\epsilon(y)(F_0f)(y)\mathrm{e}^{\mathrm{i}xy}\,\mathrm{d}y\\ &= \int (F_0g_\epsilon)(z)f(z+x)\,\mathrm{d}z\\ &= \epsilon^{-m}\int (F_0g)\left(\frac{z}{\epsilon}\right)f(z+x)\,\mathrm{d}z\\ &= \int (F_0g)(z)f(\epsilon z+x)\,\mathrm{d}z.\end{aligned}$$

If we replace here g by the function ϑ from Theorem 10.2, then it follows for $\epsilon \to 0$ (as $F_0\vartheta = \vartheta$ and $\vartheta(0) = 1$) that

$$\begin{aligned}(2\pi)^{-m/2}\int e^{ixy}(F_0 f)(y)\,dy &= \vartheta(0)(2\pi)^{-m/2}\int e^{ixy}(F_0 f)(y)\,dy\\ &= \lim_{\epsilon\to 0}(2\pi)^{-m/2}\int e^{ixy}\vartheta(\epsilon y)(F_0 f)(y)\,dy\\ &= \lim_{\epsilon\to 0}(2\pi)^{-m/2}\int \vartheta(y) f(\epsilon y + x)\,dy\\ &= (2\pi)^{-m/2} f(x)\int \vartheta(y)\,dy = f(x).\end{aligned}$$

It follows from this that F_0 is injective and F_0^{-1} has the given form. Moreover, for all $f \in S(\mathbb{R}^m)$

$$(F_0^2 f)(x) = (2\pi)^{-m/2}\int e^{-ixy}(F_0 f)(y)\,dy = f(-x),$$

and thus $F_0^4 f(x) = f(x)$, i.e., $F_0^4 = I$. Since $R(F_0) \supset R(F_0^4) = R(I) = S(\mathbb{R}^m)$, the mapping F_0 is surjective. □

In what follows we consider F_0 as an operator on $L_2(\mathbb{R}^m)$ such that $D(F_0) = S(\mathbb{R}^m)$.

Theorem 10.4. *We have $\|F_0 f\| = \|f\|$ and $\|F_0^{-1} f\| = \|f\|$ for all $f \in S(\mathbb{R}^m)$ (here $\|\,.\,\|$ denotes the norm in $L_2(\mathbb{R}^m)$). F_0 and F_0^{-1} possess uniquely determined extensions F and $\tilde{F}$ belonging to $B(L_2(\mathbb{R}^m))$. The operators F and $\tilde{F}$ are unitary, and $\tilde{F} = F^* = F^{-1}$. We have $F^4 = I$. The operator F is called the* Fourier transformation *on $L_2(\mathbb{R}^m)$.*

Proof. For $f, g \in S(\mathbb{R}^m)$ we have

$$\begin{aligned}\langle f, g\rangle &= \int f(x)^*(F_0^{-1}F_0 g)(x)\,dx\\ &= \int f(x)^*(2\pi)^{-m}\left\{\int e^{ixy}\left[\int e^{-iyz}g(z)\,dz\right]dy\right\}dx\\ &= \int (2\pi)^{-m}\left\{\int e^{-iyx}f(x)\,dx\right\}^*\left\{\int e^{-iyz}g(z)\,dz\right\}dy = \langle F_0 f, F_0 g\rangle.\end{aligned}$$

In particular, $\|F_0 f\| = \|f\|$, and thus $\|F_0^{-1} f\| = \|f\|$. Since $S(\mathbb{R}^m)$ is dense in $L_2(\mathbb{R}^m)$ (as $C_0^\infty(\mathbb{R}^m) \subset S(\mathbb{R}^m)$), there exist uniquely determined extensions F and $\tilde{F}$ of F_0 and F_0^{-1} from $B(L_2(\mathbb{R}^m))$. We obviously have $\|Ff\| = \|\tilde{F}f\| = \|f\|$ for all $f \in L_2(\mathbb{R}^m)$. If $f, g \in L_2(\mathbb{R}^m)$ and (f_n), (g_n) are sequences from $S(\mathbb{R}^m)$ such that $f_n \to f$ and $g_n \to g$, then

$$\langle Ff, g\rangle = \lim_{n\to\infty}\langle F_0 f_n, g_n\rangle = \lim_{n\to\infty}\langle f_n, F_0^{-1} g_n\rangle = \langle f, \tilde{F}g\rangle,$$

i.e., $F^* = \tilde{F}$. Moreover,

$$F\tilde{F}f = \lim_{n\to\infty} FF_0^{-1}f_n = \lim_{n\to\infty} F_0F_0^{-1}f_n = \lim_{n\to\infty} f_n = f,$$

i.e., $F\tilde{F} = I$. We can prove similarly that $\tilde{F}F = I$. Therefore, $\tilde{F} = F^{-1}$, and thus F is unitary. From $F_0^4 = I|_{S(\mathbb{R}^m)}$ it follows that $F^4 = I$. □

The mappings F_0 and F_0^{-1} can be extended to $L_1(\mathbb{R}^m)$ in a natural way. These extensions F_1 and $\tilde{F}_1$ are defined for $f \in L_1(\mathbb{R}^m)$ by the formulae,

$$(F_1 f)(x) = (2\pi)^{-m/2} \int e^{-ixy} f(y)\,dy$$

$$(\tilde{F}_1 f)(x) = (2\pi)^{-m/2} \int e^{ixy} f(y)\,dy.$$

It is easy to see that $F_1 f$ and $\tilde{F}_1 f$ are continuous functions such that $|(F_1 f)(x)| \leq (2\pi)^{-m/2}\|f\|_1$ and $|(\tilde{F}_1 f)(x)| \leq (2\pi)^{-m/2}\|f\|_1$ for all $x \in \mathbb{R}^m$.

Theorem 10.5. *The mappings F_1 and $\tilde{F}_1$ of $L_1(\mathbb{R}^m)$ into the space $C_\infty(\mathbb{R}^m)$ of continuous bounded functions defined on $\mathbb{R}^m$ are injective. For $f \in L_1(\mathbb{R}^m) \cap L_2(\mathbb{R}^m)$ we have*

$$(F_1 f)(x) = (Ff)(x) \quad \text{and} \quad (\tilde{F}_1 f)(x) = (F^{-1}f)(x)$$

almost everywhere in $\mathbb{R}^m$.

Proof. Take an f from $L_1(\mathbb{R}^m)$ for which $F_1 f = 0$ (i.e., $(F_1 f)(x) = 0$ for all $x \in \mathbb{R}^m$). We have to prove that $f = 0$. It follows from the equality $F_1 f = 0$ that

$$\int f(x)(F_0 g)(x)\,dx = (2\pi)^{-m/2} \int\int e^{-ixy} g(y) f(x)\,dy\,dx$$
$$= \int g(y)(F_1 f)(y)\,dy = 0$$

for all $g \in S(\mathbb{R}^m)$. Since $F_0 g$ runs over the whole space $S(\mathbb{R}^m)$,

$$\int f(x)h(x)\,dx = 0$$

for all $h \in S(\mathbb{R}^m)$. Then this holds also for all continuous functions h defined on $\mathbb{R}^m$ and having compact support. Define $K(n) = \{x \in \mathbb{R}^m : |x| \leq n\}$ for $n \in \mathbb{N}$, let χ_n be the characteristic function of $K(n)$, and define

$$\operatorname{sgn} f(x) = \begin{cases} f(x)|f(x)|^{-1} & \text{for} \quad f(x) \neq 0, \\ 0 & \text{for} \quad f(x) = 0. \end{cases}$$

Then $\chi_n \operatorname{sgn} f$ is measurable, $|\chi_n(x) \operatorname{sgn} f(x)| \leq 1$, and there exists a sequence (h_j) of continuous functions with supports in $K(n)$ such that

$h_j(x) \to \operatorname{sgn} f(x)$ almost everywhere in $K(n)$. We can assume without loss of generality that $|h_j(x)| \leq 1$ for all $x \in K(n)$ (since we can replace $h_j(x)$ by $h_j(x)|h_j(x)|^{-1}$ for those $x \in K(n)$ for which $|h_j(x)| > 1$). Consequently,

$$\int_{K(n)} |f(x)| \,\mathrm{d}x = \lim_{j\to\infty} \int_{K(n)} f(x) h_j^*(x) \,\mathrm{d}x = 0.$$

Therefore, $f(x) = 0$ almost everywhere in $K(n)$. As this holds for all $n \in \mathbb{N}$, we have $f(x) = 0$ almost everywhere. F_1 is then injective. The injectivity of $\tilde{F}_1$ can be proved similarly.

If $f \in L_1(\mathbb{R}^m) \cap L_2(\mathbb{R}^m)$ and $f_n = \chi_n f$, then $f_n \to f$ in $L_1(\mathbb{R}^m)$ and in $L_2(\mathbb{R}^m)$. For every $n \in \mathbb{N}$ there exists a $\varphi_n \in C_0^\infty(K(n))$ such that

$$\int_{K(n)} |f_n(x) - \varphi_n(x)|^2 \,\mathrm{d}x < \frac{1}{nV(K(n))},$$

and thus

$$\int_{K(n)} |f_n(x) - \varphi_n(x)| \,\mathrm{d}x \leq \left\{ V(K(n)) \int_{K(n)} |f_n(x) - \varphi_n(x)|^2 \,\mathrm{d}x \right\}^{1/2}$$
$$\leq \left(\frac{1}{n}\right)^{1/2},$$

where $V(K(n))$ denotes the volume of $K(n)$. Consequently, $\varphi_n \to f$ in $L_1(\mathbb{R}^m)$ and in $L_2(\mathbb{R}^m)$. Therefore,

$$(F_1 f)(x) = \lim_{n\to\infty} (2\pi)^{-m/2} \int \mathrm{e}^{-\mathrm{i}xy} \varphi_n(y) \,\mathrm{d}y$$

and

$$(Ff)(x) = \operatorname*{l.i.m.}_{n\to\infty} (2\pi)^{-m/2} \int \mathrm{e}^{-\mathrm{i}xy} \varphi_n(y) \,\mathrm{d}y.$$

Here "l.i.m. = *limit in mean*" stands for the limit in $L_2(\mathbb{R}^m)$. Hence, $(Ff)(x) = (F_1 f)(x)$ almost everywhere. We can prove in a similar way that $(F^{-1}f)(x) = (\tilde{F}_1 f)(x)$ almost everywhere. □

Theorem 10.6. *For all* $f \in L_2(\mathbb{R}^m)$

$$(Ff)(x) = \operatorname*{l.i.m.}_{n\to\infty} (2\pi)^{-m/2} \int_{K(n)} \mathrm{e}^{-\mathrm{i}xy} f(y) \,\mathrm{d}y.$$

A similar formula holds for F^{-1}.

Proof. The functions $\chi_n f$ belong to $L_1(\mathbb{R}^m) \cap L_2(\mathbb{R}^m)$, and $\chi_n f \to f$ in $L_2(\mathbb{R}^m)$. Therefore, $F(\chi_n f) \to Ff$ in $L_2(\mathbb{R}^m)$. Since by Theorem 10.5

$$(F(\chi_n f))(x) = (2\pi)^{-m/2} \int_{K(n)} \mathrm{e}^{-\mathrm{i}xy} f(y) \,\mathrm{d}y, \quad \text{almost everywhere in } \mathbb{R}^m,$$

the assertion follows. □

For $f, g \in L_2(\mathbb{R}^m)$ the *convolution* $f * g$ is defined by the integral

$$(f * g)(x) = (2\pi)^{-m/2} \int f(x-y) g(y)\, dy.$$

The integral exists for all $x \in \mathbb{R}^m$, because $f(x-.) \in L_2(\mathbb{R}^m)$ and $g(.) \in L_2(\mathbb{R}^m)$ imply $f(x-.)g(.) \in L_1(\mathbb{R}^m)$. (For the convolution of L_1-functions, see Exercise 10.1.) Moreover,

$$\begin{aligned}(f * g)(x) &= (2\pi)^{-m/2} \int f(x-y) g(y)\, dy \\ &= (2\pi)^{-m/2} \int g(x-y) f(y)\, dy = (g * f)(x).\end{aligned}$$

Theorem 10.7 (The convolution theorem). *For* $f, g \in L_2(\mathbb{R}^m)$

(a) $\tilde{F}_1(Ff \cdot Fg) = F_1(F^{-1}f \cdot F^{-1}g) = f * g$.

(b) *The following assertions are equivalent*:

(i) $Ff \cdot Fg \in L_2(\mathbb{R}^m)$,

(ii) $F^{-1}f \cdot F^{-1}g \in L_2(\mathbb{R}^m)$,

(iii) $f * g \in L_2(\mathbb{R}^m)$.

In this case

$$f * g = F^{-1}(Ff \cdot Fg) = F(F^{-1}f \cdot F^{-1}g).$$

PROOF.

(a) $\tilde{F}_1(Ff \cdot Fg)(x) = \langle (Ff)^*, h(x, .)\rangle$ with

$$\begin{aligned}h(x,y) &= (2\pi)^{-m/2} e^{ixy} (Fg)(y) \\ &= (2\pi)^{-m} \operatorname*{l.i.m.}_{n\to\infty} \int_{K(n)} e^{-iy(z-x)} g(z)\, dz \\ &= (2\pi)^{-m} \operatorname*{l.i.m.}_{n\to\infty} \int_{K(n)} e^{-iyz} g(z+x)\, dz = (2\pi)^{-m/2} (Fg_x)(y),\end{aligned}$$

where $g_x(z) = g(z+x)$. Moreover, let us set $f_-(x) = f(-x)$. Then it is obvious that $(Ff)^* = Ff_-^*$, and thus

$$\begin{aligned}\tilde{F}_1(Ff \cdot Fg)(x) &= (2\pi)^{-m/2} \langle Ff_-^*, Fg_x\rangle = (2\pi)^{-m/2} \langle f_-^*, g_x\rangle \\ &= (2\pi)^{-m/2} \int f(-z) g(z+x)\, dz = (f * g)(x).\end{aligned}$$

We can prove analogously that $F_1(F^{-1}f \cdot F^{-1}g) = f * g$.

(b) If $Ff \cdot Fg \in L_2(\mathbb{R}^m)$, then in the formula $f * g = \tilde{F}_1(Ff \cdot Fg)$ we can replace the operator $\tilde{F}_1$ by F^{-1} and obtain that $f * g \in L_2(\mathbb{R}^m)$ and $f * g = F^{-1}(Ff \cdot Fg)$. If $f * g \in L_2(\mathbb{R}^m)$, then with $h_1 = Ff \cdot Fg \in L_1(\mathbb{R}^m)$ and $h_2 = F(f * g) \in L_2(\mathbb{R}^m)$ we have

$$(\tilde{F}_1 h_1)(x) = (f * g)(x) = (F^{-1}h_2)(x) \quad \text{almost everywhere.}$$

We show that $h_1(x) = h_2(x)$ almost everywhere, and thus $Ff \cdot Fg = h_1 \in L_2(\mathbb{R}^m)$. To prove this it is sufficient to prove that $\int (h_1(x) - h_2(x))\varphi(x)\,\mathrm{d}x = 0$ for all $\varphi \in S(\mathbb{R}^m)$ (cf. the proof of Theorem 10.5). This follows from the equalities

$$\begin{aligned}\int h_1(x)\varphi(x)\,\mathrm{d}x &= \int h_1(x)\big(F_0^{-1}F_0\varphi\big)(x)\,\mathrm{d}x \\ &= \int (\tilde{F}_1 h_1)(x)(F_0\varphi)(x)\,\mathrm{d}x \\ &= \int (F^{-1}h_2)(x)(F_0\varphi)(x)\,\mathrm{d}x \\ &= \langle (F^{-1}h_2)^*, F\varphi\rangle = \langle Fh_2^*, F\varphi\rangle = \langle h_2^* \,\varphi\rangle \\ &= \int h_2(x)\varphi(x)\,\mathrm{d}x.\end{aligned}$$

The equivalence of (i) and (ii) and the second equality follow from the formula $F(h) = (F^{-1}h)_-$. □

Exercises

10.1. The convolution $(f * g)(x) = (2\pi)^{-m/2} \int f(x-y)g(y)\,\mathrm{d}y$ is defined almost everywhere for all $f, g \in L_1(\mathbb{R}^m)$, and is a function from $L_1(\mathbb{R}^m)$. We have $F_1(f * g) = (F_1 f) \cdot (F_1 g)$.

10.2. If f and $F_1 f$ are from $L_1(\mathbb{R}^m)$, then $f = \tilde{F}_1(F_1 f)$.

10.3. (a) For $f \in L_2(\mathbb{R}^m)$ and $a \in \mathbb{R}^m$ let f_a be defined by the equality $f_a(x) = f(x+a)$. Then $(Ff_a)(x) = \mathrm{e}^{ixa}(Ff)(x)$.

(b) Let $f \in L_2(\mathbb{R}^m)$ and let $(Ff)(x) \neq 0$ almost everywhere in $\mathbb{R}^m$. Then the set $\{f_a : a \in \mathbb{R}^m\}$ is total in $L_2(\mathbb{R}^m)$ (*Wiener*'s theorem).
Hint: If $g \perp \{f_a : a \in \mathbb{R}^m\}$, then $F_1(Ff \cdot Fg) = 0$.

(c) Let ϑ be as in Theorem 10.2. Then the set $\{\vartheta_a : a \in \mathbb{R}^m\}$ is total in $L_2(\mathbb{R}^m)$.

10.4. We have $(F_1 f)(x) \to 0$ as $|x| \to \infty$ for every $f \in L_1(\mathbb{R}^m)$ (*Riemann-Lebesque*).
Hint: $S(\mathbb{R}^m)$ is dense in $L_1(\mathbb{R}^m)$, and $f_n \to f$ in $L_1(\mathbb{R}^m)$ implies $F_1 f_n \to F_1 f$ uniformly in $\mathbb{R}^m$.

10.2 Sobolev spaces and differential operators on $L_2(\mathbb{R}^m)$ with constant coefficients

In what follows, for all $s \geq 0$ define

$$k_s(x) = (1+|x|^2)^{s/2} \quad \text{for} \quad x \in \mathbb{R}^m$$

and

$$L_{2,s}(\mathbb{R}^m) = \{f \in L_2(\mathbb{R}^m) : k_s f \in L_2(\mathbb{R}^m)\}.$$

$L_{2,s}(\mathbb{R}^m)$ is obviously a dense subspace of $L_2(\mathbb{R}^m)$. The equality

$$\langle f, g\rangle_{(s)} = \int f(x)^* g(x) k_s(x)^2 \, dx \quad \text{for} \quad f, g \in L_{2,s}(\mathbb{R}^m)$$

defines a scalar product on $L_{2,s}(\mathbb{R}^m)$. We denote the corresponding norm by $\|\,.\,\|_{(s)}$. The space $(L_{2,s}(\mathbb{R}^m), \langle .\,,.\rangle_{(s)})$ is a separable Hilbert space, since

$$U_s : L_{2,s}(\mathbb{R}^m) \to L_2(\mathbb{R}^m), \quad U_s f = k_s f$$

is an isomorphism of $L_{2,s}(\mathbb{R}^m)$ onto $L_2(\mathbb{R}^m)$.

The *Sobolev space* of order s is defined by the equality

$$W_{2,s}(\mathbb{R}^m) = \{f \in L_2(\mathbb{R}^m) : Ff \in L_{2,s}(\mathbb{R}^m)\} = F^{-1}L_{2,s}(\mathbb{R}^m).$$

$W_{2,s}(\mathbb{R}^m)$ is therefore a dense subspace of $L_2(\mathbb{R}^m)$, and the equality

$$\langle f, g\rangle_s = \langle Ff, Fg\rangle_{(s)} \quad \text{for} \quad f, g \in W_{2,s}(\mathbb{R}^m)$$

defines a scalar product of $W_{2,s}(\mathbb{R}^m)$. We denote the corresponding norm by $\|\,.\,\|_s$. Since F is an isomorphism of $W_{2,s}(\mathbb{R}^m)$ onto $L_{2,s}(\mathbb{R}^m)$, the space $W_{2,s}(\mathbb{R}^m)$ is also a separable Hilbert space.

First we show that the functions from $W_{2,s}(\mathbb{R}^m)$ are differentiable in a certain weak sense.

Theorem 10.8.

(a) *Let* $s \geq 1$, $w_j = (\delta_{j1}, \delta_{j2}, \ldots, \delta_{jm})$, $(M_j g)(x) = x_j g(x)$ *and* $f_{j,\epsilon}(x) = f(x + \epsilon w_j)$ *for* $j = 1, 2, \ldots, m$. *Then for all* $f \in W_{2,s}(\mathbb{R}^m)$ *and* $j = 1, 2, \ldots, m$

$$\lim_{\epsilon \to 0} \frac{1}{i\epsilon}(f_{j,\epsilon} - f) = F^{-1}M_j Ff$$

in $L_2(\mathbb{R}^m)$. *If* $f \in S(\mathbb{R}^m)$, *then this limit equals* $D^\alpha f$ *with* $\alpha = (\delta_{j1}, \delta_{j2}, \ldots, \delta_{jm})$. *We write* $D^\alpha f$ *for this limit in case* $f \in W_{2,s}(\mathbb{R}^m)$, *as well.*

(b) *If* α *is a multiindex and* $|\alpha| \leq s$, *then the derivative* $D^\alpha f$ *can be computed by iteration. The order of differentiation is irrelevant.*

(c) *If* $s \in \mathbb{N}_0$, *then* $\|\,.\,\|_s$ *is equivalent to the norms*

$$\|f\|_{s,0} = \Big\{\sum_{|\alpha| \leq s} \|D^\alpha f\|^2\Big\}^{1/2}, \quad \|f\|_{s,1} = \Big\{\|f\|^2 + \sum_{|\alpha| = s} \|D^\alpha f\|^2\Big\}^{1/2}.$$

PROOF.

(a) For all $x \in \mathbb{R}^m$

$$F\Big(\frac{1}{i\epsilon}(f_{j,\epsilon} - f)\Big)(x) = \frac{1}{i\epsilon}(e^{ix_j\epsilon} - 1)(Ff)(x),$$

$$\Big|\frac{1}{i\epsilon}(e^{ix_j\epsilon} - 1)\Big| \leq |x_j| < (1 + |x|^2)^{s/2},$$

and

$$\frac{1}{i\epsilon}(e^{ix_j\epsilon}-1) \to x_j \quad \text{as} \quad \epsilon \to 0.$$

It follows from this that $F(1/i\epsilon)(f_{j,\epsilon}-f) \to M_j Ff$, and thus

$$(1/i\epsilon)(f_{j,\epsilon}-f) \to F^{-1}M_j Ff \text{ as } \epsilon \to 0.$$

By Theorem 10.1 this limit is equal to $D^\alpha f$ for all $f \in S(\mathbb{R}^m)$.

(b) If $|\alpha|=1$, then it is obvious that $F\,D^\alpha f = x^\alpha Ff \in L_{2,s-1}(\mathbb{R}^m)$. Consequently, $D^\alpha f \in W_{2,s-1}(\mathbb{R}^m)$. If $s \geq 2$, then by part (a) we can therefore differentiate further. The commutativity of the order of differentiation follows from the equality $D^\alpha f = F^{-1}M_\alpha Ff$.

(c) We have $\|f\|_s^2 = \|k_s Ff\|^2$,

$$\|f\|_{s,0}^2 = \sum_{|\alpha| \leq s} \|M_\alpha Ff\|^2 \quad \text{and} \quad \|f\|_{s,1}^2 = \|Ff\|^2 + \sum_{|\alpha|=s} \|M_\alpha Ff\|^2.$$

Since

$$1 + \sum_{|\alpha|=s} |x^\alpha|^2 \leq C_1(1+|x|^2)^s \leq C_2 \sum_{|\alpha| \leq s} |x^\alpha|^2$$
$$\leq C_3\Big(1 + \max_{j=1,\ldots,m} x_j^{2s}\Big) \leq C_3\Big(1 + \sum_{|\alpha|=s} |x^\alpha|^2\Big)$$

with appropriate constants C_1, C_2, and C_3 (that depend only on m and s), all three norms are equivalent. □

Theorem 10.9. *Suppose $f \in W_{2,s}(\mathbb{R}^m)$ and α is a multiindex such that $|\alpha| \leq s$.*

(a) We have $\langle D^\alpha f, g\rangle = \langle f, D^\alpha g\rangle$ for all $g \in W_{2,|\alpha|}(\mathbb{R}^m)$.

(b) The element $D^\alpha f \in L_2(\mathbb{R}^m)$ is uniquely determined by the equality

$$\langle D^\alpha f, g\rangle = \langle f, D^\alpha g\rangle \quad \text{for all} \quad g \in C_0^\infty(\mathbb{R}^m).$$

PROOF.

(a) If $g \in W_{2,|\alpha|}(\mathbb{R}^m)$, then $Fg \in L_{2,|\alpha|}(\mathbb{R}^m)$, and thus Fg belongs to the domain of the operator of multiplication by x^α. Therefore,

$$\langle D^\alpha f, g\rangle = \langle F\,D^\alpha f, Fg\rangle = \langle M_\alpha Ff, Fg\rangle = \langle Ff, M_\alpha Fg\rangle$$
$$= \langle Ff, F\,D^\alpha g\rangle = \langle f, D^\alpha g\rangle.$$

(b) By part (a) we have $\langle D^\alpha f, g\rangle = \langle f, D^\alpha g\rangle$ for every $g \in C_0^\infty(\mathbb{R}^m) \subset W_{2,|\alpha|}(\mathbb{R}^m)$. If f_α is a further element from $L_2(\mathbb{R}^m)$ such that $\langle f_\alpha, g\rangle = \langle f, D^\alpha g\rangle$, then $\langle f_\alpha - D^\alpha f, g\rangle = 0$ for all $g \in C_0^\infty(\mathbb{R}^m)$. Therefore, $f_\alpha = D^\alpha f$, since $C_0^\infty(\mathbb{R}^m)$ is dense in $L_2(\mathbb{R}^m)$. □

Theorem 10.10.

(a) For every $s \geq 0$ the set $C_0^\infty(\mathbb{R}^m)$ is dense in $S(\mathbb{R}^m)$ with respect to the norm $\|\,.\,\|_s$.

(b) *For every $s > 0$ the set $C_0^\infty(\mathbb{R}^m)$ is dense in $W_{2,s}(\mathbb{R}^m)$ with respect to the norm $\|\cdot\|_s$.*

PROOF.

(a) Suppose $r \in \mathbb{N}_0$ and $r \geqslant s$. We show that $C_0^\infty(\mathbb{R}^m)$ is dense in $S(\mathbb{R}^m)$ with respect to $\|\cdot\|_{r,0}$; since $\|\cdot\|_s \leqslant C\|\cdot\|_{r,0}$, the assertion follows from this. In order to prove the former statement, let $\vartheta \in C_0^\infty(\mathbb{R})$ be such that $\vartheta(t) = 1$ for $t \leqslant 1$, $\vartheta(t) = 0$ for $t \geqslant 2$ and $0 \leqslant \vartheta(t) \leqslant 1$ for all $t \in \mathbb{R}$. For every $n \in \mathbb{N}$ let $\vartheta_n \in C_0^\infty(\mathbb{R}^m)$ be defined by the equality $\vartheta_n(x) = \vartheta(n^{-1}|x|)$ for all $x \in \mathbb{R}^m$. If $f \in S(\mathbb{R}^m)$ and we define $f_n = \vartheta_n f \in C_0^\infty(\mathbb{R}^m)$ for all $n \in \mathbb{N}$, then $\|D^\alpha f_n - D^\alpha f\| \to 0$ as $n \to \infty$ for every multi-index α. It follows from this that $\|f_n - f\|_{r,0} \to 0$ as $n \to \infty$.

(b) Because of part (a) we only have to show that $S(\mathbb{R}^m)$ is dense in $W_{2,s}(\mathbb{R}^m)$. Since F is an isomorphism of $W_{2,s}(\mathbb{R}^m)$ onto $L_{2,s}(\mathbb{R}^m)$ that maps $S(\mathbb{R}^m)$ onto itself, it is sufficient to prove that $S(\mathbb{R}^m)$ is dense in $L_{2,s}(\mathbb{R}^m)$. This is surely true, as $C_0^\infty(\mathbb{R}^m)$ is dense in $L_{2,s}(\mathbb{R}^m)$ (we can prove this the same way as we did the corresponding assertion for $L_2(\mathbb{R}^m)$, cf. Section 2.2, Example 8). $\qquad\square$

An m-variable *polynomial P of degree r* has the form

$$P(x) = \sum_{|\alpha| \leqslant r} c_\alpha x^\alpha,$$

where $c_\alpha \in \mathbb{C}$ and $c_\alpha \neq 0$ for at least one α such that $|\alpha| = r$. If P is a polynomial of degree r, then the formula

$$P(D) = \sum_{|\alpha| \leqslant r} c_\alpha D^\alpha = \sum_{|\alpha| \leqslant r} c_\alpha (-i)^{|\alpha|} \prod_{j=1}^{m} \frac{\partial^{\alpha_j}}{\partial x_j^{\alpha_j}}$$

defines a *differential form of order r*. We always assume that $r \neq 0$, i.e., we only consider non-trivial differential operators. In what follows let P be a (fixed) polynomial. The equalities

$$D(T_0) = C_0^\infty(\mathbb{R}^m), \; T_0 f = P(D)f \quad \text{for} \quad f \in C_0^\infty(\mathbb{R}^m)$$

define a *differential operator* on $L_2(\mathbb{R}^m)$ with constant coefficients. If we denote by M_g the operator of multiplication by the function g, then the following theorem holds.

Theorem 10.11.

(a) T_0 *is closable. For $T = \bar{T}_0$ we have $T = F^{-1} M_P F$. T is called the maximal differential operator with constant coefficients induced by P. The operator T_0^* is equal to the maximal differential operator induced by the conjugate polynomial P^*.*

(b) *We have $\sigma(T) = \overline{\{P(x) : x \in \mathbb{R}^m\}}$ and $(\lambda - T)^{-1} = F^{-1} M_{(\lambda - P)^{-1}} F$ for $\lambda \in \rho(T)$. The operator T has no eigenvalues.*

(c) *If* $(1+|P|)^{-1}\in L_2(\mathbb{R}^m)$, *then* $(\lambda-T)^{-1}$ *is a Carleman operator for every* $\lambda\in\rho(T)$, *and*

$$(\lambda-T)^{-1}f(x)=(2\pi)^{-m/2}\int h_\lambda(x-y)f(y)\,dy$$

with

$$h_\lambda=F^{-1}((\lambda-P)^{-1})\in L_2(\mathbb{R}^m).$$

PROOF.
(a) First we define T_1 by the equalities

$$D(T_1)=S(\mathbb{R}^m),\quad T_1f=P(\mathrm{D})f\quad\text{for}\quad f\in S(\mathbb{R}^m).$$

Then $T_0\subset T_1$ and $T_1=F^{-1}M_{P,1}F$, where $M_{P,1}$ denotes the restriction of M_P to $S(\mathbb{R}^m)$ (cf. Theorem 10.1). Since $M_{P,1}$ (as a restriction of a closed operator) is closable, the operator T_1 that is unitarily equivalent to $M_{P,1}$ is also closable. Therefore, T_0 is closable, also. Now we show that $\overline{T_0}=\overline{T_1}$ and $\overline{M_{P,1}}=M_P$. It will follow from this that

$$T=\overline{T_0}=\overline{T_1}=\overline{F^{-1}M_{P,1}F}=F^{-1}\overline{M_{P,1}}F=F^{-1}M_PF.$$

$\overline{T_0}=\overline{T_1}$: Since $T_0\subset T_1$, we have $\overline{T_0}\subset\overline{T_1}$. Consequently, it is sufficient to prove that $D(T_1)=S(\mathbb{R}^m)\subset D(\overline{T_0})$. For every $f\in S(\mathbb{R}^m)$ let us construct, as in the proof of Theorem 10.10(a), a sequence (f_n) from $C_0^\infty(\mathbb{R}^m)$ such that $\|\mathrm{D}^\alpha f_n-\mathrm{D}^\alpha f\|\to 0$ for all α. It follows from this that $\|f_n-f\|\to 0$ and $\|T_0f_n-T_1f\|\to 0$. Consequently, $f\in D(\overline{T_0})$, and thus $D(T_1)\subset D(\overline{T_0})$.

$\overline{M_{P,1}}=M_P$: For this it is sufficient to prove that $\overline{M_{P,0}}=M_P$, where $M_{P,0}$ denotes the restriction of M_P to $C_0^\infty(\mathbb{R}^m)$. In order to prove this we have to show the following: For every $f\in D(M_P)=\{f\in L_2(\mathbb{R}^m): Pf\in L_2(\mathbb{R}^m)\}$ there exists a sequence (f_n) from $C_0^\infty(\mathbb{R}^m)$ such that $f_n\to f$ and $M_Pf_n\to M_Pf$. This can also be proved the same way as in Section 2.2, Example 8 (cf. also the proof of Theorem 10.10(b)).

Finally, $T_0^*=\overline{T_0}^*=T^*=F^{-1}(M_P)^*F=F^{-1}M_{P^*}F$. (cf. Section 5.1, Example 1, (5.1)).

(b) Since T and M_P are unitarily equivalent,

$$\sigma(T)=\sigma(M_P)=\overline{\{P(x):x\in\mathbb{R}^m\}}.$$

For $\lambda\in\rho(T)$

$$\begin{aligned}(\lambda-T)^{-1}&=(\lambda-F^{-1}M_PF)^{-1}\\&=\left[F^{-1}(\lambda-M_P)F\right]^{-1}=F^{-1}(\lambda-M_P)^{-1}F.\end{aligned}$$

Because of the assumption $r>0$, the set $\{x\in\mathbb{R}^m: P(x)=s\}$ is a null set for every $s\in\mathbb{C}$ (proof!). By Section 5.2, Example 1, (5.14) the operator M_P therefore has no eigenvalue. Then the same holds for T.

(c) Assume that $(1+|P|)^{-1}\in L_2(\mathbb{R}^m)$ and $\lambda\in\rho(T)$, i.e., there is an $\eta>0$ such that $|\lambda-P(x)|\geq\eta$ for all $x\in\mathbb{R}^m$. For all $x\in\mathbb{R}^m$ such that $|P(x)|\geq 2|\lambda|+1$ we have the estimate

$$|\lambda-P(x)|\geq|P(x)|-|\lambda|=\left(\tfrac{1}{2}|P(x)|-|\lambda|\right)+\tfrac{1}{2}|P(x)|$$

$$\geq\tfrac{1}{2}+\tfrac{1}{2}|P(x)|=\tfrac{1}{2}(1+|P(x)|).$$

Consequently,

$$|\lambda-P(x)|^{-1}\leq 2(1+|P(x)|)^{-1}$$

for $x\in\mathbb{R}^m$ such that $(1+|P(x)|)^{-1}\leq\frac{1}{2}(1+|\lambda|)^{-1}$.

Since, besides, $|\lambda-P(x)|^{-1}\leq\eta^{-1}$ for all $x\in\mathbb{R}^m$, it follows that $(\lambda-P)^{-1}\in L_2(\mathbb{R}^m)$. The convolution theorem (Theorem 10.7) therefore implies that

$$(\lambda-T)^{-1}f=F^{-1}[(\lambda-P)^{-1}Ff]=h_\lambda*f,$$

where $h_\lambda=F^{-1}((\lambda-P)^{-1})$. Since $h_\lambda\in L_2(\mathbb{R}^m)$, the operator $(\lambda-T)^{-1}$ is a Carleman operator. □

REMARK. In Theorem 10.11(b) the closure is superfluous for $m=1$, as can be easily seen. This is not true for $m>1$, as the example of $P(x_1,x_2)=(1-x_1x_2)^2+x_1^2$ shows; in this case we have $\{P(x):x\in\mathbb{R}^2\}=(0,\infty)$.

Theorem 10.12. *The following assertions are equivalent:*
(i) *All coefficients c_α of P are real.*
(ii) *T_0 is symmetric.*
(iii) *T_0 is essentially self-adjoint.*
(iv) *T is self-adjoint.*

The *proof* immediately follows from Theorem 10.11(a).

Theorem 10.13. *Let T be a self-adjoint differential operator with constant coefficients induced by P, and let E denote the spectral family of T.*
(a) *For all $s\in\mathbb{R}$*

$$E(s)=F^{-1}M\chi_{\{x\in\mathbb{R}^m:\,P(x)<s\}}F.$$

(b) *If $|P(x)|\to\infty$ as $|x|\to\infty$, then $E(t)-E(s)$ is a Carleman operator for all $s,t\in\mathbb{R}$ such that $s\leq t$, and*

$$(E(t)-E(s))f(x)=(2\pi)^{-m/2}\int e_{s,t}(x-y)f(y)\,dy,$$

where $e_{s,t}=F^{-1}\chi_{\{x\in\mathbb{R}^m:\,s<P(x)\leq t\}}\in L_2(\mathbb{R}^m)$.
(c) *If $P(x)\to\infty$ as $|x|\to\infty$, then $E(s)$ is a Carleman operator for all $s\in\mathbb{R}$. The same holds for $I-E(s)$ provided that $P(x)\to-\infty$ as $|x|\to\infty$.*

PROOF.

(a) The first assertion is clear, since

$$F(s) = M\chi_{\{x \in \mathbb{R}^m \,:\, P(x) \leqslant s\}} \quad \text{for} \quad s \in \mathbb{R}$$

is the spectral family of M_P.

(b) We have

$$E(t) - E(s) = F^{-1}M\chi_{\{x \in \mathbb{R}^m \,:\, s < P(x) \leqslant t\}}F.$$

Since $|P(x)| \to \infty$ as $|x| \to \infty$, the set $\{x \in \mathbb{R}^m : s < P(x) \leqslant t\}$ is bounded. Therefore, $\chi_{\{x \in \mathbb{R}^m \,:\, s < P(x) \leqslant t\}}$ belongs to $L_2(\mathbb{R}^m)$. The assertion then follows from Theorem 10.7.

(c) If $P(x) \to \infty$ as $|x| \to \infty$, then $\{x \in \mathbb{R}^m : P(x) \leqslant s\}$ is compact, and thus $\chi_{\{x \in \mathbb{R}^m \,:\, P(x) \leqslant s\}} \in L_2(\mathbb{R}^m)$. If $P(x) \to -\infty$ as $|x| \to \infty$, then $\chi_{\{x \in \mathbb{R}^m \,:\, P(x) > s\}} \in L_2(\mathbb{R}^m)$. The assertion follows from Theorem 10.7 in both cases. $\square$

A polynomial P of degree r and the operators T_0 and T induced by P are said to be *elliptic* if there exists a $C > 0$ such that

$$1 + |P(x)| \geqslant C(1 + |x|^2)^{r/2} = Ck_r(x) \quad \text{for all} \quad x \in \mathbb{R}^m.$$

(Observe that we always have $1 + |P(x)| \leqslant C'k_r(x)$ with an appropriate choice of $C' > 0$.) The *principal part* of P is given by

$$P_r(x) = \sum_{|\alpha| = r} c_\alpha x^\alpha.$$

Correspondingly, the *principal part* of $P(D)$ is given by

$$P_r(D) = \sum_{|\alpha| = r} c_\alpha D^\alpha.$$

Theorem 10.14. *Let P be a polynomial of degree r, and let T be the maximal differential operator induced by P. Then the following statements are equivalent:*

(i) *P is elliptic.*

(ii) *The principal part of P vanishes only for $x = 0$.*

(iii) *$D(T) = W_{2,\,r}(\mathbb{R}^m)$.*

In this case the norms $\|\cdot\|_T$ and $\|\cdot\|_r$ are equivalent.

PROOF. (i) *implies* (ii): Let us assume that there is an $x_0 \in \mathbb{R}^m$, $x_0 \neq 0$ such that $P_r(x_0) = 0$. Then we also have $P_r(sx_0) = 0$ for all $s \in \mathbb{R}$. Therefore,

$$|P(sx_0)| = \Big| \sum_{|\alpha| < r} c_\alpha s^{|\alpha|} x_0^\alpha \Big| \leqslant C(1 + |sx_0|^2)^{(r-1)/2},$$

in contradiction with the ellipticity of P.

(ii) *implies* (i): Let $\eta = \min\{|P_r(x)| : x \in \mathbb{R}^m, |x| = 1\}$. Then for all $x \in \mathbb{R}^m$

$$|P(x)| = \left|P_r(x) + \sum_{|\alpha|<r} c_\alpha x^\alpha\right| \geq |x|^r\eta - C(1+|x|^2)^{(r-1)/2}.$$

The ellipticity of P follows from this.

(i) *implies* (iii): We have

$$\begin{aligned} D(T) &= F^{-1}\{f \in L_2(\mathbb{R}^m) : Pf \in L_2(\mathbb{R}^m)\} \\ &= F^{-1}\{f \in L_2(\mathbb{R}^m) : k_r f \in L_2(\mathbb{R}^m)\} = W_{2,r}(\mathbb{R}^m). \end{aligned}$$

Since

$$\|f\|_T^2 = \|f\|^2 + \|Tf\|^2 = \|Ff\|^2 + \|M_P Ff\|^2,$$

the equivalence of the norms $\|\,.\,\|_T$ and $\|\,.\,\|_r$ follows.

(iii) *implies* (i): We obtain from the equality $D(T) = W_{2,r}(\mathbb{R}^m)$ that

$$D(M_P) = FD(T) = L_{2,r}(\mathbb{R}^m) = D(M_{k_r}).$$

By Theorem 5.9 M_{k_r} is relatively bounded with respect to M_P, therefore with respect to $M_{|P|}$ as well (since $D(M_P) = D(M_{|P|})$ and $\|M_{|P|}f\| = \|M_P f\|$). The boundedness of $M_{k_r}(1+M_{|P|})^{-1}$, i.e., the boundedness of the function $k_r(1+|P|)^{-1}$, follows from this by Theorem 9.9. Consequently, P is elliptic. □

Theorem 10.15. *Let T be a self-adjoint elliptic differential operator with constant coefficients on $L_2(\mathbb{R}^m)$.*

(a) *If $m > 1$, then T is semibounded.*

(b) *If T is bounded from below, then $E(t)$ is a Carleman operator for every $t \in \mathbb{R}$.*

PROOF.

(a) Since T is self-adjoint, P is real-valued. As T is elliptic, $|P(x)| \to \infty$ as $|x| \to \infty$. Consequently, $|P(x)| > 0$ for all $|x| \geq c_0$. Because of the continuity of P it follows (due to the assumption $m > 1$) that $P(x) > 0$ for all $|x| \geq c_0$ or $P(x) < 0$ for all $|x| \geq c_0$; hence $P(x) \to \infty$ or $P(x) \to -\infty$ as $|x| \to \infty$. The boundedness from below or from above follows from this.

(b) If T is elliptic and bounded from below, then we obtain (as in the proof of part (a)) that $P(x) \to \infty$ as $|x| \to \infty$. The assertion follows by Theorem 10.13. □

Corollary. *Let T be a self-adjoint differential operator on $L_2(\mathbb{R}^m)$ with constant coefficients. Then*

1. $\sigma(T) = \mathbb{R}$ *or* $\sigma(T) = [\gamma, \infty)$ *or* $\sigma(T) = (-\infty, \gamma]$.
2. *If T is elliptic and $m > 1$, then* $\sigma(T) = [\gamma, \infty)$ *or* $\sigma(T) = (-\infty, \gamma]$.

3. *We always have* $H_p(T)=\{0\}$, $H_c(T)=L_2(\mathbb{R}^m)$, *and* $\sigma(T)=\sigma_e(T)=\sigma_c(T)$ (*cf. also Exercise* 10.7).

EXAMPLE 1. $m=1$, $P(x)=x$, $P(\mathrm{D})=(1/\mathrm{i})(\mathrm{d}/\mathrm{d}x)$. Then T is elliptic and not semibounded; $\sigma(T)=\mathbb{R}$.

EXAMPLE 2. $m\geq 1$, $P(x)=\sum_{j=1}^m x_j^2=|x|^2$, $P(\mathrm{D})=-\sum_{j=1}^m(\partial^2/\partial x_j^2)=-\Delta$. Then T is elliptic and bounded from below; $\sigma(T)=[0,\infty)$.

EXAMPLE 3. $m=2$, $P(x)=x_1^2+x_2^4$, $P(\mathrm{D})=-(\partial^2/\partial x_1^2)+(\partial^4/\partial x_2^4)$. Then T is *not* elliptic but is bounded from below; $\sigma(T)=[0,\infty]$.

EXERCISES

10.5. Let f belong to $W_{2,s}(\mathbb{R}^m)$ with $s>m/2$. Then f is Lipschitz with exponent $\delta\in(0,1]\cap(0,s-(m/2))$, i.e., there exists a $C\geq 0$ such that $|f(x)-f(y)|\leq C|x-y|^\delta$ for all $x,y\in\mathbb{R}^m$.

10.6. For $r\in\mathbb{N}_0$ such that $r<m$ the set $C_0^\infty(\mathbb{R}^m\setminus\{0\})$ is dense in $W_{2,r}(\mathbb{R}^m)$. Hint: If $\vartheta\in C_0^\infty(\mathbb{R}^m)$ is such that $\vartheta(x)=1$ for $|x|\leq 1/2$ and $\vartheta(x)=0$ for $|x|\geq 1$ and $\vartheta_n(x)=\vartheta(nx)$, then $\vartheta_n f\xrightarrow{w}0$; therefore, $(1-\vartheta_n)f\xrightarrow{w}f$ in the sense of $W_{2,r}(\mathbb{R}^m)$ (cf. Exercise 4.25).

10.7. If P is a non-constant polynomial, then M_P and $F^{-1}M_PF$ have a pure absolutely continuous spectrum.

10.3 Relatively bounded and relatively compact perturbations

In this section we first give conditions in order that an operator $M_q: W_{2,r}(\mathbb{R}^m)\to L_2(\mathbb{R}^m)$, $f\mapsto qf$ be bounded or compact. For the sake of simplicity, we only consider integers r; this is sufficient in most applications to differential operators. (Corresponding results for arbitrary r can be found in *M. Schechter* [34], Chapter 6.)

Theorem 10.16. *Let* $0\leq s<r$ (*not necessarily integral*). *Then for every* $\eta>0$ *there exists a* $C_\eta\geq 0$ *such that*

$$\|f\|_s\geq\eta\|f\|_r+C_\eta\|f\|\quad\text{for all}\quad f\in W_{2,r}(\mathbb{R}^m).$$

PROOF. The assertion is equivalent to the inequality

$$\|f\|_{(s)}\leq\eta\|f\|_{(r)}+C_\eta\|f\|\quad\text{for all}\quad f\in L_{2,r}(\mathbb{R}^m).$$

For all $N>0$ we have

$$\begin{aligned}\|f\|_{(s)}^2 &= \int |f(x)|^2(1+|x|^2)^s \, dx \\ &\leq (1+N^2)^s \int_{|x|\leq N} |f(x)|^2 \, dx + (1+N^2)^{s-r} \int_{|x|>N} |f(x)|^2(1+|x|^2)^r \, dx \\ &\leq (1+N^2)^s \|f\|^2 + (1+N^2)^{s-r} \|f\|_{(r)}^2.\end{aligned}$$

Due to the inequality $s-r<0$, the assertion follows from this if N is chosen large enough. □

A measurable function $q : \mathbb{R}^m \to \mathbb{C}$ is said to be *locally square integrable* if $\vartheta q \in L_2(\mathbb{R}^m)$ for every $\vartheta \in C_0^\infty(\mathbb{R}^m)$. This holds if and only if $|q|^2$ is integrable over every compact subset of $\mathbb{R}^m$. The set of locally square integrable functions obviously constitutes a (complex) vector space. This space will be denoted by $L_{2,\mathrm{loc}}(\mathbb{R}^m)$. For every $q \in L_{2,\mathrm{loc}}(\mathbb{R}^m)$ let us define

$$N_q(x) = \left\{ \int_{|x-y|\leq 1} |q(y)|^2 \, dy \right\}^{1/2} \quad \text{for all} \quad x \in \mathbb{R}^m.$$

N_q is obviously locally bounded, i.e., it is bounded on every compact subset. N_q is even continuous (proof!).

For every measurable function $q : \mathbb{R}^m \to \mathbb{C}$ and every $\rho \in \mathbb{R}$ let us define (the value ∞ is allowed)

$$M_{q,\rho}(x) = \begin{cases} \left\{ \int_{|x-y|\leq 1} |q(y)|^2 |x-y|^{\rho-m} \, dy \right\}^{1/2} & \text{for} \quad \rho<m,^2 \\ N_q(x) & \text{for} \quad \rho \geq m. \end{cases}$$

We denote by $M_{\rho,\mathrm{loc}}(\mathbb{R}^m)$ the vector space of measurable functions $q : \mathbb{R}^m \to \mathbb{C}$ for which $M_{q,\rho}(\,.\,)$ is a locally bounded function. $M_\rho(\mathbb{R}^m)$ denotes the subspace of those $q \in M_{\rho,\mathrm{loc}}(\mathbb{R}^m)$ for which $M_{q,\rho}(\,.\,)$ is bounded. For $q \in M_\rho(\mathbb{R}^m)$ we set

$$M_{q,\rho} = \sup\{ M_{q,\rho}(x) : x \in \mathbb{R}^m \}.$$

For all $\rho_1, \rho_2 \in \mathbb{R}$ such that $\rho_1 \leq \rho_2$ we have

$$\begin{aligned} M_{\rho_1,\mathrm{loc}}(\mathbb{R}^m) &\subset M_{\rho_2,\mathrm{loc}}(\mathbb{R}^m) \subset L_{2,\mathrm{loc}}(\mathbb{R}^m), \\ M_{\rho_1}(\mathbb{R}^m) &\subset M_{\rho_2}(\mathbb{R}^m); \end{aligned} \tag{10.1}$$

if $q \in M_{\rho_1}(\mathbb{R}^m)$, then we obviously have

$$M_{q,\rho_2} \leq M_{q,\rho_1}. \tag{10.2}$$

[2] This definition goes back to *F. Stummel* [52] for $\rho<4$.

EXAMPLE 1. Every bounded measurable function belongs to $M_\rho(\mathbb{R}^m)$ for every $\rho>0$.

EXAMPLE 2. Assume $a\in\mathbb{R}^m$, $c\geq 0$, and $0<\delta<m/2$. Assume, furthermore, that the function $q:\mathbb{R}^m\to\mathbb{C}$ is measurable and $|q(x)|\leq c|x-a|^{-\delta}$ almost everywhere in $\mathbb{R}^m$. Then $q\in M_\rho(\mathbb{R}^m)$ for all $\rho>2\delta$. This is obvious for $\rho\geq m$, since q is locally square integrable and $q(x)\to 0$ as $|x|\to\infty$. Now suppose that $2\delta<\rho<m$. Then

$$M_{q,\rho}^2(x)\leq c^2\left\{\int_{\substack{|x-y|\leq 1\\|y-a|\leq|x-y|}}+\int_{\substack{|x-y|\leq 1\\|y-a|>|x-y|}}|y-a|^{-2\delta}|x-y|^{\rho-m}\,dy\right\}$$
$$\leq c^2\left\{\int_{|y-a|\leq 1}|y-a|^{\rho-2\delta-m}\,dy+\int_{|x-y|\leq 1}|x-y|^{\rho-2\delta-m}\,dy\right\}$$
$$=2c^2\int_{|y|\leq 1}|y|^{\rho-2\delta-m}\,dy=C<\infty.$$

Of course, sums of such functions also belong to $M_\rho(\mathbb{R}^m)$.

Theorem 10.17. *Assume that $r\in\mathbb{N}$ and $\rho<2r$.*
(a) *There is a constant $C\geq 0$ such that*

$$\|qf\|\leq CM_{q,\rho}\|f\|_r \quad \textit{for all} \quad q\in M_\rho(\mathbb{R}^m) \quad \textit{and all} \quad f\in W_{2,r}(\mathbb{R}^m).$$

(b) *For every $q\in M_\rho(\mathbb{R}^m)$ and every $\eta>0$ there is a C_η such that*

$$\|qf\|\leq\eta\|f\|_r+C_\eta\|f\| \quad \textit{for all} \quad f\in W_{2,r}(\mathbb{R}^m).$$

PROOF. Take a $\vartheta\in C_0^\infty(\mathbb{R}^m)$ such that $\vartheta(0)=1$ and $\vartheta(x)=0$ for $|x|\geq 1$, and let $\vartheta_s\in C_0^\infty(\mathbb{R}^m)$ be defined by the equality

$$\vartheta_s(x)=\vartheta(s^{-1}x) \quad \text{for} \quad x\in\mathbb{R}^m \quad \text{and} \quad s\in(0,1].$$

Let Ω_m denote the unit sphere in $\mathbb{R}^m$. Then for all $f\in C_0^\infty(\mathbb{R}^m)$, $s\in(0,1]$ and $\omega\in\Omega_m$

$$f(0)=\vartheta_s(0)f(0)=-\int_0^s\frac{\partial}{\partial t}(\vartheta_s(t\omega)f(t\omega))\,dt=\cdots$$
$$=(-1)^r\int_0^s\int_{t_r}^s\cdots\int_{t_2}^s\frac{\partial^r}{\partial t_1^r}(\vartheta_s(t_1\omega)f(t_1\omega))\,dt_1\ldots dt_{r-1}\,dt_r$$
$$=(-1)^r\int_0^s\frac{\partial^r}{\partial t_1^r}(\vartheta_s(t_1\omega)f(t_1\omega))\left\{\int_0^{t_1}\cdots\int_0^{t_{r-1}}dt_r\ldots dt_2\right\}dt_1$$
$$=\frac{(-1)^r}{(r-1)!}\int_0^s\frac{\partial^r}{\partial t^r}(\vartheta_s(t\omega)f(t\omega))t^{r-1}\,dt.$$

With

$$g(y) = g(t\omega) = \frac{\partial^r}{\partial t^r}(\vartheta_s(t\omega)f(t\omega)) \quad \text{for} \quad y = t\omega, 0 < t < s$$

we therefore obtain by integration over the unit sphere Ω_m that

$$|f(0)| \leq \frac{1}{V_m(r-1)!}\int_{\Omega_m}\int_0^s |g(t\omega)|t^{r-1}\,dt\,d\omega$$

$$= C_1\int_0^s t^{r-1}\int_{\Omega_m}|g(t\omega)|\,d\omega\,dt = C_1\int_{|y|\leq s}|y|^{r-m}|g(y)|\,dy,$$

where V_m denotes the area of Ω_m and $d\omega$ the area element on Ω_m. Since

$$|g(y)| \leq C_2\left(\sum_{|\alpha|\leq r}|D^\alpha f(y)| + s^{-r}\sum_{|\alpha|<r}|D^\alpha f(y)|\right) = : h(y),$$

we obtain for all $\epsilon > 0$ that

$$|f(0)|^2 \leq C_3\int_{|y|\leq s}|y|^{\epsilon-m}\,dy\int_{|y|\leq s}|h(y)|^2|y|^{2r-\epsilon-m}\,dy$$

$$\leq C_4 s^\epsilon\int_{|y|\leq 1}|y|^{2r-\epsilon-m}\left\{\sum_{|\alpha|\leq r}|D^\alpha f(y)|^2 + s^{-2r}\sum_{|\alpha|<r}|D^\alpha f(y)|^2\right\}dy.$$

It follows similarly that for all $s \in \mathbb{R}^m$

$$|f(x)|^2 \leq C_4 s^\epsilon\left\{\int_{|x-y|\leq 1}|x-y|^{2r-\epsilon-m}\sum_{|\alpha|\leq r}|D^\alpha f(y)|^2\,dy\right.$$

$$\left.+ s^{-2r}\int_{|x-y|\leq 1}|x-y|^{2r-\epsilon-m}\sum_{|\alpha|<r}|D^\alpha f(y)|^2\,dy\right\}.$$

In order to estimate $\|qf\|$ we distinguish between two different cases.

First assume that $2r > m$. In this case, without loss of generality we can assume on the basis of (10.1) and (10.2) that $\rho > m$. Furthermore, choose $\epsilon = 2r - m$. Then

$$\int|q(x)f(x)|^2\,dx \leq C_4 s^{2r-m}\int|q(x)|^2$$

$$\times\left\{\int_{|x-y|\leq 1}\sum_{|\alpha|\leq r}|D^\alpha f(y)|^2\,dy + s^{-2r}\int_{|x-y|\leq 1}\sum_{|\alpha|<r}|D^\alpha f(y)|^2\,dy\right\}dx$$

$$\leq C_4 s^{2r-m}M_{q,\rho}^2\int\left\{\sum_{|\alpha|\leq r}|D^\alpha f(y)|^2 + s^{-2r}\sum_{|\alpha|<r}|D^\alpha f(y)|^2\right\}dy$$

$$\leq C_5 M_{q,\rho}^2\{s^{2r-m}\|f\|_r^2 + s^{-m}\|f\|_{r-1}^2\}.$$

Now let $2r < m$. Then $\rho < m$. Choose $\epsilon = 2r - \rho$. We have

$$\int |q(x)f(x)|^2 \,dx \leq C_4 s^{2r-\rho} \int |q(x)|^2 \left\{ \int_{|x-y|<1} |x-y|^{\rho-m} \sum_{|\alpha|\leq r} |D^\alpha f(y)|^2 \,dy \right.$$

$$\left. + s^{-2r} \int_{|x-y|<1} |x-y|^{\rho-m} \sum_{|\alpha|<r} |D^\alpha f(y)|^2 \,dy \right\} dx$$

$$\leq C_6 M_{q,\rho}^2 \{ s^{2r-\rho} \|f\|_r^2 + s^{-\rho} \|f\|_{r-1}^2 \}.$$

In both cases we obtain assertion (a) for $f \in C_0^\infty(\mathbb{R}^m)$ provided that we take, for example, $s = 1$.

Since by Theorem 10.16 for every $\mu > 0$ there exists a $C_\mu \geq 0$ such that

$$\|f\|_{r-1}^2 \leq \mu \|f\|_r^2 + C_\mu \|f\|^2,$$

assertion (b) follows for all $f \in C_0^\infty(\mathbb{R}^m)$ in both cases provided that first s and then μ is chosen small enough.

If $f \in W_{2,r}(\mathbb{R}^m)$, then there exists a sequence (f_n) from $C_0^\infty(\mathbb{R}^m)$ for which $f_n \to f$ in the sense of $W_{2,r}(\mathbb{R}^m)$. Since the sequence (f_n) then converges to f also in $L_2(\mathbb{R}^m)$, there is a subsequence (f_{n_k}) that converges to f almost everywhere. Then this holds for (qf_{n_k}) and qf, as well. Since the sequences $(\|f_{n_k}\|_r)$ and $(\|f_{n_k}\|)$ are bounded, $(\|qf_{n_k}\|)$ is also bounded and Fatou's lemma implies that $qf \in L_2(\mathbb{R}^m)$ and

$$\|qf\| \leq \liminf_{k\to\infty} \left\{ \int |q(x) f_{n_k}(x)|^2 \,dx \right\}^{1/2}$$

$$= \liminf_{k\to\infty} \|qf_{n_k}\| \leq \liminf_{k\to\infty} (\eta \|f_{n_k}\|_r + C_\eta \|f_{n_k}\|)$$

$$= \eta \|f\|_r + C_\eta \|f\|.$$

We can prove assertion (a) for $f \in W_{2,r}(\mathbb{R}^m)$ similarly. □

Using this result, we can now give conditions for the relative boundedness and relative compactness of the perturbations of a closed operator T such that $D(T) \subset W_{2,r}(\mathbb{R}^m)$.

Theorem 10.18. *Assume $r \in \mathbb{N}$ and T is a closed operator on $L_2(\mathbb{R}^m)$ such that $D(T) \subset W_{2,r}(\mathbb{R}^m)$. Let V be an operator on $L_2(\mathbb{R}^m)$ such that*

$$D(V) \supset W_{2,r}(\mathbb{R}^m), \quad Vf = \sum_{|\alpha|\leq r} q_\alpha D^\alpha f \quad \text{for} \quad f \in W_{2,r}(\mathbb{R}^m).$$

Let the functions q_α be measurable, and assume that q_α is bounded for $|\alpha| = r$, $\sum_{|\alpha|=r} \sup\{|q_\alpha(x)| : x \in \mathbb{R}^m\} = c$, $q_\alpha \in M_{\rho_\alpha}(\mathbb{R}^m)$ with $\rho_\alpha < 2(r - |\alpha|)$ for $|\alpha < r$.

Then V is T-bounded. If $\|f\|_r \leq d\|Tf\| + e\|f\|$ for all $f \in D(T)$, then the T-bound of V is less than or equal to dc. If $q_\alpha = 0$ for $|\alpha| = r$, then the T-bound of V is equal to 0.

PROOF. If M_{k_r} is the operator of multiplication by $k_r=(1+|\cdot|^2)^{r/2}$ on $L_2(\mathbb{R}^m)$ and $T_r=F^{-1}M_{k_r}F$, then T_r is a self-adjoint operator on $L_2(\mathbb{R}^m)$ and $D(T_r)=W_{2,r}(\mathbb{R}^m)$. By Theorem 5.9 the operator T_r is T-bounded, i.e., there exist contants d and e such that $\|T_r f\| \leqslant d\|Tf\|+e\|f\|$ for all $f \in D(T)$. Let

$$V_r=\sum_{|\alpha|=r} q_\alpha \mathrm{D}^\alpha \quad \text{and} \quad V_0=\sum_{|\alpha|<r} q_\alpha \mathrm{D}^\alpha .$$

We obviously have

$$\|V_r f\| \leqslant c\|f\|_r = c\|T_r f\| \leqslant cd\|Tf\|+ce\|f\|.$$

It is therefore sufficient to show that for every $\epsilon>0$ there exists a $C_\epsilon>0$ such that

$$\|V_0 f\| \leqslant \epsilon\|f\|_r + C_\epsilon\|f\| \quad \text{for all} \quad f\in W_{2,r}(\mathbb{R}^m).$$

Since $\mathrm{D}^\alpha f \in W_{2,r-|\alpha|}(\mathbb{R}^m)$ (because $M_\alpha Ff \in L_{2,r-|\alpha|}(\mathbb{R}^m)$), by Theorem 10.17 for every $\eta>0$ there exists a $C_\eta \geqslant 0$ such that for all $|\alpha|<r$ and $f\in W_{2,r}(\mathbb{R}^m)$

$$\|q_\alpha \mathrm{D}^\alpha f\| \leqslant \eta\|\mathrm{D}^\alpha f\|_{r-|\alpha|} + C_\eta\|\mathrm{D}^\alpha f\| \leqslant \eta\|f\|_r + C_\eta\|f\|_{|\alpha|}.$$

This gives, together with Theorem 10.16, the assertion. □

In order to prove the relative compactness of perturbations, we need the following auxiliary results.

Theorem 10.19. *Let $0\leqslant s<r$ (not necessarily integral). The mapping*

$$\Phi : W_{2,r}(\mathbb{R}^m) \to W_{2,s}(\mathbb{R}^m), \quad f \mapsto \varphi f$$

is compact for every $\varphi \in C_0^\infty(\mathbb{R}^m)$.

PROOF. The operator Φ is compact if and only if the operator K defined on $L_2(\mathbb{R}^m)$ by the formula

$$K: L_2(\mathbb{R}^m) \overset{U_r^{-1}}{\to} L_{2,r}(\mathbb{R}^m) \overset{F^{-1}}{\to} W_{2,r}(\mathbb{R}^m) \overset{\Phi}{\to} W_{2,s}(\mathbb{R}^m) \overset{F}{\to} L_{2,s}(\mathbb{R}^m) \overset{U_s}{\to} L_2(\mathbb{R}^m)$$

is compact, since the four operators at the ends are unitary (cf. Section 10.2 for U_s and U_r). By Theorem 10.7 we obviously have for $f\in L_2(\mathbb{R}^m)$ that

$$(Kf)(x) = \int (1+|x|^2)^{s/2}\psi(x-y)(1+|y|^2)^{-r/2} f(y)\,\mathrm{d}y$$

with $\psi=(2\pi)^{-m/2}F\varphi \in S(\mathbb{R}^m)$. The operator K is therefore an integral operator with kernel

$$k(x,y) = (1+|x|^2)^{s/2}\psi(x-y)(1+|y|^2)^{-r/2} \quad \text{for} \quad x,y\in\mathbb{R}^m.$$

Since $\psi \in S(\mathbb{R}^m)$, for every $l \in \mathbb{N}$ there is a B_l such that

$$|\psi(x)| \leq B_l(1+|x|)^{-l} \quad \text{for all} \quad x \in \mathbb{R}^m.$$

For $n \in \mathbb{N}$ let us set

$$k_n(x,y) = \begin{cases} k(x,y) & \text{for} \quad |y| \leq n, \\ 0 & \text{for} \quad |y| > n, \end{cases} \qquad h_n(x,y) = k(x,y) - k_n(x,y).$$

We shall prove the assertion by showing that for the operators K_n and H_n induced by k_n and h_n the following holds: K_n is a Hilbert-Schmidt operator and $\|H_n\| \to 0$ as $n \to \infty$.

For $|y| \leq n$ and $|x| \geq 2n$ we have

$$|\psi(x-y)| \leq B_l(1+|x-y|)^{-l} \leq B_l\left(1+\tfrac{1}{2}|x|\right)^{-l}.$$

It follows from this that for $l = m+1+s$ and an appropriate constant C

$$|\psi(x-y)| \leq C(1+|x|)^{-m-1-s} \quad \text{for} \quad |y| \leq n \quad \text{and all} \quad x \in \mathbb{R}^m.$$

We therefore have

$$|k_n(x,y)| \leq \begin{cases} C(1+|x|)^{-m-1}(1+|y|^2)^{-r/2} & \text{for} \quad |y| \leq n, \\ 0 & \text{for} \quad |y| > n, \end{cases}$$

and thus $k_n \in L_2(\mathbb{R}^m \times \mathbb{R}^m)$, i.e., K_n is a Hilbert-Schmidt operator.

To estimate the norm of H_n, we use the corollary to Theorem 6.24. We have

$$\begin{aligned} \int_{|y|>n} |\psi(x-y)|(1+|y|^2)^{-r/2}\,dy &\leq C_1(1+n)^{-r}\int|\psi(x-y)|dy \\ &\leq C_2(1+n)^{-r} \leq C_2(1+|x|)^{-s}(1+n)^{s-r} \end{aligned}$$

for $|x| \leq n$. Since $\psi \in S(\mathbb{R}^m)$, we have

$$\begin{aligned} \int_{|y|>n} |\psi(x-y)|(1+|y|^2)^{-r/2}\,dy &\leq C_3\int(1+|x-y|)^{-m-r}(1+|y|)^{-r}\,dy \\ &\leq C_3\left\{\int_{|x-y|<|x|/2}(1+|x-y|)^{-m-r}(1+|y|)^{-r}\,dy\right. \\ &\qquad \left. + \int_{|x-y|>|x|/2}(1+|x-y|)^{-m-r}\,dy\right\} \\ &\leq C_4(1+|x|)^{-r}\int(1+|x-y|)^{-m-r}\,dy \\ &\qquad + C_3\int_{|y|>|x|/2}(1+|y|)^{-m-r}\,dy \\ &\leq C_5(1+|x|)^{-r} \leq C_5(1+|x|)^{-s}(1+n)^{s-r} \end{aligned}$$

for $|x|>n$. Consequently, it follows for all $x\in\mathbb{R}^m$ that

$$\int|h_n(x,y)|\mathrm{d}y \leqslant C_6(1+n)^{s-r}\to 0 \quad \text{as} \quad n\to\infty.$$

Furthermore,

$$\begin{aligned}\int(1+|x|)^s|\psi(x-y)|\mathrm{d}x &\leqslant C_7\int(1+|x|)^s(1+|x-y|)^{-m-s-1}\,\mathrm{d}x\\ &= C_7\left\{\int_{|x|\leqslant 2|y|}+\int_{|x|>2|y|}(1+|x|)^s(1+|x-y|)^{-m-s-1}\,\mathrm{d}x\right\}\\ &\leqslant C_7\left\{(1+2|y|)^s\int(1+|x-y|)^{-m-s-1}\,\mathrm{d}x\right.\\ &\qquad\left.+\int(1+|x|)^s\left(1+\tfrac{1}{2}|x|\right)^{-m-s-1}\mathrm{d}x\right\}\\ &\leqslant C_8(1+|y|)^s.\end{aligned}$$

Hence, for all $y\in\mathbb{R}^m$ such that $|y|>n$ (and thus for all $y\in\mathbb{R}^m$)

$$\int|h_n(x,y)|\mathrm{d}x \leqslant C_8(1+|y|)^{s-r}\leqslant C_8(1+n)^{s-r}\to 0 \quad \text{as} \quad n\to\infty.$$

It follows therefore from the Corollary to Theorem 6.24 that $\|H_n\|\leqslant C_9(1+n)^{s-r}\to 0$ as $n\to\infty$. □

Theorem 10.20. *Assume that $s\in\mathbb{N}$, $q\in M_\rho(\mathbb{R}^m)$ for some $\rho<2s$ and $N_q(x)\to 0$ as $|x|\to\infty$. Then the mapping*

$$Q: W_{2,s}(\mathbb{R}^m)\to L_2(\mathbb{R}^m),\quad f\mapsto qf$$

is compact.

Proof. Let M_{k_s} be the operator of multiplication by the function k_s, and let $T_s=F^{-1}M_{k_s}F$. We have to prove the T_s-compactness of Q. By Theorem 10.17(b) the operator Q is obviously T_s-bounded with T_s-bound zero. Therefore, by Theorem 9.11 it is enough to prove the T_s^2-compactness of Q. This, in turn, is equivalent to the compactness of

$$Q: W_{2,2s}(\mathbb{R}^m)\to L_2(\mathbb{R}^m),\quad f\mapsto qf.$$

We shall prove this in what follows.

If $\rho>m$, then set $\tau=\rho$. Then

$$M_{q,\tau}(x) = M_{q,\rho}(x) = N_q(x)\to 0 \quad \text{as} \quad |x|\to\infty.$$

If $\rho<m$, then we choose a τ for which $\rho<\tau<\min(m,2s)$. Then Hölder's inequality with the exponents $p=(m-\rho)/(\tau-\rho)$ and $p'=(m-\rho)/(m-\tau)$

gives

$$M_{q,\tau}(x)^2 = \int_{|x-y|<1} \{|q(y)|^{2/p}\}\{|q(y)|^{2/p'}|x-y|^{\tau-m}\}\,dy$$
$$\leq \left\{\int_{|x-y|<1} |q(y)|^2\,dy\right\}^{1/p}\left\{\int_{|x-y|<1} |q(y)|^2|x-y|^{\rho-m}\,dy\right\}^{1/p'}$$
$$= N_q(x)^{2/p} M_{q,\rho}(x)^{2/p'} \to 0 \quad \text{as} \quad |x| \to \infty.$$

Consequently, in each case we have found a τ for which

$$\tau < 2s \quad \text{and} \quad M_{q,\tau}(x) \to 0 \quad \text{as} \quad |x| \to \infty.$$

Now let $\varphi \in C_0^\infty(\mathbb{R}^m)$ be such that $\varphi(x) = 1$ for $|x| \leq 1$, $\varphi(x) = 0$ for $|x| \geq 2$ and $0 \leq \varphi(x) \leq 1$ for all $x \in \mathbb{R}^m$. Let the functions $\varphi_n \in C_0^\infty(\mathbb{R}^m)$ be defined by the equality $\varphi_n(x) = \varphi(n^{-1}x)$ for $n \in \mathbb{N}$ and $x \in \mathbb{R}^m$. Then by Theorem 10.19 the operator

$$\Phi_n : W_{2,2s}(\mathbb{R}^m) \to W_{2,s}(\mathbb{R}^m), \quad f \mapsto \varphi_n f$$

is compact for all $n \in \mathbb{N}$. Since the mapping $Q : W_{2,s}(\mathbb{R}^m) \to L_2(\mathbb{R}^m)$ is bounded by Theorem 10.17, the compactness of

$$Q\Phi_n : W_{2,2s}(\mathbb{R}^m) \to L_2(\mathbb{R}^m), \quad f \mapsto q\varphi_n f$$

follows. As $M_{q,\tau}(x) \to 0$ when $|x| \to \infty$, we obviously have $M_{(1-\varphi_n)q,\tau} \to 0$ when $n \to \infty$. It follows from this for the operators

$$A_n : W_{2,2s}(\mathbb{R}^m) \to L_2(\mathbb{R}^m), \quad f \mapsto q(1-\varphi_n)f$$

that $\|A_n\| \to 0$ (cf. Theorem 10.17). This implies that $Q = \lim_{n\to\infty} Q\Phi_n$. Hence, Q is compact. □

Theorem 10.21. *Let r, T, and V be defined as in Theorem 10.18.*
(a) *If $q_\alpha = 0$ for $|\alpha| = r$ and*

$$N_{q_\alpha}(x) \to 0 \quad \text{as} \quad |x| \to \infty \quad \text{for} \quad |\alpha| < r,$$

then V is T-compact.
(b) *If T is self-adjoint, $D(T^p) \subset W_{2,s}(\mathbb{R}^m)$ for some $p > 1$ and $s > r$, and*

$$q_\alpha(x) \to 0 \quad \text{for} \quad |x| \to \infty \quad \text{and} \quad |\alpha| = r,$$
$$N_{q_\alpha}(x) \to 0 \quad \text{for} \quad |x| \to \infty \quad \text{and} \quad |\alpha| < r,$$

then V is T^t-compact for every $t > 1$.

Proof. The mappings

$$W_{2,r}(\mathbb{R}^m) \to W_{2,r-|\alpha|}(\mathbb{R}^m), \quad f \mapsto D^\alpha f$$

are bounded, and by Theorem 10.20 the mappings

$$W_{2,r-|\alpha|}(\mathbb{R}^m) \to L_2(\mathbb{R}^m), \quad g \mapsto q_\alpha g$$

are compact for $|\alpha|<r$. The compactness of the operators

$$V_\alpha : W_{2,r}(\mathbb{R}^m) \to L_2(\mathbb{R}^m), \quad f \mapsto q_\alpha D^\alpha f$$

follows from this for all α such that $|\alpha|<r$. This gives the T-compactness of V for part (a) of the assertion.

We can show the compactness of

$$\tilde{V}_\alpha : W_{2,s}(\mathbb{R}^m) \to L_2(\mathbb{R}^m), \quad f \mapsto q_\alpha D^\alpha f$$

for $|\alpha|=r$ in an analogous way. Since $D(T^p) \subset W_{2,s}(\mathbb{R}^m)$, this gives the T^p-compactness of V. As V is T-bounded by Theorem 10.18, we obtain the T^t-compactness of V for every $t>1$ by Theorem 9.11. This proves part (b). □

EXERCISES

10.8. (a) Assume that $q \in L_{p,\text{loc}}(\mathbb{R}^m)$ for some $p \geq 2$ (i.e., q is measurable and $|q|^p$ is integrable over every compact subset of $\mathbb{R}^m$). If $\rho > 2m/p$ for $p>2$ and $\rho \geq m$ for $p=2$, then $q \in M_{\rho,\text{loc}}(\mathbb{R}^m)$.
(b) Prove a corresponding result for $M_\rho(\mathbb{R}^m)$.

10.9. Let T be defined by the equalities $D(T)=W_{2,2}(\mathbb{R})$, $Tf=-\Delta f+qf$, where q is a continuous real-valued function with compact support. If $\int_{\mathbb{R}} q(t)\,dt<0$, then T has at least one negative eigenvalue.
Hint: Theorem 6.33, 7.26(b) and 10.21(a).

10.10. Let T be a self-adjoint operator on $L_2(\mathbb{R}^m)$.
(a) If $D(T) \subset W_{2,r}(\mathbb{R}^m)$ for some $r>m/2$, then $(\lambda - T)^{-1}$ is a Carleman operator for every $\lambda \in \rho(T)$. The operator $E(b)-E(a)$ is a Carleman operator for all $a, b \in \mathbb{R}$.
(b) If $D(T^n) \subset W_{2,r}(\mathbb{R}^m)$ for some $r>m/2$, then $E(b)-E(a)$ is a Carleman operator for all $a, b \in \mathbb{R}$.

10.4 Essentially self-adjoint Schrödinger operators

In this section we consider operators on $L_2(\mathbb{R}^m)$ that are induced by differential forms

$$\tau f = \sum_{j=1}^{m} (D_j - b_j)^2 f + qf,$$

with $D_j = (1/i)(\partial/\partial x_j)$ and with real-valued functions $b_j \in C^1(\mathbb{R}^m)$ and $q \in L_{2,\text{loc}}(\mathbb{R}^m)$ (we have used the notation $(D_j - b_j)^2 f = (D_j - b_j)[(D_j - b_j)f]$). The operator T defined on $L_2(\mathbb{R}^m)$ by

$$D(T) = C_0^\infty(\mathbb{R}^m) \quad \text{and} \quad Tf = \tau f \quad \text{for} \quad f \in D(T) \tag{10.3}$$

is obviously symmetric. For $f \in D(T)$

$$Tf = -\Delta f - 2\sum_{j=1}^{m} b_j D_j f + (b^2 + \mathrm{i}\,\mathrm{div} b + q) f \tag{10.4}$$

with $b = (b_1, b_2, \ldots, b_m)$ and $b^2 = \sum_{j=1}^{m} b_j^2$.

In non-relativistic quantum mechanics operators of this form occur as *Schrödinger operators* of systems consisting of finitely many charged particles in an electromagnetic field. For $m = 3$ we encounter the *Schrödinger operator of one particle* in an electric field with potential q and a magnetic field with the vector potential $b = (b_1, b_2, b_3)$,

$$Tf = \sum_{j=1}^{3} (D_j - b_j)^2 f + qf. \tag{10.5}$$

For $m = 3N$ the *Schrödinger operator of a system of N charged particles* has this form: If in this case we write

$$\mathbb{R}^{3N} \ni x = (x_{1,1}, x_{1,2}, x_{1,3}, x_{2,1}, \ldots, x_{N-1,3}, x_{N,1}, x_{N,2}, x_{N,3}),$$

where $x_j = (x_{j,1}, x_{j,2}, x_{j,3})$ are the coordinates of the jth particle, then

$$Tf(x) = \sum_{j=1}^{N} \sum_{k=1}^{3} (D_{jk} - b_{jk}(x_j))^2 f(x) + q(x) f(x) \tag{10.6}$$

with

$$q(x) = \sum_{l=2}^{N} \sum_{j=1}^{l-1} q_{jl}(x_j - x_l) + \sum_{j=1}^{N} q_j(x_j) \tag{10.7}$$

and

$$b_{jk} = e_j b_k \quad \text{for} \quad j = 1, 2, \ldots, N \quad \text{and} \quad k = 1, 2, 3,$$

where the factor e_j depends on the charge of the jth particle. (Here we have replaced by 1 all physical quantities that are irrelevant for the properties studied here.)

Theorem 10.22. *Let the operator T be defined as in* (10.3). *Assume that $b_1, \ldots, b_m \in C^1(\mathbb{R}^m)$ are bounded with bounded derivatives, and $q \in M_\rho(\mathbb{R}^m)$ for some $\rho < 4$. Then T is essentially self-adjoint and $D(\overline{T}) = W_{2,2}(\mathbb{R}^m)$. The operator T (and thus also $\overline{T}$) is bounded from below.*

The *proof* can be obtained immediately from Theorems 5.28, 10.18, and 9.1 if we consider $-\Delta$ as the unperturbed operator (cf. the representation (10.4) of T).

The operators under (10.5) and (10.6) satisfy the assumptions of Theorem 10.22 in almost all physically realizable cases.

In this section we actually prove that under much more general assumptions these operators still remain essentially self-adjoint, while $D(\overline{T})$ is then generally no longer equal to $W_{2,2}(\mathbb{R}^m)$. The following theorem is due to *T. Ikebe* and *T. Kato* [44] in a somewhat more general form. Our presentation relies on a proof of *C.G. Simader* [51] for a somewhat more general result.

For $r \in \mathbb{N}$ let $W_{2,r,\text{loc}}(\mathbb{R}^m)$ denote the space of (equivalence classes of) the functions $f: \mathbb{R}^m \to \mathbb{C}$ for which $\vartheta f \in W_{2,r}(\mathbb{R}^m)$ for all $\vartheta \in C_0^\infty(\mathbb{R}^m)$. For every $f \in W_{2,r,\text{loc}}(\mathbb{R}^m)$ and $j \in \{1, 2, \ldots, m\}$ there is exactly one $f_j \in W_{2,r-1,\text{loc}}(\mathbb{R}^m)$ such that for all $k > 0$

$$\chi_k f_j = \lim_{\epsilon \to 0} \chi_k \frac{1}{i\epsilon}(f_{j,\epsilon} - f) \quad \text{in the sense of} \quad L_2(\mathbb{R}^m),$$

where $f_{j,\epsilon}$ is defined as in Theorem 10.8 and χ_k denotes the characteristic function of $M_k = \{x \in \mathbb{R}^m : |x| < k\}$ (on M_k the function f_j can be defined by the equality $\chi_k f_j = \chi_k D_j(\vartheta f)$ where $\vartheta \in C_0^\infty(\mathbb{R}^m)$ and $\vartheta(x) = 1$ for $x \in M_k$; this definition of $\chi_k f_j$ is independent of ϑ). For $f \in W_{2,r}(\mathbb{R}^m)$ we have $f_j = D_j f$. Consequently, for $f \in W_{2,r,\text{loc}}(\mathbb{R}^m)$ we define $D_j f = f_j$. If $\vartheta \in C_0^\infty(\mathbb{R}^m)$ is such that $\vartheta(x) = 1$ for $x \in M_k$, then

$$D_j(\vartheta f)(x) = D_j f(x) \quad \text{almost everywhere in} \quad M_k.$$

Since $D_j f \in W_{2,r-1,\text{loc}}(\mathbb{R}^m)$, we can define successively all derivatives $D^\alpha f$ for $|\alpha| \leq r$; we have $D^\alpha f \in W_{2,r-|\alpha|,\text{loc}}(\mathbb{R}^m)$. The function hf belongs to $W_{2,1,\text{loc}}(\mathbb{R}^m)$ provided that $h \in C^1(\mathbb{R}^m)$ and $f \in W_{2,1,\text{loc}}(\mathbb{R}^m)$. In particular, the expression $\sum_{j=1}^m (D_j - b_j)^2 f$ is meaningful for $b_j \in C^1(\mathbb{R}^m)$ and $f \in W_{2,2,\text{loc}}(\mathbb{R}^m)$.

Theorem 10.23. *Let the operator T be defined as in* (10.3). *Let* $b_1, b_2, \ldots, b_m \in C^1(\mathbb{R}^m)$ *and let q belong to* $M_{\rho,\text{loc}}(\mathbb{R}^m)$ *for some* $\rho < 4$. *Moreover, assume that* $q = q_1 + q_2$ *with*

$$q_1 \in M_\rho(\mathbb{R}^m) \qquad \text{for some} \quad \rho < 4,$$

$$q_2(x) \geq -C|x|^2 \quad \text{for some} \quad C \geq 0 \quad \text{and all} \quad x \in \mathbb{R}^m.$$

Then T is essentially self-adjoint. We have

$$D(\overline{T}) = \{f \in L_2(\mathbb{R}^m) \cap W_{2,2,\text{loc}}(\mathbb{R}^m) : \tau f \in L_2(\mathbb{R}^m)\},$$

$$\overline{T} f = \sum_{j=1}^m (D_j - b_j)^2 f + qf \quad \text{for} \quad f \in D(\overline{T}).$$

For the proof we need the following auxiliary theorems.

Auxiliary theorem 10.24. *If* $q \in M_\rho(\mathbb{R}^m)$, *then* $|q|^{1/2} \in M_\sigma(\mathbb{R}^m)$ *for every* $\sigma > \rho/2$. (*A similar result holds for* $M_{\rho,\text{loc}}(\mathbb{R}^m)$ *and* $M_{\sigma,\text{loc}}(\mathbb{R}^m)$).

PROOF. If $\rho \geq 2m$, then it follows by the Schwarz inequality that

$$\int_{|x-y|<1} |q(y)| \mathrm{d}y \leq \left\{ C \int_{|x-y|<1} |q(y)|^2 \, \mathrm{d}y \right\}^{1/2} \leq C^{1/2} M_{q,\rho}$$

for all $x \in \mathbb{R}^m$, i.e., $|q|^{1/2} \in M_\sigma(\mathbb{R}^m)$ for $\sigma \geq m$, and thus for all $\sigma > \rho/2$.

If $m \leqslant \rho < 2m$ and $\sigma > \rho/2 \geqslant m/2$, then

$$\int_{|x-y|\leqslant 1} |q(y)||x-y|^{\sigma-m}\,dy$$
$$\leqslant \left\{ \int_{|x-y|\leqslant 1} |q(y)|^2\,dy \int_{|x-y|\leqslant 1} |x-y|^{2\sigma-2m}\,dy \right\}^{1/2} \leqslant CM_{q,\rho},$$

because $2\sigma - 2m > -m$.

If $\rho < m$ and $\sigma > \rho/2$, then it follows for $\epsilon = \sigma - (\rho/2) > 0$ that

$$\int_{|x-y|\leqslant 1} |q(y)||x-y|^{\sigma-m}\,dy$$
$$= \int_{|x-y|\leqslant 1} |x-y|^{\epsilon-(m/2)}\big[|q(y)||x-y|^{\sigma-\epsilon-(m/2)}\big]\,dy$$
$$\leqslant \left\{ \int_{|x-y|\leqslant 1} |x-y|^{2\epsilon-m}\,dy \int_{|x-y|\leqslant 1} |q(y)|^2|x-y|^{\rho-m}\,dy \right\}^{1/2} \leqslant CM_{q,\rho}.$$

Consequently, the assertion is proved in each case. □

Auxiliary theorem 10.25. *Let the functions $b_j (j = 1, 2, \ldots, m)$ and q_1 be the same as in Theorem 10.23. For every $\eta > 0$ there exists a $C_\eta > 0$ such that*

$$\langle |q_1| f, f\rangle \leqslant \eta \sum_{j=1}^{m} \|(D_j - b_j) f\|^2 + C_\eta \|f\|^2$$

for all $f \in C_0^\infty(\mathbb{R}^m)$.

PROOF. $|q_1|^{1/2}$ belongs to $M_\sigma(\mathbb{R}^m)$ for some $\sigma < 2$ by Auxiliary theorem 10.24. Take a $\psi \in C_0^\infty(\mathbb{R}^m)$ such that $\psi(x) = 1$ for $x \in \operatorname{supp} f$, and set $h = |\operatorname{grad}\psi| + |\psi|$. Set, furthermore, $\varphi_\epsilon(x) = (|f(x)|^2 + \epsilon)^{1/2}$ for every $\epsilon > 0$. Then by Theorem 10.17

$$\langle |q_1| f, f\rangle = \big\||q_1|^{1/2} f\big\|^2 \leqslant \big\|\,|q_1|^{1/2}\varphi_\epsilon\psi\big\|^2 \leqslant \frac{\eta}{2}\|\varphi_\epsilon\psi\|_1^2 + C_1\|\varphi_\epsilon\psi\|^2$$
$$\leqslant \eta \sum_{j=1}^{m} \int |\psi(x) D_j\varphi_\epsilon(x)|^2\,dx + C_2\big\||\operatorname{grad}\psi|\varphi_\epsilon + |\psi|\varphi_\epsilon\big\|^2$$
$$= \frac{\eta}{4}\sum_{j=1}^{m} \int \varphi_\epsilon(x)^{-2}|\psi(x)|^2|f(x)^* D_j f(x) - f(x)(D_j f(x))^*|^2\,dx$$
$$+ C_2\|h\varphi_\epsilon\|^2$$
$$= \frac{\eta}{4}\sum_{j=1}^{m} \int \varphi_\epsilon^{-2}|f^*(D_j f - b_j f) - f(D_j f - b_j f)^*|^2\,dx + C_2\|h\varphi_\epsilon\|^2$$
$$\leqslant \eta \sum_{j=1}^{m} \int \varphi_\epsilon^{-2}|f|^2|D_j f - b_j f|^2\,dx + C_2\|h\varphi_\epsilon\|^2$$
$$\leqslant \eta \sum_{j=1}^{m} \int |D_j f(x) - b_j(x) f(x)|^2\,dx + C_2\|h\varphi_\epsilon\|^2.$$

The assertion follows by letting ϵ tend to 0, because $h(x) = 1$ for $x \in \operatorname{supp} f$. □

Auxiliary theorem 10.26. *Let T be defined as in Theorem 10.23. For $n \in \mathbb{N}$ let*

$$R_n = \{x \in \mathbb{R}^m : n \leq |x| \leq 2n\} \quad \text{and} \quad Q_n = \left\{x \in \mathbb{R}^m : \frac{n}{2} \leq |x| \leq 3n\right\}.$$

There exists a $C > 0$ such that for all $f \in C_0^\infty(\mathbb{R}^m)$ and all $n \in \mathbb{N}$

$$\sum_{j=1}^m \int_{R_n} |(D_j - b_j)f|^2 \, dx \leq 2\int_{Q_n} |Tf|^2 \, dx + Cn^2 \int_{Q_n} |f|^2 \, dx$$

and

$$\sum_{j=1}^m \int_{|x| \leq n} |(D_j - b_j)f|^2 \, dx \leq 2\int_{|x| \leq 2n} |Tf|^2 \, dx + Cn^2 \int_{|x| \leq 2n} |f|^2 \, dx.$$

PROOF. Let $\eta \in C_0^\infty(\mathbb{R}^m)$ be such that

$$0 \leq \eta(x) \leq 1 \text{ for all } x \in \mathbb{R}^m, \eta(x) = \begin{cases} 1 & \text{for} \quad 1 \leq |x| \leq 2, \\ 0 & \text{for} \quad |x| \leq \frac{1}{2} \text{ and } |x| \geq 3. \end{cases}$$

Moreover, for all $n \in \mathbb{N}$ set $\eta_n(x) = \eta(n^{-1}x)$. Since $\eta_n \in C_0^\infty(\mathbb{R}^m)$ and

$$\operatorname{supp} \eta_n \subset Q_n, \operatorname{supp}(\operatorname{grad} \eta_n) \subset Q_n \setminus R_n,$$

by Auxiliary theorem 10.25 we have for all $f \in C_0^\infty(\mathbb{R}^m)$ that

$$\begin{aligned}
\|\eta_n Tf\|^2 + \|\eta_n f\|^2 &\geq 2\mathrm{Re}\langle \eta_n Tf, \eta_n f\rangle = 2\mathrm{Re}\langle Tf, \eta_n^2 f\rangle \\
&= 2\mathrm{Re}\left\{ \sum_{j=1}^m \langle (D_j - b_j)f, (D_j - b_j)(\eta_n^2 f)\rangle + \langle q\eta_n f, \eta_n f\rangle \right\} \\
&= 2\mathrm{Re}\left\{ \sum_{j=1}^m \|\eta_n(D_j - b_j)f\|^2 + 2\sum_{j=1}^m \langle \eta_n(D_j - b_j)f, (D_j\eta_n)f\rangle \right. \\
&\qquad \left. + \langle q_1\eta_n f, \eta_n f\rangle + \langle q_2\eta_n f, \eta_n f\rangle \right\} \\
&\geq 2\left\{ \sum_{j=1}^m \|\eta_n(D_j - b_j)f\|^2 - \frac{1}{4}\sum_{j=1}^m \|\eta_n(D_j - b_j)f\|^2 \right. \\
&\qquad - 4\sum_{j=1}^m \|(D_j\eta_n)f\|^2 - \frac{1}{4}\sum_{j=1}^m \|(D_j - b_j)\eta_n f\|^2 \\
&\qquad \left. - C_1\|\eta_n f\|^2 - C_2 n^2 \|\eta_n f\|^2 \right\}.
\end{aligned}$$

Since

$$\begin{aligned}
\|(D_j - b_j)\eta_n f\|^2 &= \|(D_j\eta_n)f + \eta_n(D_j - b_j)f\|^2 \\
&\leq 2\{\|(D_j\eta_n)f\|^2 + \|\eta_n(D_j - b_j)f\|^2\},
\end{aligned}$$

and

$$|D_j\eta_n(x)| \leq C_3\frac{1}{n} \quad \text{for all} \quad n \in \mathbb{N} \quad \text{and all} \quad x \in \mathbb{R}^m,$$

we can further estimate:

$$\|\eta_n Tf\|^2 + \|\eta_n f\|^2 \geqslant \frac{1}{2}\sum_{j=1}^{m} \|\eta_n(\mathrm{D}_j - b_j)f\|^2 - C_4 n^2 \int_{Q_n} |f(x)|^2\,\mathrm{d}x$$

$$\geqslant \frac{1}{2}\sum_{j=1}^{m} \int_{R_n} |(\mathrm{D}_j - b_j(x))f(x)|^2\,\mathrm{d}x - C_4 n^2 \int_{Q_n} |f(x)|^2\,\mathrm{d}x.$$

The first assertion follows from this immediately. The second assertion can be proved analogously if we replace η_n by a function $\zeta_n \in C_0^\infty(\mathbb{R}^m)$ such that $\zeta_n(x) = 1$ for $|x| \leqslant n$ and $\zeta_n(x) = 0$ for $|x| \geqslant 2n$. □

Auxiliary theorem 10.27. *Let T be defined as in Theorem 10.23. For every $\vartheta \in C_0^\infty(R^m)$ there is a $C > 0$ such that*

$$\|\Delta(\vartheta f)\| \leqslant C\{\|Tf\| + \|f\|\} \quad \textit{for all} \quad f \in C_0^\infty(\mathbb{R}^m).$$

Proof. By (10.4) and Auxiliary theorem 10.26 we have for $f \in C_0^\infty(\mathbb{R}^m)$ that

$$\|\Delta(\vartheta f)\| = \left\| T(\vartheta f) + 2\sum_{j=1}^{m} b_j\,\mathrm{D}_j(\vartheta f) - (b^2 + \mathrm{i}\,\mathrm{div}\,b + q)\vartheta f \right\|$$

$$= \left\| \vartheta Tf - (\Delta\vartheta)f + 2\sum_{j=1}^{m} (b_j\vartheta + \mathrm{D}_j\vartheta)\,\mathrm{D}_j f - (b^2 + \mathrm{i}\,\mathrm{div}\,b + q)\vartheta f \right\|$$

$$\leqslant \|\vartheta Tf\| + \|(\Delta\vartheta)f\| + 2\left\| \sum_{j=1}^{m} (b_j\vartheta + \mathrm{D}_j\vartheta)\,\mathrm{D}_j f \right\|$$

$$+ \|(b^2 + \mathrm{i}\,\mathrm{div}\,b)\vartheta f\| + \|q\vartheta f\|$$

$$\leqslant C_1\{\|Tf\| + \|f\|\} + \|q\vartheta f\|,$$

where C_1 depends on ϑ (more precisely, on the supremum of $|\vartheta|$, $|\mathrm{grad}\,\vartheta|$, $|\Delta\vartheta|$ and on the supremum of b^2 and $|\mathrm{div}\,b|$ on $\mathrm{supp}\,\vartheta$). If we set $q_0(x) = q(x)$ for $x \in \mathrm{supp}\,\vartheta$ and $q_0(x) = 0$ for $x \notin \mathrm{supp}\,\vartheta$, then q_0 obviously belongs to $M_\rho(\mathbb{R}^m)$ for some $\rho < 4$, and by Theorem 10.17 and by the inequality $\|g\|_2 \leqslant \|\Delta g\| + \|g\|$

$$\|q(\vartheta f)\| = \|q_0(\vartheta f)\| \leqslant \tfrac{1}{2}\|\Delta(\vartheta f)\| + C_2\|\vartheta f\|,$$

and thus

$$\|\Delta(\vartheta f)\| \leqslant C_1\{\|Tf\| + \|f\|\} + \tfrac{1}{2}\|\Delta(\vartheta f)\| + C_2\|f\|.$$

The assertion follows from this. □

Proof of Theorem 10.23.

(a) First we prove the essential self-adjointness of T. Take a $\varphi \in C_0^\infty(\mathbb{R}^m)$ such that

$$0 \leqslant \varphi(x) \leqslant 1, \quad \varphi(x) = \begin{cases} 1 & \text{for} \quad |x| \leqslant 1, \\ 0 & \text{for} \quad |x| \geqslant 2. \end{cases}$$

For every $n \in \mathbb{N}$ let $\varphi_n(x) = \varphi(n^{-1}x)$, and let T_n be defined similarly to T provided that we replace the functions b_j and q_l by the functions

$$b_{j,n}(x) = \varphi_{3n}(x) b_j(x), \quad q_{l,n}(x) = \varphi_{3n}(x) q_l(x).$$

Then it is obvious that for all $n \in \mathbb{N}$ and $f \in C_0^\infty(\mathbb{R}^m)$

$$\varphi_n T f = \varphi_n T_n f \quad \text{and} \quad T(\varphi_n f) = T_n(\varphi_n f).$$

Now assume that $g \in R(\mathrm{i} - T)^\perp$. We have to show that $g = 0$. Since the operators T_n are essentially self-adjoint by Theorem 10.22, for every $n \in \mathbb{N}$ there exists an $f_n \in C_0^\infty(\mathbb{R}^m)$ for which

$$\|\varphi_n g - (\mathrm{i} - T_n) f_n\| < \frac{1}{n}.$$

Then we have in particular that (cf. Theorem 5.18)

$$\|f_n\| \leq \|(\mathrm{i} - T_n) f_n\| \leq \|\varphi_n g\| + \frac{1}{n}.$$

Therefore, the sequence (f_n) is bounded. Auxiliary theorem 10.26 implies that

$$\begin{aligned}
\sum_{j=1}^m \int_{R_n} |(\mathrm{D}_j - b_j) f_n|^2 \, \mathrm{d}x &\leq 2 \int_{Q_n} |T f_n|^2 \, \mathrm{d}x + C_1 n^2 \int_{Q_n} |f_n|^2 \, \mathrm{d}x \\
&= 2 \int_{Q_n} |T_n f_n|^2 \, \mathrm{d}x + C_1 n^2 \int_{Q_n} |f_n|^2 \, \mathrm{d}x \\
&= 2 \int_{Q_n} |(\mathrm{i} - T_n) f_n - \varphi_n g + (\varphi_n g - \mathrm{i} f_n)|^2 \, \mathrm{d}x \\
&\quad + C_1 n^2 \int_{Q_n} |f_n|^2 \, \mathrm{d}x \\
&\leq 4n^{-2} + 4 \int_{Q_n} |\varphi_n g - \mathrm{i} f_n|^2 \, \mathrm{d}x + C_1 n^2 \int_{Q_n} |f_n|^2 \, \mathrm{d}x \\
&\leq 4n^{-2} + 8\|\varphi_n g\|^2 + C_2 n^2 \|f_n\|^2 \\
&\leq C_3 \{1 + n^2 \|\varphi_n g\|^2\} \leq C_3 (1 + n\|\varphi_n g\|)^2,
\end{aligned}$$

and thus (because $|\mathrm{D}_j \varphi_n| \leq K n^{-1}$ with $K = \sup\{|\operatorname{grad} \varphi(x)| : x \in \mathbb{R}^m\}$)

$$\begin{aligned}
\left\| \sum_{j=1}^m (\mathrm{D}_j \varphi_n)(\mathrm{D}_j - b_{j,n}) f_n \right\| &= \left\| \sum_{j=1}^m (\mathrm{D}_j \varphi_n)(\mathrm{D}_j - b_j) f_n \right\| \\
&= \left\{ \int_{R_n} \left| \sum_{j=1}^m (\mathrm{D}_j \varphi_n)(\mathrm{D}_j - b_j) f_n \right|^2 \mathrm{d}x \right\}^{1/2} \\
&\leq C_4 \frac{1}{n} \left\{ \sum_{j=1}^m \int_{R_n} |(\mathrm{D}_j - b_j) f_n|^2 \mathrm{d}x \right\}^{1/2} \\
&\leq C_5 \left\{ \frac{1}{n} + \|\varphi_n g\| \right\}.
\end{aligned}$$

Consequently, because of the equality $\langle g, (\mathrm{i}-T_n)(\varphi_n f_n)\rangle = \langle g, (\mathrm{i}-T)(\varphi_n f_n)\rangle = 0$ we obtain the estimate

$$\begin{aligned}
\|\varphi_n g\|^2 &= \langle \varphi_n g, (\mathrm{i}-T_n)f_n + [\varphi_n g - (\mathrm{i}-T_n)f_n]\rangle \\
&\leq |\langle g, \varphi_n(\mathrm{i}-T_n)f_n\rangle| + \frac{1}{n}\|\varphi_n g\| \\
&= \left|\langle g, (\mathrm{i}-T_n)(\varphi_n f_n) + 2\sum_{j=1}^{m}(\mathrm{D}_j\varphi_n)(\mathrm{D}_j - b_{j,n})f_n - (\Delta\varphi_n)f_n\rangle\right| \\
&\quad + \frac{1}{n}\|\varphi_n g\| \\
&\leq \left\{\int_{R_n}|g|^2\,\mathrm{d}x\right\}^{1/2}\left\{2\left\|\sum_{j=1}^{m}(\mathrm{D}_j\varphi_n)(\mathrm{D}_j - b_{j,n})f_n\right\| + \|(\Delta\varphi_n)f_n\|\right\} \\
&\quad + \frac{1}{n}\|\varphi_n g\| \\
&\leq C_6\left\{\int_{R_n}|g|^2\,\mathrm{d}x\right\}^{1/2}\left\{\frac{1}{n} + \|\varphi_n g\| + \frac{1}{n^2}\|f_n\|\right\} + \frac{1}{n}\|\varphi_n g\|.
\end{aligned}$$

Since the left side tends to $\|g\|^2$ and the right side tends to 0 as $n\to\infty$, it follows from this that $g=0$. Hence, $R(\mathrm{i}-T)^{\perp} = \{0\}$. We can prove similarly that $R(-\mathrm{i}-T)^{\perp} = \{0\}$. The operator T is therefore essentially self-adjoint.

(b) If $f\in L_2(\mathbb{R}^m)\cap W_{2,2,\mathrm{loc}}(\mathbb{R}^m)$ and $\sum_{j=1}^m(\mathrm{D}_j - b_j)^2 f + qf \in L_2(\mathbb{R}^m)$, then $f\in D(T^*) = D(\overline{T})$ and

$$T^*f = \overline{T}f = \sum_{j=1}^{m}(\mathrm{D}_j - b_j)^2 f + qf,$$

because for every $g\in C_0^\infty(\mathbb{R}^m) = D(T)$ with $\operatorname{supp} g \subset \{x\in\mathbb{R}^m : |x| < k\}$ and for any $\vartheta\in C_0^\infty(\mathbb{R}^m)$ with $\vartheta(x) = 1$ for $|x|<k$ we have

$$\begin{aligned}
\langle f, Tg\rangle = \langle \vartheta f, Tg\rangle &= \langle \sum_{j=1}^{m}(\mathrm{D}_j - b_j)^2(\vartheta f) + q\vartheta f, g\rangle \\
&= \langle \sum_{j=1}^{m}(\mathrm{D}_j - b_j)^2 f + qf, g\rangle.
\end{aligned}$$

Now let $f\in D(\overline{T})$. There exists a sequence (f_n) from $C_0^\infty(\mathbb{R}^m)$ for which $f_n\to f$ and $Tf_n\to\overline{T}f$. In particular, (f_n) and (Tf_n) are Cauchy sequences. Then $(\Delta(\vartheta f_n))$ is also a Cauchy sequence for every $\vartheta\in C_0^\infty(\mathbb{R}^m)$ by Auxiliary theorem 10.27. Since $\|h\|_2 \leq \|\Delta h\| + \|h\|$ for $h\in C_0^\infty(\mathbb{R}^m)$, the sequence (ϑf_n) is also a Cauchy sequence in $W_{2,2}(\mathbb{R}^m)$. Let g be the limit of the sequence (ϑf_n) in $W_{2,2}(\mathbb{R}^m)$. Since $\vartheta f_n\to\vartheta f$ in $L_2(\mathbb{R}^m)$, we then have $\vartheta f = g\in W_{2,2}(\mathbb{R}^m)$.

For an arbitrary $r>0$ let $\vartheta \in C_0^\infty(\mathbb{R}^m)$ be such that $\vartheta(x)=1$ for $|x| \leq r$. Then for all $g \in C_0^\infty(\{x \in \mathbb{R}^m : |x|<r\})$

$$\langle \bar{T}f, g\rangle = \langle T^* f, g\rangle = \langle f, Tg\rangle = \langle \vartheta f, Tg\rangle$$
$$= \langle \sum_{j=1}^m (\mathrm{D}_j - b_j)^2(\vartheta f) + q\vartheta f, g\rangle$$
$$= \langle \vartheta\left\{\sum_{j=1}^m (\mathrm{D}_j - b_j)^2 f + qf\right\}, g\rangle.$$

As the set of these g is dense in $L_2(\{x \in \mathbb{R}^m : |x| \leq r\})$, it follows that

$$\bar{T}f(x) = \sum_{j=1}^m (\mathrm{D}_j - b_j(x))^2 f(x) + q(x)f(x)$$

almost everywhere in $\{x \in \mathbb{R}^m : |x| \leq r\}$. Since r was arbitrary, this equality holds almost everywhere in $\mathbb{R}^m$. In particular, $\Sigma_{j=1}^m(\mathrm{D}_j - b_j)^2 f + qf$ belongs to $L_2(\mathbb{R}^m)$. □

REMARK. The operator T of Theorem 10.23 is still essentially self-adjoint if $q = q_1 + q_2$ with

$$q_1 \in M_\rho(\mathbb{R}^m) \quad \text{for some} \quad \rho < 4, \quad q_2 \in L_{2,\mathrm{loc}}(\mathbb{R}^m),$$
$$q_2(x) \geq -C|x|^2 \quad \text{for some} \quad C \geq 0 \quad \text{and all} \quad x \in \mathbb{R}^m.$$

The above proof can then be employed without change if we first show that the corresponding operators T_n are essentially self-adjoint. A proof is given, for example, by *C.G. Simader* [51]. The reader can find further references concerning this circle of problems there.

In order to apply the results of Section 9.2, it is useful to have criteria for the T-compactness and the T^2-compactness of a perturbation of T. We shall prove such a criterion now. If A and B are operators on $L_2(\mathbb{R}^m)$, then we say that A is *B-small at infinity* if A is B-bounded and for every $\epsilon>0$ there exists an $r(\epsilon) \geq 0$ such that

$$\|Af\| \leq \epsilon(\|Bf\| + \|f\|)$$

for all $f \in D(B)$ such that $f(x)=0$ for $|x|<r(\epsilon)$.

Theorem 10.28. *Let T be defined as in Theorem 10.23. If V is closed and T-small at infinity, then V is $\bar{T}^2$-compact. If, in addition, the T-bound of V is zero, then V is $\bar{T}$-compact.*

PROOF. If V is T-bounded with T-bound zero, then it is also $\bar{T}$-bounded with $\bar{T}$-bound zero. Consequently, the second statement follows from the first with the aid of Theorem 9.11d. It remains to prove the first statement.

First step: The operator A defined by the formulae

$$D(A) = D(\overline{T}) \quad \text{and} \quad Af = \vartheta f \quad \text{for} \quad f \in D(A)$$

is $\overline{T}$-compact for every $\vartheta \in C_0^\infty(\mathbb{R}^m)$.

Proof of the first step. It is sufficient to prove that A is T-compact. Let ψ be an element of $C_0^\infty(\mathbb{R}^m)$ such that $\psi(x) = 1$ for $x \in \operatorname{supp} \vartheta$. The operator B defined by the equalities

$$D(B) = D(T) \quad \text{and} \quad Bf = \Delta(\psi f) \quad \text{for} \quad f \in D(B)$$

is T-bounded by Auxiliary theorem 10.27. Since $Af = A(\psi f)$ and since A is Δ-compact by Theorem 10.21, the operator A is B-compact, and thus also T-compact (cf. the proposition preceding Theorem 9.7).

Second step: The operator C defined by the equalities

$$D(C) = D(\overline{T}) \quad \text{and} \quad Cf = \vartheta \overline{T} f - \overline{T}(\vartheta f) \quad \text{for} \quad f \in D(\overline{T})$$

is $\overline{T}$-compact for every $\vartheta \in C_0^\infty(\mathbb{R}^m)$.

Proof of the second step. Again, it is sufficient to show that C is T-compact. Since C is a differential operator of the first order having continuous coefficients with compact support, the B-compactness and thus also the T-compactness of C follow as in the first step.

Third step: V is $\overline{T}^2$-compact.

Proof of the third step. It is clear that V is also $\overline{T}$-small at infinity. Let $\epsilon > 0$ be given, and let $r(\epsilon)$ be chosen according to the definition of $\overline{T}$-smallness at infinity. Moreover, let ϑ be an element of $C_0^\infty(\mathbb{R}^m)$ such that $\vartheta(x) = 1$ for $|x| \leqslant r(\epsilon)$ and $0 \leqslant \vartheta(x) \leqslant 1$ for all $x \in \mathbb{R}^m$. Since V is $\overline{T}$-bounded by assumption, we have for all $f \in D(\overline{T})$ with appropriate $a, b \geqslant 0$ that

$$\begin{aligned} \|Vf\| &\leqslant \|V(\vartheta f)\| + \|V[(1-\vartheta)f]\| \\ &\leqslant a\|\vartheta f\| + b\|\overline{T}(\vartheta f)\| + \epsilon\{\|(1-\vartheta)f\| + \|\overline{T}[(1-\vartheta)f]\|\} \\ &\leqslant a\|\vartheta f\| + b\|\vartheta \overline{T} f\| + (b+\epsilon)\|\vartheta \overline{T} f - \overline{T}(\vartheta f)\| \\ &\quad + \epsilon\{\|(1-\vartheta)f\| + \|(1-\vartheta)\overline{T} f\|\}. \end{aligned}$$

If we replace here f by a sequence (f_n) from $D(\overline{T}^2)$ for which $f_n \xrightarrow{w} 0$ and $\overline{T}^2 f_n \xrightarrow{w} 0$, then we also have $\overline{T} f_n \xrightarrow{w} 0$, and by steps 1 and 2 the first three terms of the right side converge to zero. The fourth term is bounded by $\epsilon \sup (\|f_n\| + \|\overline{T} f_n\|)$. Since $\epsilon > 0$ was arbitrary, it follows that $\|Vf_n\| \to 0$, and thus V is $\overline{T}^2$-compact. □

Exercise

10.11. Let T be the self-adjoint Schrödinger operator from Theorem 10.23 for $m \leqslant 3$. Then $(z-T)^{-1}$ is a Carleman operator for every $z \in \rho(T)$. The operator $E(b) - E(a)$ is a Carleman operator for all $a, b \in \mathbb{R}$.
Hint: $M_\vartheta(z-T)^{-1}$ is a Carleman operator for $\vartheta \in C_0^\infty(\mathbb{R}^m)$.

10.5 Spectra of Schrödinger operators

In this section we prove some properties of the spectra of self-adjoint operators of the form considered in the previous section. The results can be applied equally well to one particle and several particle Schrödinger operators (Theorems 10.29(a), 10.30 and 10.33).

In what follows let T always have the form

$$D(T) = C_0^\infty(\mathbb{R}^m), \qquad Tf = \sum_{j=1}^{m} (\mathrm{D}_j - b_j)^2 f + qf \quad \text{for} \quad f \in D(T), \tag{10.8}$$

where the coefficients satisfy the assumptions of Theorem 10.23. Therefore, T is essentially self-adjoint. We denote by S the closure of T (that is, the uniquely determined self-adjoint extension of T). Moreover, let the self-adjoint operator S_0 be defined by (cf. Section 10.2)

$$D(S_0) = W_{2,2}(\mathbb{R}^m) \quad \text{and} \quad S_0 f = -\Delta f \quad \text{for} \quad f \in D(S_0), \tag{10.9}$$

and let V be defined by the formulae

$$\begin{aligned} &D(V) = D(S) \cap W_{2,2}(\mathbb{R}^m), \\ &Vf = Sf - S_0 f = -2 \sum_{j=1}^{m} b_j \, \mathrm{D}_j f + (b^2 + \mathrm{i} \operatorname{div} b + q) f. \end{aligned} \tag{10.10}$$

Theorem 10.29.

(a) *If $q_- \in M_\rho(\mathbb{R}^m)$ for some $\rho < 4$, then S is bounded from below (here $q_-(x) = \max(-q(x), 0)$). If $q \geq 0$, then S is non-negative.*

(b) *If $b^2 \in M_\rho(\mathbb{R}^m)$, $\operatorname{div} b \in M_\rho(\mathbb{R}^m)$, $q \in M_\rho(\mathbb{R}^m)$ for some $\rho < 4$, and*

$$\int_{|x-y|<1} (b^4(y) + |\operatorname{div} b(y)|^2 + |q(y)|^2) \, \mathrm{d}y \to 0 \quad \text{as} \quad |x| \to \infty,$$

then V is relatively compact with respect to S_0, and $\sigma_e(S) = [0, \infty)$.

PROOF.

(a) By Auxiliary theorem 10.25 with $\eta = 1$ we have for all $f \in C_0^\infty(\mathbb{R}^m)$ that

$$\begin{aligned} \langle f, Sf \rangle &= \sum_{j=1}^{m} \|(\mathrm{D}_j - b_j) f\|^2 + \langle f, qf \rangle \\ &\geq \sum_{j=1}^{m} \|(\mathrm{D}_j - b_j) f\|^2 - \langle q_- f, f \rangle \geq -C_1 \|f\|^2. \end{aligned}$$

This then holds for all $f \in D(S)$, since $C_0^\infty(\mathbb{R}^m)$ is a core of S. If $q \geq 0$, then $q_- = 0$. The above estimate then gives that $S \geq 0$.

(b) We have $b_j \in M_\sigma(\mathbb{R}^m)$ for some $\sigma < 2$ $(j = 1, 2, \ldots, m)$ by Auxiliary theorem 10.24. Consequently, the assertion follows from Theorem 10.21 together with (10.10) and Theorem 10.11. □

The spectrum of the operator S_0 is $[0, \infty)$; the operator S_0 has no eigenvalue. We can actually prove for a large class of operators S of the form given above (without magnetic field) that no eigenvalues lie in $[0, \infty)$.

Theorem 10.30. *Assume that $b_j = 0$ for $j = 1, 2, \ldots, m$, $q \in M_\rho(\mathbb{R}^m)$ for some $\rho < 4$, and*

$$q(ax) = a^{-\gamma} q(x) \quad \textit{for} \quad x \in \mathbb{R}^m \setminus \{0\}$$

with some $\gamma \in (0, 2)$. If f is an eigenelement of S belonging to the eigenvalue λ, then

$$2\langle f, -\Delta f\rangle + \gamma\langle f, qf\rangle = 0 \quad (\textit{virial theorem}).$$

The interval $[0, \infty)$ contains no eigenvalue of S. (For a somewhat more general result we refer to J. Weidmann [55].)

Proof. If f is an eigenelement of S belonging to the eigenvalue λ, then

$$-\Delta f(x) = \lambda f(x) - q(x) f(x). \tag{10.11}$$

It follows from this for every $a > 0$ using the notation $f_a(x) = f(ax)$ that

$$\begin{aligned} -\Delta f_a(x) &= -a^2(\Delta f)(ax) = a^2(\lambda f(ax) - q(ax) f(ax)) \\ &= a^2 \lambda f_a(x) - a^{2-\gamma} q(x) f_a(x). \end{aligned} \tag{10.12}$$

It follows from (10.11) and (10.12) that

$$\begin{aligned} 0 &= \langle -\Delta f, f_a\rangle - \langle f, -\Delta f_a\rangle \\ &= \lambda\langle f, f_a\rangle - \langle qf, f_a\rangle - a^2\lambda\langle f, f_a\rangle + a^{2-\gamma}\langle qf, f_a\rangle \\ &= (1 - a^2)\lambda\langle f, f_a\rangle + (a^{2-\gamma} - 1)\langle qf, f_a\rangle. \end{aligned}$$

For $a \neq 1$ we obtain by dividing by $(a-1)$ that

$$(a+1)\lambda\langle f, f_a\rangle = \frac{a^{2-\gamma} - 1}{a - 1}\langle qf, f_a\rangle.$$

By letting a tend to 1, it follows from this that

$$2\lambda\|f\|^2 = (2-\gamma)\langle f, qf\rangle.$$

Since $-\Delta f + qf = \lambda f$, this implies

$$2\langle f, -\Delta f\rangle = -2\langle f, qf\rangle + 2\lambda\|f\|^2 = -\gamma\langle f, qf\rangle$$

(this is the virial theorem) and

$$\gamma\lambda\|f\|^2 = (2-\gamma)\langle f, qf - \lambda f\rangle = (2-\gamma)\langle f, \Delta f\rangle.$$

It follows from the last equality that $\lambda < 0$ provided that $f \neq 0$. □

Remark. The assumptions of Theorem 10.30 are satisfied in particular by every Schrödinger operator without magnetic field with pure Coulomb interaction (this also holds for many-particle operators).

The operators of Theorem 10.29(b) also satisfy the assumptions of part (a); the negative part of the spectrum therefore consists of at most countably many eigenvalues of finite multiplicity that are bounded from below and can only cluster at 0. The following theorem shows that in many interesting cases there actually exist infinitely many negative eigenvalues.

Theorem 10.31. *Let S be as in Theorem 10.29(b). Assume further that there exist constants $C \geq 0$, $\epsilon > 0$ and $r \geq 0$ such that*

$$q(x) \leq -C|x|^{-2+\epsilon} \quad \textit{for all} \quad x \in \mathbb{R}^m \quad \textit{with} \quad |x| \geq r,$$

and

$$p(t) := \sup_{|x| \geq t} |x|^{2-\epsilon}(b^2(x) + |\operatorname{div} b(x)|) \to 0 \quad \textit{for} \quad t \to \infty.$$

Then S has infinitely many negative eigenvalues accumulating at zero.

Proof. We only have to prove that S has infinitely many negative eigenvalues. According to Theorem 7.26(b) it is sufficient to find an infinite-dimensional subspace M of $D(S)$ with $\langle f, Sf\rangle < 0$ for every non-vanishing $f \in M$. Let $\vartheta \in C_0^\infty(\mathbb{R}^m)$ such that $\|\vartheta\| = 1$ and $\operatorname{supp}\vartheta \subset \{x \in \mathbb{R}^m : 1 < |x| < 2\}$. Then for the function $\vartheta_t(x) = t^{m/2}\vartheta(t^{-1}x)$ we have $\|\vartheta_t\| = 1$ and $\operatorname{supp}\vartheta_t \subset \{x \in \mathbb{R}^m : t < |x| < 2t\}$. Therefore it follows for $t > r$ that

$$\begin{aligned}
\langle \vartheta_t, S\vartheta_t\rangle &= \langle \vartheta_t, q\vartheta_t - \Delta\vartheta_t + (b^2 + \mathrm{i}\operatorname{div} b)\vartheta_t - 2\sum_{j=1}^m b_j D_j \vartheta_t\rangle \\
&\leq -Ct^{-2+\epsilon}\|\vartheta_t\|^2 + t^{-2}\langle\vartheta, -\Delta\vartheta\rangle + t^{-2+\epsilon}p(t)\|\vartheta_t\|^2 \\
&\quad + 2t^{-1+(\epsilon/2)}p(t)^{1/2}\sum_{j=1}^m \langle \vartheta_t, D_j\vartheta_t\rangle \\
&\leq -Ct^{-2+\epsilon} + t^{-2}\langle\vartheta, -\Delta\vartheta\rangle + t^{-2+\epsilon}p(t) \\
&\quad + 2t^{-2+(\epsilon/2)}p(t)^{1/2}\sum_{j=1}^m \langle\vartheta, D_j\vartheta\rangle.
\end{aligned}$$

(Here we have used that $|b_j(x)| \leq t^{-1+(\epsilon/2)}p(t)^{1/2}$ for $|x| \geq t$, which follows from the definition of p.) Therefore there exists a $t_0 \geq 0$ such that

$$\langle\vartheta_t, S\vartheta_t\rangle < 0 \quad \text{for} \quad t \geq t_0.$$

Let now $f_n = \vartheta_{2^n t_0}$, $n \in \mathbb{N}_0$. These functions have mutually disjoint supports. Therefore $M = L\{f_n : n \in \mathbb{N}_0\}$ is infinite-dimensional and $\langle f, Sf\rangle < 0$ for all $f \in M, f \neq 0$. This implies the theorem. □

Now we want to show that the smallest eigenvalue of a Schrödinger operator (without magnetic field) is always simple, i.e., that the system has a uniquely determined ground state. The proof is essentially taken from a work of *W. Faris* [41]. We need some preparation.

In the following for an element $f \in L_2(\mathbb{R}^m)$ we write $f \geqslant 0$ ($f > 0$) provided that f is real and $f(x) \geqslant 0$ ($f(x) > 0$) almost everywhere in $\mathbb{R}^m$. Such elements are said to be *non-negative* (*positive*). A bounded operator A on $L_2(\mathbb{R}^m)$ is said to be *positivity preserving* if $f \geqslant 0$ implies $Af \geqslant 0$. It is said to be *positivity improving* if $f \geqslant 0$ and $f \neq 0$ imply $Af > 0$.

Theorem 10.32. *Let $A \in B(L_2(\mathbb{R}^m))$ be a real*[3] *positivity improving self-adjoint operator. Assume that $\|A\|$ is an eigenvalue of A. Then the multiplicity of the eigenvalue $\|A\|$ equals 1 and there is an $f > 0$ that spans the eigenspace $N(\|A\| - A)$.*

PROOF. Assume that $f \neq 0$ and $Af = \|A\| f$. Since A is real, we may assume that f is real (otherwise we could replace f by $\operatorname{Re} f$ or $\operatorname{Im} f$, because $A(\operatorname{Re} f) = A(f + Kf)/2 = (Af + KAf)/2 = (\|A\| f + K\|A\| f)/2 = \|A\|(\operatorname{Re} f)$ and $A(\operatorname{Im} f) = \|A\|(\operatorname{Im} f)$). From the inequality $\pm f \leqslant |f|$ it follows that $\pm Af \leqslant A|f|$. Therefore, $|Af| \leqslant A|f|$, and thus

$$\langle f, Af\rangle \leqslant \langle |f|, |Af|\rangle \leqslant \langle |f|, A|f|\rangle.$$

This implies that

$$\|A\|\,\|f\|^2 = \langle f, Af\rangle \leqslant \langle |f|, A|f|\rangle \leqslant \|A\|\,\||f|\|^2 = \|A\|\,\|f\|^2,$$

i.e., that

$$\langle f, Af\rangle = \langle |f|, A|f|\rangle.$$

Let us define f_+ and f_- by the equalities

$$f_+(x) = \max\{0, f(x)\}, \quad f_- = f_+ - f.$$

Then $|f| = f_+ + f_-$. Consequently,

$$\langle f_+, Af_-\rangle = \tfrac{1}{4}\{\langle |f|, A|f|\rangle - \langle f, Af\rangle\} = 0.$$

Hence we have $f_+ = 0$ or $f_- = 0$, since $f_+ \neq 0$ and $f_- \neq 0$ imply that $Af_- > 0$, and thus that $\langle f_+, Af_-\rangle \neq 0$. Consequently, we have proved that $f \geqslant 0$ or $f \leqslant 0$. We can assume, without loss or generality, that $f \geqslant 0$. Since $f = \|A\|^{-1} Af$ and $f \neq 0$, it then follows that we even have $f > 0$, because A is positivity improving.

The theorem will be proved if we show that f spans the space $N(\|A\| - A)$. For every element g of $N(\|A\| - A)$ the functions $\operatorname{Re} g$ and $\operatorname{Im} g$ do not change sign. Such an element can only be orthogonal to the positive element f if $g = 0$. Therefore, $N(\|A\| - A) = L(f)$. □

Theorem 10.33. *Let S be defined as above with $b_j = 0$ ($j = 1, 2, \ldots, m$), $q \in M_{\rho,\mathrm{loc}}(\mathbb{R}^m)$ and $q_- \in M_\rho(\mathbb{R}^m)$ for some $\rho < 4$. Then S is bounded from below. If the lowest point of $\sigma(S)$ is an eigenvalue, then it is simple.*

[3] Here "real" refers to the natural conjugation K on $L_2(\mathbb{R}^m)$, $(Kf)(x) = f(x)^*$ (cf. Section 8.1, Example 1).

PROOF. By Theorem 10.29(a) the operators S and $S_0 - q_-$ are bounded from below. The lower bound of $S_0 - q_-$ is, at the same time, a lower bound of the operators S_n and $S - Q_n$ used in steps 2 and 3. These operators therefore have a common lower bound, so that Theorem 9.18(b) can be applied.

First step: Let S_0 be defined as above. Then the operator $\exp(-tS_0)$ is order improving for all $t > 0$.

Proof of the first step. With $\vartheta_t(x) = \exp(-t|x|^2)$ for $x \in \mathbb{R}^m$ we have

$$\exp(-tS_0) = F^{-1}M_{\vartheta_t}F,$$

where M_{ϑ_t} denotes the operator of multiplication by ϑ_t. Hence, by the convolution theorem (Theorem 10.7) the operator $\exp(-tS_0)$ is equal to the operator of convolution by the function $F^{-1}\vartheta_t$. With $\vartheta(x) = \exp(-|x|^2/2)$ we obtain from Theorem 10.2 that

$$\begin{aligned}(F^{-1}\vartheta_t)(x) &= (2\pi)^{-m/2}\int e^{ixy}\vartheta_t(y)\,dy\\ &= (2\pi)^{-m/2}\int e^{ixy}\vartheta(\sqrt{2t}\,y)\,dy\\ &= (2t)^{-m/2}(2\pi)^{-m/2}\int \exp\left(i\frac{x}{\sqrt{2t}}z\right)\vartheta(z)\,dz\\ &= (2t)^{-m/2}(F^{-1}\vartheta)\left(\frac{x}{\sqrt{2t}}\right) = (2t)^{-m/2}\vartheta\left(\frac{x}{\sqrt{2t}}\right) > 0\end{aligned}$$

for all $x \in \mathbb{R}^m$ and $t > 0$. Since the operator of convolution by a positive function is obviously positivity improving, the assertion follows.

Second step: $\exp(-tS)$ is positivity preserving for all $t \geq 0$.

Proof of the second step. For every $n \in \mathbb{N}$ let q_n be defined by the equality

$$q_n(x) = \begin{cases} q(x), & \text{if } \ |q(x)| \leq n\\ 0, & \text{if } \ |q(x)| > n.\end{cases}$$

Let S_n be the operator defined by q_n instead of q. By Theorem 7.41

$$\exp(-tS_n) = s - \lim_{k\to\infty}\left[\exp\left(-\frac{t}{k}S_0\right)\exp\left(-\frac{t}{k}Q_n\right)\right]^k$$

for all $t \geq 0$, where Q_n denotes the operator of multiplication by q_n. Since every term on the right side is order preserving, it follows from this that $\exp(-tS_n)$ is also positivity preserving. Since by Theorem 9.16(i) (with $D_0 = C_0^\infty(\mathbb{R}^m)$) we have $(i - S_n)^{-1} \xrightarrow{s} (i - S)^{-1}$, it follows by Theorem 9.18(b) that

$$\exp(-tS_n) \xrightarrow{s} \exp(-tS) \quad \text{for all} \quad t \geq 0.$$

Therefore, $\exp(-tS)$ is also positivity preserving for all $t \geq 0$.

Third step: If $f \geqslant 0, f \neq 0, g \geqslant 0$ and $g \neq 0$, then there exists a $t \geqslant 0$ such that $\langle f, \exp(-tS)g\rangle > 0$.

Proof of the third step. It is sufficient to prove that if $f \geqslant 0$ and $f \neq 0$, then

$$K(f) = \{g \in L_2(\mathbb{R}^m) : g \geqslant 0, \langle g, \exp(-tS)f\rangle = 0 \quad \text{for all} \quad t \geqslant 0\}$$

contains only the zero element. The set $K(f)$ is closed. It is mapped into itself by $\exp(-sS)$ for $s \geqslant 0$, since $s, t \geqslant 0$ and $g \in K(f)$ imply that $\langle \exp(-sS)g, \exp(-tS)f\rangle = \langle g, \exp[-(s+t)S]f\rangle = 0$. This then holds for $\exp(sQ_n)$, as well: It follows from $f \geqslant 0$, $g \geqslant 0$, $\langle g, \exp(-tS)f\rangle = 0$ and $\exp(-tS)f \geqslant 0$ (cf. step 2) that $g(x)[\exp(-tS)f](x) = 0$ almost everywhere; we also have then that $\langle \exp(sQ_n)g, \exp(-tS)f\rangle = 0$. Since

$$\exp(-t(S - Q_n)) = s - \lim_{k\to\infty}\left[\exp\left(-\frac{t}{k}S\right)\exp\left(\frac{t}{k}Q_n\right)\right]^k,$$

the operator $\exp(-t(S - Q_n))$ also maps the closed set $K(f)$ into itself. Since, moreover,

$$\exp(-t(S - Q_n)) \xrightarrow{s} \exp(-tS_0) \quad \text{for all} \quad t \geqslant 0,$$

this follows also for $\exp(-tS_0)$. If $g \in K(f)$, then we therefore have $\langle g, \exp(-tS_0)f\rangle = 0$ for all $t \geqslant 0$. Because $\exp(-tS_0)$ is positivity improving, it follows from this that $g = 0$.

Fourth step: If $\lambda \in \mathbb{R}$ is smaller than the lower bound of S, then $(S-\lambda)^{-1}$ is positivity improving.

Proof of the fourth step. If γ is the lower bound of S, then

$$\|e^{-t(S-\lambda)}\| \leqslant e^{-t(\gamma-\lambda)} \quad \text{for} \quad t \geqslant 0.$$

Consequently, the following integrals exist. For all $f, g \in L_2(\mathbb{R}^m)$

$$\begin{aligned}\int_0^\infty e^{\lambda t}\langle g, e^{-tS}f\rangle\, dt &= \int_0^\infty \langle g, e^{-t(S-\lambda)}f\rangle\, dt \\ &= \int_0^\infty \int_{[\lambda,\infty)} e^{-t(s-\lambda)}\, d_s\langle g, E(s)f\rangle\, dt \\ &= \int_{[\lambda,\infty)}\int_0^\infty e^{-t(s-\lambda)}\, dt\, d_s\langle g, E(s)f\rangle \\ &= \int_{[\lambda,\infty)} (s-\lambda)^{-1}\, d_s\langle g, E(s)f\rangle = \langle g, (S-\lambda)^{-1}f\rangle.\end{aligned}$$

If $f \geqslant 0, f \neq 0, g \geqslant 0$ and $g \neq 0$, then the second and third steps imply that $\langle g, e^{-tS}f\rangle \geqslant 0$ for all $t \geqslant 0$ and $\langle g, e^{-t_0S}f\rangle > 0$ for some $t_0 \geqslant 0$. Since the function $t \mapsto \langle g, e^{-tS}f\rangle$ is continuous on $[0, \infty)$, it follows from this that $\langle g, (S-\lambda)^{-1}f\rangle > 0$. Therefore, $(S-\lambda)^{-1}$ is positivity improving (if we had $(S-\lambda)^{-1}f \not> 0$ for some $f \geqslant 0$ such that $f \neq 0$, then there would be a $g \geqslant 0$ such that $g \neq 0$ and $g(x) = 0$ for all x with $(S-\lambda)^{-1}f(x) \neq 0$; we would therefore have $\langle g, (S-\lambda)^{-1}f\rangle = 0$).

Fifth step: The smallest eigenvalue of S is simple.

Proof of the fifth step. Since the lowest point γ of $\sigma(S)$ is an eigenvalue, $\|(S-\lambda)^{-1}\| = (\gamma-\lambda)^{-1}$ is an eigenvalue of the positive operator $(S-\lambda)^{-1}$. This eigenvalue is simple by Theorem 10.32 and the fourth step. Consequently, the assertion follows for S. □

EXERCISE

10.12. Let T be defined as in Theorem 10.23, and assume that $q(x)\to\infty$ as $|x|\to\infty$. Then $\bar{T}$ has a pure discrete spectrum.
Hint: The identity operator is T-small at infinity and its T-bound is equal to 0. The compactness of $(\mathrm{i}-\bar{T})^{-1}$ follows from this by Theorems 10.28, 9.11(d) and 9.10.

10.6 Dirac operators

In this section we consider the Hilbert space

$$L_2(\mathbb{R}^3)^4 = L_2(\mathbb{R}^3) \oplus L_2(\mathbb{R}^3) \oplus L_2(\mathbb{R}^3) \oplus L_2(\mathbb{R}^3).$$

The elements of this space are the 4-tuples $f=(f_1, f_2, f_3, f_4)$ of elements $f_j \in L_2(\mathbb{R}^3)$, and the scalar product is defined by the equality

$$\langle f, g\rangle = \sum_{j=1}^{4} \int f_j(x)^* g_j(x)\,\mathrm{d}x.$$

The elements of $L_2(\mathbb{R}^3)^4$ may also be considered as equivalence classes of functions $f: \mathbb{R}^3 \to \mathbb{C}^4$; then

$$\langle f, g\rangle = \int (f(x), g(x))\,\mathrm{d}x,$$

where $(.\,,.)$ is the usual scalar product in $\mathbb{C}^4$ ($(\xi,\eta)=\sum_{j=1}^4 \xi_j^*\eta_j$ for $\xi,\eta\in \mathbb{C}^4$; the corresponding norm in $\mathbb{C}^4$ will be denoted by $|\,.\,|$). For any continuously differentiable function $f: \mathbb{R}^3 \to \mathbb{C}^4$, $f(x) = (f_1(x), f_2(x), f_3(x), f_4(x))$ let us define

$$(\mathrm{D}_j f)(x) = \frac{1}{i}\left(\frac{\partial}{\partial x_j} f_1(x), \frac{\partial}{\partial x_j} f_2(x), \frac{\partial}{\partial x_j} f_3(x), \frac{\partial}{\partial x_j} f_4(x)\right) \qquad j = 1, 2, 3.$$

For an arbitrary function $f: \mathbb{R}^3 \to \mathbb{C}^4$ and a 4×4 matrix-valued function γ we define γf by the equality

$$(\gamma f)(x) = \gamma(x) f(x) = \left(\sum_{k=1}^{4} \gamma_{1k}(x) f_k(x), \ldots, \sum_{k=1}^{4} \gamma_{4k}(x) f_k(x)\right);$$

we use a similar definition if γ is a constant matrix. The operator norm of 4×4 matrices corresponding to the norm $|\,.\,|$ on $\mathbb{C}^4$ will be denoted by $|\,.\,|$, as well.

The *free Dirac operator* (which describes the free electron in relativistic quantum mechanics) is defined on $L_2(\mathbb{R}^3)^4$ by the differential form

$$\tau f(x) = \sum_{j=1}^{3} \alpha_j D_j f(x) + \beta f(x)$$

with the matrices

$$\alpha_j = \begin{pmatrix} 0 & \sigma_j \\ \sigma_j & 0 \end{pmatrix} \quad \text{for } j = 1, 2, 3, \qquad \beta = \begin{bmatrix} 1 & 0 & 0 & 0 \\ 0 & 1 & 0 & 0 \\ 0 & 0 & -1 & 0 \\ 0 & 0 & 0 & -1 \end{bmatrix}$$

$$\sigma_1 = \begin{pmatrix} 0 & 1 \\ 1 & 0 \end{pmatrix}, \ \sigma_2 = \begin{pmatrix} 0 & -i \\ i & 0 \end{pmatrix}, \ \sigma_3 = \begin{pmatrix} 1 & 0 \\ 0 & -1 \end{pmatrix} \quad \text{(the \textit{Pauli} matrices).}$$

Here we have made all physical constants 1 again. Since the matrices α_j and β are Hermitian, the operator T_0 defined by

$$D(T_0) = C_0^\infty(\mathbb{R}^3)^4 \quad \text{and} \quad T_0 f = \tau f \quad \text{for} \quad f \in D(T_0)$$

is symmetric on $L_2(\mathbb{R}^3)^4$, as a simple computation shows.

Theorem 10.34. *The operator T_0 defined above is essentially self-adjoint. For $T = \overline{T}_0$ we have:*
(i) *$D(T) = W_{2,1}(\mathbb{R}^3)^4$ and $\|Tf\| = \|f\|_1$ for $f \in D(T)$,*[4]
(ii) *$\sigma(T) = (-\infty, -1] \cup [1, \infty)$,*
(iii) *T has no eigenvalue.*

Proof.
(i) Let T_1 be defined by

$$D(T_1) = S(\mathbb{R}^3)^4 \quad \text{and} \quad T_1 f = \tau f \quad \text{for} \quad f \in D(T_1).$$

It is clear that $T_1 \subset \overline{T}_0$ (cf. the proof of Theorem 10.11(a)), and thus $T = \overline{T}_0 = \overline{T}_1$. On the other hand,

$$F T_1 F^{-1} = M_{P,1},$$

where

$$D(M_{P,1}) = S(\mathbb{R}^3)^4, \quad M_{P,1} f = Pf,$$

and

$$P(x) = \begin{bmatrix} 1 & 0 & x_3 & x_1 - i x_2 \\ 0 & 1 & x_1 + i x_2 & -x_3 \\ x_3 & x_1 - i x_2 & -1 & 0 \\ x_1 + i x_2 & -x_3 & 0 & -1 \end{bmatrix}.$$

[4] In what follows we also write $\|\cdot\|_p$ and $\|\cdot\|_{(p)}$ for the norms in $W_{2,p}(\mathbb{R}^m)^4$ and $L_{2,p}(\mathbb{R}^m)^4$, respectively.

(Here the Fourier transformation is applied to the elements of $L^2(\mathbb{R}^3)^4$ component-wise.) $\overline{M_{P,1}} = M_P$, where M_P denotes the maximal operator of multiplication by the matrix-valued function P (cf. the proof of Theorem 10.11(a)):

$$D(M_P) = \{f \in L_2(\mathbb{R}^3)^4 : Pf \in L_2(\mathbb{R}^3)^4\}, \quad M_P f = Pf.$$

The reader can easily verify that the eigenvalues $p_j(x)$ and the corresponding orthogonal eigenelements $\tilde{e}_j(x)$ of the matrix $P(x)$ are given by the formulae (observe that $\sigma_j \sigma_k + \sigma_k \sigma_j = 2\,\delta_{jk}\begin{pmatrix} 1 & 0 \\ 0 & 1 \end{pmatrix}$)

$$p_1(x) = p_2(x) = -p_3(x) = -p_4(x) = (1+|x|^2)^{1/2},$$

$$\tilde{e}_1(x) = \begin{bmatrix} \begin{pmatrix} 1 \\ 0 \end{pmatrix} \\ \dfrac{\sigma_1 x_1 + \sigma_2 x_2 + \sigma_3 x_3}{1+(1+|x|^2)^{1/2}} \begin{pmatrix} 1 \\ 0 \end{pmatrix} \end{bmatrix}, \qquad \tilde{e}_2(x) = \begin{bmatrix} \begin{pmatrix} 0 \\ 1 \end{pmatrix} \\ \dfrac{\sigma_1 x_1 + \sigma_2 x_2 + \sigma_3 x_3}{1+(1+|x|^2)^{1/2}} \begin{pmatrix} 0 \\ 1 \end{pmatrix} \end{bmatrix},$$

$$\tilde{e}_3(x) = \begin{bmatrix} -\dfrac{\sigma_1 x_1 + \sigma_2 x_2 + \sigma_3 x_3}{1+(1+|x|^2)^{1/2}} \begin{pmatrix} 1 \\ 0 \end{pmatrix} \\ \begin{pmatrix} 1 \\ 0 \end{pmatrix} \end{bmatrix}, \qquad \tilde{e}_4(x) = \begin{bmatrix} -\dfrac{\sigma_1 x_1 + \sigma_2 x_2 + \sigma_3 x_3}{1+(1+|x|^2)^{1/2}} \begin{pmatrix} 0 \\ 1 \end{pmatrix} \\ \begin{pmatrix} 0 \\ 1 \end{pmatrix} \end{bmatrix}.$$

The corresponding normalized eigenelements are

$$e_j(x) = \left[2+2|x|^2+2(1+|x|^2)^{1/2}\right]^{1/2}\left[2+|x|^2+2(1+|x|^2)^{1/2}\right]^{-1/2}\tilde{e}_j(x).$$

Consequently,

$$(M_P f)(x) = \sum_{j=1}^{4} p_j(x)(e_j(x), f(x))e_j(x) \quad \text{for all } f \in D(M_P).$$

It follows that

$$D(M_P) = \left\{ f \in L_2(\mathbb{R}^3)^4 : \left[\sum_{j=1}^{4} p_j(x)^2 |(e_j(x), f(x))|^2 \right]^{1/2} \in L_2(\mathbb{R}^3) \right\}$$

$$= \{ f \in L_2(\mathbb{R}^3)^4 : (1+|\,.\,|^2)^{1/2} f \in L_2(\mathbb{R}^3)^4 \} = L_{2,1}(\mathbb{R}^3)^4,$$

$$D(T) = F^{-1}D(M_P) = F^{-1}L_{2,1}(\mathbb{R}^3)^4 = W_{2,1}(\mathbb{R}^3)^4,$$

$$Tf = F^{-1}M_P Ff \quad \text{for } f \in D(T),$$

and

$$\|Tf\| = \|PFf\| = \|(1+|\,.\,|^2)^{1/2}Ff\| = \|Ff\|_{(1)} = \|f\|_1.$$

(ii) If $\lambda > 1$, then there is an $x_0 \in \mathbb{R}^3$ for which $p_1(x_0) = \lambda$. Since p_1 is continuous, for every $\epsilon > 0$ there is a ball K_ϵ around x_0 such that $|p_1(x) - \lambda| < \epsilon$ for all $x \in K_\epsilon$. If $\eta \in L_2(\mathbb{R}^3)$ is such that $\eta \neq 0$ and

$\eta(x)=0$ for $x \notin K_\epsilon$, then for $f=\eta e_1 \in L_2(\mathbb{R}^3)^4$

$$\|(\lambda - M_p)f\|^2 = \int |\lambda - p_1(x)|^2 |\eta(x)|^2 \, dx \leq \epsilon^2 \|\eta\|^2 = \epsilon^2 \|f\|^2.$$

Since $\epsilon > 0$ was arbitrary, $\lambda - M_p$ is not continuously invertible. If $\lambda \leq -1$, then we can prove this analogously if we use p_3 and e_3 instead of p_1 and e_1. Therefore, $(-\infty, -1] \cup [1, \infty) \subset \sigma(M_p) = \sigma(T)$.

If now $\lambda \in (-1, 1)$, then

$$\begin{aligned} \|(\lambda - M_p)f\|^2 &= \sum_{j=1}^4 \int |\lambda - p_j(x)|^2 |(e_j(x), f(x))|^2 \, dx \\ &\geq (1-|\lambda|)^2 \sum_{j=1}^4 \int |(e_j(x), f(x))|^2 \, dx = (1-|\lambda|)^2 \|f\|^2 \end{aligned}$$

for all $f \in L_2(\mathbb{R}^3)^4$. Consequently, $\lambda - M_p$ is continuously invertible. Hence, $(-1, 1) \subset \rho(T) = \rho(M_p)$, and thus $\sigma(T) = (-\infty, -1] \cup [1, \infty)$.

(iii) If $(\lambda - M_p)f = 0$, then we must have

$$(\lambda - p_j(x))(e_j(x), f(x)) = 0 \quad \text{almost everywhere} \quad (j = 1, 2, 3, 4).$$

As the set of zeros of $\lambda - p_j(x)$ is a null set (it is either empty, consists of the origin, or is a sphere), we must have

$$(e_j(x), f(x)) = 0 \quad \text{almost everywhere} \quad (j = 1, 2, 3, 4).$$

It follows from this (as $e_j(x)$ is a basis in $\mathbb{C}^4$) that $f(x) = 0$ almost everywhere; therefore, $f = 0$. Consequently, no λ can be an eigenvalue of M_p (and hence of T). □

The Dirac operator of an electron in the electric field with potential q has the form

$$S = T + Q,$$

where Q is the operator of multiplication by a function $q : \mathbb{R}^3 \to \mathbb{R}$ and $Qf = qf = (qf_1, qf_2, qf_3, qf_4)$. Because of physical reasons, the Coulomb potential $q(x) = c(1/|x|)$ is particularly interesting. As this q does not belong to $M_\rho(\mathbb{R}^3)$ for $\rho < 2$, the results of Section 10.3 cannot be applied to it. The following auxiliary theorem enables us to treat such potentials. We prove it with somewhat more generality than we need.

Auxiliary theorem 10.35. *For $f \in C_0^\infty(\mathbb{R}^m)$ $(m \geq 3)$ and for $f \in C_0^\infty(\mathbb{R} \setminus \{0\})$ $(m=1)$*

$$\int |x|^{-2} |f(x)|^2 \, dx \leq \frac{4}{(m-2)^2} \int \sum_{j=1}^m \left| \frac{\partial}{\partial x_j} f(x) \right|^2 dx.$$

For $f \in C_0^\infty(\mathbb{R}^2 \setminus \{0\})$

$$\int |x|^{-2}|f(x)|^2 \, dx \leq 4 \int |\ln|x||^2 \sum_{j=1}^{2} \left|\frac{\partial}{\partial x_j} f(x)\right|^2 dx.$$

PROOF. Let Ω_m denote the unit sphere in $\mathbb{R}^m$ and let $d\omega_m$ be the area element on Ω_m. For $m \geq 3$ we have

$$\int |x|^{-2}|f(x)|^2 \, dx = \int_0^\infty r^{m-3} \int_{\Omega_m} |f(r\omega)|^2 \, d\omega_m \, dr$$

$$= \lim_{\epsilon \to 0} \int_\epsilon^\infty r^{m-3} \int_{\Omega_m} |f(r\omega)|^2 \, d\omega_m \, dr$$

$$= \lim_{\epsilon \to 0} \left\{ \frac{1}{m-2} \epsilon^{m-2} \int_{\Omega_m} |f(\epsilon\omega)|^2 \, d\omega_m \right.$$

$$\left. - \frac{2}{m-2} \int_\epsilon^\infty r^{m-2} \operatorname{Re} \int_{\Omega_m} f(r\omega)^* \frac{\partial}{\partial r} f(r\omega) \, d\omega_m \, dr \right\}.$$

Here the first term vanishes as $\epsilon \to 0$, since the integral tends to $f(0)\int d\omega_m$ as $\epsilon \to 0$. Therefore,

$$\int |x|^{-2}|f(x)|^2 \, dx \leq \frac{2}{m-2} \int |x|^{-1}|f(x)| \left\{ \sum_{j=1}^{m} \left| \frac{\partial}{\partial x_j} f(x) \right|^2 \right\}^{1/2} dx$$

$$\leq \frac{2}{m-2} \left\{ \int |x|^{-2}|f(x)|^2 \, dx \int \sum_{j=1}^{m} \left| \frac{\partial}{\partial x_j} f(x) \right|^2 dx \right\}^{1/2}.$$

The assertion follows from this for $m \geq 3$. The case $m = 1$ can be proved similarly though with certain simplifications. In the case $m = 2$ a logarithmic term arises after the integration by parts; everything else goes as above. □

Theorem 10.36. *Assume that* $q = q_1 + q_2$, *where* q_1 *and* q_2 *are measurable Hermitian* 4×4 *matrix-valued functions such that*

$$|q_1(x)| \leq C \frac{1}{|x|} \quad \text{and} \quad q_2(\,.\,) \in M_\rho(\mathbb{R}^3) \quad \text{for some} \quad \rho < 2.$$

Then Q *is* T*-bounded with* T*-bound less than or equal to* $2C$. *If* $C < 1/2$, *then* $S = T + Q$ *is essentially self-adjoint on* $C_0^\infty(\mathbb{R}^3)^4$ *and self-adjoint on* $W_{2,1}(\mathbb{R}^3)^4$.

PROOF. By Auxiliary theorem 10.35

$$\|q_1 f\|^2 \leqslant C^2 \int |x|^{-2} \sum_{j=1}^{4} |f_j(x)|^2 \, dx \leqslant 4C^2 \sum_{k=1}^{3} \sum_{j=1}^{4} \int \left| \frac{\partial}{\partial x_k} f_j(x) \right|^2 dx$$

$$= 4C^2 \sum_{k=1}^{3} \sum_{j=1}^{4} \int |x_k (Ff_j)(x)|^2 \, dx$$

$$= 4C^2 \int \| |x| (Ff)(x) \|^2 \, dx \leqslant 4C^2 \| M_p Ff \|^2 = 4C^2 \| Tf \|^2,$$

for all $f \in C_0^\infty(\mathbb{R}^3)^4$. The operator Q_1 induced by q_1 is therefore T-bounded with T-bound $\leqslant 2C$. The operator Q_2 induced by q_2 has T-bound 0, since $D(T) = W_{2,1}(\mathbb{R}^3)^4$ (cf. Theorem 10.18). Consequently, the T-bound of $Q = Q_1 + Q_2$ is not greater than 2C. The remaining assertions follow from Theorem 5.28. □

Theorem 10.37. *Let q be as in Theorem 10.36. Assume, moreover, that $N_{|q|}(x) \to 0$ as $|x| \to \infty$. Then Q is T^2-compact. If $C < 1/2$, then $\sigma_e(S) = \sigma_e(T) = (-\infty, -1] \cup [1, \infty)$. If, in addition, $q \leqslant 0$, then the eigenvalues of S in $(-1, 1)$ can only accumulate at 1; if $q \geqslant 0$, then the eigenvalues of S in $(-1, 1)$ can only accumulate at -1.*

PROOF. We have $T^2 = F^{-1} M_p^2 F$. Therefore,

$$D(T^2) = F^{-1} D(M_p^2) = F^{-1} L_{2,2}(\mathbb{R}^3)^4 = W_{2,2}(\mathbb{R}^3)^4.$$

Since $|q| = |q_1 + q_2| \in M_\rho(\mathbb{R}^3)$ for some $\rho < 4$, the operator Q is T^2-compact (cf. Theorem 10.21). (We can show in the same way that the operator Q_2 induced by q_2 is T-compact.) The remaining assertions follow from Theorems 9.13, 9.14 and 10.36. □

In analogy with the virial theorem (Theorem 10.30), we can now prove yet the following result.

Theorem 10.38. *Let q be as in Theorem 10.36 with some $C < 1/2$. Moreover, assume that $q(ax) = q(x)/a$ for all $a > 0$ and $x \in \mathbb{R}^3 \setminus \{0\}$. Then $S = T + Q$ has no eigenvalue in $(-\infty, -1) \cup (1, \infty)$. If $q(x)$ is diagonal, then -1 and 1 are not eigenvalues of S, either.*

PROOF. If $(\lambda - S)f = 0$, then the function $f_a(x) = f(ax)$ obviously belongs to $D(S)$ and

$$\begin{aligned}(Sf_a)(x) &= \sum_{j=1}^{3} \alpha_j D_j f(ax) + (\beta + q(x)) f(ax)\\ &= a \sum_{j=1}^{3} \alpha_j (D_j f)(ax) + (\beta + q(x)) f(ax)\\ &= a \left\{ \sum_{j=1}^{3} \alpha_j (D_j f)(ax) + (\beta + q(ax)) f(ax) \right\} + \beta(1-a) f(ax)\\ &= a(Sf)(ax) + \beta(1-a) f_a(x)\\ &= a\lambda f(ax) + \beta(1-a) f_a(x) = a\lambda f_a(x) + \beta(1-a) f_a(x).\end{aligned}$$

It follows from this that

$$\begin{aligned}0 &= \langle Sf, f_a\rangle - \langle f, Sf_a\rangle = \lambda\langle f, f_a\rangle - a\lambda\langle f, f_a\rangle - (1-a)\langle f, \beta f_a\rangle\\ &= (1-a)\langle f, (\lambda - \beta) f_a\rangle.\end{aligned}$$

For $a \neq 1$ we can divide by $(1-a)$. Taking the limit as $a \to 1$ gives

$$0 = \langle f, (\lambda - \beta) f\rangle.$$

For $\lambda \in (-\infty, -1) \cup (1, \infty)$ the matrix $\lambda - \beta$ is strictly positive or strictly negative; this equality is therefore possible only if $f = 0$.

If $\lambda = 1$ and q is diagonal, then it follows first that $f_3 = f_4 = 0$. The eigenvalue equality $(1 - S)f = 0$ then gives

$$\begin{pmatrix} x_3 & x_1 - ix_2 \\ x_1 + ix_2 & -x_3 \end{pmatrix} \begin{pmatrix} f_1(x) \\ f_2(x) \end{pmatrix} = 0.$$

This implies that $f_1 = f_2 = 0$, and thus $f = 0$. We can argue similarly if $\lambda = -1$. □

To conclude this section, we give a general criterion for the essential self-adjointness of Dirac operators (cf. also Exercise 10.13).

Theorem 10.39. *Assume that $q = q_1 + q_2$, where q_1 and q_2 are measurable Hermitian 4×4 matrix-valued functions such that*

$$|q_1(x)| \leq C \frac{1}{|x|} \quad \text{for some} \quad C < \frac{1}{2},$$

$$|q_2(\,.\,)| \in M_{\rho, \mathrm{loc}}(\mathbb{R}^3) \quad \text{for some} \quad \rho < 2.$$

Then $S = T + Q$ is essentially self-adjoint on $C_0^\infty(\mathbb{R}^3)^4$. The closure $\bar{S}$ is given by the formulae

$$D(\bar{S}) = \{ f \in L_2(\mathbb{R}^3)^4 \cap W_{2,1,\mathrm{loc}}(\mathbb{R}^3)^4 : \tau f + qf \in L_2(\mathbb{R}^3)^4 \},$$
$$\bar{S}f = \tau f + qf \quad \textit{for} \quad f \in D(\bar{S}).$$

PROOF. Let φ_n be as in the proof of Theorem 10.23, let $\tilde{q}_n = \varphi_{3n} q$, and let $S_n = T + Q_n$, where Q_n is the operator of multiplication by $\tilde{q}_n$. The function $\tilde{q}_n$ satisfies the assumption of Theorem 10.36 for every $n \in \mathbb{N}$; the operator S_n is therefore essentially self-adjoint on $C_0^\infty(\mathbb{R}^3)^4$.

The rest of the proof follows the proof of Theorem 10.23; however, it is essentially simpler : Let $g \in R(\mathrm{i} - S)^\perp$. Since S_n is essentially self-adjoint, for every $n \in \mathbb{N}$ there exists an f_n such that

$$\|\varphi_n g - (\mathrm{i} - S_n) f_n\| \leqslant \frac{1}{n}.$$

Hence, $\|f_n\| \leqslant \|(\mathrm{i} - S_n) f_n\| \leqslant \|\varphi_n g\| + (1/n)$. It follows from this that

$$\begin{aligned} \|\varphi_n g\|^2 &= \langle \varphi_n g, (\mathrm{i} - S_n) f_n + [\varphi_n g - (\mathrm{i} - S_n) f_n] \rangle \\ &\leqslant |\langle g, \varphi_n (\mathrm{i} - S_n) f_n \rangle| + \frac{1}{n} \|\varphi_n g\| \\ &\leqslant |\langle g, (\mathrm{i} - S)(\varphi_n f_n) \rangle| + C_1 \|g\| \sum_{j=1}^{3} \|f_n \, D_j \varphi_n\| + \frac{1}{n} \|\varphi_n g\| \\ &\leqslant C_2 \frac{1}{n} (\|g\| \, \|f_n\| + \|\varphi_n g\|) \to 0 \quad \text{as} \quad n \to \infty. \end{aligned}$$

This implies that $g = 0$. Consequently, $\overline{R(\mathrm{i} - S)} = L_2(\mathbb{R}^3)^4$. The equality $\overline{R(-\mathrm{i} - S)} = L_2(\mathbb{R}^3)^4$ follows analogously. The rest of the proof is analogous to part (b) of the proof of Theorem 10.23. □

EXERCISE

10.13. (a) In the first part of Theorem 10.36 the assumption on q_1 can be replaced by the assumption $|q_1(x)| \leqslant \sum_{j=1}^N c_j |x - a_j|^{-1}$ with N different points $a_j \in \mathbb{R}^3$.

(b) In the second part of Theorem 10.36, in Theorem 10.37, and in Theorem 10.39 the assumption on q_1 can be replaced by the assumption $|q_1(x)| \leqslant \sum_{j=1}^N c_j |x - a_j|^{-1}$ with N different points $a_j \in \mathbb{R}^3$ and $0 \leqslant c_j < \frac{1}{2}$ for $j = 1, \ldots, N$.

(c) In Theorem 10.39 we can allow an infinite sum of the above form for the bound of q_1 provided that the a_j have no accumulation point.
Hint: If $\vartheta \in C_0^\infty(\mathbb{R}^m)$, $\vartheta(x) = 1$ for $|x| \leqslant s/2$, $\vartheta(x) = 0$ for $|x| \geqslant s$, and $0 \leqslant \vartheta(x) \leqslant 1$ for all $x \in \mathbb{R}^3$, then there is a $C > 0$ such that

$$\int |x|^{-2} |\vartheta(x) f(x)|^2 \, dx \leqslant 4 \int_{|x| < s} \sum_{j=1}^{3} |D_j f(x)|^2 \, dx + C \int_{|x| < s} |f(x)|^2 \, dx.$$

Scattering theory 11

11.1 Wave operators

The theory of *wave operators* provides a useful means of studying the absolutely continuous spectrum. We wish to present this theory briefly in what follows. For any two self-adjoint operators T_1 and T_2 on the complex Hilbert space H we define $\Omega_{\pm}(T_2, T_1)$ by the equalities

$$D(\Omega_{\pm}(T_2, T_1)) = \left\{ f \in H: \lim_{t \to \pm\infty} e^{itT_2} e^{-itT_1} f \text{ exists} \right\},$$

$$\Omega_{\pm}(T_2, T_1) f = \lim_{t \to \pm\infty} e^{itT_2} e^{-itT_1} f \quad \text{for} \quad f \in D(\Omega_{\pm}(T_2, T_1)).$$

It is obvious that $\Omega_{\pm}(T_2, T_1)$ are linear operators. Since $e^{itT_2}e^{-itT_1}$ is unitary for all $t \in \mathbb{R}$, the operators $\Omega_{\pm}(T_2, T_1)$ are obviously isometric (as mappings from $D(\Omega_{\pm}(T_2, T_1))$ into H).

Theorem 11.1. *The subspaces $D(\Omega_{\pm}(T_2, T_1))$ are closed and reduce T_1. The subspaces $R(\Omega_{\pm}(T_2, T_1))$ are closed and reduce T_2. We have $R(\Omega_{\pm}(T_2, T_1)) = D(\Omega_{\pm}(T_1, T_2))$ and $\Omega_{\pm}(T_1, T_2) = \Omega_{\pm}(T_2, T_1)^{-1}$. Moreover, for all $s \in \mathbb{R}$*

$$\Omega_{\pm}(T_2, T_1)\, e^{isT_1} = e^{isT_2} \Omega_{\pm}(T_2, T_1)$$

and

$$\Omega_{\pm}(T_2, T_1) E_1(s) = E_2(s) \Omega_{\pm}(T_2, T_1),$$

where E_1 and E_2 are the spectral families of T_1 and T_2, respectively. We have $\Omega_{\pm}(T_2, T_1)u(T_1) = u(T_2)\Omega_{\pm}(T_2, T_1)$ for every bounded continuous function $u : \mathbb{R} \to \mathbb{R}$ ("intertwining" property).

PROOF. First we show that $D(\Omega_+(T_2, T_1))$ is closed. Let $f \in \overline{D(\Omega_+(T_2, T_1))}$. We have to show that for every $\epsilon > 0$ there exists a $t_0 \in \mathbb{R}$ such that

$$\|e^{i\,tT_2} e^{-i\,tT_1}f - e^{i\,sT_2} e^{-i\,sT_1}f\| < \epsilon$$

for all $s, t \geq t_0$. In order to prove this, we take an $f_0 \in D(\Omega_+(T_2, T_1))$ such that $\|f - f_0\| < \epsilon/3$. Since $\lim_{t\to\infty} e^{i\,tT_2} e^{-i\,tT_1}f_0$ exists, there is a $t_0 \in \mathbb{R}$ such that

$$\|e^{i\,tT_2} e^{-i\,tT_1}f_0 - e^{i\,sT_2} e^{-i\,sT_1}f_0\| < \tfrac{1}{3}\epsilon$$

for all $s, t \geq t_0$. Consequently, for all $s, t \geq t_0$

$$\|e^{i\,tT_2} e^{-i\,tT_1}f - e^{i\,sT_2} e^{-i\,sT_1}f\| < \|e^{i\,tT_2} e^{-i\,tT_1}(f - f_0)\|$$
$$+ \|e^{i\,sT_2} e^{-i\,sT_1}(f_0 - f)\| + \|e^{i\,tT_2} e^{-i\,tT_1}f_0 - e^{i\,sT_2} e^{-i\,sT_1}f_0\| < \epsilon.$$

We can show in a similar way that $D(\Omega_-(T_2, T_1))$ is closed. As the operators $\Omega_\pm(T_2, T_1)$ are isometric, the ranges $R(\Omega_\pm(T_2, T_1))$ are also closed.

If $g = \Omega_+(T_2, T_1)f \in R(\Omega_+(T_2, T_1))$ with some $f \in D(\Omega_+(T_2, T_1))$, then

$$\|e^{i\,tT_1} e^{-i\,tT_2}g - f\| = \|g - e^{i\,tT_2} e^{-i\,tT_1}f\| \to 0$$

as $t \to \infty$, i.e., $g \in D(\Omega_+(T_1, T_2))$ and $\Omega_+(T_1, T_2)g = f$. If $g \in D(\Omega_+(T_1, T_2))$ and $f = \Omega_+(T_1, T_2)g$, then we can show similarly that $f \in D(\Omega_+(T_2, T_1))$ and $g = \Omega_+(T_2, T_1)f$. Therefore, $R(\Omega_+(T_2, T_1)) = D(\Omega_+(T_1, T_2))$ and $\Omega_+(T_1, T_2) = \Omega_+(T_2, T_1)^{-1}$. We can show analogously that $R(\Omega_-(T_2, T_1)) = D(\Omega_-(T_1, T_2))$ and $\Omega_-(T_1, T_2) = \Omega_-(T_2, T_1)^{-1}$.

If $f \in D(\Omega_+(T_2, T_1))$ and $g = \Omega_+(T_2, T_1)f$, then for every $s \in \mathbb{R}$

$$e^{i\,tT_2} e^{-i\,tT_1}(e^{i\,sT_1}f) = e^{i\,sT_2} e^{i(t-s)T_2} e^{-i(t-s)T_1}f \to e^{i\,sT_2}g \quad \text{as} \quad t \to \infty.$$

Consequently, $e^{i\,sT_1}f \in D(\Omega_+(T_2, T_1))$ for all $s \in \mathbb{R}$. By Theorem 7.39 the operator T_1 is therefore reduced by $D(\Omega_+(T_2, T_1))$. We can prove this assertion for $D(\Omega_-(T_2, T_1))$ similarly. Since $R(\Omega_\pm(T_2, T_1)) = D(\Omega_\pm(T_1, T_2))$, the operator T_2 is reduced by $R(\Omega_\pm(T_2, T_1))$.

For every $s \in \mathbb{R}$ the operator $e^{i\,sT_1}$ maps the space $D(\Omega_+(T_2, T_1))$ onto itself (cf. Theorem 7.39). Since for all $f \in D(\Omega_+(T_2, T_1))$ and all $s \in \mathbb{R}$

$$\Omega_+(T_2, T_1) e^{i\,sT_1}f = \lim_{t\to\infty} e^{i\,tT_2} e^{-i(t-s)T_1}f$$

$$= e^{i\,sT_2} \lim_{t\to\infty} e^{i(t-s)T_2} e^{-i(t-s)T_1}f = e^{i\,sT_2}\Omega_+(T_2, T_1)f,$$

it follows that $\Omega_+(T_2, T_1) e^{i\,sT_1} = e^{i\,sT_2}\Omega_+(T_2, T_1)$. A corresponding equality holds for $\Omega_-(T_2, T_1)$.

For $j=1, 2$, $\operatorname{Im} z>0$, and all $f, g \in H$

$$\begin{aligned} -\mathrm{i}\int_0^{\infty} \mathrm{e}^{\mathrm{i} zt}\langle f, \mathrm{e}^{-\mathrm{i} tT_j} g\rangle \,\mathrm{d}t &= -\mathrm{i}\int_0^{\infty} \mathrm{e}^{\mathrm{i} zt}\left\{\int \mathrm{e}^{-\mathrm{i} ts}\,\mathrm{d}\langle f, E_j(s)g\rangle\right\}\mathrm{d}t \\ &= -\mathrm{i}\int\left\{\int_0^{\infty} \mathrm{e}^{\mathrm{i} t(z-s)}\,\mathrm{d}t\right\}\mathrm{d}\langle f, E_j(s)g\rangle \\ &= \int (z-s)^{-1}\,\mathrm{d}\langle f, E_j(s)g\rangle \\ &= \langle f, (z-T_j)^{-1} g\rangle. \end{aligned}$$

Similarly, for $\operatorname{Im} z<0$

$$\mathrm{i}\int_{-\infty}^{0} \mathrm{e}^{\mathrm{i} zt}\langle f, \mathrm{e}^{-\mathrm{i} tT_j} g\rangle \,\mathrm{d}t = \langle f, (z-T_j)^{-1} g\rangle.$$

Together with the representation (7.22) of the spectral family, it follows therefore that $\Omega_{\pm}(T_2, T_1)E_1(s)=E_2(s)\Omega_{\pm}(T_2, T_1)$ for all $s \in \mathbb{R}$. The last assertion follows from this, because $u(T_j)=\int u(t)\,\mathrm{d}E_j(t)$. □

Remark. If f is an eigenelement of T_1 for the eigenvalue λ, then $f \in D(\Omega_+(T_2, T_1))$ if and only if f is also an eigenelement of T_2 for the same eigenvalue. The same holds for $\Omega_-(T_2, T_1)$.

Proof. If $f \in D(T_2)$ and $T_2 f=\lambda f$, then

$$\mathrm{e}^{\mathrm{i} tT_2}\,\mathrm{e}^{-\mathrm{i} tT_1} f = \mathrm{e}^{\mathrm{i} tT_2}\,\mathrm{e}^{-\mathrm{i} t\lambda} f = \mathrm{e}^{\mathrm{i} t\lambda}\,\mathrm{e}^{-\mathrm{i} t\lambda} f = f$$

for all $t \in \mathbb{R}$. Consequently, $f \in D(\Omega_{\pm}(T_2, T_1))$. If $f \in D(\Omega_+(T_2, T_1))$, then for every $s \in \mathbb{R}$

$$\|\mathrm{e}^{\mathrm{i} s(T_2-\lambda)} f - f\| = \|\mathrm{e}^{\mathrm{i}(t+s)T_2}\mathrm{e}^{-\mathrm{i}(t+s)\lambda} f - \mathrm{e}^{\mathrm{i} tT_2}\,\mathrm{e}^{-\mathrm{i} t\lambda} f\| \to 0 \quad \text{as} \quad t\to\infty$$

(since $\mathrm{e}^{\mathrm{i}(t+s)T_2}\mathrm{e}^{-\mathrm{i}(t+s)\lambda} f$ and $\mathrm{e}^{\mathrm{i} tT_2}\mathrm{e}^{-\mathrm{i} t\lambda} f$ converge to $\Omega_+(T_2, T_1)f$). Hence,

$$\mathrm{e}^{\mathrm{i} s(T_2-\lambda)} f = f \quad \text{for all} \quad s \in \mathbb{R}.$$

This implies that $f \in D(T_2-\lambda)=D(T_2)$ and

$$(T_2-\lambda)f = \lim_{s\to 0}\frac{-\mathrm{i}}{s}\left(\mathrm{e}^{\mathrm{i} s(T_2-\lambda)} f - f\right) = 0,$$

i.e., that $T_2 f=\lambda f$. □

In what follows we write $M \subset_{T_1} D(\Omega_{\pm}(T_2, T_1))$ if M is a closed subspace of $D(\Omega_{\pm}(T_2, T_1))$ that reduces T_1. If $M \subset_{T_1} D(\Omega_{\pm}(T_2, T_1))$ and P denotes the orthogonal projection onto M, then we define

$$W_{\pm}(T_2, T_1, P) = \Omega_{\pm}(T_2, T_1)P.$$

These operators are called *generalized wave operators*. In particular, we write $W_{\pm}(T_2, T_1)$ for $W_{\pm}(T_2, T_1, I)$ provided that $D(\Omega_{\pm}(T_2, T_1))=H$. The operators $W_\pm(T_2, T_1)$ are called *wave operators*. In the sequel we say that $W_\pm(T_2, T_1, P)$ exists if $R(P) \subset_{T_1} D(\Omega_\pm(T_2, T_1))$. (Some of the following statements still hold when T_1 is not reduced by $R(P)$.) The operators $W_\pm(T_2, T_1, P)$ are obviously partially isometric with initial domain $R(P)$ and final domain $R(W_\pm(T_2, T_1, P))$. Now we prove a few simple properties of the generalized wave operators.

Theorem 11.2. *Assume that T_1 and T_2 are self-adjoint operators on the complex Hilbert space H, $M_1 \subset_{T_1} D(\Omega_+(T_2, T_1))$, P_1 is the orthogonal projection onto M_1, and $M_2 = R(W_+(T_2, T_1, P_1))$. Then $M_2 \subset_{T_2} D(\Omega_+(T_1, T_2))$, and with the notation $W_+ = W_+(T_2, T_1, P_1)$ we have*

$$e^{isT_2}W_+ = W_+ e^{isT_1} \quad \textit{for all} \quad s \in \mathbb{R}$$

and

$$T_2 W_+ = W_+ T_1 P_1 \supset W_+ T_1.$$

If P_2 is the orthogonal projection onto M_2, then

$$T_1 W_+^* = W_+^* T_2 P_2 \supset W_+^* T_2.$$

The operators $T_1|_{M_1}$ and $T_2|_{M_2}$ are unitarily equivalent. If $M_1 \subset H_c(T_1)$ or $M_1 \subset H_{ac}(T_1)$, then $M_2 \subset H_c(T_2)$ or $M_2 \subset H_{ac}(T_2)$, respectively. A similar result holds for $W_-(T_2, T_1, P_1)$.

Proof. Since M_1 is closed and is contained in $D(\Omega_+(T_2, T_1))$, the subspace $M_2 = \Omega_+(T_2, T_1)M_1$ is also closed. By Theorem 11.1

$$\begin{aligned} e^{isT_2}W_+ &= e^{isT_2}\Omega_+(T_2, T_1)P_1 = \Omega_+(T_2, T_1)\, e^{isT_1}P_1 \\ &= \Omega_+(T_2, T_1)P_1\, e^{isT_1} = W_+ e^{isT_1} \end{aligned}$$

for all $s \in \mathbb{R}$. In particular, $e^{isT_2}g \in M_2$ for all $g = W_+ f \in M_2$ and all $s \in \mathbb{R}$, i.e., M_2 reduces T_2(cf. Theorem 7.39). The equality $e^{isT_2}W_+ = W_+ e^{isT_1}$ implies that

$$e^{isT_2}P_2 = e^{isT_2}W_+ W_+^* = W_+ e^{isT_1}W_+^* = W_+ e^{isT_1}P_1 W_+^*.$$

The restrictions of e^{isT_1} and e^{isT_2} to M_1 and M_2, respectively, hence are unitarily equivalent. This then follows also for their infintesimal generators, i.e., $T_2P_2 = W_+ T_1 P_1 W_+^*$, and thus

$$T_2 W_+ = T_2 P_2 W_+ = W_+ T_1 P_1 W_+^* W_+ = W_+ T_1 P_1 \supset W_+ T_1.$$

By taking adjoints, we obtain via Theorem 4.19 that

$$W_+^* T_2 = W_+^* T_2^* \subset (T_2 W_+)^* \subset (W_+ T_1)^* = T_1^* W_+^* = T_1 W_+^*.$$

On the basis of what we have proved so far, $T_2|_{M_2}$ and $T_1|_{M_1}$ are unitarily equivalent. If $M_1 \subset H_c(T_1)$ or $M_1 \subset H_{ac}(T_1)$, then $T_1|_{M_1}$ has a pure continuous or a pure absolutely continuous spectrum, correspondingly. Then this also holds for the operator $T_2|_{M_2}$, unitarily equivalent to $T_1|_{M_1}$, i.e., $M_2 \subset H_c(T_2)$ or $M_2 \subset H_{ac}(T_2)$, respectively. □

Theorem 11.3. *Assume that T_1 and T_2 are self-adjoint operators on the complex Hilbert space H, $M_1 \subset_{T_1} D(\Omega_+(T_2, T_1))$, P_1 is the orthogonal projection onto M_1, $M_2 = R(W_+(T_2, T_1, P_1))$, and P_2 is the orthogonal projection onto M_2. Then with the notation $W_+ = W_+(T_2, T_1, P_1)$ we have, as $t \to \infty$, that*

$$e^{-itT_2}W_+ - e^{-itT_1}P_1 \xrightarrow{s} 0, \tag{11.1}$$

$$e^{itT_1}e^{-itT_2}W_+ \xrightarrow{s} P_1, \tag{11.2}$$

$$e^{itT_1}e^{-itT_2}P_2 \xrightarrow{s} W_+^*, \tag{11.3}$$

$$(W_+ - I)\,e^{-itT_1}P_1 \xrightarrow{s} 0, \tag{11.4}$$

$$(W_+^* - I)\,e^{-itT_1}P_1 \xrightarrow{s} 0, \tag{11.5}$$

$$e^{itT_1}W_+\,e^{-itT_1} \xrightarrow{s} P_1, \tag{11.6}$$

$$e^{itT_1}W_+^*\,e^{-itT_1} \xrightarrow{s} P_1, \tag{11.7}$$

$$(I - P_2)\,e^{-itT_1}P_1 \xrightarrow{s} 0. \tag{11.8}$$

We have $W_+(T_1, T_2, P_2) = W_+(T_2, T_1, P_1)^*$. Similar results hold for $W_- = W_-(T_2, T_1, P_1)$ as $t \to -\infty$.

Proof. (11.1) follows from the definition of W_+ by multiplication by e^{-itT_2}. (11.2) follows from (11.1) by multiplication by e^{itT_1}. If we multiply (11.2) by W_+^* from the right, then we obtain (11.3). Relation (11.4) follows from (11.1), because $e^{-itT_2}W_+ = W_+\,e^{-itT_1} = W_+P_1\,e^{-itT_1} = W_+\,e^{-itT_1}P_1$. If we multiply (11.4) by W_+^*, then (11.5) follows, because $W_+^*W_+\,e^{-itT_1}P_1 = P_1\,e^{-itT_1}P_1 = e^{-itT_1}P_1$. Relations (11.6) and (11.7) follow from (11.4) and (11.5), respectively, by multiplication by e^{itT_1}. Finally, (11.8) follows from the equalities

$$\|(I - P_2)\,e^{-itT_1}P_1 f\| = \|e^{itT_2}(I - P_2)\,e^{-itT_1}P_1 f\|$$
$$= \|(I - P_2)\,e^{itT_2}\,e^{-itT_1}P_1 f\| \to \|(I - P_2)W_+ f\| = 0.$$

The equlaity $W_+(T_1, T_2, P_2) = W_+(T_2, T_1, P_1)^*$ follows from (11.3). □

The following *chain rule* is useful in many investigations.

Theorem 11.4. *Assume that T_1, T_2, and T_3 are self-adjoint operators on the complex Hilbert space H, $M_1 \subset_{T_1} D(\Omega_+(T_2, T_1))$, $M_2 \subset_{T_2} D(\Omega_+(T_3, T_2))$, P_j is the orthogonal projection onto M_j $(j = 1, 2)$, and $R(W_+(T_2, T_1, P_1)) \subset M_2$.*

Then $M_1 \subset D(\Omega_+(T_3, T_1))$ *and*

$$W_+(T_3, T_1, P_1) = W_+(T_3, T_2, P_2)W_+(T_2, T_1, P_1).$$

A similar result holds for W_-.

PROOF. For $f \in M_1$

$$\begin{aligned} e^{itT_3}e^{-itT_1}f &= (e^{itT_3}e^{-itT_2})(e^{itT_2}e^{-itT_1})f \\ &= e^{itT_3}e^{-itT_2}\Omega_+(T_2, T_1)f \\ &\quad + e^{itT_3}e^{-itT_2}\left[e^{itT_2}e^{-itT_1}f - \Omega_+(T_2, T_1)f\right] \\ &\to \Omega_+(T_3, T_2)\Omega_+(T_2, T_1)f \quad \text{as} \quad t \to \infty, \end{aligned}$$

since $\Omega_+(T_2, T_1)f = W_+(T_2, T_1, P_1)f \in M_2$ and

$$e^{itT_2}e^{-itT_1}f - \Omega_+(T_2, T_1)f \to 0.$$

Consequently, $M_1 \subset D(\Omega_+(T_3, T_1))$ and

$$\Omega_+(T_3, T_1)f = \Omega_+(T_3, T_2)\Omega_+(T_2, T_1)f \quad \text{for} \quad f \in M_1.$$

Because of the inclusion $\Omega_+(T_2, T_1)M_1 \subset M_2$ we have

$$\begin{aligned} W_+(T_3, T_1, P_1) &= \Omega_+(T_3, T_1)P_1 = (\Omega_+(T_3, T_2)\Omega_+(T_2, T_1)P_1 \\ &= \Omega_+(T_3, T_2)P_2\Omega_+(T_2, T_1)P_1 \\ &= W_+(T_3, T_2, P_2)W_+(T_2, T_1, P_1). \end{aligned}$$

□

Theorem 11.5. *Assume that* T_1 *and* T_2 *are self-adjoint operators on the complex Hilbert space* H, M_1 *and* M_2 *are closed subspaces,* P_1 *and* P_2 *are the orthogonal projections onto* M_1 *and* M_2, *respectively. If* $M_1 \subset_{T_1} D(\Omega_+(T_2, T_1))$, $M_2 \subset_{T_2} D(\Omega_+(T_1, T_2))$, $R(W_+(T_2, T_1, P_1)) \subset M_2$ *and* $R(W_+(T_1, T_2, P_2)) \subset M_1$, *then*

$$W_+(T_1, T_2, P_2) = W_+(T_2, T_1, P_1)^*$$

and

$$R(W_+(T_2, T_1, P_1)) = M_2, \quad R(W_+(T_1, T_2, P_2)) = M_1.$$

PROOF. We obviously have $W_+(T_1, T_1, P_1) = P_1$ and $W_+(T_2, T_2, P_2) = P_2$. It therefore follows from Theorem 11.4 that

$$\begin{aligned} W_+(T_1, T_2, P_2)W_+(T_2, T_1, P_1) &= P_1, \\ W_+(T_2, T_1, P_1)W_+(T_1, T_2, P_2) &= P_2. \end{aligned}$$

Because of the formulae $R(W_+(T_1, T_2, P_2)) \subset M_1$ and $P_1 = W_+(T_2, T_1, P_1)^* W_+(T_2, T_1, P_1)$ it follows that

$$\begin{aligned} W_+(T_1, T_2, P_2) &= P_1W_+(T_1, T_2, P_2) \\ &= W_+(T_2, T_1, P_1)^* W_+(T_2, T_1, P_1)W_+(T_1, T_2, P_2) \\ &= W_+(T_2, T_1, P_1)^* P_2. \end{aligned}$$

Because $W_+(T_2, T_1, P_1)^*$ is a partial isometry with initial domain $R(W_+(T_2, T_1, P_1)) \subset M_2$, we have

$$W_+(T_1, T_2, P_2) = W_+(T_2, T_1, P_1)^* P_2 = W_+(T_2, T_1, P_1)^*.$$

The equality $R(W_+(T_2, T_1, P_2)) = M_2$ follows from the fact that $W_+(T_2, T_1, P_1) = W_+(T_1, T_2, P_2)^*$ is a partial isometry with final domain M_2. The equality $R(W_+(T_1, T_2, P_2)) = M_1$ follows from $W_+(T_1, T_2, P_2) = W_+(T_2, T_1, P_1)^*$ correspondingly. □

On the basis of the remark after the end of Theorem 11.1, an eigenelement f of T_1 belongs to $D(\Omega_+(T_2, T_1))$ only if f is also an eigenelement of T_2 for the same eigenvalue. Consequently, it is unrealistic to expect that an eigenelement belongs to $D(\Omega_+(T_2, T_1))$. We shall therefore always assume in the sequel that $M_1 \subset H_c(T_1)$. If $M_1 \subset H_c(T_1) \cap D(\Omega_+(T_2, T_1))$ and P_1 is the orthogonal projection onto M_1, then $R(W_+(T_2, T_1, P_1)) \subset H_c(T_2)$ by Theorem 11.2. Actually, for technical reasons we shall consider only subspaces M_1 of $H_{ac}(T_1)$. Particularly, we obtain from Theorems 11.2 and 11.5:

Theorem 11.6. *Let T_1 and T_2 be self-adjoint operators on the complex Hilbert space H, and let $P_{j,ac}$ be the orthogonal projections onto $H_{ac}(T_j)$ for $j = 1, 2$. If $H_{ac}(T_1) \subset D(\Omega_+(T_2, T_1))$ and $H_{ac}(T_2) \subset D(\Omega_+(T_1, T_2))$, then*

$$R(W_+(T_2, T_1, P_{1,ac})) = H_{ac}(T_2), \quad R(W_+(T_1, T_2, P_{2,ac})) = H_{ac}(T_1)$$

and

$$W_+(T_1, T_2, P_{2,ac}) = W_+(T_2, T_1, P_{1,ac})^*;$$

the absolutely continuous parts of T_1 and T_2 are unitarily equivalent. The corresponding results hold for W_-. (In the assertions of this theorem $P_{j,ac}$ and $H_{ac}(T_j)$ can be replaced by $P_{j,c}$ and $H_c(T_j)$.)

Corollary. *Assume that $W_+(T_2, T_1, P_{1,ac})$ exists. Then $R(W_+(T_2, T_1, P_{1,ac})) = R(P_{2,ac})$ if and only if $W_+(T_1, T_2, P_{2,ac})$ also exists.*

11.2 The existence and completeness of wave operators

Useful existence results for $W_\pm(T_2, T_1, P)$ are known only in the case where $P = P_{1,ac}$ (or $P \leqslant P_{1,ac}$). Consequently, we shall consider only this case in what follows. If, for example, T_1 is a (non-trivial) differential operator on $L_2(\mathbb{R}^m)$ with constant coefficients, then $P_{1,ac} = I$. Therefore in many cases there is no loss of generality in assuming that $P \leqslant P_{1,ac}$.

The wave operator $W_+(T_2, T_1, P_{1,ac})$ is said to be *complete* if $R(W_+(T_2, T_1, P_{1,ac})) = H_{ac}(T_2)$. By the Corollary to Theorem 11.6 this is the case if and only if $W_+(T_1, T_2, P_{2,ac})$ also exists; this wave operator is then also complete, and the absolutely continuous parts $T_{1,ac}$ and $T_{2,ac}$ are unitarily equivalent. A similar result holds for $W_-(T_2, T_1, P_{1,ac})$ and $W_-(T_1, T_2, P_{2,ac})$. If $R(W_-(T_2, T_1, P_{1,ac})) = R(W_+(T_2, T_1, P_{1,ac}))$, then the *scattering operator* $S(T_2, T_1) = W_+(T_2, T_1, P_{1,ac})^* W_-(T_2, T_1, P_{1,ac})$ is obviously a unitary operator on $H_{ac}(T_1)$; this holds in particular if $W_+(T_2, T_1, P_{1,ac})$ *and* $W_-(T_2, T_1, P_{1,ac})$ are complete.

The purpose of this section is to prove a few abstract criteria for the existence and completeness of $W_\pm(T_2, T_1, P_{1,ac})$. These will then be applied to differential operators in the next section. The reader can find further references, for example, in *T. Kato* [45].

Assertions similar to those occuring in the following theorem are known in the literature as *Cook's lemma.*

Theorem 11.7. *Let T_1 and T_2 be self-adjoint operators on the complex Hilbert space H. If $e^{-itT_1}f \in D(T_1) \cap D(T_2)$ for all $t \in \mathbb{R}$, the function*

$$\mathbb{R} \to H,\ t \mapsto (T_2 - T_1)\, e^{-itT_1}f$$

is continuous, and

$$\int_{-\infty}^{\infty} \|(T_2 - T_1)\, e^{-itT_1}f\|\, dt < \infty,$$

then $f \in D(\Omega_\pm(T_2, T_1))$.

Proof. Since $e^{-itT_1}f \in D(T_2) \cap D(T_1)$, the function $\Omega(t)f = e^{itT_2}\, e^{-itT_1}f$ is differentiable for all $t \in \mathbb{R}$, and its derivative is

$$\frac{d}{dt}(\Omega(t)f) = i\, e^{itT_2}(T_2 - T_1)\, e^{-itT_1}f.$$

This derivative is continuous by assumption. Therefore,

$$\Omega(t)f - \Omega(s)f = i\int_s^t e^{iuT_2}(T_2 - T_1)e^{-iuT_1}f\, du,$$

$$\|\Omega(t)f - \Omega(s)f\| \leq \int_s^t \|(T_2 - T_1)\, e^{-iuT_1}f\|\, du.$$

Since the integral over $(-\infty, \infty)$ is bounded, the limits $\Omega_\pm(T_2, T_1)f = \lim_{t\to\pm\infty}\Omega(t)f$ exist. □

Theorem 11.8. *Let T_1 and T_2 be self-adjoint operators on the complex Hilbert space H, and let E be the spectral family of T_1. If $f \in H_{ac}(T_1)$, $H_f = L\{E(t)f : t \in \mathbb{R}\}$, P_f is the orthogonal projection onto H_f, $H_f \subset D(T_1) \cap D(T_2)$, and $(T_2 - T_1)P_f \in B_1(H)$, then $H_f \subset D(\Omega_\pm(T_2, T_1))$.*

Proof. The proof is in three steps.

First step: Without loss of generality we can assume that $H_f = L_2(J)$ with J a bounded interval and that the restriction $T_{1,f}$ of T_1 to the reducing subspace $H_f = L_2(J)$ is equal to the operator of multiplication by the variable.

Proof of the first step. The subspace H_f reduces T_1, because it obviously reduces $E(t)$ for all $t \in \mathbb{R}$ (cf. Theorem 7.28). The operator $T_{1,f}$ is self-adjoint on H_f and is defined on the whole of H_f. Therefore, $T_{1,f}$ is bounded (Theorem 5.6). Hence there exists an interval (t_1, t_2) such that $\|E(t)f\| = 0$ for $t < t_1$ and $\|E(t)f\| = \|f\|$ for $t \geqslant t_2$.

If we set $\rho_f(t) = \|E(t)f\|^2$ for $t \in \mathbb{R}$, then there exists (cf. the proof of Theorem 7.16) a unitary mapping $V_f : H_f \to L_2(\mathbb{R}, \rho_f)$ for which $V_f T_{1,f} V_f^{-1}$ is the multiplication operator on $L_2(\mathbb{R}, \rho_f)$:

$$\left(V_f T_{1,f} V_f^{-1} g\right)(x) = x g(x) \quad \text{for} \quad g \in L_2(\mathbb{R}, \rho_f).$$

Because $f \in H_{ac}(T_1)$, the function ρ_f is absolutely continuous, i.e., there exists a $\sigma \in L_1(\mathbb{R})$ such that

$$\rho_f(t) = \int_{-\infty}^{t} \sigma(s)\, ds = \int_{t_1}^{t} \sigma(s)\, ds \quad \text{for} \quad t \in \mathbb{R}.$$

Denote $S_f = \{t \in \mathbb{R} : \sigma(t) \neq 0\}$. (Since σ is uniquely determined only up to a Lebesgue null set, S_f is uniquely determined only up to a null set; the reader can verify that this plays no role in what follows.) The set S_f is measurable and can obviously be chosen to be bounded. The formula

$$W_f : L_2(\mathbb{R}, \rho_f) \to L_2(S_f), \quad g \mapsto \sigma^{1/2} g$$

defines a unitary mapping. With the notation $U_f = W_f V_f$ the operator $U_f T_{1,f} U_f^{-1}$ is the multiplication operator on $L_2(S_f)$. Let us extend the operator U_f to the operator

$$U : H \to H_f^{\perp} \oplus L_2(S_f), \quad g \mapsto \left((I - P_f)g, U_f P_f g\right).$$

U is obviously unitary; U maps H_f onto $L_2(S_f)$ and transforms $T_{1,f}$ to the multiplication operator on $L_2(S_f)$. The subspace $H_f^{\perp}$ and the restriction of T_1 to $H_f^{\perp}$ remain invariant under U. Consequently, we may assume without loss of generality that $H_f = L_2(S)$ with some bounded measureable subset S of $\mathbb{R}$ and that $T_{1,f}$ is the multiplication operator on $L_2(S)$.

Let J be a bounded open interval with $S \subset J$, and denote $\check{S} = J \setminus S$ and $\hat{H} = H \oplus L_2(\check{S}) = H_f^{\perp} \oplus L_2(J)$. The operators T_j on H $(j = 1, 2)$ will be extended to self-adjoint operators $\hat{T}_j$ on $\hat{H}$ if we define

$$D(\hat{T}_j) = D(T_j) \oplus L_2(\check{S}),$$
$$\hat{T}_j(g + h) = T_j g + Mh \quad \text{for} \quad g \in D(T_j), h \in L_2(\check{S}),$$

where M denotes the multiplication operator on $L_2(\hat{S})$. The restriction of $\hat{T}_1$ to $L_2(J)$ is then equal to the multiplication operator on $L_2(J)$. We therefore have $L_2(J) \subset H_{ac}(\hat{T}_1)$. If we choose an $\hat{f} \in L_2(J)$ such that $\hat{f}(x) = 1$ for all $x \in J$, then $\hat{f} \in H_{ac}(\hat{T}_1)$ and

$$H_{\hat{f}} := \overline{L\{F(t)\hat{f} : t \in \mathbb{R}\}} = L_2(J),$$

where F denotes the spectral family of $\hat{T}_1$. Since $\hat{T}_2 - \hat{T}_1$ vanishes on $L_2(\hat{S})$, the operator $(\hat{T}_2 - \hat{T}_1)P_{\hat{f}}$ belongs to $B_1(\hat{H})$, where $P_{\hat{f}}$ denotes the orthogonal projection onto $H_{\hat{f}}$. If we prove that $H_{\hat{f}} \subset D(\Omega_{\pm}(\hat{T}_2, \hat{T}_1))$, then it follows that $H_f \subset D(\Omega_{\pm}(\hat{T}_2, \hat{T}_1))$. Since $e^{itT_2}e^{-itT_1}g = e^{it\hat{T}_2}e^{-it\hat{T}_1}g$ for $g \in H_f$, we also have then that $H_f \subset D(\Omega_{\pm}(T_2, T_1))$.

Second step: Let $V = T_2 - T_1$. The assertion of the theorem holds if VP_f has the form

$$VP_f g = \sum_{j=1}^{n} c_j \langle g_j, g \rangle h_j \quad \text{for all} \quad g \in H,$$

where $g_j \in C_0^\infty(J)$, $h_j \in H$ and $\|g_j\| = \|h_j\| = 1$. Then for all $g \in C_0^\infty(J)$ (with $\Omega(t) = e^{itT_2}e^{-itT_1}$)

$$\|\Omega(t)g - \Omega(s)g\|^2 \leqslant 8\pi \|g\|_\infty \sum_{j=1}^{n} |c_j| \left[\left\{ \int_t^\infty |F(g_j^* g)(u)|^2 \, du \right\}^{1/2} + \left\{ \int_s^\infty |F(g_j^* g)(u)|^2 \, du \right\}^{1/2} \right] \tag{11.9}$$

(here F is the Fourier transformation on $L_2(\mathbb{R})$ and $g_j^* g$ is taken to be equal to zero outside J). A similar estimate holds for the integrals over $(-\infty, t)$ and $(-\infty, s)$.

Proof of the second step. $e^{-itT_1}g \in H_f \subset D(T_2)$ for all $g \in H_f = L_2(J)$ and $t \in \mathbb{R}$. The function

$$\mathbb{R} \to H, \; t \mapsto (T_2 - T_1)e^{-itT_1}g = VP_f e^{-itT_1}g$$

is therefore continuous for all $g \in H_f$. Moreover, for $g \in H_f = L_2(J)$

$$\|VP_f e^{-itT_1}g\| = \left\| \sum_{j=1}^{n} c_j \int_J g_j^*(x) e^{-itx} g(x) \, dx h_j \right\|$$

$$\leqslant \sum_{j=1}^{n} |c_j| \, \|h_j\| (2\pi)^{1/2} |F(g_j^* g)(t)|.$$

If $g \in C_0^\infty(J)$, then the (extended) function $g_j^* g$ belongs to $C_0^\infty(\mathbb{R})$, and thus $F(g_j^* g) \in S(\mathbb{R})$. Hence, in this case

$$\int_{-\infty}^{\infty} \| VP_f e^{-itT_1} g \| \, dt < \infty \quad \text{for all} \quad g \in C_0^\infty(J).$$

Therefore, $C_0^\infty(J) \subset D(\Omega_\pm(T_2, T_1))$ by Theorem 11.7. Since the subspaces $D(\Omega_\pm(T_2, T_1))$ are closed by Theorem 11.1, it follows that $H_f = L_2(J) \subset D(\Omega_\pm(T_2, T_1))$.

With $\Omega_+ = \Omega_+(T_2, T_1)$ we have for $g \in C_0^\infty(J)$ that

$$\begin{aligned}
\|\Omega_+ g - \Omega(t)g\|^2 &= \|\Omega_+ g\|^2 - 2\,\mathrm{Re}\langle \Omega_+ g, \Omega(t)g\rangle + \|\Omega(t)g\|^2 \\
&= 2\|\Omega_+ g\|^2 - 2\,\mathrm{Re}\langle \Omega_+ g, \Omega(t)g\rangle \\
&= 2\,\mathrm{Re}\langle \Omega_+ g, \Omega_+ g - \Omega(t)g\rangle \\
&= 2\,\mathrm{Re}\, i\int_t^\infty \langle \Omega_+ g, e^{isT_2} VP_f\, e^{-isT_1} g\rangle \, ds \\
&= -2\,\mathrm{Im}\int_t^\infty \langle \Omega_+ e^{-isT_1} g, VP_f e^{-isT_1} g\rangle \, ds \quad \text{(cf. Theorem 11.1)} \\
&= -2\,\mathrm{Im}\sum_{j=1}^n c_j \int_t^\infty \langle \Omega_+ e^{-isT_1} g, h_j\rangle \langle g_j, e^{-isT_1} g\rangle \, ds \\
&\leqslant 2\sum_{j=1}^n |c_j| \left\{ \int_t^\infty |\langle \Omega_+ e^{-isT_1} g, h_j\rangle|^2 \, ds \int_t^\infty |\langle g_j, e^{-isT_1} g\rangle|^2 \, ds \right\}^{1/2}.
\end{aligned}$$

Since $e^{-isT_1} g \in H_f$, we have

$$\begin{aligned}
\int_t^\infty |\langle \Omega_+ e^{-isT_1} g, h_j\rangle|^2 \, ds &= \int_t^\infty |\langle \Omega_+ P_f\, e^{-isT_1} g, h_j\rangle|^2 \, ds \\
&= \int_t^\infty |\langle e^{-isT_1} g, (\Omega_+ P_f)^* h_j\rangle|^2 \, ds \\
&= \int_t^\infty \left| \int e^{isx} g(x)^* [(\Omega_+ P_f)^* h_j](x) \, dx \right|^2 ds \\
&\leqslant 2\pi \| F^{-1}(g^*(\Omega_+ P_f)^* h_j)\|^2 = 2\pi \| g^*(\Omega_+ P_f)^* h_j\|^2 \\
&\leqslant 2\pi \| g\|_\infty^2 \|(\Omega_+ P_f)^* h_j\|^2 \leqslant 2\pi \| g\|_\infty^2
\end{aligned}$$

and

$$\int_t^\infty |\langle g_j, e^{-isT_1} g\rangle|^2 \, ds = 2\pi \int_t^\infty |F(g_j^* g)(s)|^2 \, ds.$$

Consequently, we obtain that

$$\|\Omega_+ g - \Omega(t)g\|^2 \leqslant 4\pi \| g\|_\infty \sum_{j=1}^n |c_j| \left\{ \int_t^\infty |F(g_j^* g)(s)|^2 \, ds \right\}^{1/2},$$

and hence the assertion follows from the inequalities

$$\begin{aligned}
\|\Omega(t)g - \Omega(s)g\|^2 &\leqslant (\|\Omega(t)g - \Omega_+ g\| + \|\Omega_+ g - \Omega(s)g\|)^2 \\
&\leqslant 2(\|\Omega_+ g - \Omega(t)g\|^2 + \|\Omega_+ g - \Omega(s)g\|^2).
\end{aligned}$$

Third step: The assertion of the theorem holds in general.

Proof of the third step. Let the operator V_0 be defined by the equality

$$V_0 = (VP_f)^* + VP_f - P_fVP_f.$$

V_0 is obviously self-adjoint and belongs to $B_1(H)$. For all $g \in H_f$ and $h \in H$

$$\begin{aligned}\langle h, V_0 g\rangle &= \langle h, (VP_f)^* g + Vg - P_f Vg\rangle \\ &= \langle VP_f h - VP_f h, g\rangle + \langle h, Vg\rangle = \langle h, Vg\rangle,\end{aligned}$$

i.e., $V_0 P_f = VP_f$.

The operator $V_0 \in B_1(H)$ has the form

$$V_0 g = \sum_{j=1}^{\infty} c_j \langle g_j, g\rangle g_j \quad \text{for} \quad g \in H,$$

where $g_j \in H$, $\|g_j\| = 1$ and $\sum_{j=1}^{\infty}|c_j| < \infty$. For all $j \in \mathbb{N}$ we choose sequences $(g_{j,n})_{n\in\mathbb{N}}$ from H with the properties $P_f g_{j,n} \in C_0^\infty(J)$, $\|g_{j,n}\| \leqslant 1$ and $g_{j,n} \to g_j$ as $n \to \infty$. With

$$V_n g = \sum_{j=1}^{n} c_j \langle g_{j,n}, g\rangle g_{j,n} \quad \text{for} \quad g \in H,\ n \in \mathbb{N}$$

it follows that $\|V_n - V_0\| \to 0$, since for every $N \in \mathbb{N}$ and $n \geqslant N$

$$\|V_n - V_0\| \leqslant 2\sum_{j=1}^{N} |c_j|\, \|g_{j,n} - g_j\| + 2\sum_{j=N+1}^{\infty} |c_j|.$$

If we now set $T_{2,n} = T_2 + (V_n - V_0)$, then by Theorem 9.16 and Theorem 9.18.

$$e^{itT_{2,n}} \xrightarrow{s} e^{itT_2} \quad \text{as} \quad n \to \infty,\ t \in \mathbb{R}.$$

If we set $\Omega_n(t) = e^{itT_{2,n}}e^{-itT_1}$, then

$$\Omega_n(t) \xrightarrow{s} \Omega(t) \quad \text{as} \quad n \to \infty,\ t \in \mathbb{R}.$$

The second step can be applied to $T_{2,n}$ in place of T_2, since

$$(T_{2,n} - T_1)Pf = (V + V_n - V_0)P_f = V_n P_f,$$

and $V_n P_f$ has the form required in the second step:

$$V_n P_f g = \sum_{j=1}^{n} c_j \langle P_f g_{j,n}, g\rangle g_{j,n} \quad \text{for} \quad g \in H.$$

Therefore, (11.9) holds for all $g \in C_0^\infty(J)$ with $\Omega_n(\,.\,)$ in place of $\Omega(\,.\,)$ and with $P_f g_{j,n}$ in place of g_j. We obtain from this for $n \to \infty$ that

$$\begin{aligned}\|\Omega(t)g - \Omega(s)g\|^2 \leqslant 8\pi\|g\|_\infty \sum_{j=1}^{\infty} |c_j| \Bigg[&\left\{ \int_t^\infty |F((P_f g_j)^* g)(u)|^2\, du \right\}^{1/2} \\ &+ \left\{ \int_s^\infty |F((P_f g_j)^* g)(u)|^2\, du \right\}^{1/2} \Bigg]; \qquad (11.10)\end{aligned}$$

for the proof we notice that $F((P_f g_{j,n})^* g)$ converges to $F((P_f g_j)^* g)$ in $L_2(\mathbb{R})$ as $n \to \infty$. Since the right side of (11.10) tends to zero as $t, s \to \infty$, we have $C_0^\infty(J) \subset D(\Omega_+(T_2, T_1))$, and thus $H_f = L_2(J) \subset D(\Omega_+(T_2, T_1))$. We can prove similarly that $H_f \subset D(\Omega_-(T_2, T_1))$.

This completes the proof of Theorem 11.8. □

Theorem 11.9. *Let T_1 and T_2 be self-adjoint operators on the complex Hilbert space H, and let E be the spectral family of T_1.*

(a) *If J is a bounded interval, $R(E(J)P_{1,ac}) \subset D(T_2)$, and $(T_2 - T_1)E(J)P_{1,ac} \in B_1(H)$, then the wave operators $W_\pm(T_2, T_1, E(J)P_{1,ac})$ exist.*

(b) *If the assumption of* (a) *is satisfied for every bounded interval J, then the wave operators $W_\pm(T_2, T_1, P_{1,ac})$ exist.*

PROOF.

(a) If $f \in R(E(J)P_{1,ac})$ then $H_f \subset R(E(J)P_{1,ac})$, and thus $(T_2 - T_1)P_f \in B_1(H)$. By Theorem 11.8 $H_f \subset D(\Omega_\pm(T_2, T_1))$; therefore, $f \in D(\Omega_\pm(T_2, T_1))$. As this holds for all $f \in R(E(J)P_{1,ac})$, we have $R(E(J)P_{1,ac}) \subset D(\Omega_\pm(T_2, T_1))$, i.e., $W_\pm(T_2, T_1, E(J)P_{1,ac})$ exist.

(b) $R(E(J)P_{1,ac}) \subset D(\Omega_\pm(T_2, T_1))$ for every bounded interval J, by part (a). Since the linear hull of these spaces (for all bounded intervals J) is dense in $R(P_{1,ac})$, the assertion follows. □

Theorem 11.10. *Let T_1 and T_2 be self-adjoint operators on the complex Hilbert space H. The wave operators $W_\pm(T_2, T_1, P_{1,ac})$ and $W_\pm(T_1, T_2, P_{2,ac})$ exist and are complete provided that $R(E_1(J)) \subset D(T_2)$, $R(E_2(J)) \subset D(T_1)$ and $(T_2 - T_1)E_j(J) \in B_1(H)$ for $j = 1, 2$ and for every bounded interval J. This holds in particular if one of the following assumptions is satisfied*:

(i) *There is an operator $V \in B_1(H)$ such that $T_2 = T_1 + V$* (Kato-Rosenblum).

(ii) *$D(T_1) = D(T_2)$ and $(T_2 - T_1)(z - T_1)^{-2} \in B_1(H)$ for some $z \in \rho(T_1)$.*

(iii) *$D(T_1) = D(T_2)$ and $(T_2 - T_1)(z - T_2)^{-2} \in B_1(H)$ for some $z \in \rho(T_2)$.*

(iv) *$D(T_1) = D(T_2)$, and there exists an $n \in \mathbb{N}$ such that $(T_2 - T_1)(z - T_1)^{-n} \in B_1(H)$ for some $z \in \rho(T_1)$ and $(T_2 - T_1)(z - T_2)^{-n} \in B_1(H)$ for some $z \in \rho(T_2)$.*

(v) *There exists an $n \in \mathbb{N}$ such that $D(T_2) \supset D(T_1^n) \supset R(E_2(J))$ for every bounded interval J and $(T_2 - T_1)(z - T_1)^{-n} \in B_1(H)$ for some $z \in \rho(T_1)$.*

PROOF. If $(T_2 - T_1)E_j(J) \in B_1(H)$ for $j = 1, 2$ and for every bounded interval J, then the wave operators $W_\pm(T_2, T_1, P_{1,ac})$ and $W_\pm(T_1, T_2, P_{2,ac})$ exist by Theorem 11.9(b). These wave operators are complete by the Corollary to Theorem 11.6.

(i) obviously implies (iv) for all $n \in \mathbb{N}$. The assumptions (ii) and (iii) are equivalent according to (9.7) (observe that the assumptions of (9.7) are fulfilled. (iv) follows from (ii) and (iii). In order to prove that (i), (ii), (iii), or (iv) implies the assertion, it is sufficient to prove that (iv) implies the assertion: For every bounded interval J

$$(T_2-T_1)E_1(J)=(T_2-T_1)(z-T_1)^{-n}(z-T_1)^nE_1(J)\in B_1(H).$$

The formula $(T_2-T_1)E_2(J)\in B_1(H)$ follows similarly.

It remains to prove that (v) implies the assertion: For this, we prove $(T_2-T_1)E_1(J)\in B_1(H)$ as above. Since $R(E_2(J))\subset D((z-T_1)^n)$, the operator $(z-T_1)^nE_2(J)$ is bounded, and thus

$$(T_2-T_1)E_2(J)=(T_2-T_1)(z-T_1)^{-n}(z-T_1)^nE_2(J)\in B_1(H). \qquad \square$$

In the case $D(T_1)=D(T_2)$ Theorem 11.10 can be essentially sharpened.

Theorem 11.11. *Let T_1 and T_2 be self-adjoint operators on the complex Hilbert space H. Assume that $D(T_1)=D(T_2)$ and $(T_2-T_1)E_1(J)\in B_1(H)$ for every bounded interval J. Then the wave operators $W_\pm(T_2,T_1)$ exist and are complete.*

I am indebted to Dr. R. Colgen for the proof given here. The case where $D(T_1^{1/2})=D(T_2^{1/2})$ can be also treated without many changes. First we prove the following auxiliary theorem.

Auxiliary theorem 11.12. *Let T_1 and T_2 be self-adjoint operators on the complex Hilbert space H. Assume that $D(T_1)=D(T_2)$.*

(a) *$\|(I-E_1((-n,n)))E_2(J)\|\to 0$ as $n\to\infty$ for every bounded interval J.*

(b) *If the limits $s-\lim_{t\to\pm\infty}E_1((-n,n))e^{itT_1}e^{-itT_2}E_2(J)P_{2,ac}$ exist for all $n\in\mathbb{N}$, then the limits*

$$s-\lim_{t\to\pm\infty} e^{itT_1}e^{-itT_2}E_2(J)P_{2,ac}$$

also exist.

PROOF.

(a) It is obvious that $(I-E_1((-n,n)))E_2(J)=(I-E_1((-n,n)))(T_1+i)^{-1}\times(T_1+i)(T_2+i)^{-1}(T_2+i)E_2(J)$ for all $n\in\mathbb{N}$. Since $(T_2+i)E_2(J)$ and $(T_1+i)(T_2+i)^{-1}$ are bounded and since $\|(I-E_1((-n,n)))(T_1+i)^{-1}\|\leq(1/n)$, the assertion follows.

(b) This assertion immediately follows from

$$\begin{aligned}\|e^{itT_1}e^{-itT_2}E_2(J)P_{2,ac}&-E_1((-n,n))e^{itT_1}e^{-itT_2}E_2(J)P_{2,ac}\|\\&=\|e^{itT_1}(I-E_1((-n,n)))E_2(J)e^{-itT_2}P_{2,ac}\|\\&\leq\|(I-E_1((-n,n)))E_2(J)\|\to 0\quad\text{as}\quad n\to\infty.\end{aligned}$$

$\square$

PROOF OF THEOREM 11.11. The existence of $W_{\pm}(T_2, T_1)$ follows from Theorem 11.9(b), because $R(E_1(J)) \subset D(T_1) = D(T_2)$ for every bounded interval J. It therefore remains to prove the existence of

$$W_{\pm}(T_1, T_2, E_2(J)P_{2,ac}) = s - \lim_{t\to\pm\infty} e^{itT_1} e^{-itT_2} E_2(J) P_{2,ac}$$

for every bounded interval J.

Let J be a bounded interval. As $(T_2 - T_1)E_1(J) \in B_1(H)$, the operator

$$V_0 := (T_2 - T_1)E_1(J) + ((T_2 - T_1)E_1(J))^* - E_1(J)(T_2 - T_1)E_1(J)$$

is a self-adjoint element of $B_1(H)$ and

$$V_0 E_1(J) = (T_2 - T_1)E_1(J),$$
$$E_1(J)V_0 = (V_0 E_1(J))^* \supset E_1(J)T_2 - T_1 E_1(J).$$

Define $T_3 = T_2 - V_0$. Since $R(E_3(J)) \subset D(T_3) = D(T_2)$, we then have

$$E_1(J)V_0 E_3(J) = (E_1(J)T_2 - T_1 E_1(J))E_3(J),$$

and thus

$$T_1 E_1(J)E_3(J) - E_1(J)T_3 E_3(J) = (T_1 E_1(J) - E_1(J)T_2 + E_1(J)V_0)E_3(J) = 0.$$

If we define $A : \mathbb{R} \to B(H)$ by the equality

$$A(t) = E_1(J)e^{itT_1} e^{-itT_3} E_3(J) P_{3,ac},$$

then

$$\frac{d}{dt}\langle u, A(t)v\rangle = i\langle u, e^{itT_1}(T_1 E_1(J)E_3(J) - E_1(J)T_3 E_3(J))\, e^{-itT_3} P_{3,ac} v\rangle = 0$$

for all $u, v \in H$. A is therefore constant, i.e., $A(t) = A(0) = E_1(J)E_3(J)P_{3,ac}$ for all $t \in \mathbb{R}$.

Since $T_3 = T_2 - V_0$ with $V_0 \in B_1(H)$, the operators

$$W_{\pm}(T_3, T_2, E_2(J)P_{2,ac}) = s - \lim e^{itT_3} e^{-itT_2} E_2(J) P_{2,ac}$$

exist by Theorem 11.9(a), and

$$W_{\pm}(T_3, T_2, E_2(J)P_{2,ac}) = E_3(J)P_{3,ac} W_{\pm}(T_3, T_2, E_2(J)P_{2,ac})$$

by Theorem 11.2. This implies

$$s - \lim_{t\to\pm\infty} (I - E_3(J)P_{3,ac})\, e^{itT_3} e^{-itT_2} E_2(J) P_{2,ac} = 0,$$

and thus the existence of

$$\begin{aligned} s - \lim_{t\to\pm\infty} & E_1(J)\, e^{itT_1} e^{-itT_2} E_2(J) P_{2,ac} \\ &= s - \lim_{t\to\pm\infty} E_1(J)\, e^{itT_1} e^{-itT_3} e^{itT_3} e^{-itT_2} E_2(J) P_{2,ac} \\ &= s - \lim_{t\to\pm\infty} E_1(J)\, e^{itT_1} e^{-itT_3} E_3(J) P_{3,ac}\, e^{itT_3} e^{-itT_2} E_2(J) P_{2,ac} \\ &= s - \lim_{t\to\pm\infty} A(0)\, e^{itT_3} e^{-itT_2} E_2(J) P_{2,ac}. \end{aligned}$$

As this holds for all bounded intervals J (in particular, for $J=(-n, n)$, as well), the existence of

$$s-\lim_{t\to\pm\infty} E_1((-n,n))\, e^{itT_1}\, e^{-itT_2}E_2(J)P_{2,ac}$$

also follows for all $n\in\mathbb{N}$. Consequently, the assertion of the theorem is obtained by Auxiliary theorem 11.12(b). □

In what follows we would like to prove a version of the so-called *invariance principle* for wave operators (cf. also *T. Kato* [22], X.4).

Theorem 11.13 (Invariance principle). *Let the assumptions of Theorem* 11.8 *be satisfied (let H_f and P_f be defined as there). Assume that the function $\vartheta:\mathbb{R}\to\mathbb{R}$ is twice continuously differentiable and $\vartheta'(x)>0$ (respectively $\vartheta'(x)<0$) for all $x\in\mathbb{R}$. Then $H_f\subset D(\Omega_\pm(\vartheta(T_2),\vartheta(T_1)))$, and $\Omega_\pm(\vartheta(T_2),\vartheta(T_1))g=\Omega_\pm(T_2,T_1)g$ (respectively $\Omega_\pm(\vartheta(T_2),\vartheta(T_1))g=\Omega_\mp(T_2,T_1)g$) for all $g\in H_f$.*

For the proof we need an auxiliary theorem.

Auxiliary theorem 11.14. *Let $\vartheta:\mathbb{R}\to\mathbb{R}$ be twice continuously differentiable. Assume that $\vartheta'(x)>0$ (respectively $\vartheta'(x)<0$) for all $x\in\mathbb{R}$. Then for all $g\in L_2(\mathbb{R})$*

$$\|g\|^2\geqslant\int_0^\infty|F(e^{-is\vartheta(\cdot)}g)(u)|^2\,du\to 0$$

as $s\to\infty$ (respectively $s\to-\infty$). (A similar result holds for $\int_{-\infty}^0$ if we exchange "$s\to\infty$" with "$s\to-\infty$".)

Proof. The inequality is clear since $F:L_2(\mathbb{R})\to L_2(\mathbb{R})$ is unitary and $\|e^{-is\vartheta(\cdot)}g\|=\|g\|$ for all $s\in\mathbb{R}$. It is therefore enough to prove the convergence relation for all g from a total subset of $L_2(\mathbb{R})$, for example, for characteristic functions of bounded intervals. We consider the case $\vartheta'>0$. Then for $g=\chi_{[a,b)}$, $s>0$, and $u>0$

$$\begin{aligned}F(e^{-is\vartheta(\cdot)}g)(u)&=(2\pi)^{-1/2}\int_a^b e^{-iux}\,e^{-is\vartheta(x)}\,dx\\&=i(2\pi)^{-1/2}\int_a^b(u+s\vartheta'(x))^{-1}\frac{\partial}{\partial x}e^{-iux-is\vartheta(x)}\,dx\\&=i(2\pi)^{-1/2}\left\{\left[\frac{e^{-iux-is\vartheta(x)}}{u+s\vartheta'(x)}\right]_a^b\right.\\&\qquad\left.+\int_a^b\frac{s\vartheta''(x)}{(u+s\vartheta'(x))^2}e^{-iux-is\vartheta(x)}\,dx\right\}.\end{aligned}$$

There exist $C_1>0$ and $C_2\geqslant 0$ such that $\vartheta'(x)\geqslant C_1$ and $|\vartheta''(x)|\leqslant C_2$ for all

$x\in[a, b]$. Consequently, for all $u>0$ and $s>0$

$$|F(e^{-is\vartheta(\cdot)}g)(u)| \leq C_3(u+s)^{-1},$$

and thus

$$\int_0^\infty |F(e^{-is\vartheta(\cdot)}g)(u)|^2\,du \leq C_3^2 s^{-1} \quad \text{for all} \quad s>0.$$

The case $\vartheta'<0$ can be treated in a similar way if we consider $u>0$ and $s<0$. □

PROOF OF THEOREM 11.13. As in the proof of Theorem 11.8, we can assume without loss of generality that $H_f = L_2(J)$ with some bounded interval J and that the restriction of T_1 to $L_2(J)$ is equal to the operator of multiplication by the variable.

Let $\vartheta'>0$. We consider Ω_+: First we remark that (11.10) obviously holds for $g\in C(J)$, as well. If we replace g by $e^{-is\vartheta(T_1)}g = e^{-is\vartheta(\cdot)}g(\,.\,)$, then we obtain for $s=0$ and $t\to\infty$ that

$$\begin{aligned}\|\Omega_+(T_2, T_1)\, e^{-is\vartheta(T_1)}g - e^{-is\vartheta(T_1)}g\|^2 \\ \leq 8\pi\|g\|_\infty \sum_{j=1}^\infty |c_j|\Big\{\int_0^\infty |F((P_f g_j)^* e^{-is\vartheta(\cdot)}g)(u)|^2\,du\Big\}^{1/2}.\end{aligned}$$

By Auxiliary theorem 11.14, the expressions in the braces are bounded by $\|g\|_\infty^2\|g_j\|^2 = \|g\|_\infty^2$ and tend to 0 as $s\to\infty$. Consequently,

$$\Omega_+(T_2, T_1)\, e^{-is\vartheta(T_1)}g - e^{-is\vartheta(T_1)}g \to 0 \quad \text{as} \quad s\to\infty$$

for all $g\in C_0^\infty(J)$. Then this holds for all $g\in L_2(J)$, as well. Hence, for all $g\in L_2(J)$

$$e^{-is\vartheta(T_2)}\Omega_+(T_2, T_1)g - e^{-is\vartheta(T_1)}g \to 0 \quad \text{as} \quad s\to\infty$$

(cf. Theorem 11.1). Multiplication by $e^{is\vartheta(T_2)}$ gives assertion in this case. The other cases can be handled similarly. □

Theorem 11.15. *Let T_1 and T_2 be self-adjoint operators on the complex Hilbert space H, and let E denote the spectral family of T_1. Assume that $R(E(J)P_{1,ac})\subset D(T_2)$ and $(T_2-T_1)E(J)P_{1,ac}\in B_1(H)$ for every bounded interval J. Assume, furthermore, that the function $\vartheta: \mathbb{R}\to\mathbb{R}$ is twice continuously differentiable and $\vartheta'(x)>0$ (or $\vartheta'(x)<0$) for all $x\in\mathbb{R}$. Then $H_{ac}(\vartheta(T_1)) = H_{ac}(T_1)$, the wave operators $W_\pm(\vartheta(T_2), \vartheta(T_1), P_{1,ac})$ exist, and $W_\pm(\vartheta(T_2), \vartheta(T_1), P_{1,ac}) = W_\pm(T_2, T_1, P_{1,ac})$ (or $W_\pm(\vartheta(T_2), \vartheta(T_1), P_{1,ac}) = W_\mp(T_2, T_1, P_{1,ac})$, respectively).*

PROOF. A subset N of $\mathbb{R}$ is a null set if and only if $\vartheta(N)$ is a null set. If F is the spectral family of $\vartheta(T_1)$, then $F(S) = E(\vartheta^{-1}(S))$ for every Borel set S (cf. Section 7.3, Proposition 6). It follows that $H_{ac}(\vartheta(T_1)) = H_{ac}(T_1)$. The rest of the assertion follows from Theorem 11.13 (cf. the proof of Theorem 11.9). □

Theorem 11.16. *Let T_1 and T_2 be self-adjoint operators on the complex Hilbert space H. Assume that $T_1 \geqslant \gamma$, $T_2 \geqslant \gamma$, and $(T_1-\lambda)^{-p}-(T_2-\lambda)^{-p} \in B_1(H)$ for some $\lambda \in \mathbb{R}$ such that $\lambda < \gamma$ and some $p \in \mathbb{N}$. Then the wave operators $W_{\pm}(T_2, T_1, P_{1,ac})$ exist and are complete.*

PROOF. Let $\vartheta : \mathbb{R} \to \mathbb{R}$ be a twice continuously differentiable function such that $\vartheta(t)=\lambda+t^{-1/p}$ for $t \geqslant (\gamma-\lambda)^{-p}$ and $\vartheta'(t)<0$ for all $t \in \mathbb{R}$ (hence, $\vartheta((T_j-\lambda)^{-p})=T_j$ for $j=1, 2$). The wave operators

$$W_{\pm}(T_2, T_1, P_{1,ac}) = W_{\mp}((T_2-\lambda)^{-p}, (T_1-\lambda)^{-p}, P_{1,ac})$$

and

$$W_{\pm}(T_1, T_2, P_{2,ac}) = W_{\mp}((T_1-\lambda)^{-p}, (T_2-\lambda)^{-p}, P_{2,ac})$$

therefore exist according to Theorems 11.10 and 11.15. The assertion thus follows. □

11.3 Applications to differential operators on $L_2(\mathbb{R}^m)$

If T_1 is a self-adjoint differential operator on $L_2(\mathbb{R}^m)$ with constant coefficients, then in many cases the existence of $W_{\pm}(T_2, T_1)$ can be proved without using Theorem 11.8 or any of its consequences. We only have to apply Theorem 11.7 and appropriate estimates of $\exp(-\mathrm{i}\, tT_1)f$ that can be proved with the aid of the Fourier transformation. Here we prove a special case of a result of *K. Veselić* and *J. Weidmann* [53], [54] (compare also with *L. Hörmander* [43]).

Auxiliary theorem 11.17. *Let $A \subset \mathbb{R}^m$ be a closed set. Assume that the function $h : \mathbb{R}^m \to \mathbb{R}$ is infinitely many times continuously differentiable in $\mathbb{R}^m \setminus A$ and* grad $h(x) \neq 0$ *for $x \in \mathbb{R}^m \setminus A$. Then for every function $g \in C_0^\infty(\mathbb{R}^m \setminus A)$ and for every $\rho \geqslant 0$ there is a constant C such that*

$$\left|\int e^{\mathrm{i}\, xy}\, e^{-\mathrm{i}\, th(y)} g(y)\, dy\right| \leqslant C|t|^{-\rho}(1+|x|)^{\rho}$$

for all $x \in \mathbb{R}^m$ and $t \in \mathbb{R} \setminus \{0\}$ (here g is considered as a function defined on $\mathbb{R}^m$ with $g(x)=0$ for $x \in A$).

PROOF. For every $y \in \mathbb{R}^m \setminus A$ there is a $j \in \{1, 2, \ldots, m\}$ for which $\partial_j h(y) = (\partial/\partial y_j)h(y) \neq 0$. Since $\partial_j h$ is continuous, there is a neighborhood U_y of y in $\mathbb{R}^m \setminus A$ such that $\partial_j h(z) \neq 0$ for all $z \in U_y$. The compact set supp g can be covered by finitely many such neighborhoods $U_{y_1}, \ldots, U_{y_n}$. Then there are functions $\vartheta_j \in C_0^\infty(\mathbb{R}^m)$ such that supp $\vartheta_j \subset U_{y_j}$ and $\sum_{j=1}^n \vartheta_j(y)=1$ for all $y \in$ supp g (partition of unity; cf. *W. Rudin* [33], Theorem 6.20). It is obviously sufficient to prove the assertion for the functions $\vartheta_j g$ in place of g, i.e., we can assume that a $j\in\{1, 2, \ldots, m\}$ exists for which $\partial_j h(y)\neq 0$ for all $y\in\operatorname{supp} g$. (In many concrete cases there is a $j\in\{1, 2, \ldots, m\}$ such that $\partial_j h(y)\neq 0$ for all $y\in\mathbb{R}^m\setminus A$; then the foregoing step of the proof is superfluous.)

Hence, let $\partial_j h(y)\neq 0$ for all $y\in\operatorname{supp} g$. Then by k-fold partial integration with respect to y_j we obtain for all $k\in\mathbb{N}$ and $t\neq 0$ that

$$\begin{aligned}\int e^{i\,xy-i\,th(y)}g(y)\,dy &= \int e^{-i\,th(y)}\partial_j h(y)\frac{e^{i\,xy}g(y)}{\partial_j h(y)}\,dy\\ &= (i\,t)^{-1}\int e^{-i\,th(y)}\partial_j\left[\frac{e^{ixy}g(y)}{\partial_j h(y)}\right]dy\\ &= \cdots = (i\,t)^{-k}\int e^{-i\,th(y)}\partial_j\left[\frac{1}{\partial_j h(y)}\partial_j\left[\cdots\partial_j\left[\frac{e^{i\,xy}g(y)}{\partial_j h(y)}\right]\cdots\right]\right]dy\\ &= (i\,t)^{-k}\int e^{i\,xy-i\,th(y)}\chi(x_j, y)\,dy,\end{aligned}$$

where

$$\chi(x_j, y) = \sum_{p=0}^{k}\vartheta_p(y)x_j^p$$

and the supports of the functions $\vartheta_p\in C_0^\infty(\mathbb{R}^m\setminus A)$ are contained in supp g. Therefore,

$$\int e^{i\,xy-i\,th(y)}g(y)\,dy = (i\,t)^{-k}\sum_{p=0}^{k}x_j^p\int e^{i\,xy-i\,th(y)}\vartheta_p(y)\,dy.$$

We consequently obtain for every $k\in\mathbb{N}$ that

$$\left|\int e^{i\,xy-i\,th(y)}g(y)\,dy\right|\leq C_k|t|^{-k}(1+|x|)^k.$$

This estimate evidently holds for $k=0$, as well.

Now let $k\in\mathbb{N}_0$ be such that $k\leq\rho<k+1$. Then

$$\begin{aligned}|t|^{-\rho}(1+|x|)^\rho &\geq |t|^{-k}(1+|x|)^k &&\text{for}\quad |t|^{-1}(1+|x|)\geq 1,\\ |t|^{-\rho}(1+|x|)^\rho &> |t|^{-(k+1)}(1+|x|)^{k+1} &&\text{for}\quad |t|^{-1}(1+|x|)<1.\end{aligned}$$

If we apply the above result with k and $k-1$, then we obtain the assertion. □

In what follows, for a real-valued measurable function h defined on $\mathbb{R}^m$ let M_h denote the maximal operator of multiplication by h on $L_2(\mathbb{R}^m)$, and let $T_h = F^{-1}M_hF$. If h is a polynomial, then T_h is a differential operator with constant coefficients.

Theorem 11.18. *Let us assume that with a closed set $A \subset \mathbb{R}^m$ of measure zero we have*

$$h \in C^\infty(\mathbb{R}^m \setminus A) \quad \text{and} \quad \operatorname{grad} h(x) \neq 0 \quad \text{for} \quad x \notin A.$$

Let V be a symmetric operator on $L_2(\mathbb{R}^m)$ such that $S(\mathbb{R}^m) \subset D(V)$, and assume that there exist a $p \in \mathbb{N}_0$, a $\Theta > 1$, and a $C \geq 0$ with the property that for all $r \geq 0$

$$\|Vf\| \leq C(1+r)^{-\Theta}\|f\|_p{}^{1} \text{ for all } f \in S(\mathbb{R}^m) \text{ such that } f(x) = 0 \text{ for } |x| < r.$$

Then for every self-adjoint extension T of $T_h + V$ (provided that any exist) the wave operators $W_\pm(T, T_h)$ exist.

In the following we shall not prove this theorem but a somewhat more general one that also considers operators on $L_2(\mathbb{R}^m)^M$ (for example, Dirac operators). For this let H be a measurable function defined on $\mathbb{R}^m$ whose values are $M \times M$ Hermitian matrices (such a function is said to be *measurable* if the entries of the matrix are measurable functions). Let M_H again be the maximal "operator of multiplication" by H on $L_2(\mathbb{R}^m)^M$ and let $T_H = F^{-1}M_H F$. The operator T_H is evidently self-adjoint. We denote by $h_1(x), \ldots, h_M(x)$ the M (not necessarily different) eigenvalues of $H(x)$ and by $e_1(x), \ldots, e_M(x)$ the corresponding normalized eigenelements. (There is a great deal of freedom [especially if multiple eigenvalues occur] in the choice of these functions; in what follows it will be possible to choose them in such a way that the functions h_j and e_j are infinitely many times differentiable.) The operator T_H is a differential operator if all entries of the matrix function H are polynomials; however, the functions h_j and e_j are in general then not polynomials (cf., for example, the Dirac operator, Section 10.6).

A function H defined on $\mathbb{R}^m$ whose values are Hermitian $M \times M$ matrices is said to be *permissible* if the functions h_j and e_j can be chosen such that there exists a closed set $A \subset \mathbb{R}^m$ of measure zero for which

$$h_j \in C^\infty(\mathbb{R}^m \setminus A),\ e_j \in C^\infty(\mathbb{R}^m \setminus A)^M,$$
$$\operatorname{grad} h_j(x) \neq 0 \text{ for } x \notin A \ (j = 1, \ldots, M).$$

Theorem 11.19. *Let H be a permissible function. Let V be a symmetric operator on $L_2(\mathbb{R}^m)^M$ with the properties: $S(\mathbb{R}^m)^M \subset D(V)$, and there exists a $p \in \mathbb{N}_0$, a $\Theta > 1$, and a $C \geq 0$ such that for all $r \geq 0$*

$$\|Vf\| \leq C(1+r)^{-\Theta}\|f\|_p{}^{2} \text{ for all } f \in S(\mathbb{R}^m)^M \text{ such that } f(x) = 0 \text{ for } |x| < r.$$

[1] $\|\cdot\|_p$ is the norm of $W_{2,p}(\mathbb{R}^m)$.
[2] Here $\|\cdot\|_p$ denotes the norm in $W_{2,p}(\mathbb{R}^m)^M$.

Then the wave operators $W_\pm(T, T_H)$ exist for every self-adjoint extension T of $T_H + V$.

Proof. We prove that if $Ff \in C_0^\infty(\mathbb{R}^m \setminus A)^M$, then f satisfies the assumptions of Theorem 11.7. Since $C_0^\infty(\mathbb{R}^m \setminus A)^M$ is dense in $L_2(\mathbb{R}^m)^M = L_2(\mathbb{R}^m \setminus A)^M$, this will prove the assertion.

Let $Ff \in C_0^\infty(\mathbb{R}^m \setminus A)^M$. Then, in particular, $f \in S(\mathbb{R}^m)^M$. The assumption on V (for $r = 0$) implies that

$$\begin{aligned}\|V(e^{-isT_H} - e^{-itT_H})f\| &\leq C\|(e^{-isT_H} - e^{-itT_H})f\|_p \\ &= C\|(e^{-isH(.)} - e^{-itH(.)})Ff\|_{(p)} \\ &= C\left\{\int |e^{isH(x)} - e^{-itH(x)}|^2 |(Ff)(x)|^2 (1+|x|^2)^p \, dx\right\}^{1/2} \\ &\to 0 \quad \text{as} \quad s \to t.\end{aligned}$$

The function $t \mapsto Ve^{-itT_H}f$ is therefore continuous on $\mathbb{R}$.

Now let $\vartheta \in C^\infty(\mathbb{R})$ be such that $0 \leq \vartheta(s) \leq 1$, $\vartheta(s) = 1$ for $s \leq 0$ and $\vartheta(s) = 0$ for $s \geq 1$. With some $\mu \in (1/\Theta, 1)$ let $\vartheta_t \in C_0^\infty(\mathbb{R}^m)$ be defined by the equality $\vartheta_t(x) = \vartheta(|x| - |t|^\mu)$ for all $x \in \mathbb{R}^m$ and $t \in \mathbb{R}$.

The assumption on V (for $r = 0$) implies (cf. Theorem 10.8(c)) that

$$\begin{aligned}\|V\vartheta_t e^{-itT_H}f\| &\leq C_1 \max_{|\alpha| \leq p} \left\{\int_{|x| \leq |t|^\mu + 1} |(D^\alpha e^{-itT_H}f)(x)|^2 \, dx\right\}^{1/2} \\ &= C_1 \max_{|\alpha| \leq p} \left\{\int_{|x| \leq |t|^\mu + 1} (2\pi)^{-m/2} \left|\int e^{ixy} e^{-itH(y)} y^\alpha (Ff)(y) \, dy\right|^2 dx\right\}^{1/2}.\end{aligned}$$

The operator induced on $\mathbb{C}^M$ by the matrix $e^{-itH(y)}$ can be written in the form

$$e^{-itH(y)}\xi = \sum_{j=1}^M e^{-ith_j(y)}(e_j(y), \xi)e_j(y) \quad \text{for} \quad \xi \in \mathbb{C}^M,$$

where $(.\,,.)$ is the scalar product in $\mathbb{C}^M$. Consequently,

$$\int e^{ixy} e^{-itH(y)} y^\alpha (Ff)(y) \, dy = \sum_{j=1}^M \int e^{ixy} e^{-ith_j(y)} (e_j(y), y^\alpha (Ff)(y)) e_j(y) \, dy.$$

Since $Ff \in C_0^\infty(\mathbb{R}^m \setminus A)^M$, we also have $y^\alpha Ff \in C_0^\infty(\mathbb{R}^m \setminus A)^M$. Therefore, the functions $y \mapsto (e_j(y), y^\alpha (Ff)(y))e_j(y)$ also belong to $C_0^\infty(\mathbb{R}^m \setminus A)^M$. Hence we can apply Auxiliary theorem 11.17 to the last integral. We thus obtain for all $\rho > 0$ and $|t| \geq 1$ that

$$\begin{aligned}\|V\vartheta_t e^{-itT_H}f\| &\leq C_2 \left\{\int_{|x| < |t|^\mu + 1} (1+|x|)^{2\rho} |t|^{-2\rho} \, dx\right\}^{1/2} \\ &\leq C_3 |t|^{\rho(\mu - 1) + (m\mu/2)}.\end{aligned}$$

We obtain similarly from the assumption on V (for $r=|t|^{\mu}$) that

$$\|V(1-\vartheta_t)e^{-itT_H}f\| \leqslant C_4(1+|t|^{\mu})^{-\Theta} \max_{|\alpha|<p} \|e^{-itT_H}D^{\alpha}f\|$$
$$\leqslant C_4 \max_{\alpha<p} \|D^{\alpha}f\|\, |t|^{-\mu\Theta} = C_5|t|^{-\mu\Theta}.$$

If we choose ρ so large that $\rho(\mu-1)+(m\mu/2) \leqslant -\mu\Theta < -1$, then

$$\|V e^{-itT_H}f\| \leqslant \|V\vartheta_t e^{-itT_H}f\| + \|V(1-\vartheta_t)e^{-itT_H}f\| \leqslant C_6|t|^{-\mu\Theta}$$

for $|t|\geqslant 1$. Since $t\mapsto Ve^{-itT_H}f$ is continuous, the integrability of $t\mapsto \|V e^{-itT_H}f\|$ follows from this. □

Theorem 11.20.
(a) *Let T_1 be the self-adjoint operator defined by the formulae $D(T_1) = W_{2,2}(\mathbb{R}^m)$, $T_1 f = -\Delta f$ for $f\in D(T_1)$. Let V be defined by the equality*

$$Vf(x) = \sum_{|\alpha|\leqslant k} c_{\alpha}(x)D^{\alpha} f(x) \quad \textit{for} \quad f\in S(\mathbb{R}^m),$$

and assume that $|c_{\alpha}(x)| \leqslant C(1+|x|)^{-1-\epsilon}$ for some $C\geqslant 0$ and some $\epsilon>0$. Then the wave operators $W_{\pm}(T_2, T_1)$ exist for every self-adjoint extension T_2 of T_1+V.
(b) *A corresponding result holds if T_1 is the free Dirac operator on $L_2(\mathbb{R}^3)^4$ and the $c_{\alpha}(x)$ are 4×4 matrices.*

PROOF.
(a) We have $T_1=T_h$ with $h(x)=|x|^2$. Since grad $h(x)\neq 0$ for $x\neq 0$, the assumption of Theorem 11.18 is obviously satisfied with $p=k$.
(b) The functions h_j and e_j are known from Section 10.6 and satisfy the assumptions of Theorem 11.19 with $A=\{0\}$. Theorem 11.19 is therefore applicable if we choose $p=k$. □

The assumptions of Theorem 11.20 on the coefficients c_{α} can be essentially weakened (cf. Exercise 11.1).

In Theorems 11.18 to 11.20 we showed the existence of $W_{\pm}(T_2, T_1)$ in many cases where T_1 is a differential operator with constant coefficients. Since in these cases $H_{ac}(T_1)=H$ (cf. Exercise 10.7; this is clear for $-\Delta$ and the unperturbed Dirac operator), this is equivalent to the existence of $W_{\pm}(T_2, T_1, P_{1,ac})$. We cannot expect that the completeness (i.e., the equality $R(W_{\pm}(T_2, T_1)) = H_{ac}(T_2)$) can be proved under such general assumptions. A few simple assertions can be proved with the aid of the results of Section 11.2. In order to be able to apply them we need the following auxiliary theorems.

Auxiliary theorem 11.21. *If $r, s\geqslant 0$, $r-s>m/2$ and $\varphi\in C_0^{\infty}(\mathbb{R}^m)$, then the operator*

$$\Phi : W_{2,r}(\mathbb{R}^m) \to W_{2,s}(\mathbb{R}^m), \quad f\mapsto \varphi f$$

belongs to $B_2(W_{2,r}(\mathbb{R}^m), W_{2,s}(\mathbb{R}^m))$. *If* $r-s>m$, *then* $\Phi \in B_1(W_{w,r}(\mathbb{R}^m), W_{2,s}(\mathbb{R}^m))$.

PROOF. We recall the proof of Theorem 10.19. Φ is a Hilbert-Schmidt operator if and only if the integral operator $\dot{K}$ on $L_2(\mathbb{R}^m)$ defined there is a Hilbert-Schmidt operator. The kernel of K is given by the formula

$$k(x,y)=(1+|x|^2)^{s/2}\psi(x-y)(1+|y|^2)^{-r/2} \quad \text{for} \quad x,y\in\mathbb{R}^m$$

with $\psi = F\varphi \in S(\mathbb{R}^m)$. Because of the inequality $|\psi(x)| \leq C(1+|x|^2)^{-l}$ we have

$$\begin{aligned}|k(x,y)| &\leq C(1+|x-y|^2)^{-l}(1+|x|^2)^{s/2}(1+|y|^2)^{-r/2}\\ &\leq \begin{cases} C_1(1+|x|^2)^{-l+s/2}(1+|y|^2)^{-r/2} & \text{for} \quad |y|\leq \frac{1}{2}|x|,\\ C_1(1+|x-y|^2)^{-l}(1+|y|^2)^{(s/2)-(r/2)} & \text{for} \quad |y|>\frac{1}{2}|x|.\end{cases}\end{aligned}$$

If we choose $l>(m/4)+(s/2)$, then $k\in L_2(\mathbb{R}^m\times\mathbb{R}^m)$; the operator K is therefore a Hilbert-Schmidt operator.

Now assume that $r-s>m$, $s_1=(r+s)/2$ and $\varphi_1\in C_0^\infty(\mathbb{R}^m)$ with $\varphi_1(x)=1$ for $x\in\operatorname{supp}\varphi$. Then Φ can be considered the product of the mappings

$$\Phi' : W_{2,r}(\mathbb{R}^m)\to W_{2,s_1}(\mathbb{R}^m),\quad f\mapsto \varphi_1 f,$$

$$\Phi'' : W_{2,s_1}(\mathbb{R}^m)\to W_{2,s}(\mathbb{R}^m),\quad f\mapsto \varphi f.$$

As both of these operators are Hilbert-Schmidt, Φ belongs to $B_1(W_{2,r}(\mathbb{R}^m), W_{2,s}(\mathbb{R}^m))$ (cf. Theorem 7.9). □

Auxiliary theorem 11.22. *Let T be a self-adjoint operator on $L_2(\mathbb{R}^m)$. Assume that $D(T^n)\subset W_{2,t}(\mathbb{R}^m)$ for some $n\in\mathbb{N}$ and some $t>m$. Let V be a symmetric operator on $L_2(\mathbb{R}^m)$ such that $D(V)\supset W_{2,s}(\mathbb{R}^m)$ for some $s\in[0, t-m)$. Assume that there exist a $C\geq 0$ and a $\Theta>m$ such that for all $r\geq 0$*

$$\|Vf\| \leq C(1+r)^{-\Theta}\|f\|_s$$

for all $f\in W_{2,s}(\mathbb{R}^m)$ such that $f(x)=0$ for $|x|<r$. Then

$$V(\mathrm{i}-T)^{-n}\in B_1(L_2(\mathbb{R}^m)).$$

(An analogous result holds in $L_2(\mathbb{R}^m)^M$.)

PROOF. Let $\delta=\delta_1\in C_0^\infty(\mathbb{R}^m)$ be defined as in Section 2.2, Example 8 (i.e., $0\leq\delta(x)$, $\delta(x)=0$ for $|x|\geq 1$ and $\int\delta(x)\,\mathrm{d}x=1$). For all m-tuples $\gamma=(\gamma_1,\ldots,\gamma_m)\in\mathbb{Z}^m$ let

$$Q_\gamma=\{x\in\mathbb{R}^m : \gamma_j\leq x_j<\gamma_j+1,\quad j=1,2,\ldots,m\},$$

$$\varphi_\gamma(x)=\int\delta(x-y)\chi_{Q_\gamma}(y)\,\mathrm{d}y.$$

Then we obviously have

$$\sum_{\gamma \in \mathbb{Z}^m} \varphi_\gamma(x) = 1 \quad \text{for all} \quad x \in \mathbb{R}^m,$$

where the sum is finite for every $x \in \mathbb{R}^m$ (with at most 3^m summands). By Auxiliary theorem 11.21 the operator

$$\Phi_\gamma : W_{2,t}(\mathbb{R}^m) \to W_{2,s}(\mathbb{R}^m), \quad f \mapsto \varphi_\gamma f$$

belongs to $B_1(W_{2,t}(\mathbb{R}^m), W_{2,s}(\mathbb{R}^m))$ for every $\gamma \in \mathbb{Z}^m$, and the trace norm $\|\Phi_\gamma\|_{B_1(W_{2,t}, W_{2,s})}$ does not depend on γ (we have $\Phi_\gamma = \tau_\gamma \Phi_0 \tau_{-\gamma}$, where τ_γ denotes the operator of translation by $\gamma : (\tau_\gamma f)(x) = f(x - \gamma)$).

The operator $(\mathrm{i} - T)^{-n}$ is a continuous mapping from $L_2(\mathbb{R}^m)$ into $W_{2,t}(\mathbb{R}^m)$: As in the proof of Theorem 10.18 we can prove that $\|\,.\,\|_{T^n} \geqslant C_1\|\,.\,\|_t$ for some $C_1 > 0$, and thus

$$\|(\mathrm{i} - T)^{-n} f\|_t^2 \leqslant C_1^{-2}\{\|T^n(\mathrm{i} - T)^{-n} f\|^2 + \|(\mathrm{i} - T)^{-n} f\|^2\} \leqslant C_2\|f\|^2.$$

Now let $\tilde{Q}_\gamma = \{x \in \mathbb{R}^m : \gamma_j - 1 \leqslant x_j < \gamma_j + 2 \ \text{ for } j = 1, 2, \ldots, m\}$, $\tilde{\varphi}_\gamma = \int \delta(x - y)\chi_{\tilde{Q}_\gamma}(y)\,\mathrm{d}y$, and

$$\tilde{\Phi}_\gamma : W_{2,s}(\mathbb{R}^m) \to W_{2,s}(\mathbb{R}^m), \quad f \mapsto \tilde{\varphi}_\gamma f.$$

Then $\tilde{\Phi}_\gamma \Phi_\gamma = \Phi_\gamma$, and $V\tilde{\Phi}_\gamma$, as an operator from $W_{2,s}(\mathbb{R}^m)$ into $L_2(\mathbb{R}^m)$, is, by assumption, bounded by

$$\|V\tilde{\Phi}_\gamma\|_{B(W_{2,s}, L_2)} \leqslant C_3(1 + |\gamma|)^{-\theta}.$$

Observe now that $(B_1, \|\,.\,\|_1)$ is a Banach space by Exercise 7.10 and that by Theorem 7.8(c) $\|AB\|_1 \leqslant \|A\|_1\|B\|$ and $\|CA\|_1 \leqslant \|C\|\,\|A\|_1$ for bounded B and C. It therefore follows that

$$\begin{aligned}\|V(\mathrm{i} - T)^{-n}\|_1 &= \Big\|\sum_{y \in \mathbb{Z}^m} V\tilde{\Phi}_\gamma \Phi_\gamma (\mathrm{i} - T)^{-n}\Big\|_1 \leqslant \sum_{y \in \mathbb{Z}^m} \|V\tilde{\Phi}_\gamma \Phi_\gamma (\mathrm{i} - T)^{-n}\|_1 \\ &\leqslant \sum_{y \in \mathbb{Z}^m} \|V\tilde{\Phi}_\gamma\|_{B(W_{2,s}, L_2)} \|\Phi_\gamma\|_{B_1(W_{2,t}, W_{2,s})} \|(\mathrm{i} - T)^{-n}\|_{B(L_2, W_{2,t})} \\ &\leqslant C_4 \sum_{y \in \mathbb{Z}^m} (1 + |\gamma|)^{-\theta} \leqslant C_5 \int_{\mathbb{R}^m} (1 + |\xi|)^{-\theta}\,\mathrm{d}\xi < \infty.\end{aligned}$$

□

From Auxiliary theorem 11.22 we can immediately derive criteria for the existence and completeness of wave operators with the aid of Theorem 11.10. (These existence statements are weaker than those contained in Theorem 11.20). Here we only give a typical result. Actually, for $m \geqslant 2$ much better results can be proved using entirely different methods. We shall not consider them here (compare with, for example, *S.T. Kuroda* [47, 48] and *M. Schechter* [49, 50] for further references).

Theorem 11.23. *Let T_1 be equal to $-\Delta$, with $D(T_1) = W_{2,2}(\mathbb{R}^m)$. Let V be symmetric such that $D(V) \supset D(T_1)$, and let $T_2 = T_1 + V$ be self-adjoint. Assume that there exist an $s \geq 0$, a $\Theta \geq m$ and a $C \geq 0$ such that for all $r \geq 0$*

$$\|Vf\| \leq C(1 + r)^{-\Theta} \|f\|_s$$

for all $f \in W_{2,s}(\mathbb{R}^m) \cap D(V)$ such that $f(x) = 0$ for $|x| < r$. Then the wave operators $W_{\pm}(T_2, T_1)$ exist and are complete.

PROOF. Since $D(T_1)^n = W_{2,2n}(\mathbb{R}^m)$, by Auxiliary theorem 11.22 $V(i - T_1)^{-n} \in B_1(L_2(\mathbb{R}^m))$ for every $n \in \mathbb{N}$ such that $2n - s > m$, and thus

$$VE_1(J) \in B_1(L_2(\mathbb{R}^m)) \quad \text{for every bounded interval } J.$$

The assertion therefore follows from Theorem 11.11.   □

REMARK. The assumptions of Theorem 11.23 hold in particular if V is a differential operator of order ≤ 2 whose coefficients decrease as $|x|^{-\Theta}$ for some $\Theta > m$. An analogous result can be proved for Dirac operators.

EXERCISES

11.1. The assertion of Theorem 11.20 holds also if the functions c_α are locally square integrable and $N_{c_\alpha}(x) \leq C(1 + |x|)^{-1-\epsilon}$ for some $C \geq 0$ and some $\epsilon > 0$. Hint: Choose $p > k + (m/2)$ in the proof.

11.2. Assume $m \leq 3$, $T_1 = -\Delta$, $D(T_1) = W_{2,2}(\mathbb{R}^m)$ and $q \in L_2(\mathbb{R}^m) \cap L_1(\mathbb{R}^m)$. If V is the operator of multiplication by q, then V is T_1-bounded with T_1-bound zero; consequently, $T_2 = T_1 + V$ is self-adjoint and $D(T_2) = D(T_1)$. The wave operators $W_{\pm}(T_2, T_1)$ exist and are complete.
Hint: $(T_1 + s)^{-1} - (T_2 + s)^{-1} = (T_1 + s)^{-1}V(T_2 + s)^{-1} \in B_1(L_2(\mathbb{R}^m))$ for sufficiently large s, since

$$|V|^{1/2}(T_1 + s)^{-1} \text{ and } |V|^{1/2}(T_2 + s)^{-1} = |V|^{1/2}(T_1 + s)^{-1}(T_1 + s)(T_2 + s)^{-1}$$

are Hilbert-Schmidt operators (Theorem 11.16).

Appendix A

Lebesgue integration

In this appendix we shall compile and prove a few results of Lebesgue integration theory that are used in several places in this book. We essentially follow the presentation of *F. Riesz* and *B. Sz.-Nagy* [31]; but notice that only the measure induced by the volume function is studied in detail there. For complete presentations of the theory of measure and integration we refer the reader to, for example, *E. Hewitt* and *K. Stromberg* [18] or *W. Rudin* [32].

A.1 Definition of the integral

Let $\mathcal{J} = \mathcal{J}(\mathbb{R}^m)$[1] be the set of bounded *intervals* $J = J_1 \times J_2 \times \ldots \times J_m$ in $\mathbb{R}^m$, where the J_j are arbitrary open, half-open, closed, one-point or empty intervals in $\mathbb{R}$. Let $\tilde{\mathcal{J}} = \tilde{\mathcal{J}}(\mathbb{R}^m)$ be the set of finite unions of intervals from $\mathcal{J}$. It is obvious that every $M \in \tilde{\mathcal{J}}$ can be written as a union of finitely many mutually disjoint intervals from $\mathcal{J}$.

A mapping $\rho : \mathcal{J} \to \mathbb{R}$ is called a *function of an interval* or a *interval function* on $\mathbb{R}^m$ if we have:

(A1) *Monotonicity*: $J_1, J_2 \in \mathcal{J}$ and $J_1 \subset J_2$ imply $\rho(J_1) \leq \rho(J_2)$.

(A2) *Additivity*: $J_1, J_2 \in \mathcal{J}$, $J_1 \cap J_2 = \emptyset$ and $J_1 \cup J_2 \in \mathcal{J}$ imply $\rho(J_1 \cup J_2) = \rho(J_1) + \rho(J_2)$.

The mapping ρ can be extended to $\tilde{\mathcal{J}}$ by the formulae

$$\rho(M) = \sum_{j=1}^{n} \rho(J_j) \quad \text{for} \quad M = \bigcup_{j=1}^{n} J_j \quad \text{with} \quad J_j \cap J_k = \emptyset \quad \text{for} \quad j \neq k.$$

[1]In the following we omit $\mathbb{R}^m$ if no confusion is possible.

This definition is obviously independent of the choice of the representation $M = \bigcup_{j=1}^{n} J_j$. This extended mapping is also monotone and additive.

The monotonicity and additivity immediately imply that

$$\rho(M) \geqslant 0 \quad \text{for all} \quad M \in \mathfrak{J}; \quad \rho(\varnothing) = 0. \tag{A3}$$

An interval function ρ is said to be *regular* if we have:

(A4) For every $J \in \mathfrak{J}$ and every $\epsilon > 0$ there exists an open interval $\tilde{J}$ such that $J \subset \tilde{J}$ and $\rho(\tilde{J}) \leqslant \rho(J) + \epsilon$.

It is easy to see that ρ is regular if and only if we have:

(A5) For every $J \in \mathfrak{J}$ and every $\epsilon > 0$ there exists a closed interval $\hat{J}$ such that $\hat{J} \subset J$ and $\rho(\hat{J}) \geqslant \rho(J) - \epsilon$.

A regular interval function on $\mathbb{R}^m$ will be called a *measure* in the sequel.

EXAMPLE 1. The *volume function*

$$\lambda(J) = \prod_{j=1}^{m} (b_j - a_j) \quad \text{for} \quad J = \{x \in \mathbb{R}^m : a_j \lesseqgtr x_j \lesseqgtr b_j\}$$

is a measure. λ is called the *Lebesgue measure.*

EXAMPLE 2. If $f : \mathbb{R} \to \mathbb{R}$ is a right continuous non-decreasing function, then the formula

$$\rho_f(J) = \begin{cases} f(b) - f(a) & \text{for} \quad J = (a, b], \\ f(b) - f(a-) & \text{for} \quad J = [a, b], \\ f(b-) - f(a) & \text{for} \quad J = (a, b), \\ f(b-) - f(a-) & \text{for} \quad J = [a, b) \end{cases}$$

defines a measure.

EXAMPLE 3. If ρ_1 is a measure on $\mathbb{R}^p$ and ρ_2 is a measure on $\mathbb{R}^q$, then the equality

$$\rho(J_1 \times J_2) = \rho_1(J_1)\rho_2(J_2) \quad \text{for} \quad J_1 \in \mathfrak{J}(\mathbb{R}^p), \ J_2 \in \mathfrak{J}(\mathbb{R}^q)$$

defines a measure on $\mathbb{R}^{p+q}$.

In what follows let ρ always be a measure on $\mathbb{R}^m$. A set $N \subset \mathbb{R}^m$ is called a *ρ-null set* if for every $\epsilon > 0$ there exists a sequence (J_n) from $\mathfrak{J}$ such that

$$N \subset \bigcup_{n \in \mathbb{N}} J_n \quad \text{and} \quad \sum_{n \in \mathbb{N}} \rho(J_n) < \epsilon.$$

Since ρ is regular, these intervals can always be chosen to be open.

EXAMPLE 4. All finite and countable subsets of $\mathbb{R}^m$ are λ-null sets.

Example 5. Every subset of a ρ-null set is a ρ-null set.

Example 6. Countable unions of ρ-null sets are ρ-null sets: If the $N_k (k \in \mathbb{N})$ are ρ-null sets, then for every $\epsilon > 0$ there are sequences $(J_{k,n})_{n\in\mathbb{N}}$ from $\mathcal{J}$ for which $N_k \subset \bigcup_{n\in\mathbb{N}} J_{k,n}$ and $\Sigma_{n\in\mathbb{N}}\rho(J_{k,n}) < \epsilon 2^{-n}$. This implies that $\bigcup_{k\in\mathbb{N}} N_k \subset \bigcup_{n,k\in\mathbb{N}} J_{k,n}$ and $\Sigma_{n,k\in\mathbb{N}}\rho(J_{k,n}) < \epsilon$.

In what follows we say that a certain assertion holds *ρ-almost everywhere* (or for *ρ-almost all* x) if there exists a ρ-null set N such that the given assertion holds for all $x \in \mathbb{R}^m \setminus N$. In particular, we write $f \underset{\rho}{=} g$ or $f_n \underset{\rho}{\to} g$ if there exists a ρ-null set N such that $f(x) = g(x)$ or $f_n(x) \to g(x)$ for all $x \in \mathbb{R}^m \setminus N$, respectively.

A function $f: \mathbb{R}^m \to \mathbb{C}$ is called a *step function* if

$$f = \sum_{j=1}^{n} c_j \chi_{J_j} \quad \text{with} \quad J_j \in \mathcal{J} \quad \text{for} \quad j = 1, \ldots, n$$

(here χ_M denotes the characteristic function of M). Of course, the intervals J_j are not uniquely determined by f. However, we can choose the intervals J_j to be mutually disjoint. We denote the set of step functions defined on $\mathbb{R}^m$ by $T = T(\mathbb{R}^m)$. The set $T(\mathbb{R}^m)$ is a complex vector space.

For an $f \in T$ we define the *ρ-integral* by the equality

$$\int f \, d\rho = \int f(x) \, d\rho(x) = \sum_{j=1}^{n} c_j \rho(J_j) \quad \text{for} \quad f = \sum_{j=1}^{n} c_j \chi_{J_j}.$$

This definition is independent of the choice of the representation of f as a linear combination of characteristic functions.

The following properties of the ρ-integral of step functions are obvious:

(A6) If $f, g \in T$ are real-valued and $f \leqslant g$, then $\int f \, d\rho \leqslant \int g \, d\rho$.

(A7) $|\int f \, d\rho| \leqslant \int |f| \, d\rho$ for every $f \in T$.

(A8) The mapping $T \to \mathbb{C}$, $f \mapsto \int f \, d\rho$ is linear.

Concerning the extension of this notion of an integral to a wider class of functions we need a few preliminary remarks. First we consider only real-valued functions.

Auxiliary theorem A1. *If $M \in \tilde{\mathcal{J}}$ and (I_n) is a sequence from $\mathcal{J}$ such that $M \subset \bigcup_{n\in\mathbb{N}} I_n$, then $\rho(M) \leqslant \Sigma_{n\in\mathbb{N}}\rho(I_n)$.*

Proof. Let $M = \bigcup_{j=1}^{k} J_j$ with pairwise disjoint intervals J_j. Since ρ is regular, for every $\epsilon > 0$ there exist closed intervals $\hat{J}_j$ for which $\hat{J}_j \subset J_j$ and $\rho(\hat{J}_j) \geqslant \rho(J_j) - \epsilon/(2k)$ and open intervals $\tilde{I}_n$ for which $I_n \subset \tilde{I}_n$ and $\Sigma_{n\in\mathbb{N}}\rho(\tilde{I}_n) \leqslant \Sigma_{n\in\mathbb{N}}\rho(I_n) + \epsilon/2$. The set $\hat{M} = \bigcup_{j=1}^{k} \hat{J}_j$ is a compact subset of M and $\rho(\hat{M}) \geqslant \rho(M) - \epsilon/2$. Consequently, there is an $N \in \mathbb{N}$ such that $\hat{M} \subset$

$\cup_{n=1}^{N}\tilde{I}_n$, and thus

$$\rho(M) < \rho(\hat{M}) + \frac{\epsilon}{2} < \sum_{n=1}^{N} \rho(\tilde{I}_n) + \frac{\epsilon}{2} < \sum_{n\in\mathbb{N}} \rho(I_n) + \epsilon.$$

As this holds for all $\epsilon > 0$, the assertion follows. □

Auxiliary theorem A2. *If (f_n) is a non-increasing sequence from T and $f_n \geq 0$, $f_n \to 0$, then $\int_\rho f_n \, d\rho \to 0$.*

PROOF. Let N_0 be the ρ-null set on which the sequence (f_n) does not converge to zero, let $K \in \mathbb{R}$ be chosen such that we have $K \geq f_1$ (then we also have $K \geq f_n$ for all $n \in \mathbb{N}$), and let I_0 be an interval such that $f_1(x) = 0$ for $x \in \mathbb{R}^m \setminus I_0$ (then $f_n(x) = 0$ for $x \in \mathbb{R}^m \setminus I_0$ and all $n \in \mathbb{N}$).

For a given $\epsilon > 0$ let $\epsilon' = \epsilon(K + \rho(I_0))^{-1}$. Let $\mathfrak{S}_n$ be the set of mutually disjoint intervals where f_n assumes constant values not smaller than ϵ'. Then with the notation $M_n = \cup_{J\in\mathfrak{S}} J$ we have

$$M_1 \supset M_2 \supset \cdots \quad \text{and} \quad \bigcap_{n\in\mathbb{N}} M_n \subset N_0,$$

because of the monotonicity of the sequence (f_n). This implies that

$$M_k = \bigcup_{n=k}^{\infty} (M_n \setminus M_{n+1}) \cup \bigcap_{n\in\mathbb{N}} M_n \subset \bigcup_{n=k}^{\infty} (M_n \setminus M_{n+1}) \cup N_0. \tag{A9}$$

For every $n \in \mathbb{N}$ let $\mathfrak{T}_n$ be a finite set of mutually disjoint intervals for which $M_n \setminus M_{n+1} = \cup_{J\in\mathfrak{T}_n} J$. Since $M_1 \supset \cup_{n=1}^{k-1}(M_n \setminus M_{n+1})$ for $k \in \mathbb{N}$ and $(M_n \setminus M_{n+1}) \cap (M_m \setminus M_{m+1}) = \varnothing$ for $n \neq m$, we have

$$\rho(M_1) \geq \rho\left(\bigcup_{n=1}^{k-1} (M_n \setminus M_{n+1})\right) = \rho\left(\bigcup_{n=1}^{k-1} \bigcup_{J\in\mathfrak{T}_n} J\right) = \sum_{n=1}^{k-1} \sum_{J\in\mathfrak{T}_n} \rho(J),$$

and therefore

$$\sum_{n=1}^{\infty} \sum_{J\in\mathfrak{T}_n} \rho(J) \leq \rho(M_1) < \infty.$$

Consequently, there exists a $k_0 \in \mathbb{N}$ such that

$$\sum_{n=k}^{\infty} \sum_{J\in\mathfrak{T}_n} \rho(J) < \frac{\epsilon'}{2} \quad \text{for} \quad k \geq k_0.$$

Since the ρ-null set N_0 can be covered by countably many intervals of total measure $< \epsilon'/2$, by (A9) the set M_k can be covered by countably many intervals of total measure $< \epsilon'$. By Auxiliary theorem A1 we therefore have $\rho(M_k) < \epsilon'$ for $k \geq k_0$, and thus

$$\int f_k \, d\rho \leq K\rho(M_k) + \epsilon'\rho(I_0) \leq \epsilon'(K + \rho(I_0)) = \epsilon \quad \textit{for} \quad k \geq k_0. \quad □$$

Auxiliary theorem A3. *If (f_n) is a non-decreasing sequence from T and the sequence of integrals $(\int f_n\, d\rho)$ is bounded, then there exists a function $f: \mathbb{R}^m \to \mathbb{R}$ for which $f_n \underset{\rho}{\to} f$.*

PROOF. Without loss of generality we may assume that the functions f_n are non-negative (otherwise we would consider the sequence $(f_n - f_1)$). Let N_0 be the set of those $x \in \mathbb{R}^m$ for which the sequence $(f_n(x))$ diverges. We have to show that N_0 is a ρ-null set.

Let $\epsilon > 0$ be given. Let $C \geq 0$ be such that $\int f_n\, d\rho \leq C$ for all $n \in \mathbb{N}$. Let $\mathfrak{T}_n$ be the set of disjoint intervals on which f_n assumes constant values not smaller than C/ϵ, and define

$$N_n = \bigcup_{J \in \mathfrak{T}_n} J = \left\{ x \in \mathbb{R}^m : f_n(x) \geq \frac{C}{\epsilon} \right\} \quad \text{for} \quad n \in \mathbb{N}.$$

If $N = \{x \in \mathbb{R}^m : f_n(x) \geq C/\epsilon \text{ for some } n \in \mathbb{N}\}$, then $N_0 \subset N = \bigcup_{n \in \mathbb{N}} N_n$ and $N_n \subset N_{n+1}$. We can therefore choose a sequence (J_k) of disjoint intervals so that

$$N_n = \bigcup_{k=1}^{k(n)} J_k \quad \text{for all} \quad n \in \mathbb{N} \quad \text{and thus} \quad N = \bigcup_{k=1}^{\infty} J_k \supset N_0.$$

It therefore follows for all $n \in \mathbb{N}$ that

$$\sum_{k=1}^{k(n)} \rho(J_k) \frac{C}{\epsilon} = \sum_{J \in \mathfrak{T}_n} \rho(J) \frac{C}{\epsilon} \leq \int f_n\, d\rho \leq C,$$

and hence

$$\sum_{k=1}^{\infty} \rho(J_k) \leq \epsilon,$$

i.e., N_0 is a ρ-null set. □

Auxiliary theorem A4. *Let (f_n) and (g_n) be non-decreasing sequences from T such that $f_n \underset{\rho}{\to} f$, $g_n \underset{\rho}{\to} g$ and $f \underset{\rho}{\leq} g$. Assume that the sequences of integrals $(\int f_n\, d\rho)$ and $(\int g_n\, d\rho)$ are bounded. Then*

$$\lim_{n\to\infty} \int f_n\, d\rho \leq \lim_{n\to\infty} \int g_n\, d\rho.$$

PROOF. For every $m \in \mathbb{N}$ the sequence $(f_m - g_n)_{n \in \mathbb{N}}$ is non-increasing and

$$\rho - \lim_{n\to\infty} (f_m - g_n) \underset{\rho}{=} f_m - g \underset{\rho}{\leq} f - g \underset{\rho}{\leq} 0.$$

Consequently, for every $m \in \mathbb{N}$ the sequence $((f_m - g_n)_+)_{n \in \mathbb{N}}$ is non-increasing and ρ-converges to 0 (here $h_+ = \max\{0, h\}$). By Auxiliary theorem

A2 we therefore have $\int (f_m - g_n)_+ \, d\rho \to 0$ as $n \to \infty$, and thus

$$\int f_m \, d\rho - \lim_{n\to\infty} \int g_n \, d\rho = \lim_{n\to\infty} \int (f_m - g_n) \, d\rho$$

$$\leqslant \lim_{n\to\infty} \int (f_m - g_n)_+ \, d\rho = 0 \quad \text{for all} \quad m \in \mathbb{N}.$$

The assertion follows from this if we let m tend to ∞. □

In what follows let $T_1 = T_1(\mathbb{R}^m, \rho)$ be the set of the functions $f : \mathbb{R}^m \to \mathbb{R}$ for which there exists a non-decreasing sequence (f_n) from T for which $f_n \underset{\rho}{\to} f$ and the sequence of integrals is bounded. Since the sequence of integrals $\left(\int f_n \, d\rho\right)$ is non-decreasing and bounded, it is convergent and we can define

$$\int f \, d\rho = \lim_{n\to\infty} \int f_n \, d\rho.$$

By Auxiliary theorem A4 this definition is independent of the choice of the sequence (f_n) (having the required properties). If $f, g \in T_1$ and $a, b \geqslant 0$, then $af + bg$ obviously also belongs to T_1 and

$$\int (af + bg) \, d\rho = a \int f \, d\rho + b \int g \, d\rho.$$

Now let $T_2 = T_2(\mathbb{R}^m, \rho)$ be the *real* vector space that is spanned by T_1, i.e., $T_2 = \{f = f_1 - f_2 : f_1, f_2 \in T_1\}$. On T_2 let us define the ρ-integral by the equality

$$\int f \, d\rho = \int f_1 \, d\rho - \int f_2 \, d\rho \quad \text{for} \quad f = f_1 - f_2 \quad \text{with} \quad f_j \in T_1.$$

This definition is independent of the choice of the functions f_1 and f_2, since $f = f_1 - f_2 = g_1 - g_2$ with $f_1, f_2, g_1, g_2 \in T_1$ implies $f_1 + g_2 = g_1 + f_2$, $\int f_1 \, d\rho + \int g_2 \, d\rho = \int g_1 \, d\rho + \int f_2 \, d\rho$, and thus

$$\int f_1 \, d\rho - \int f_2 \, d\rho = \int g_1 \, d\rho - \int g_2 \, d\rho.$$

The elements of $T_2(\mathbb{R}^m, \rho)$ are called ρ-integrable functions (observe that only real functions have been considered so far).

Theorem A5.
(a) *The mapping* $T_2 \to \mathbb{R}$, $f \mapsto \int f \, d\rho$ *is linear.*
(b) $f, g \in T_2$ *and* $f \underset{\rho}{\leqslant} g$ *imply* $\int f \, d\rho \leqslant \int g \, d\rho$.
(c) *If* $f \in T_2$, *then* $|f|, f_+, f_- \in T_2$.
(d) $|\int f \, d\rho| \leqslant \int |f| \, d\rho$ *for every* $f \in T_2$.

PROOF.

(a) This assertion is evident.

(b) It is obviously sufficient to consider the case $f=0$ (g can be replaced by $g-f$). Let $g=g_1-g_2$ with $g_1, g_2 \in T_1$. Then $g_2 \underset{\rho}{\leq} g_1$, and hence $\int g_2 \, d\rho \leq \int g_1 \, d\rho$ by Auxiliary theorem A4, i.e., $\int g \, d\rho \geq 0$.

(c) Let $f=f_1-f_2$ with $f_1, f_2 \in T_1$. Then $\max\{f_1, f_2\} \in T_1$ and $\min\{f_1, f_2\} \in T_1$ (proof!). The assertion then follows from the formulae

$$|f| = \max\{f_1, f_2\} - \min\{f_1, f_2\},$$
$$f_+ = \max\{f_1, f_2\} - f_2, \; f_- = \max\{f_1, f_2\} - f_1.$$

(d) We have

$$\left|\int f \, d\rho\right| = \left|\int f_+ \, d\rho - \int f_- \, d\rho\right|$$
$$\leq \int f_+ \, d\rho + \int f_- \, d\rho = \int (f_+ + f_-) \, d\rho = \int |f| \, d\rho. \qquad \square$$

Theorem A6. *For every function $f \in T_2$ there exists a sequence (f_n) from T such that $f_n \underset{\rho}{\rightarrow} f$ and $\int |f_n - f| \, d\rho \rightarrow 0$. In particular, $\int f_n \, d\rho \rightarrow \int f \, d\rho$.*

PROOF. Let $f=f_1-f_2$ with $f_1, f_2 \in T_1$. Then there exist non-decreasing sequences $(f_{j,n})_{n \in \mathbb{N}}$ from T such that $f_{j,n} \underset{\rho}{\rightarrow} f_j$ and $\int f_{j,n} \, d\rho \rightarrow \int f_j \, d\rho$ for $j=1, 2$. For $f_n = f_{1,n} - f_{2,n}$ we have $f_n \underset{\rho}{\rightarrow} f$ and $|\int f_n \, d\rho - \int f \, d\rho| \leq \int |f_n - f| \, d\rho$ $= \int |f_{1,n} - f_{2,n} - (f_1 - f_2)| \, d\rho \leq \int (f_1 - f_{1,n}) \, d\rho + \int (f_2 - f_{2,n}) \, d\rho \rightarrow 0$ as $n \rightarrow \infty$. $\square$

A.2 Limit theorems

The following theorem asserts that the extension process of the previous section (which took us from T over T_1 to T_2) does not lead from T_2 to any wider class of functions. In the rest of this section we shall prove theorems which show, under which assumptions the passage to the limit and the integration are exchangeable. These theorems show the essential advantage of the Lebesgue integral when compared with the Riemann integral.

Theorem A7 (B. Levi). *Let (f_n) be a monotone sequence (non-decreasing or non-increasing) from T_2 for which the sequence of integrals $(\int f_n \, d\rho)$ is bounded. Then there is an $f \in T_2$ for which*

$$f_n \underset{\rho}{\rightarrow} f \quad \textit{and} \quad \int f_n \, d\rho \rightarrow \int f \, d\rho.$$

Proof. We may assume without loss of generality that (f_n) is non-decreasing (otherwise we consider $(-f_n)$).

First step: There are non-decreasing sequences (g_n) and (h_n) from T_1 with bounded sequences of integrals for which $f_n = g_n - h_n$ (i.e., it is sufficient to prove the theorem for a sequence (f_n) from T_1).

Proof of the first step. Define $k_1 = f_1$ and $k_j = f_j - f_{j-1}$ for $j \geq 2$. Then $f_n = \Sigma_{j=1}^n k_j$ and $k_n \geq 0$ for all $n \in \mathbb{N}$. We have $k_j = k_{j,1} - k_{j,2}$ with $k_{j,1}, k_{j,2} \in T_1$. For every $j \in \mathbb{N}$ there exists an $l_j \in T$ such that $l_j \leq k_{j,2}$ and $\int (k_{j,2} - l_j)\, d\rho \leq 2^{-j}$. If we set $\hat{k}_{j,2} = k_{j,2} - l_j$ and $\hat{k}_{j,1} = k_{j,1} - l_j$, then $\hat{k}_{j,1}, \hat{k}_{j,2} \in T_1$ and $k_j = \hat{k}_{j,1} - \hat{k}_{j,2}$ for all $j \in \mathbb{N}$. The series $\Sigma_{j=1}^\infty \int \hat{k}_{j,2}\, d\rho$ is convergent (since $\int \hat{k}_{j,2}\, d\rho < 2^{-j}$). If C is a bound of the sequence $(\int f_n\, d\rho)$, then

$$\sum_{j=1}^{n} \int k_j \, d\rho = \int f_n \, d\rho \leq C \quad \text{for all} \quad n \in \mathbb{N}.$$

Then the series $\Sigma_{j=1}^\infty \int \hat{k}_{j,1}\, d\rho$ is also convergent. The functions

$$g_n = \sum_{j=1}^{n} \hat{k}_{j,1} \quad \text{and} \quad h_n = \sum_{j=1}^{n} \hat{k}_{j,2}$$

have the required properties.

Second step: The theorem holds for any sequence (f_n) from T_1.

Proof of the second step. For every $n \in \mathbb{N}$ there is a non-decreasing sequence $(g_{n,m})_{m \in \mathbb{N}}$ from T with a bounded sequence of integrals such that $g_{n,m} \underset{\rho}{\to} f_n$. Define $g_m = \max\{ g_{n,m} : n \leq m\}$. Then (g_m) is a non-decreasing sequence from T. Since $g_{n,m} \underset{\rho}{\leq} f_n \underset{\rho}{\leq} f_m$ for $n \leq m$, we have $g_m \underset{\rho}{\leq} f_m$ and $\int g_m\, d\rho \leq \int f_m\, d\rho \leq C$ for all $m \in \mathbb{N}$. By Auxiliary Theorem A3 there exists an $f \in T_1$ for which $g_n \underset{\rho}{\to} f$ and $\int g_n\, d\rho \to \int f\, d\rho$. From the inequality $g_{n,m} \underset{\rho}{\leq} g_m$ for all $n \leq m$ we obtain, by letting $m \to \infty$, that $f_n \underset{\rho}{\leq} f$, and thus $g_n \underset{\rho}{\leq} f_n \underset{\rho}{\leq} f$. Consequently, $f_n \underset{\rho}{\to} f$ and $\int f_n\, d\rho \to \int f\, d\rho$. □

Theorem A8 (Lebesgue's dominated convergence theorem). *Let (f_n) be a sequence from T_2 for which $f_n \underset{\rho}{\to} f$. Assume that there exists a $g \in T_2$ such that $|f_n| \underset{\rho}{\leq} g$ for all $n \in \mathbb{N}$. Then f also belongs to T_2 and $\int f_n\, d\rho \to \int f\, d\rho$.*

Proof. For all $n \in \mathbb{N}$ let $g_n = \sup\{f_n, f_{n+1}, \ldots\}$. Then g_n belongs to T_2: Since $\max\{f_n, f_{n+1}\} = (f_n - f_{n+1})_+ + f_{n+1} \in T_2$, it follows by induction that $\max\{f_n, f_{n+1}, \ldots, f_{n+k}\} \in T_2$. Moreover, since

$$\int \max\{f_n, f_{n+1}, \ldots, f_{n+k}\}\, d\rho \leq \int g \, d\rho \quad \text{for all} \quad k \in \mathbb{N}$$

and $\max\{f_n, f_{n+1}, \ldots, f_{n+k}\} \to g_n$ as $k \to \infty$, the function g_n indeed belongs to T_2 by B. Levi's theorem. The sequence (g_n) is non-increasing and

$\int g_n \, d\rho \geq -\int g \, d\rho$. Therefore, $f = \rho - \lim g_n \in T_2$ and $\int g_n \, d\rho \to \int f \, d\rho$ by B. Levi's theorem.

If we define $h_n = \inf\{f_n, f_{n+1}, \ldots\}$, then we can show analogously that $h_n \underset{\rho}{\to} f$ and $\int h_n \, d\rho \to \int f \, d\rho$. The assertion then follows, because $h_n \leq f_n \leq g_n$. □

Theorem A9. *Let (f_n) be a sequence from T_2 and assume that $f_n \underset{\rho}{\to} f$. If there exists a $g \in T_2$ such that $|f| \underset{\rho}{\leq} g$, then f also belongs to T_2.*

Proof. Let $\tilde{f}_n = \min\{g, \max\{f_n, -g\}\}$ for all $n \in \mathbb{N}$. Then $\tilde{f}_n \in T_2$. $|\tilde{f}_n| \leq g$ for all $n \in \mathbb{N}$, and $\tilde{f}_n \underset{\rho}{\to} f$. The assertion therefore follows from Lebesgue's theorem. □

Theorem A10 (Fatou's lemma). *Let (f_n) be a non-negative sequence from T_2 for which the sequence of the integrals is bounded and $f_n \underset{\rho}{\to} f$. Then $f \in T_2$ and*

$$\int f \, d\rho \leq \liminf_{n\to\infty} \int f_n \, d\rho.$$

Proof. Let $h_n = \inf\{f_n, f_{n+1}, \ldots\}$ for all $n \in \mathbb{N}$. The sequence (h_n) is non-decreasing and $h_n \underset{\rho}{\to} f$. The inequalities $h_n \leq f_{n+k}$ for all $n, k \in \mathbb{N}$ imply

$$\int h_n \, d\rho \leq \liminf_{k\to\infty} \int f_{n+k} \, d\rho = \liminf_{k\to\infty} \int f_k \, d\rho, \quad n \in \mathbb{N}.$$

Consequently, B. Levi's theorem implies that $f \in T_2$ and

$$\int f \, d\rho = \lim_{n\to\infty} \int h_n \, d\rho \leq \liminf_{n\to\infty} \int f_n \, d\rho. \qquad \square$$

A.3 Measurable functions and sets

In the sequel let ρ be a measure on $\mathbb{R}^m$. A function $f: \mathbb{R}^m \to \mathbb{C}$ is said to be *ρ-measurable* if there exists a sequence (f_n) from T for which $f_n \underset{\rho}{\to} f$. It is obvious that every continuous function defined on $\mathbb{R}^m$ is ρ-measurable. The sum, product, and quotient (if the denominator $\underset{\rho}{\neq} 0$) of two ρ-measurable functions are ρ-measurable. Along with f, the function $g \circ f$ is also ρ-measurable for every continuous function $g: \mathbb{C} \to \mathbb{C}$. In particular, $|f|$ is ρ-measurable if f is ρ-measurable. Every $f \in T_2(\mathbb{R}^m, \rho)$ is ρ-measurable (cf. Theorem A6).

Theorem A11.
(a) *If $f: \mathbb{R}^m \to \mathbb{R}$ is ρ-measurable and there exists a $g \in T_2$ such that $|f| \leq g$, then f also belongs to T_2.*
(b) *If (f_n) is a sequence of ρ-measurable functions such that $f_n \underset{\rho}{\to} f$, then f is ρ-measurable.*

PROOF.
(a) This immediately follows from Theorem A9.
(b) Let $h \in T_2$ be such that $h(x) > 0$ for all $x \in \mathbb{R}^m$ (the reader is advised to prove the existence of such a function). With $g_n = (h + |f_n|)^{-1} h f_n$ for all $n \in \mathbb{N}$ we have $g_n \underset{\rho}{\to} g = (h + |f|)^{-1} h f$. Since all g_n are measurable and since $|g_n| \leq h$, we obtain that all g_n belong to T_2. Consequently, it follows from Lebesgue's theorem that $g \in T_2$. The function g is therefore ρ-measurable. This then holds for $f = (h - |g|)^{-1} h g$, as well. □

A subset M of $\mathbb{R}^m$ is said to be *ρ-measurable* if its characteristic function χ_M is ρ-measurable. If χ_M is ρ-integrable, then the *measure* $\rho(M)$ of M is defined by the equality $\rho(M) = \int \chi_M \, d\rho$. If M is ρ-measurable and χ_M is not ρ-integrable, then we set $\rho(M) = \infty$. All sets $M \in \tilde{\mathfrak{J}}(\mathbb{R}^m)$ are obviously ρ-measurable and have the finite measure $\rho(M)$; for these M the definition coincides with the earlier one. Every ρ-null set N is ρ-measurable with $\rho(N) = 0$.

Theorem A12. *Countable unions and intersections of ρ-measurable sets, as well as the complement of a ρ-measurable set are ρ-measurable. If $M_1, M_2, \ldots$ are disjoint ρ-measurable sets, then*

$$\rho\Big(\bigcup_{n \in \mathbb{N}} M_n\Big) = \sum_{n \in \mathbb{N}} \rho(M_n).$$

If $M_1 \supset M_2 \supset \ldots$ are ρ-measurable sets and $\rho(M_1) < \infty$, then

$$\rho\Big(\bigcap_{n \in \mathbb{N}} M_n\Big) = \lim_{n \to \infty} \rho(M_n).$$

The *proof* of this theorem can be left to the reader. One has to consider the characteristic functions χ_{M_n} and apply the previous theorems. The cases where infinite measures occur have to be treated carefully.

Theorem A13. *A function $f: \mathbb{R}^m \to \mathbb{R}$ is ρ-measurable if and only if the set $M_s = \{x \in \mathbb{R}^m : f(x) \geq s\}$ is ρ-measurable for every $s \in \mathbb{R}$. The same holds for the sets $\{x \in \mathbb{R}^m : f(x) \leq s\}$, $\{x \in \mathbb{R}^m : f(x) > s\}$ and $\{x \in \mathbb{R}^m : f(x) < s\}$.*

Proof. Let f be ρ-measurable. We can assume without loss of generality that $s=1$ (otherwise we replace f by $f-s+1$). The function $g=\min\{1, \max\{0,f\}\}$ is ρ-measurable and $g^n \to \chi_{M_s}$ as $n\to\infty$. The function χ_{M_s} is therefore ρ-measurable.

Let M_s now be ρ-measurable for all $s\in\mathbb{R}$. Then the sets

$$\{x\in\mathbb{R}^m : f(x)<t\} = \mathbb{R}^m \setminus M_t \quad \text{and} \quad M(s,t) = \{x\in\mathbb{R}^m : s\leq f(x)<t\}$$

are also ρ-measurable. The functions

$$f_n(x) = \begin{cases} kn^{-1} \text{ for } x\in M((k-1)n^{-1}, kn^{-1}),\ k=-n^2+1, -n^2+2,\ldots, n^2, \\ 0 \text{ otherwise} \end{cases}$$

are therefore ρ-measurable for all $n\in\mathbb{N}$. The assertion follows, because $f_n\to f$. □

The family of *Borel sets* in $\mathbb{R}^m$ is the smallest family of subsets of $\mathbb{R}^m$ that contains all intervals and is closed with respect to taking complements and countable unions and intersections. It is obvious that all open and all closed subsets of $\mathbb{R}^m$ are Borel sets. A function $f:\mathbb{R}^m\to\mathbb{R}$ is said to be *Borel measurable* if the set $\{x\in\mathbb{R}^m : f(x)\geq s\}$ is a Borel set for every $s\in\mathbb{R}$. (A function $f:\mathbb{R}^m\to\mathbb{C}$ is said to be Borel measurable if $\operatorname{Re} f$ and $\operatorname{Im} f$ are Borel measurable.)

If ρ is a measure on $\mathbb{R}^m$, then every Borel set is ρ-measurable and every Borel-measurable function defined on $\mathbb{R}^m$ is ρ-measurable. (The measures considered here are hence called *Borel measures*, as well.)

Now we extend the notion of the integral to complex valued functions. A function $f:\mathbb{R}^m\to\mathbb{C}$ is said to be *ρ-integrable* if $\operatorname{Re} f, \operatorname{Im} f\in T_2$. We define

$$\int f\,d\rho = \int \operatorname{Re} f\,d\rho + i\int \operatorname{Im} f\,d\rho.$$

The set of ρ-integrable functions is a complex vector space. It will be denoted by $\mathcal{L}_1(\mathbb{R}^m,\rho)$. The mapping $\mathcal{L}_1(\mathbb{R}^m,\rho)\to\mathbb{C}$, $f\mapsto\int f\,d\rho$ is obviously linear.

Theorem A14.

(a) *If $f:\mathbb{R}^m\to\mathbb{C}$ is ρ-measurable and there exists a ρ-integrable function g such that $|f| \underset{\rho}{\leq} g$, then f is ρ-integrable and $|\int f\,d\rho|\leq\int g\,d\rho$.*

(b) $|\int f\,d\rho|\leq\int|f|\,d\rho$ *for every $f\in\mathcal{L}_1(\mathbb{R}^m,\rho)$.*

(c) *For every $f\in\mathcal{L}_1(\mathbb{R}^m,\rho)$ there exists a sequence (f_n) from T for which $\int|f_n-f|\,d\rho\to 0$.*

(d) *For any ρ-measurable function $f:\mathbb{R}^m\to\mathbb{C}$ we have $f \underset{\rho}{=} 0$ if and only if $f\in\mathcal{L}_1(\mathbb{R}^m,\rho)$ and $\int|f|\,d\rho=0$.*

PROOF.
(a) The functions $\operatorname{Re} f$ and $\operatorname{Im} f$ are ρ-measurable, and $|\operatorname{Re} f| \underset{\rho}{\leq} g$ and $|\operatorname{Im} f| \underset{\rho}{\leq} g$. The functions $\operatorname{Re} f$ and $\operatorname{Im} f$ are therefore ρ-integrable by Theorem A11(a), i.e., $f \in \mathfrak{L}_1(\mathbb{R}^m, \rho)$. The inequality follows from (b).
(b) Let $a = \operatorname{sgn}(\int f\, d\rho)^*$. Then by Theorem A5(d)

$$\left|\int f\, d\rho\right| = a \int f\, d\rho = \int \operatorname{Re}(af)\, d\rho \leq \int |af|\, d\rho = \int |f|\, d\rho.$$

(c) This follows by applying Theorem A6 to $\operatorname{Re} f$ and $\operatorname{Im} f$.
(d) The equality $f \underset{\rho}{=} 0$ implies $|f| \underset{\rho}{=} 0$, $f \in \mathfrak{L}_1(\mathbb{R}^m, \rho)$ (since the sequence whose members are all zero ρ-converges to f), and $\int |f|\, d\rho = 0$. If $\int |f|\, d\rho = 0$, then B. Levi's theorem can be applied to the sequence (g_n) with $g_n = n|f|$. Hence there exists a $g \in T_2$ for which $g_n = n|f| \underset{\rho}{\rightarrow} g$. This is possible only if $f \underset{\rho}{=} 0$. □

Theorem A15 (Lebesgue). *Let (f_n) be a sequence from $\mathfrak{L}_1(\mathbb{R}^m, \rho)$ such that $f_n \underset{\rho}{\rightarrow} f$. Assume that there exists a $g \in \mathfrak{L}_1(\mathbb{R}^m, \rho)$ such that $|f_n| \leq g$ for all $n \in \mathbb{N}$. Then $f \in \mathfrak{L}_1(\mathbb{R}^m, \rho)$ and $\int f_n\, d\rho \rightarrow \int f\, d\rho$.*

PROOF. We can obviously apply Theorem A8 to the sequences $(\operatorname{Re} f_n)$ and $(\operatorname{Im} f_n)$. This gives the assertion. □

Corollary. *If (f_n) is a sequence from $\mathfrak{L}_1(\mathbb{R}^m, \rho)$ such that $\sum_{n=1}^{\infty} \int |f_n|\, d\rho < \infty$, then there exists an $f \in \mathfrak{L}_1(\mathbb{R}^m, \rho)$ such that $\sum_{j=1}^{n} f_j \underset{\rho}{\rightarrow} f$ and $\int f\, d\rho = \sum_{n=1}^{\infty} \int f_n\, d\rho$.*

PROOF. We can apply B. Levi's theorem to the sequence $(\sum_{j=1}^{n} |f_j|)_{n \in \mathbb{N}}$ and obtain a $g \in T_2 \subset \mathfrak{L}_1(\mathbb{R}^m, \rho)$ for which $\sum_{j=1}^{n} |f_j| \underset{\rho}{\rightarrow} g$. Consequently, there exists a function $f: \mathbb{R}^m \rightarrow \mathbb{C}$ such that $\sum_{j=1}^{n} f_j \underset{\rho}{\rightarrow} f$ as $n \rightarrow \infty$. Since $|\sum_{j=1}^{n} f_j| \leq g$ for all $n \in \mathbb{N}$, the assertion follows from Lebesgue's theorem. □

Let M be a ρ-measurable subset of $\mathbb{R}^m$. A function $f: M \rightarrow \mathbb{C}$ is said to be *ρ-measurable* (*ρ-integrable*) if the function

$$\tilde{f}: \mathbb{R}^m \rightarrow \mathbb{C}, \quad \tilde{f}(x) = \begin{cases} f(x) & \text{for } x \in M \\ 0 & \text{for } x \notin M \end{cases}$$

is ρ-measurable (ρ-integrable). If $f: M \rightarrow \mathbb{C}$ is ρ-integrable, then we define

$$\int_M f\, d\rho = \int \tilde{f}\, d\rho.$$

If $f: \mathbb{R}^m \rightarrow \mathbb{C}$ is ρ-measurable (ρ-integrable), then the restriction $f|_M$ of f to M is ρ-measurable (ρ-integrable), since $\widetilde{f|_M} = \chi_M f$. If $f: \mathbb{R}^m \rightarrow \mathbb{C}$ is ρ-

measurable and $f|_M$ is ρ-integrable, then we define

$$\int_M f\, d\rho = \int_M f|_M\, d\rho = \int \chi_M f\, d\rho.$$

We denote the vector space of ρ-integrable functions $f: M \to \mathbb{C}$ by $\mathfrak{L}_1(M, \rho)$.

Theorem A16.

(a) *If $f: M \to \mathbb{C}$ is ρ-integrable, (M_n) is a sequence of mutually disjoint ρ-measurable subsets of $\mathbb{R}^m$, and $M = \bigcup_{n \in \mathbb{N}} M_n$, then $\int_M f\, d\rho = \sum_{n \in \mathbb{N}} \int_{M_n} f\, d\rho$.*

(b) *If $f: M \to \mathbb{R}$ is ρ-integrable and $\int_K f\, d\rho \leq a\rho(K)$ ($\int_K f\, d\rho \geq a\rho(K)$) for every ρ-measurable subset K of M, then $f \underset{\rho}{\leq} a$ ($f \underset{\rho}{\geq} a$).*

(c) *If $f: \mathbb{R}^m \to \mathbb{C}$ is ρ-measurable, $f \in \mathfrak{L}_1(J, \rho)$, and $\int_J f\, d\rho = 0$ for all bounded intervals J, then $f \underset{\rho}{=} 0$.*

PROOF.

(a) Apply the above corollary to the sequence $(\chi_{M_n} f)$.

(b) Let $K_a = \{x \in M : f(x) > a\}$. Then $\chi_{K_a}(f - a) \geq 0$ and $\int_M \chi_{K_a}(f - a)\, d\rho = 0$. By Theorem A14(d) we therefore have $\chi_{K_a}(f - a) \underset{\rho}{=} 0$. This gives the assertion.

(c) It is sufficient to consider real f. It is obvious that $\int g f\, d\rho = 0$ for every $g \in T$. If I is a bounded interval and $M \subset I$ is ρ-measurable, then there exists a sequence (g_n) from T such that $0 \leq g_n \leq 1$, $g_n(x) = 0$ for $x \notin I$, and $g_n \underset{\rho}{\to} \chi_M$. Then it follows from Lebesgue's theorem for the sequence $(g_n f)$ that $\int_M f\, d\rho = \int \chi_M f\, d\rho = 0$. By part (b) we obtain that $f \underset{\rho}{=} 0$. □

A.4 The Fubini-Tonelli theorem

In what follows let ρ_1 and ρ_2 be measures on $\mathbb{R}^p$ and $\mathbb{R}^q$, respectively. Let ρ denote the *product measure* on $\mathbb{R}^{p+q}$ (cf. A1, Example 3).

Auxiliary Theorem A17. *If N is a ρ-null set in $\mathbb{R}^{p+q}$, then for ρ_1-almost all $x \in \mathbb{R}^p$ the set $\{y \in \mathbb{R}^q : (x, y) \in N\}$ is a ρ_2-null set, i.e.,*

$$N_1 = \{x \in \mathbb{R}^p : \{y \in \mathbb{R}^q : (x, y) \in N\} \text{ is not a } \rho_2\text{-null set}\}$$

is a ρ_1-null set.

PROOF. Since N is a ρ-null set, there is a sequence (J_k) of intervals for which $J_k = J_{1,k} \times J_{2,k}$, $N \subset \bigcup_{k \in \mathbb{N}} J_k$, $\sum_{k \in \mathbb{N}} \rho(J_k) < \infty$, and each $z \in N$ is covered by infinitely many J_k (we choose the union of infinitely many covers by intervals with total measures $2^{-1}, 2^{-2}, 2^{-3}, \ldots$). We have

$$\sum_{k\in\mathbf{N}} \int \chi_{J_{1,k}}\, d\rho_1 \int \chi_{J_{2,k}}\, d\rho_2 = \sum_{k\in\mathbf{N}} \rho_1(J_{1,k})\rho_2(J_{2,k}) = \sum_{k\in\mathbf{N}} \rho(J_k) < \infty.$$

We can therefore apply B. Levi's theorem to the sequence of step functions

$$\left(\sum_{k=1}^{n} \chi_{J_{1,k}}(x) \int \chi_{J_{2,k}}\, d\rho_2\right)_{n\in\mathbf{N}}.$$

Consequently, there is a ρ_1-null set F_1 such that

$$\sum_{k\in\mathbf{N}} \chi_{J_{1,k}}(x) \int \chi_{J_{2,k}}\, d\rho_2 < \infty \quad \text{for all} \quad x \in \mathbf{R}^p \setminus F_1.$$

It remains to prove that $N_1 \subset F_1$. Let $x_0 \in \mathbf{R}^p \setminus F_1$. Then

$$\sum_{k\in\mathbf{N}} \int_{\mathbf{R}^q} \chi_{J_{1,k}}(x_0)\chi_{J_{2,k}}(y)\, d\rho_2(y) < \infty.$$

For every $y \in \mathbf{R}^q$ such that $(x_0, y) \in N$ the element (x_0, y) belongs to infinitely many $J_k = J_{1,k} \times J_{2,k}$. The non-decreasing sequence

$$\left(\sum_{k=1}^{n} \chi_{J_{1,k}}(x_0)\chi_{J_{2,k}}(y)\right)_{n\in\mathbf{N}}$$

is therefore divergent. Since the corresponding sequence of integrals with respect to y is bounded, it follows from B. Levi's theorem (or from Auxiliary theorem A3) that $\{y \in \mathbf{R}^q : (x_0, y) \in N\}$ is a ρ_2-null set. Therefore, $x_0 \notin N_1$, and thus $N_1 \subset F_1$. □

Theorem A18 (Fubini). *Let ρ_1, ρ_2 and ρ be as above, and let $f \in \mathfrak{L}_1(\mathbf{R}^{p+q}, \rho)$. Then we have: For ρ_1-almost all $x \in \mathbf{R}^p$ the function $f(x, .)$ belongs to $\mathfrak{L}_1(\mathbf{R}^q, \rho_2)$. The function F defined by the equality*

$$F(x) = \begin{cases} \int_{\mathbf{R}^q} f(x, y)\, d\rho_2(y) & \text{if } f(x, .) \in \mathfrak{L}_1(\mathbf{R}^q, \rho_2), \\ 0 & \text{otherwise} \end{cases}$$

belongs to $\mathfrak{L}_1(\mathbf{R}^p, \rho_1)$ and

$$\int f\, d\rho = \int_{\mathbf{R}^p} F\, d\rho_1.$$

A similar result holds if we exchange the roles of x and y. To express the content of this theorem, we briefly write

$$\int f\, d\rho = \int_{\mathbf{R}^p} \left\{ \int_{\mathbf{R}^q} f(x, y)\, d\rho_2(y) \right\} d\rho_1(x)$$
$$= \int_{\mathbf{R}^q} \left\{ \int_{\mathbf{R}^p} f(x, y)\, d\rho_1(x) \right\} d\rho_2(y). \tag{A10}$$

PROOF. Since $f=f_1-f_2+\mathrm{i}f_3-\mathrm{i}f_4$ with $f_j\in T_1$, it is enough to study the case $f\in T_1$. Then there exists a non-decreasing sequence (f_n) from T and a ρ-null set N such that $f_n(z)\to f(z)$ for all $z\in\mathbb{R}^{p+q}\setminus N$ and $\int f_n\,\mathrm{d}\rho\to\int f\,\mathrm{d}\rho$. Formula (A10) is evident for the step functions f_n. Define $g_n(x)=\int f_n(x,y)\,\mathrm{d}\rho_2(y)$. Then (g_n) is a non-decreasing sequence from $T(\mathbb{R}^p)$ and $\int g_n\,\mathrm{d}\rho_1\leqslant\int f\,\mathrm{d}\rho$. By B. Levi's theorem there hence exists a $g\in\mathfrak{L}_1(\mathbb{R}^p,\rho_1)$ for which $g_n\underset{\rho_1}{\to}g$ and $\int g_n\,\mathrm{d}\rho_1\to\int g\,\mathrm{d}\rho_1$. Consequently, by Auxiliary theorem A17 the set

$$N_0=\{x\in\mathbb{R}^p:\{y\in\mathbb{R}^q:(x,y)\in N\}\text{ is not a }\rho_2\text{-null set}\}$$
$$\cup\{x\in\mathbb{R}^p:(g_n(x))\text{ is not convergent}\}$$

is a ρ_1-null set. For $x\notin N_0$ the non-decreasing sequence $(f_n(x,.)$ has the properties

$$f_n(x,.)\underset{\rho_2}{\to}f(x,.)\quad\text{and}\quad\int_{\mathbb{R}^q}f_n(x,y)\,\mathrm{d}\rho_2(y)=g_n(x)\leqslant g(x)<\infty.$$

By B. Levi's theorem $f(x,.)$ therefore belongs to $\mathfrak{L}_1(\mathbb{R}^q,\rho_2)$ for all $x\notin N_0$ and

$$\int_{\mathbb{R}^q}f(x,y)\,\mathrm{d}\rho_2(y)=\lim_{n\to\infty}\int_{\mathbb{R}^q}f_n(x,y)\,\mathrm{d}\rho_2(y)=\lim_{n\to\infty}g_n(x)=g(x).$$

Since g is ρ_1-integrable and $g\underset{\rho_1}{=}F$, the function F is also ρ_1-integrable and

$$\int F\,\mathrm{d}\rho_1=\int g\,\mathrm{d}\rho_1=\lim_{n\to\infty}\int g_n\,\mathrm{d}\rho_1$$
$$=\lim_{n\to\infty}\int_{\mathbb{R}^p}\left\{\int_{\mathbb{R}^q}f_n(x,y)\,\mathrm{d}\rho_2(y)\right\}\mathrm{d}\rho_1(x)=\lim_{n\to\infty}\int f_n\,\mathrm{d}\rho=\int f\,\mathrm{d}\rho.$$

□

Theorem A19 (Tonelli). *Let ρ_1, ρ_2 and ρ be as above, and let $f:\mathbb{R}^{p+q}\to\mathbb{C}$ be ρ-measurable. Assume that $f(x,.)\in\mathfrak{L}_1(\mathbb{R}^q,\rho_2)$ for ρ_1-almost all $x\in\mathbb{R}^p$ and that the function $\tilde{F}$ defined by the formula*

$$\tilde{F}(x)=\begin{cases}\int_{\mathbb{R}^q}|f(x,y)|\,\mathrm{d}\rho_2(y), & \text{if } f(x,.)\in\mathfrak{L}_1(\mathbb{R}^q,\rho_2),\\ 0 & \text{otherwise}\end{cases}$$

belongs to $\mathfrak{L}_1(\mathbb{R}^p,\rho_1)$. Then $f\in\mathfrak{L}_1(\mathbb{R}^{p+q},\rho)$.

PROOF. For every $n\in\mathbb{N}$ let

$$M_n=\{(x,y)\in\mathbb{R}^{p+q}:|f(x,y)|\leqslant n,|(x,y)|\leqslant n\},\quad f_n=\chi_{M_n}f.$$

Since f is ρ-measurable, every f_n is ρ-integrable by Theorems A13 and

A14(a). The sequence $(|f_n|)$ is non-decreasing and $|f_n| \to |f|$. An application of Fubini's theorem to $|f_n|$ gives

$$\int |f_n| \, d\rho = \int_{\mathbb{R}^p} \left\{ \int_{\mathbb{R}^q} |f_n(x, y)| \, d\rho_2(y) \right\} d\rho_1(x)$$
$$\leq \int_{\mathbb{R}^p} \left\{ \int_{\mathbb{R}^q} |f(x, y)| \, d\rho_2(y) \right\} d\rho_1(x) < \infty.$$

Consequently, the ρ-integrability of $|f|$ follows from B. Levi's theorem. Since f is ρ-measurable, $f \in \mathfrak{L}_1(\mathbb{R}^{p+q}, \rho)$ by Theorem A14(a). □

A.5 The Radon-Nikodym theorem

Let ρ and μ be two measures on $\mathbb{R}^m$. The measure μ is said to be *absolutely continuous with respect to* ρ (in symbols: $\mu \ll \rho$) if every ρ-null set is also a μ-null set. (Then every ρ-measurable set is μ-measurable, as well).

Theorem A20 (Radon-Nikodym). *Let ρ and μ be two measures on $\mathbb{R}^m$. We have $\mu \ll \rho$ if and only if there exists a ρ-measurable non-negative function $h : \mathbb{R}^m \to \mathbb{R}$ such that $\chi_J h \in \mathfrak{L}_1(\mathbb{R}^m, \rho)$ for every bounded interval J and $\mu(M) = \int \chi_M h \, d\rho$ for every ρ-measurable set M (here we consider the integral to be equal to ∞ if $\chi_M h$ is not ρ-integrable). Every ρ-measurable function is also μ-measurable. If $f : \mathbb{R}^m \to \mathbb{C}$ is ρ-measurable and μ-integrable, then*

$$\int f \, d\mu = \int fh \, d\rho.$$

Proof. If μ has the above form, then $\mu \ll \rho$ obviously holds. Now let $\mu \ll \rho$ and let J be an arbitrary bounded interval in $\mathbb{R}^m$. Let us consider the Hilbert space $L_2(J, \tau)$ with the measure $\tau = \rho + \mu$. The mapping

$$L_2(J, \tau) \to \mathbb{C}, \quad f \mapsto \int_J f \, d\mu$$

is a continuous linear functional, since $|\int_J f \, d\mu| \leq \int_J |f| \, d\tau \leq \tau(J)^{1/2} \|f\|$.

By the Riesz representation theorem (Theorem 4.8) there exists a $g \in L_2(J, \tau)$ (more precisely, a $g \in \mathfrak{L}_2(J, \tau)$) such that

$$\int_J f \, d\mu = \int_J gf \, d\tau \quad \text{for} \quad f \in L_2(J, \tau). \tag{A11}$$

If here we replace f by χ_M, where M is an arbitrary ρ-measurable subset of J, then we obtain

$$\mu(M) = \int_J \chi_M \, d\mu = \int_J g\chi_M \, d\tau = \int_M g \, d\tau.$$

Since $\mu(M) \leqslant \tau(M)$, it follows that

$$0 \leqslant \int_M g \, d\tau \leqslant \tau(M).$$

It follows from this by Theorem A16(b) that

$$0 \leqslant g(x) \leqslant 1 \quad \text{for} \quad \rho\text{-almost all } x \in J.$$

We obtain from (A11) that

$$\int_J (1-g)f \, d\mu = \int_J gf \, d\rho \quad \text{for} \quad f \in L_2(J, \tau). \tag{A12}$$

Now set $N = \{x \in J : g(x) = 1\}$ and $L = J \setminus N$. Then (A12) implies for $f = \chi_N$ that

$$\rho(N) = \int \chi_N \, d\rho = \int g\chi_N \, d\rho = \int (1-g)\chi_N \, d\mu = 0,$$

and thus $\mu(N) = 0$, as well, because $\mu \ll \rho$. If in (A12) we set $f = (1 + g + g^2 + \cdots + g^n)\chi_N$, then it follows for all $n \in \mathbb{N}$ that

$$\int_M (1 - g^{n+1}) \, d\mu = \int_M g(1 + g + \cdots + g^n) \, d\rho.$$

The integrands of both sides constitute non-decreasing sequences the integrals of which are bounded by $\mu(M)$. The left integrand converges to χ_L, and hence the left side tends to $\mu(M \cap L)$. By B. Levi's theorem there exists an $h \in L_1(J, \rho)$ such that $g(1 + g + \cdots + g^n) \underset{\rho}{\to} h$ as $n \to \infty$ and $\int_M g(1 + g + \cdots + g^n) \, d\rho \to \int_M h \, d\rho$. Consequently, it follows that

$$\mu(M) = \mu(L \cap M) + \mu(N \cap M) = \mu(L \cap M) = \int_M h \, d\rho.$$

Since $g \underset{\rho}{\geqslant} 0$, we also have $h \underset{\rho}{\geqslant} 0$. Without loss of generality, we can choose $h \geqslant 0$.

Now let (J_n) be a sequence of disjoint intervals for which $\mathbb{R}^m = \bigcup_{n \in \mathbb{N}} J_n$. Let h_n be functions such that

$$\mu(M) = \int_M h_n \, d\rho \quad \text{for every} \quad \rho\text{-measurable set} \quad M \subset J_n.$$

Let $h : \mathbb{R}^m \to \mathbb{R}$ be defined by the equalities $h(x) = h_n(x)$ for $x \in J_n$. Then h has all properties required, since for every ρ-measurable subset M of $\mathbb{R}^m$

$$\mu(M) = \sum_{n \in \mathbb{N}} \mu(M \cap J_n) = \sum_{n \in \mathbb{N}} \int_{M \cap J_n} h \, d\rho = \int_M h \, d\rho.$$

If f is ρ-measurable, then there is a sequence (f_n) from T such that $f_n \underset{\rho}{\to} f$. Consequently, we also have $f_n \underset{\mu}{\to} f$, and thus f is μ-measurable.

It remains to prove the last assertion for non-negative functions f. For every $n \in \mathbb{N}$ let

$$M_{n,k} = \{x \in \mathbb{R}^m : (k-1)2^{-n} \leq f(x) < k2^{-n}\} \quad \text{for} \quad k = 1, 2, \ldots, 2^{2n}$$

and

$$f_n = \sum_{k=1}^{2^{2n}} (k-1)2^{-n} \chi_{M_{n,k}}.$$

Then $f_n \in \mathfrak{L}_1(\mathbb{R}^m, \mu)$ for all $n \in \mathbb{N}$ and

$$\int f_n \, d\mu = \sum_{k=1}^{2^{2n}} (k-1)2^{-n} \mu(M_{n,k}) = \sum_{k=1}^{2^{2n}} (k-1)2^{-n} \int_{M_{n,k}} h \, d\rho = \int f_n h \, d\rho.$$

It follows from this by B. Levi's theorem that $\int f \, d\mu = \int fh \, d\rho$. □

A function $F : \mathbb{R} \to \mathbb{C}$ is said to be *absolutely continuous* if there exists a λ-measurable (i.e., Lebesgue measurable) function $f : \mathbb{R} \to \mathbb{C}$ that is λ-integrable (Lebesgue integrable) over every bounded interval and

$$F(x) = F(0) + \int_0^x f(t) \, d\lambda(t).^2$$

This function f is called the *derivative* of F. (It is possible to show that F is λ-almost everywhere differentiable and $F'(x) \underset{\lambda}{=} f(x)$.) The derivative f is uniquely determined by F. (This follows from Theorem A16(c).)

If ρ is a measure on $\mathbb{R}$, then $\rho \ll \lambda$ if and only if the function

$$f : \mathbb{R} \to \mathbb{R}, \quad f(x) = \begin{cases} \rho((0, x]) & \text{for} \quad x > 0 \\ -\rho((x, 0]) & \text{for} \quad x \leq 0 \end{cases}$$

is absolutely continuous, i.e., if ρ is induced by an absolutely continuous function f in the sense of Section A1, Example 2.

Let F and G be absolutely continuous functions on $\mathbb{R}$, and denote by f and g their respective derivatives. If $F(0) = G(0) = 0$, then we obtain by Fubini's theorem that

$$\begin{aligned} \int_0^x (Fg + fG) \, d\lambda &= \int_0^x \left\{ g(t) \int_0^t f(s) \, d\lambda(s) + f(t) \int_0^t g(s) \, d\lambda(s) \right\} d\lambda(t) \\ &= \int_0^x \left\{ f(s) \int_s^x g(t) \, d\lambda(t) + g(s) \int_s^x f(t) \, d\lambda(t) \right\} d\lambda(s) \\ &= 2F(x)G(x) - \int_0^x (fG + gF) \, d\lambda, \end{aligned}$$

and thus

$$F(x)G(x) = \int_0^x (Fg + fG) \, d\lambda.$$

[2]Here we set $\int_0^x = -\int_x^0$ for $x < 0$.

If we do not necessarily have $F(0) = G(0) = 0$, then

$$F(x)G(x) - F(0)G(0) = \int_0^x (Fg + fG)\, d\lambda.$$

Consequently, for $-\infty < a < b < \infty$ we have the formula of *integration by parts*

$$\int_a^b Fg\, d\lambda = F(b)G(b) - F(a)G(a) - \int_a^b fG\, d\lambda. \tag{A13}$$

Appendix B

A representation theorem for holomorphic functions with values in a half-plane

A function $h : \mathbb{R} \to \mathbb{C}$ is said to be of *bounded variation* when it can be written in the form $h = h_1 - h_2 + \mathrm{i}h_3 - \mathrm{i}h_4$, where the functions $h_j : \mathbb{R} \to \mathbb{R}$ are non-decreasing bounded functions. (We can show that h is of this form if and only if there is a $C > 0$ such that $\Sigma_n |h(b_n) - h(a_n)| < C$ for every sequence $((a_n, b_n])$ of disjoint intervals. The smallest C of this kind is called the *variation* of h. We do not need this result here.) If h is a right continuous function of bounded variation, then the integral

$$\int_{-\infty}^{\infty} (z-t)^{-1}\, \mathrm{d}h(t) \quad \text{for} \quad z \in \mathbb{C} \setminus \mathbb{R}$$

can be considered as a Riemann-Stieltjes integral. We will retain this notation in the sequel. We can also view this integral as a linear combination of the corresponding integrals with respect to the measures ρ_{h_j} (cf. Section A1, Example 2). Consequently, the theorems of Appendix A are at our disposal.

Theorem B1. *Assume that* $w : \mathbb{R} \to \mathbb{R}$ *is right continuous and of bounded variation,* $\lim_{t \to -\infty} w(t) = 0$, *and*

$$f(z) = \int_{-\infty}^{\infty} (z-t)^{-1}\, \mathrm{d}w(t) \quad \textit{for} \quad z \in G = \{z \in \mathbb{C} : \operatorname{Im} z > 0\}.$$

(a) *For all* $t \in \mathbb{R}$ *we have the* Stieltjes inversion formula

$$w(t) = \lim_{\delta \to 0_+} \lim_{\epsilon \to 0_+} \frac{-1}{\pi} \int_{-\infty}^{t+\delta} \operatorname{Im} f(s + \mathrm{i}\,\epsilon)\, \mathrm{d}s.$$

(b) *If* $f(z) = 0$ *for all* $z \in G$, *then* $w(t) = 0$ *for all* $t \in \mathbb{R}$.

Proof. Part (b) follows from part (a). Consequently, it is sufficient to prove (a). Since w is real-valued,

$$\begin{aligned}\operatorname{Im} f(s+i\epsilon) &= \int_{-\infty}^{\infty} \operatorname{Im}\left[(s+i\epsilon-u)^{-1}\right] dw(u)\\ &= -\epsilon\int_{-\infty}^{\infty}\left[(s-u)^2+\epsilon^2\right]^{-1} dw(u)\end{aligned}$$

for every $\epsilon > 0$. It follows from this by Fubini's theorem that

$$\begin{aligned}\int_{-\infty}^{r} \operatorname{Im} f(s+i\epsilon)\, ds &= -\int_{-\infty}^{\infty}\int_{-\infty}^{r} \frac{\epsilon}{(s-u)^2+\epsilon^2}\, ds\, dw(u)\\ &= -\int_{-\infty}^{\infty}\left[\arctan\frac{r-u}{\epsilon}+\frac{\pi}{2}\right] dw(u).\end{aligned}$$

Since

$$\left|\arctan\frac{r-u}{\epsilon}+\frac{\pi}{2}\right| \leq \pi \quad \text{for all} \quad r\in\mathbb{R}$$

and

$$\arctan\frac{r-u}{\epsilon}+\frac{\pi}{2} \to \begin{cases}\pi & \text{for} \quad r>u,\\ \frac{\pi}{2} & \text{for} \quad r=u, \quad \textit{as} \quad \epsilon\to 0,\\ 0 & \text{for} \quad r<u,\end{cases}$$

Lebesgue's theorem implies that

$$\begin{aligned}&\lim_{\epsilon\to 0_+}\int_{-\infty}^{r} \operatorname{Im} f(s+i\epsilon)\, ds\\ &= -\int_{(-\infty,r)} \pi\, dw(u) - \int_{\{r\}} \frac{\pi}{2}\, dw(u) - \int_{(r,\infty)} 0\, dw(u)\\ &= -\pi w(r-) - \frac{\pi}{2}\left[w(r)-w(r-)\right] = -\frac{\pi}{2}\left[w(r)+w(r-)\right].\end{aligned}$$

(In order to be able to apply Lebesgue's theorem, we write $\int\cdots dw = \int\cdots dw_1 - \int\cdots dw_2$, where w_1 and w_2 are non-decreasing right continuous functions, $w=w_1-w_2$, and $\lim_{u\to-\infty} w_1(u) = \lim_{u\to-\infty} w_2(u) = 0$.) If we set $r=t+\delta$ with $\delta>0$ and let δ tend to zero, then the assertion follows. □

Theorem B2. *Assume that* $w: \mathbb{R}\to\mathbb{C}$ *is right continuous, of bounded variation, and* $\lim_{t\to-\infty} w(t) = 0$. *If* $\int_{-\infty}^{\infty}(z-t)^{-1}\, dw(t) = 0$ *for all* $z\in\mathbb{C}\setminus\mathbb{R}$ *then* $w(t)=0$ *for all* $t\in\mathbb{R}$.

Proof. For $z\in G=\{z\in\mathbb{C} : \operatorname{Im} z>0\}$ we have

$$\int_{-\infty}^{\infty}(z-t)^{-1}\, dw(t) = 0$$

and

$$\int_{-\infty}^{\infty}(z-t)^{-1}\, dw^*(t) = \left\{\int_{-\infty}^{\infty}(z^*-t)^{-1}\, dw(t)\right\}^* = 0.$$

Therefore,

$$\int_{-\infty}^{\infty} (z-t)^{-1}\,\mathrm{d}[\operatorname{Re} w(t)] = \int_{-\infty}^{\infty} (z-t)^{-1}\,\mathrm{d}[\operatorname{Im} w(t)] = 0$$

for all $z \in G$. It then follows from Theorem B1(b) that $\operatorname{Re} w(t) = \operatorname{Im} w(t) = 0$, and thus $w(t) = 0$ for all $t \in \mathbb{R}$. □

Theorem B3 (Herglotz). *Let* $G = \{z \in \mathbb{C} : \operatorname{Im} z > 0\}$, *and let* $f: G \to \mathbb{C}$ *be holomorphic such that* $\operatorname{Im} f(z) \leq 0$ *and* $|f(z) \operatorname{Im} z| \leq M$ *for all* $z \in G$. *Then there exists a unique right continuous non-decreasing function* $w: \mathbb{R} \to \mathbb{R}$ *for which* $w(t) \to 0$ *as* $t \to -\infty$ *and*

$$f(z) = \int_{-\infty}^{\infty} (z-t)^{-1}\,\mathrm{d}w(t) \quad \textit{for all} \quad z \in G.$$

For all $t \in \mathbb{R}$ *we have* $w(t) \leq M$ *and*

$$w(t) = \lim_{\delta \to 0_+} \lim_{\epsilon \to 0_+} \frac{-1}{\pi} \int_{-\infty}^{t+\delta} \operatorname{Im} f(s + \mathrm{i}\,\epsilon)\,\mathrm{d}s.$$

PROOF. The last equality will follow from Theorem B1(a) if we prove the existence of a function w having the remaining properties.

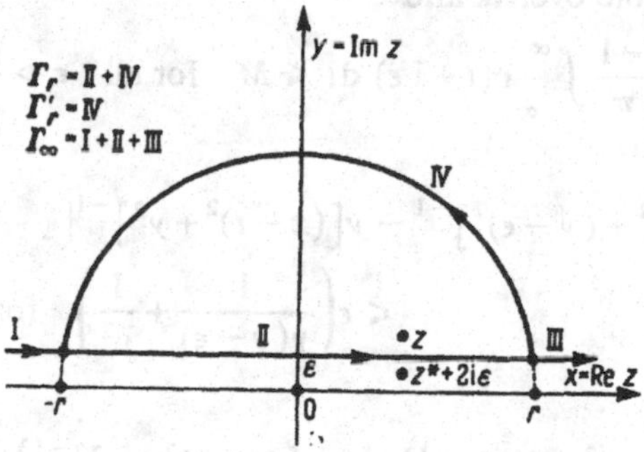

For $0 < \epsilon < r$ let the paths Γ_r, Γ_r', and Γ_∞ be defined as the above figure shows. For $z = x + \mathrm{i}\,y \in G$ such that $\operatorname{Im} z = y > \epsilon$ and for $r > |z|$ the point $z^* + 2\mathrm{i}\,\epsilon$ lies outside Γ_r. Therefore, by the Cauchy integral formula

$$\begin{aligned} f(z) &= \frac{1}{2\pi\,\mathrm{i}} \int_{\Gamma_r} (\zeta - z)^{-1} f(\zeta)\,\mathrm{d}\zeta \\ &= \frac{1}{2\pi\,\mathrm{i}} \int_{\Gamma_r} [(\zeta - z)^{-1} - (\zeta - (z^* + 2\,\mathrm{i}\,\epsilon))^{-1}] f(\zeta)\,\mathrm{d}\zeta \\ &= \frac{1}{2\pi\,\mathrm{i}} \int_{\Gamma_r} (z - z^* - 2\,\mathrm{i}\,\epsilon)[(\zeta - z)(\zeta - z^* - 2\,\mathrm{i}\,\epsilon)]^{-1} f(\zeta)\,\mathrm{d}\zeta \\ &= \frac{1}{\pi} \int_{\Gamma_r} (y - \epsilon)[(\zeta - z)(\zeta - z^* - 2\,\mathrm{i}\,\epsilon)]^{-1} f(\zeta)\,\mathrm{d}\zeta. \end{aligned}$$

For $\zeta \in \Gamma_r'$ we have for fixed z that

$$|f(\zeta)| \leq \epsilon^{-1}M \quad \text{and} \quad |(y-\epsilon)|[(\zeta - z)(\zeta - z^* - 2\,\mathrm{i}\,\epsilon)]^{-1} \leq Cr^{-2}.$$

The integral over Γ_r' therefore tends to 0 as $r \to \infty$, and there remains

$$\begin{aligned} f(z) &= \frac{1}{\pi}\int_{\Gamma\infty} (y-\epsilon)[(\zeta - z)(\zeta - z^* - 2\,\mathrm{i}\,\epsilon)]^{-1} f(\zeta)\, d\zeta \\ &= \frac{1}{\pi}\int_{-\infty}^{\infty} (y-\epsilon)[(t+\mathrm{i}\,\epsilon - z)(t - \mathrm{i}\,\epsilon - z^*)]^{-1} f(t+\mathrm{i}\,\epsilon)\, dt \\ &= \frac{1}{\pi}\int_{-\infty}^{\infty} (y-\epsilon)[(x-t)^2 + (y-\epsilon)^2]^{-1} f(t+\mathrm{i}\,\epsilon)\, dt. \end{aligned}$$

If we set $v(z) = \operatorname{Im} f(z)$, then it follows for $0 < \epsilon < \operatorname{Im} z = y$ that

$$v(z) = \frac{1}{\pi}\int_{-\infty}^{\infty} (y-\epsilon)[(x-t)^2 + (y-\epsilon)^2]^{-1} v(t+\mathrm{i}\,\epsilon)\, dt.$$

The inequalities $|yv(z)| \leq |f(z) \operatorname{Im} z| \leq M$ imply for $0 < \epsilon < y$ that

$$\left|\frac{1}{\pi}\int_{-\infty}^{\infty} (y-\epsilon)^2[(x-t)^2 + (y-\epsilon)^2]^{-1} v(t+\mathrm{i}\,\epsilon)\, dt\right| = |(y-\epsilon)v(z)| \leq M.$$

By letting $y \to \infty$, we obtain from Fatou's lemma (observe that $v \leq 0$) that $v(\,.+\mathrm{i}\,\epsilon)$ is integrable over $\mathbb{R}$ and

$$0 \leq \frac{-1}{\pi}\int_{-\infty}^{\infty} v(t+\mathrm{i}\,\epsilon)\, dt \leq M \quad \text{for all} \quad \epsilon > 0.$$

Since

$$\begin{aligned} &|(y-\epsilon)[(x-t)^2 + (y-\epsilon)^2]^{-1} - y[(x-t)^2 + y^2]^{-1}| \\ &\qquad \leq \epsilon\left(\frac{1}{y(y-\epsilon)} + \frac{1}{y^2}\right) \quad \text{for} \quad 0 < \epsilon < y, \end{aligned}$$

it follows that

$$\int_{-\infty}^{\infty} \{(y-\epsilon)[(x-t)^2 + (y-\epsilon)^2]^{-1} - y[(x-t)^2 + y^2]^{-1}\} v(t+\mathrm{i}\,\epsilon)\, dt \to 0$$

as $\epsilon \to 0+$. Therefore, for all $z \in G$

$$v(z) = \lim_{\epsilon \to 0_+} \frac{1}{\pi}\int_{-\infty}^{\infty} y[(x-t)^2 + y^2]^{-1} v(t+\mathrm{i}\,\epsilon)\, dt.$$

In what follows let

$$\vartheta_\epsilon(t) = \frac{-1}{\pi}\int_{-\infty}^{t} v(s+\mathrm{i}\,\epsilon)\, ds \quad \text{for} \quad t \in \mathbb{R},\ \epsilon > 0.$$

The functions ϑ_ϵ are all non-decreasing and bounded, $0 \leq \vartheta_\epsilon(t) \leq M$ for all $t \in \mathbb{R}$.[1] Let us construct, with the aid of the diagonal process, a positive null

[1]The following steps can be much shortened if we make use of the fact that the family of measures induced by $\{\vartheta_\epsilon : 0 < \epsilon < 1\}$ is compact in the "vague" topology.

sequence (ϵ_n) in such a way that $(\vartheta_{\epsilon_n}(t))$ is convergent for all rational numbers t. If we set

$$\vartheta(t) = \lim_{n\to\infty} \vartheta_{\epsilon_n}(t) \quad \text{for rational } t,$$

then $\vartheta(s) \leq \vartheta(t)$ for all rational s and t such that $s < t$. If we extend ϑ to a function $\vartheta : \mathbb{R}\to\mathbb{R}$ by defining

$$\vartheta(t) = \inf\{\vartheta(s) : s > t, s \text{ rational}\} \text{ for irrational } t,$$

then ϑ is obviously non-decreasing, and $\lim_{t\to\infty}(\vartheta(t)-\vartheta(-t)) \leq M$.

We show that in the sense of the Riemann–Stieltjes integral

$$v(z) = -\int_{-\infty}^{\infty} y\left[(x-t)^2+y^2\right]^{-1} \mathrm{d}\vartheta(t) \quad \text{for} \quad z \in G.$$

Since

$$\begin{aligned} v(z) &= \lim_{\epsilon\to 0+} \frac{1}{\pi}\int_{-\infty}^{\infty} y\left[(x-t)^2+y^2\right]^{-1} v(t+\mathrm{i}\,\epsilon)\,\mathrm{d}t \\ &= -\lim_{\epsilon\to 0+} \int_{-\infty}^{\infty} y\left[(x-t)^2+y^2\right]^{-1} \frac{\mathrm{d}}{\mathrm{d}t}\vartheta_\epsilon(t)\,\mathrm{d}t \\ &= -\lim_{\epsilon\to 0+} \int_{-\infty}^{\infty} y\left[(x-t)^2+y^2\right]^{-1} \mathrm{d}\vartheta_\epsilon(t), \end{aligned}$$

this assertion is equivalent to the equality

$$\lim_{n\to\infty} \int_{-\infty}^{\infty} y\left[(x-t)^2+y^2\right]^{-1} \mathrm{d}\vartheta_{\epsilon_n}(t) = \int_{-\infty}^{\infty} y\left[(x-t)^2+y^2\right]^{-1} \mathrm{d}\vartheta(t).$$

For the proof of this equality we notice that if we wish to approximate this Riemann-Stieltjes integral (with a continuous integrand) by Riemann sums, then it is enough to consider only partitions of $(-\infty, \infty)$ with rational division points. For every such *rational* partition P and for fixed $z = x + \mathrm{i}\,y$ let U_P, L_P, $U_{P,n}$, and $L_{P,n}$ be the upper and lower sums of the integrals

$$J = \int_{-\infty}^{\infty} y\left[(x-t)^2+y^2\right]^{-1} \mathrm{d}\vartheta(t)$$

respectively

$$J_n = \int_{-\infty}^{\infty} y\left[(x-t)^2+y^2\right]^{-1} \mathrm{d}\vartheta_{\epsilon_n}(t)$$

that correspond to P. For every rational partition P we obviously have $U_{P,n}\to U_P$ and $L_{P,n}\to L_P$. For every $\delta > 0$ there exists a rational partition P for which $U_P - L_P < \delta/2$. For this P there is an $n_0 \in \mathbb{N}$ such that $|U_{P,n} - U_P| < \delta/2$ and $|L_{P,n} - L_P| < \delta/2$ for all $n \geq n_0$. Since $L_{P,n} \leq J_n \leq U_{P,n}$ and $L_P \leq J \leq U_P$, it follows that $|J - J_n| \leq \delta$ for $n \geq n_0$. Therefore, $J_n \to J$ as $n\to\infty$.

Consequently, we have shown that for $z \in G$

$$\operatorname{Im} f(z) = v(z) = -\int_{-\infty}^{\infty} y\left[(x-t)^2 + y^2\right]^{-1} \mathrm{d}\vartheta(t)$$
$$= \operatorname{Im} \int_{-\infty}^{\infty} (z-t)^{-1} \mathrm{d}\vartheta(t).$$

Since f and $z \mapsto \int_{-\infty}^{\infty}(z-t)^{-1}\,\mathrm{d}\vartheta(t)$ are holomorphic in G and have the same imaginary part, it follows that

$$f(z) = \int_{-\infty}^{\infty} (z-t)^{-1} \mathrm{d}\vartheta(t) + C \quad \text{with some} \quad C \in \mathbb{R}.$$

Because $|f(z) \operatorname{Im} z| \leqslant M$ and

$$|(\operatorname{Im} z)\int_{-\infty}^{\infty} (z-t)^{-1} \mathrm{d}\vartheta(t)| \leqslant \int_{-\infty}^{\infty} 1\, \mathrm{d}\vartheta(t) \leqslant M \quad \text{for} \quad z \in G,$$

we must also have $|C \operatorname{Im} z| \leqslant 2M$, and thus $C = 0$.

If we now define

$$\tilde{\vartheta}(t) = \lim_{\delta \to 0_+} \vartheta(t+\delta) \quad \text{for} \quad t \in \mathbb{R}$$

and

$$w(t) = \tilde{\vartheta}(t) - \lim_{s \to -\infty} \tilde{\vartheta}(x) \quad \text{for} \quad t \in \mathbb{R},$$

then w has the required properties: The passage from ϑ to $\tilde{\vartheta}$ does not change anything in the integral formula we have just proved, since $\tilde{\vartheta}$ has at most countably many points of discontinuity and they can be avoided during the formation of the partitions. The passage from $\tilde{\vartheta}$ to w does not influence the integral formula. □

References

(A) Textbooks and monographs

[1] Achieser, N. I.; Glasman, I. M.: Theorie der linearen Operatoren im Hilbertraum. 5. Aufl. Berlin 1968

[2] Bourbaki, N.: Eléments de mathématique, Livre V; Espace vectoriels topologiques. Paris 1967

[3] Berberian, S. K.: Lectures in functional analysis and operator theory. New York-Heidelberg-Berlin 1974

[4] Dunford, N.; Schwartz, J. T.: Linear operators, Part I: General theory. New York-London-Sydney 1958/1964

[5] Dunford, N.; Schwartz, J. T.: Linear operators, Part II: Spectral theory, self adjoint operators in Hilbert space. New York-London-Sydney 1963

[6] Dunford, N.; Schwartz, J. T.: Linear operators, Part III: Spectral operators. New York-London-Sydney-Toronto 1971

[7] Faris, W. G.: Self-adjoint operators. Berlin-Heidelberg-New York 1975

[8] Foias, C.; Sz.-Nagy, B.: Harmonic analysis of operators on Hilbert space. Budapest-Amsterdam-London 1970

[9] Friedrichs, K. O.: Spectral theory of operators in Hilbert space. Berlin-Heidelberg-New York 1973

[10] Gohberg, I. C.; Krein, M. G.: Introduction to the theory of non-self-adjoint operators in Hilbert space. Providence, R. I. 1969

[11] Goldberg, S.: Unbounded linear operators, theory and applications. New York-St. Louis-San Francisco-Toronto-London-Sydney 1966

[12] Halmos, P. R.: Introduction to Hilbert space and the theory of spectral multiplicity. New York 1951

[13] Halmos, P. R.: Finite dimensional vector spaces. Princeton, N. J.-Toronto-New York-London 1958

[14] Halmos, P. R.: A Hilbert space problem book. New York-Heidelberg 1974

[15] Hellwig, G.: Differentialoperatoren der mathematischen Physik. Berlin-Göttingen-Heidelberg 1964

[16] Helmberg, G.: Introduction to spectral theory in Hilbert space. Amsterdam-London 1969
[17] Heuser, H.: Funktionalanalysis. Stuttgart 1975
[18] Hewitt, E.; Stromberg, K.: Real and abstract analysis. Berlin-Heidelberg-New York 1965
[19] Jörgens, K.: Lineare Integraloperatoren. Stuttgart 1970
[20] Jörgens, K.; Rellich, F.: Eigenwerttheorie gewöhnlicher Differentialgleichungen. Berlin-Heidelberg-New York 1976
[21] Jörgens, K.; Weidmann, J.: Spectral properties of Hamiltonian operators. Berlin-Heidelberg-New York 1973
[22] Kato, T.: Perturbation theory for linear operators. Berlin-Heidelberg-New York. Second Edition 1976
[23] Knobloch, H. W.; Kappel, F.: Gewöhnliche Differentialgleichungen. Stuttgart 1974
[24] Maurin, K.: Methods of Hilbert spaces. Warschau 1967
[25] Maurin, K.: General eigenfunction expansions and unitary representations of topological groups. Warschau 1968
[26] Nelson, E.: Topics in dynamics I: Flows. Princeton, N. J. 1969
[27] Neumark, M. A.: Lineare Differentialoperatoren. Berlin 1960
[28] Putnam, C. R.: Commutation properties of Hilbert space operators. Berlin-Heidelberg-New York 1967
[29] Reed, M.; Simon, B.: Methods of modern mathematical physics I: Functional analysis. New York-London 1972
[30] Reed, M.; Simon, B.: Methods of modern mathematical physics II: Fourier analysis, self-adjointness. New York-London 1975
[31] Riesz, F.; Sz.-Nagy, B.: Vorlesungen über Funktionalanalysis. Berlin 1956
[32] Rudin, W.: Real and complex analysis. New York 1966
[33] Rudin, W.: Functional analysis. New York 1973
[34] Schechter, M.: Spectra of partial differential operators. Amsterdam-London 1971
[35] Stone, M. H.: Linear transformations in Hilbert space and their applications to analysis. New York 1932
[36] Sz.-Nagy, B.: Spektraldarstellung linearer Transformationen des Hilbertschen Raumes. Berichtigter Nachdruck. Berlin-Heidelberg-New York 1967
[37] Triebel, H.: Höhere Analysis. Berlin 1972
[38] Voigt, A.; Wloka, J.: Hilberträume und elliptische Differentialoperatoren. Mannheim-Wien-Zürich 1975
[39] Wloka, J.: Funktionalanalysis und Anwendungen. Berlin-New York 1971
[40] Yosida, K.: Functional analysis. Berlin-Heidelberg-New York 1965

(B) Articles referred to in the book

[41] Faris, W. G.: Invariant cones and uniqueness of the ground state for fermion systems. J. Mathematical Phys. **13** (1972) 1285–1290
[42] Hess, P.; Kato, T.: Perturbation of closed operators and their adjoints. Comment. Math. Helv. **45** (1970) 524–529
[43] Hörmander, L.: The existence of wave operators in scattering theory. Math. Z. **146** (1976) 69–91

[44] Ikebe, T.; Kato, T.: Uniqueness of the self-adjoint extension of singular elliptic differential operators. Arch. Rational Mech. Anal. **9** (1962) 77–92

[45] Kato, T.: Scattering theory and perturbation of continuous spectra. Actes, Congrès intern. math., 1970. Band 1, 135–140

[46] Korotkov, V. B.: Integral operators with Carleman kernels. Dokl. Akad. Nauk SSSR **165** (1965) 1496–1499 (russisch)

[47] Kuroda, S. T.: Scattering theory for differential operators, I, operator theory. J. Math. Soc. Japan **25** (1973) 75–104

[48] Kuroda, S. T.: Scattering theory for differential operators, II, self adjoint elliptic operators. J. Math. Soc. Japan **25** (1973) 222–234

[49] Schechter, M.: Scattering theory for elliptic operators of arbitrary order. Comment. Math. Helv. **49** (1974) 84–113

[50] Schechter, M.: A unified approach to scattering. J. Math. Pures Appl. **53** (1974) 373–396

[51] Simader, C. G.: Bemerkung über Schrödinger-Operatoren mit stark singulären Potentialen. Math. Z. **138** (1974) 53–70

[52] Stummel, F.: Singuläre elliptische Differentialoperatoren in Hilbertschen Räumen. Math. Ann. **132** (1956) 150–176

[53] Veselić, K.; Weidmann, J.: Existenz der Wellenoperatoren für eine allgemeine Klasse von Operatoren. Math. Z. **134** (1973) 255–274

[54] Veselić, K.; Weidmann, J.: Asymptotic estimates of wave functions and the existence of wave operators. J. Functional Analysis **17** (1974) 61–77

[55] Weidmann, J.: The virial theorem and its application to spectral theory of Schrödinger operators. Bull. Amer. Math. Soc. **73** (1967) 452–456

[56] Weyl, H.: Über gewöhnliche Differentialgleichungen mit Singularitäten und die zugehörigen Entwicklungen willkürlicher Funktionen. Math. Ann. **68** (1909) 220–269

[57] Wüst, R.: Generalisations of Rellich's theorem on perturbations of (essentially) selfadjoint operators. Math. Z. **119** (1971) 276–280

[58] Wüst, R.: Holomorphic operator families and stability of selfadjointness. Math. Z. **125** (1972) 349–358

Index of symbols

Author and subject index

Printed in the United States
By Bookmasters